THE HOLOGRAPHIC ANTHROPIC MULTIVERSE

Formalizing the Complex Geometry of Reality

K&E Series on Knots and Everything — Vol. 43

THE HOLOGRAPHIC ANTHROPIC MULTIVERSE

Formalizing the Complex Geometry of Reality

Richard L. Amoroso

Noetic Advanced Studies Institute, USA

Elizabeth A. Rauscher

Tecnic Research Laboratories, USA

World Scientific

NEW JERSEY • LONDON • SINGAPORE • BEIJING • SHANGHAI • HONG KONG • TAIPEI • CHENNAI

Published by

World Scientific Publishing Co. Pte. Ltd.

5 Toh Tuck Link, Singapore 596224

USA office: 27 Warren Street, Suite 401-402, Hackensack, NJ 07601

UK office: 57 Shelton Street, Covent Garden, London WC2H 9HE

Library of Congress Cataloging-in-Publication Data

Amoroso, Richard L.

The holographic anthropic multiverse : formalizing the complex geometry of reality / by Richard L Amoroso & Elizabeth A Rauscher.

p. cm. -- (Series on knots and everything ; v. 43)

Includes bibliographical references and index.

ISBN-13: 978-981-283-930-5 (hardcover : alk. paper)

ISBN-10: 981-283-930-5 (hardcover : alk. paper)

1. Quantum cosmology. 2. Anthropic Principle. 3. Supersymmetry. I. Rauscher, Elizabeth A. II. Title.

QB991.Q36A46 2009

523.1--dc22

2009017158

British Library Cataloguing-in-Publication Data

A catalogue record for this book is available from the British Library.

Printed in Singapore by World Scientific Printers

For Juliette –

May the hope we have for the entelechies of our children's incarnation parallel the conformal scale-invariant hope for the evolution of human consciousness this little volume might engender as reflected in the following musing…

THE GREATEST POEM EVER WRITTEN

This Is The Greatest Poem Ever Written
Should it win a Pulitzer Prize?
Emerson says, 'Beauty lights the beholder's eyes'
This rule cannot be bent by critic's cries.

This Is The Greatest Poem Ever Written
No court of law, Miranda oath, legal imposition
Can deny this lofty disposition.

This Is The Greatest Poem Ever Written
And as you read, oh awestruck 'mirer,
We give you leave to hone your senses
While you listen t'what our plain defense is.

This Is The Greatest Poem Ever Written
Though Sheldrake's morphogenic field finds it so
And fundamental Platonic form illuminates it's glow;

This Is The Greatest Poem Ever Written
Simply because we've plumed it so
Thus this stanza or the next will find you smitten.

This Is The Greatest Poem Ever Written
The prior stanza profounds the next
Deny this Muse be *Voo-doo* hexed.

This Is The Greatest Poem Ever Written
It needn't be popular to make it so
Listen to the repetition metered incantation.

This Is The Greatest Poem Ever Written
No need to use subliminal suggestion
To hypno-muse myopic perception.
This Is The Greatest Poem Ever Written
Some say moot, we easily refute !
This common bane of artistic creation.

This Is The Greatest Poem Ever Written
As joy now surely swells your breast
You know our Muse and so attest.

This Is The Greatest Poem Ever Written
It makes no difference if nonsense fills these lines
For as recorded everywhere and every-when
This Is The Greatest Poem Ever Written.

Oh, Calliope Queen of Muses
Of sister Goddesses, Muses nine
You inspire us with your pensive rays.

Oh, Clio, Oh Erat, Euterpe and Melpomene
Oh, Polymnia, Terpsichore, Urania and Thalia
You infuse our verse with all divine.

Voilá le Plus Poĕme Jamais Ĕcrit Magnifique
Sentez les mots encore comme ils passent vos lèvres [1]
Voilá le plus poĕme jamais écrit magnifique.

This Is The Greatest Poem Ever Written
Do you realize why at last?
It's simply Love;
It's Love that makes it so.

This Is The Greatest Poem Ever Written
Loving you, you know
How could love not be promulgated so…[1]

- Juliette & Richard Amoroso

Reference

[1] Amoroso, J.R.S. & Amoroso, R.L. (2006) The greatest poem ever written, in Rendevous at the Temple of Love: Poems of Peace, Love & Whimsy, pp. 105-106, Oakland: Noetic Press.

[1] French for: Feel the words again as they pass your lips.

Preface

...the right hand side includes all that cannot be described so far in the Unified Field Theory, of course, not for a fleeting moment, have I had any doubt that such a formulation is just a temporary answer, undertaken to give General Relativity some closed expression. This formulation has been in essence nothing more than the theory of the gravitational field which has been separated in a somewhat artificial manner from the unified field of a yet unknown nature - Einstein.

What kind of a book is this? Does it live up to its hype? We suppose critics will have a field day with it as has always been the case when someone goes out on a limb further than the current *conquistadores* are wont to. This is not a book for those as it were 'geocentric nay sayers' who vehemently oppose anything new or who myopically adhere to the status quo. A roiling plethora of physicists these days are deeply troubled by the morose perversity of the conundrum experienced as a result of such ratiocinations; perhaps hundreds or even thousands struggle with additional dimensions, string theory → M-Theory → F-Theory and myriad other issues. Recently a colleague remarked 'he was finally willing to pay the price to embrace the Everett Many World's model'; I wanted to answer that I cared nothing for what was considered politically correct and would rather embrace the true intellectual freedom entailed in the fundamental epistemological foundations of science itself, seek only for the truth and apply due diligence to the tiniest indicia if even a scintilla of evidence warranted searching a seemingly spurious path to find it. In that guise 'we let our hair down' so to speak and explore avenues that have remained unexplored for far too long to their penultimate conclusions.

We hope thereby dear reader that you might keep an open mind if for no other reason than as an exercise in epistemology. It was the famous paradox or antinomy, $R = \{A | A \notin A\}$ of noted philosopher Bertrand Russell who proved that 4 + 5 = 7 [1]. Contrariwise we use no logical trickery but take the usual empirical evidence to follow 'another' path of interpretation. For example Hubble discovered cosmological redshift not a Doppler expansion of the universe which has fostered the 'political' mindset of the time.

We mentioned the struggle with 'political pressure' in the scientific community that even one of the greatest physicists of all time gave in to and withdrew (what was considered at the time *ad hoc*) his concept of the cosmological constant calling it 'his greatest blunder'. We disagree, his greatest blunder was giving in to what was thought to be politically correct. It is said there has been little new in physics in the last hundred years. We insist as Smolin and others have said that this has occurred because of the pressure to be 'mainstream' and the severe punishment (being marginalized - no career, no publications, no funding) of those that veer off the beaten track. If science is allowed to be science, this may not have occurred because the seeds of all we relate in this volume have been available for planting for most of this 100 years war.

Admittedly this volume's contents are *avante guarde* in 2009, hopefully by 2012 they will be mainstay science if empirical tests are performed. Let us elaborate. This book has been pitched as a 'Copernican class volume'. How do we know if we have deluded ourselves? More importantly, how can you the reader judge tommyrot and twaddle from truth? Too many of our friends and colleagues insist that the universe is only 4D. One said, 'of course if you add more degrees of freedom you can do anything!' Two of the major contributions of this volume are a design for universal bulk quantum computing (Chap. 11) based on a model for surmounting the quantum uncertainty principle (Chap. 9). History has generally shown that when a 'correct' theory is found it is elegant, logical and internally consistent and has broad explanatory power. Most saliently this little tome is full of explanatory power. But the point we wish to make regards empirical testing which is the main pragmatic task of science. Moore's Law named after the founder of Intel, has shown for over 40 years that every 18 months the number of transistors on a CPU doubles and the processing speed doubles. Moore's Law has never been wrong. In Chap. 11 you will find a graph projecting Moore's Law into the quantum domain. This seems to occur about 2011 or 2012. The chink in making this prediction more precisely is that the transition from the usual technology to quantum technology may involve a phase change like the one that occurs when one puts a stick into a pool of water – The image of the stick is bent by a certain small angle. Does such a transition angle apply here? In any case we challenge the quantum computing community to 'immediately' build a bulk universal quantum computing prototype using the empirically testable model presented here. Do not let Gordon Moore down. Do not let us down either for that matter because we have use for a special class of quantum computer required to develop new medical technologies.

We state our case(s) here matter-of-factly as axiomatic elements of the new noetic paradigm presented without too much in the way of humble apologetic mumbo jumbo 'this is highly speculative', or theoretical etc. That's too boring for us and now should be obvious since stated up front. Pretty much all of our heretical views are empirically testable in the near term so we take liberty to play with your minds a little. Do the experiments then pick on us or not as the case may be because by then we would doubly deserve it. Why attempt to start such a puerile brouhaha? For two reasons: 1) As Smolin has said 'we are in an era where brilliant young scientists are not given jobs, tenure or funding unless they rigidly adhere to current thinking' [2]. Some of us are sick of this and we personally hope this volume finally has enough *chutzpah* to knock 'Humpty Dumpty' permanently off the wall. 2) Gandhi said, 'first they ignore you, then they laugh at you, then they fight with you, then you win'. So put up your dukes...

It is uncommon to be so brazen, so why do we do it? We want to make certain our point does not go unnoticed. Copernicus and Galileo were almost killed for their ideas; and the modern form of this 'murder' is reflected in Smolin's manifesto noted above. Worse yet, it is said that it took about 150 years before Copernicus' ideas became generally accepted. More recently for Einstein's introduction of the photoelectric effect (initially considered absurd) it took ~15 years before the empirical work was performed. 'Things' seem to occur asymptotically quicker in current times. We suggest this can be so and hope to force the issue – $150 \to 15 \to 1.5$ years which would be about the 2011 or 2012 predicted by Moore's Law! Again please do not disappoint Mr. Moore; it would be breaking the law!

The main 'claim to fame' of this little tome's is the semiformal introduction of a new cosmological paradigm called the Holographic

Anthropic Multiverse (HAM) which we have done our best to make 'logically coherent and internally consistent'. In spite of its flaws which stem from our inability to have performed better at this point in time; there is no denying the breath of its explanatory power:

- Empirical model for surmounting the quantum uncertainty principle
- Design for immediate implementation of bulk universal quantum computing
- Redshift-CMBR as black body cavity-QED exciplex equilibrium
- QSO luminosity as gravitational shock waves
- Integration of G & EM in a manner supporting the new cosmology
- Design for de Broglie matter-wave antiballistic defense shields
- Putative new empirically testable derivation of M-Theory tension
- New protocol for SETI success that calls for a new kind of telescope
- A simple unified field equation with implications for the nature of the observer, brane tension and CMBR/redshift
- Reference to a companion volume that delineates the physical basis of awareness from the mind-body side of the anthropic cosmology

This is a reasonable list of accomplishments for any book and as we mentioned our bodacious claims are attenuated in the fact that most are empirically testable in the near term.

't Hooft said "nature is much more crazy at the Planck scale than even string theorists could have imagined" [3]. This volume could just as easily have been called 'The Nature of the Singularity' because that is really what it is all about or 'Demise of the Big Bang' instead of being called the Holographic Anthropic Multiverse; but the chosen title has broader scope. In the sense of the above bulleted list we wish we could call it 'Nobel Prize Giveaway Manual'. We are curious to see how many it spawns. Again don't let us down; And yes, we have that much confidence, which may seem surprising in the face of the 2006 Nobel prize in Physics to John C Mather and George F Smoot:

> "...for recording faint echoes of the birth of the universe. Their precise satellite measurements of the cosmic background radiation, remnants of the sea of light emitted by the new universe, have confirmed fundamental predictions arising from the Big Bang theory, leading to its further acceptance as the standard model of cosmology."

Never has so much intelligence and pride been expended on something so wrong. "There are some ideas so wrong that only a very intelligent person could believe them." - George Orwell. Certainly Occam's Razor can only be applied if the choices it is applied to is correct. Hubble discovered a cosmological redshift relation not expansion of the universe. Certainly COBE has been a boon to cosmology but it would have been better if the Nobel Committee stuck with their statement 'They have been honored for their discovery of the blackbody form and anisotropy of the cosmic microwave background radiation'. Smoot went so far as to quote Kiyoshi Shiraishi in saying "It is impossible that the Big Bang is wrong" [4].

> Consensus discourages dissent... It is the enemy of science, just as it is the triumph of politics. A theory accepted by 99 percent of scientists may be wrong. Committees... that decide which projects shall be funded are inevitably run by scientists who are at peace with the dominant theory. Changing the consensus on cosmology will be an arduous task, like turning a supertanker with a broken rudder. ...the competition of theories has been the driving force behind scientific progress. Isolated individuals and private companies have been the most fruitful sources of this advance [5].

We have decided to throw caution to the wind and not hold anything back. Parts are informal and highly speculative but included because they follow directly from the new model in unique and interesting ways. Other parts are as pragmatic, rigorous and as empirically testable as the current state of our abilities and understanding allow. It is said that the bigger the step and the farther ahead of its time the greater the challenge for acceptance. Nevertheless we are a product of the current *Zeitgeist* suggesting that this book's radical stance is timely. We know that to some it will be long overdue. What is it that drives the evolution of human consciousness? That could be a subject for a multi-volume series. The one-word answer is 'necessity'.

If a Static Universe model had been continuously embraced since first introduced by Newton in the 17th century in the same way the Big Bang has; it is possible we would not be in a different place now. We consider this unlikely because a lot of intellect would have been expended exploring avenues neglected for no reason other than political myopia:

- The Dirac Covariant Polarized Vacuum
- Photon Mass Anisotropy
- Anthropic Principle / Teleology
- Nature Of The Observer
- Physical Basis Of Awareness/Consciousness
- Completion Of The Tools Of Epistemology
- Extended Electromagnetic Theory
- De Broglie-Bohm Ontology To Complete Quantum Theory
- Additional Dimensionality
- Alternative To Quantum Gravity And Higgs Mechanism

This volume is a full spectrum of theoretical insights based on application of a new cosmological paradigm and the insights that naturally drop out of it - from more rigorous theoretical considerations on quantum theory to the raw speculations on SETI research. But there is another wrinkle going on from which we take license to speculate. We wish to take a step toward completing the tools of human epistemology as stated formally in the last chapter of a companion volume [6]. Epistemology has evolved from superstition to logic to empiricism. The remaining tool first proposed by Plato is the utility of transcendence in theory formation. Plato said, 'noetic insight is the highest form of knowing; no matter how broad ones knowledge base or how great ones intelligence noetic insight comes form beyond the individual'.

This volume is our attempt to institute a Galilean or Copernican class revolution. The subtitle of this volume summarizes much of our purpose in writing it. Specifically we make the case that cosmology takes the form of a holographic anthropic multiverse. Avenues of our approach have been unpopular during the last sixty years. While the terms holographic, anthropic and multiverse have each been around for various degrees of time; we believe we are first to merge them into a unified cosmology in a formal way. In the hope that the views presented are inspired we do not take a conservative approach. Where required to facilitate development of the model our approach is axiomatic. We justify our radical stance by evidentiary conclusions from the history of science; and in that sense hope our approach is bold enough to take part in implementing a second Galilean class revolution in cosmology. By this we mean that the history of science has shown that when the correct theory is found it has elegance, internal logic, simplicity and broad explanatory power.

In simple terms we build the Holographic Anthropic Multiverse cosmology by taking alternative interpretations of all the purported pillars of the Big Bang. We ask the reader to indulge us with an open mind especially if the 'chains of myopia' have tethered the minds eye rigidly to the currently popular Big Bang. At the time of Galileo it was inherently obvious to the same degree and built on what was considered sound logical deduction that 'the heavier object would fall faster in a gravitational field'. We hope to show that today even with hundreds of years of sophisticated experimental development; we are prone to the same logical errors. As we have professed it is possible by using the tenets of anthropic cosmology to complete the tools of epistemology so that we are less and less likely to make the same kind of errors in the future in spite of our personal biases.

A surprising number of contemporary physical scientists do not accept dimensionality beyond four. The Euclidean line is deemed the real line because it is what our 'eyes' observe; but even in 3D dimensionality cannot be adequately proven. Newton gave us three; then Einstein introduced a fourth. String theory has struggled with thirty-two to twenty-six, eleven to ten and back again with M-Theory settling for eleven, as the parade continues with the recent addition of F-Theory cast in twelve for which we make a formal case for the ultimate basis of a Holographic Anthropic Multiverse (HAM).

While a reasonably large number of papers are published each year in cosmology, astrophysics, string theory and various areas of extended physical theory, not so many books are written and few of these have attempted to condense the arena and organize pertinent aspects into a coherent whole. This is due in part to the fact that the associated fields are relatively new and vibrant with evolution. Most workers confine themselves to narrow areas of research and typically spend little effort considering a larger framework. This suggests the time is ripe for monographs with the capacity to present order to the field. Our main purpose is not to present a review of recent thinking in order to survey and connect disparate pieces for the sake of adding coherence; but rather an attempt at engendering a grander new step forward based on numerous breakthroughs in our research and that of others related to the holographic cosmological model. The inherent purpose of an Anthropic Multiverse is life and consciousness, therefore intelligence in the cosmos is the evolutionary rule not the accidental statistical exception.

Not long ago cosmology was not considered a science; it was at best a form of philosophical/theological rumination. Some still say ‘first comes speculation, then speculation squared followed by cosmology’. To this critic that is what Big Bangers have done and continue to do. Of course somehow this is what we all try to do to preserve our theories especially if the alternative threatens to shatter our world view. Aristotle thought experimentation was flawed and foolish; that only logic could lead to the truth. Why should cosmology ultimately have Multiverse, Holographic and/or Anthropic properties; and especially the integrated Holographic-Anthropic-Multiverse form we promote here? This is what we attempt in the volume.

One of our boldest premises is the suggestion that there is no quantum gravity. This is not a deal breaker in our view for the holographic principle because ‘t Hooft’s motivation for intruding the principle to aid the development of quantum gravity is only entailed in the quest for a fundamental unified theory. Without a quantum gravity this quest still exists; but in a different form. Feynman said:

> ...maybe we should not try to quantize gravity. Is it possible that gravity is not quantized and all the rest of the world is?...Now the postulate defining quantum mechanical behavior is that there is an amplitude for different processes. It cannot be that a particle which is described by an amplitude, such as an electron, has an interaction which is not described by an amplitude but by a probability...it seems that it should be impossible to destroy the quantum nature of fields. In spite of these arguments, we should like to keep an open mind. It is still possible that quantum theory does not absolutely guarantee that gravity has to be quantized.

We use standard abbreviations for acronyms such as QT for quantum theory; we mention this here because we took the most liberty with terms for dimensionality or dimension, D with usage such as 3D, 4D, XD, HD which spelled out would add pages to the volume.

Shortcomings - we could try to be sufficiently arrogant to pass ourselves off as string theorists, but we don’t really want to be as it’s a life-long career path. So we merely dabble to make certain points because ultimately we have another time consuming agenda which will appear in future volumes.

The SETI work in Chap. 13 is our most speculative, but it falls right out of the anthropic portion of the HAM all on its own. We hope someone will finish inventing the interdimensional Q-Telescope proposed by the time the SETI-I program is considered a failure.

The QSO luminosity work is reasonably OK, needs a little more work for full rigor. Gravity is after all a classical theory and needs to be extended; not as quantum gravity which our theory says doesn't exist. The integration of the two principles is at unitarity not with each other.

The defense shield, maybe we got away without putting in actual engineering diagrams; but the perceptive reader will notice that all these threads are based on the very same principles of manipulating vacuum topology. You understand how to get one; you get all the rest of them. Programming the vacuum for the defense shield isn't more difficult than ontologically programming a quantum computer; only that one needs the additional nanoscale programmable matter substrate to imbed it in a more clever L.O.V.E.R. (Laser Oscillated Vacuum Energy Resonator) configuration (see Chap. 9) to get sufficiently Gödelized[1] [7], whereas in universal quantum computing the resonance hierarchy for surmounting the uncertainty principle is simpler to arrange. And in the defense shield case (see acknowledgement at the end of Chap. 12) we're not sure just anyone should be able to build it. We believe all scientific discovery comes as 'revelation from God' and we wanted to leave a little wiggle room for God to play his hand in the *Zeitgeist*.

We had a sense of humor but use it up here to write our own review of this book: 'This insidious volume is a conspiracy by the international psychoanalytic community to drum up business during a troubled world economy; if you read it you will need psychoanalysis for the rest of your life...'

Finally we would like to thank Lou Kauffman, the Knot Series editor, for his confidence in us and hope we have not misbehaved to the degree that he needs to wear a 'Flak Jacket', at least until after the ink sufficiency dries.

---

[1] Gödelization – according to Gödel's incompleteness theorem a system cannot be completely understood in terms of itself. In this case cannot be sufficiently controlled from within its own limits.

References

[1] Russell, B. (1908) Mathematical logic as based on the theory of types, Am J Mathematics, 30, 222-262.
[2] Smolin, L. (2006) The Trouble With Physics: The Rise of String Theory, the Fall of a Science, and What Comes Next, New York: Houghton Mifflin.
[3] ’t Hooft, G. (1993) Dimensional reduction in quantum gravity, arXiv:gr-qc/9310026v1 1993
[4] Smoot, G. & Davidson, K. (1993) Wrinkles in Time, quotation by Kiyoshi Shiraishi: *It's impossible that the Big Bang is wrong*, p. 69, New York: W. Morrow & Co.
[5] Bethell, T. (2005) The Politically Incorrect Guide to Science, Washington: Regnery Publishing.
[6] Amoroso, R.L. & Pribram, K.H. (2009) The Complementarity of Mind and Body: Realizing the Dream of DesCartes, Einstein and Eccles, Cambridge: MIT University Press.
[7] Smullyan, R.M. (1992) Gödel’s Incompleteness Theorems, Oxford: Oxford University Press.

Richard L. Amoroso
Noetic Advanced Studies Institute
Oakland, CA USA
April, 2009

Contents

Chapter 1

Demise of the Big Bang - A Philosophical Conundrum

> *Perhaps never in the history of science has so much quality evidence accumulated against a model so widely accepted within a field. Even the most basic elements of the theory, the expansion of the universe and the fireball remnant radiation, remain interpretations with credible alternative explanations. One must wonder why, in this circumstance, four good alternative models are not even being comparatively discussed by most astronomers* - T. Van Flandern [1].

> *The crucial discoveries needed to break away from current dogma will only be communicated in alternative journals, conferences and books such as the present one, where investigators can speak frankly about fundamental issues* - Halton C. Arp [2].

Critics of the Big Bang have said every time the Big Bang fails an adjustment to the theory is made or a new parameter added to fix it. While a valid criticism, it misses the mark in that this is the business of science, constant tinkering until truth is eventually found. Our complaint is in the quotes above. Others have said that there is no truth in science because a theory can never be proven true only falsified. However there is another wrinkle in that respect; when a 'best' theory is finally found and later falsified what remains is Absolute Truth [3]. Absolute truth in science refers to a finite regime described by a theory that has been falsified. This somewhat rarefied condition is best said about Newton's theories. They remain absolute truth in reference to the finite classical regimes they describe [4]. The aim of this volume is to provide sufficient insight that Big Bang cosmology may finally be falsified.

1.1 Philosophical Overview

The Ptolemaic system with the Earth as center of the universe (geocentric model) lasted for 2,000 years after the Greek Ionian school first postulated a model with the sun as the center of the universe (heliocentric) in 300 BC because of the forceful persona of Hipparchus, considered the greatest mathematician and astronomer of antiquity, in 150 BC and the fervor with which he defended it. It is interesting that Hipparchus first tried to prove the sun as the center, but the contemporary Platonic idea that the circle or sphere were the only perfect shapes and therefore divine and thus the only orbit a planet could have became a significant fact in changing world history. Even though the planetary orbits are nearly circular his calculations were so precise that he totally abandoned the heliocentric system. If he had abandoned circularity instead, another construct based on incorrect religious dogma, the correct heliocentric system would have been accepted 2,000 years earlier. The Ptolemaic system was called the greatest intellectual achievement of ancient astronomy and lasted until the time of Copernicus. These early ideas of perfection and the belief of an eternal universe led to the Newtonian static cosmological model. Einstein assumed that the universe was uniform which came to be called the cosmological principle and later generalized to the perfect cosmological principle for steady-state cosmologies.

Sometimes ‘truth’ takes awhile to uncover because the avenue leading to it might be unpopular or seem in apparent violation of ‘Occam’s razor’ (All other things being equal, the simplest solution is the best.). Before the 20th Century scientific cosmology was little more than philosophy; and it appears that Einstein’s motivation for a static universe model may have been theological. The Big Bang hallowed for over eighty years was motivated by the antithesis of that condition.

Aristotle insisted that ‘logic was superior to experiment’; but at the time of Galileo pure logic failed giving rise to empiricism as the dominant pragmatic test. Cosmological data is the most difficult to acquire with precision. It is fascinating to realize that we are on the cusp of another Galilean class revolution in that empiricism has failed epistemologically, not in and of itself, but in the Aristotelian sense that unscientific bias for a preferred theory has demanded errors in interpretation. Indeed not only have Mather and Smoot won the Nobel prize for the Big Bang, but Smoot quoted Kiyoshi Shiraishi in saying: “It is impossible that the Big Bang is wrong" [5].

The 2006 Nobel Prize in Physics was awarded to J.C. Mather and G.F. Smoot "for their discovery of the blackbody form and anisotropy of the cosmic microwave background radiation" from measurements made by the COBE satellite launched by NASA in 1989. All that is fine; but Hubble discovered a redshift distance law not expansion of the universe, so we have trouble with this statement by the Swedish Academy: 'The COBE results provided increased support for the Big Bang scenario for the origin of the Universe, as this is the only scenario that predicts the kind of cosmic microwave background radiation measured by COBE'.

It is said that COBE measurements also marked the inception of cosmology as a precise empirical science. According to Big Bang theory, the cosmic microwave background radiation is a relic of the earliest hot phase of the Universe immediately after the big bang which has gradually cooled as the Universe has expanded which today corresponds to a blackbody temperature 2.75 degrees above absolute zero (equivalent to -273.15° C or -459° F). The COBE measurements revealed a perfect blackbody spectrum for the microwave background radiation. But this is the same scenario as what one would expect from a cosmological QED blackbody cavity without an initial Big Bang singularity for a static universe model for example. And this temperature is precisely what Eddington was able to calculate from fundamental parameters [6].

> The microwave 'background' makes more sense as the limiting temperature of space heated by starlight than as the remnant of a fireball - Sir Arthur Eddington [6].

History has repeated itself. Not in hundreds of years have such conditions existed in science. Copernicus and Galileo were nearly executed for their views. Today the 'murder' has taken a different approach: Young scientists are not given funding or tenure if they try to pursue research avenues not considered politically correct [7].

The cosmological principle which the Big Bang is based on states that the universe is homogeneous, isotropic and time dependent wherein 4D spacetime is described by the Friedmann-Robertson-Walker metric [8,9]

$$ds^2 = -dt^2 + a^2(t)\left[\frac{dr^2}{1-kr^2} + r^2\left(d\theta^2 + \sin^2\theta d\varphi^2\right)\right] \qquad (1.1)$$

which is an exact solution to Einstein's field equations

$$G_{\mu\nu} + \Lambda g_{\mu\nu} = \frac{8\pi G}{c^4} T_{\mu\nu} . \tag{1.2}$$

Only one of the Friedmann solutions to Einstein's field equations is stationery which Einstein chose for his static cosmological model by introducing a cosmological constant, Λ equal to $\Lambda_E = 4\pi G\rho / c^2$ with G Newton's constant and ρ the cosmic matter energy density. Einstein then added the cosmological constant to General Relativity in order to counteract the effects of gravity which in a universe full of matter would cause the universe to collapse. By putting $\dot{a} = \ddot{a} = 0$ in the Friedmann equation the Einstein radius of curvature, R_E for a static universe is

$$R_E = \Lambda_E^{-1/2} \frac{c}{\sqrt{4\pi G\rho}} . \tag{1.3}$$

A number of other values of cosmological constants have been proposed by various authors and the value zero was particularly popular before 1998. The zero value of cosmological constant predicts a decelerating expansion of the universe. After 1998, when observations established beyond any reasonable doubt that the expansion of the universe seems to be accelerating, the value zero had to be given up and a quest for establishing a real value started and is still going on.

Big Bangers consider the static solution unphysical because of their interpretation of the Hubble redshift as a Doppler shift indicative of an expanding universe. Einstein's field equations do allow the possibility of singularities allowing for the putative occurrence of Big Bang singularities and black holes. The Big Bang's main strengths have been interpreting a Hubble redshift distance relation that appears to coincide with the age of the universe derived from it. Although of course each time the model ran into trouble new parameters like inflation and quintessence were added to fix the problem. The other main pillar is the Cosmic Microwave Background Radiation (CMBR) which is theorized to have cooled to be sufficiently isotropic with a black body temperature spectrum to support the model. There are of course other unpopular interpretations for these two parameters that will be addressed in this volume. A number of problems remain that are sufficiently threatening to warrant the exploration of alternative considerations.

Perhaps never in the history of science has so much quality evidence accumulated against a model so widely accepted within a field. Even the most basic elements of the theory, the expansion of the universe and the fireball remnant radiation, remain interpretations with credible alternative explanations. One must wonder why, in this circumstance, four good alternative models are not even being comparatively discussed by most astronomers [1].

THE HORIZON PROBLEM

According to the Big Bang the CMBR received in the current epoch originated after the primordial explosion at the time, T_d when matter and radiation 'decoupled' for a cosmological temperature considered to be $T_d \approx 3,000^\circ K$. The decoupling time, t_d is calculated by the formula

$$\frac{T_0}{T_d} = \frac{2.73^\circ K}{3,000^\circ K} = \frac{a(t_d)}{a(t_0)} = \left(\frac{t_d}{t_0}\right)^{2/3} \tag{1.4}$$

yielding a $t_d \approx 200,000 \quad h^{-1}$ years which in this scenario corresponds to a distance the CMBR photons traveled since emission of

$$a(t_0)\int_{t_d}^{t_0}\frac{dt'}{a(t')}3t_0\left[1-\left(\frac{t_d}{t_0}\right)^{2/3}\right] \approx 3t_0 \approx 6,000 \quad h^{-1}Mpc \tag{1.5}$$

coinciding with the present particle horizon size [10].

The problem is that this decupling horizon allows the sky to split into $\approx$ 14,000 causally separated patches sending light to us. The difficulty arising is how can the black body radiation temperature from all these patches be so well tuned [10]?

THE FLATNESS PROBLEM

The current energy density of the universe is observed to be asymptotically flat corresponding to a matter dominated universe. The unsolved question surrounding this issue is why should the initial energy density of the universe be so finely tuned as to be equal to its critical value, $k = 0$ in Eq. (1.1).

DENSITY FLUCTUATION

In order to have structure formation in a Big Bang cosmos there must be a primordial density perturbation, $\delta\rho / \rho$ at all scales and an explanation of the causality violation coinciding with the horizon problem above. Additional difficulties are that in a matter dominated inflationary universe expansion separates particles slowing structure formation and in a radiation dominated universe there is no structure formation [10]. The Big Bang offers no explanation for how these primordial density fluctuations originate.

EX NIHILO CREATION PROBLEM

How could the universe arise from nothing? What triggered the creation process [11]? Something arising from nothing is a logical contradiction. The Greek and Hebrew terms for creation suggest 'built from or organized from existing materials', but what is the cause of the original instability? *Creatio ex Dios?*

Steinhardt and Turok have proposed a cosmology where space and time always existed [12]. By using string theory they claim the Big Bang was a bridge to a pre-existing universe. Using this idea they speculate that individual creations could undergo eternal successions, with trillions of years of evolution between each Big Crunch and Big Bang.

LARGE SCALE STRUCTURE & AGE OF UNIVERSE PROBLEM

Recent observations have shown that the size of large galactic structures in the 'Great Wall' are far too large to have formed in 10 to 20 billion years. The new data shows a universe full of super-structures and companion super-voids with a scale of about 10^9 light years that could require over 100 billion years to form by gravitational attraction. This problem, by itself, should be strong enough to discard the Big Bang model and replace it with a new one.

COSMOLOGICAL CONSTANT – DARK ENERGY PROBLEM

Until recently cosmologists assumed the cosmological constant to have a value of zero because it predicted a decelerating expansion of the universe. But the discovery of acceleration caused a revival of the cosmological constant as a mechanism for explaining dark energy. Dark energy in Big Bang cosmology has recently been recast as a scalar field called Quintessence [13] to explain the cause of the observed acceleration of the universe. Quintessence is sold as a way of replacing the cosmological constant with a negative energy pressure of magnitude equal to the positive energy density

$$p = -\rho c^2 \tag{1.6}$$

This Quintessence replacement energy is derived by

$$\rho \to \rho + \frac{\Lambda c^2}{8\pi G}; \quad p \to p - \frac{\Lambda c^4}{8\pi G}. \tag{1.7}$$

where to create an accelerating expansion term the scalar Quintessence field must be

$$p < -\rho c^2 / 3 \text{ [13].} \tag{1.8}$$

QUASAR REDSHIFT-LUMINOSITY PROBLEM

The redshift and luminosity of some Quasars is far greater than would be expected by their distance or compared to the galaxy they are located within [14]. Some are purported to be at the limit of observation suggesting they are both too old to have formed in the time since the Big Bang and also farther away than the calculated Hubble radius.

More troubling problems occurred in the spectrum of blueshifted objects like ESO 323-G077 that contain three times more iron than possible for their age [15].

OTHER PROBLEMS

- Galactic rotation speeds suggest 45 to 60 rotations since the Big Bang which is not sufficient time to achieve a spiral shape. Many spiral galaxies are observed at large distances for times closer to the Big Bang indicating time for even fewer rotations. Recent Hubble images show spiral galaxies within 5% of the big bang time leaving time for only 2 or 3 rotations at the Milky Way galaxy's rotation rate [16].

- Can galaxies collide if they are flying away from each other?

- Galactic redshift surveys reveal a regular spacing a quarter of the way to the time of the putative Big Bang origin, but Big Bang theory says they should be closer together the closer they are to the time of the Big Bang [16,17].

- Old galaxies are observed near the time of Big Bang origin with insufficient time to evolve.

- High energy cosmic rays are observed at energies beyond a theoretical cutoff for a hot Big Bang conflicting with the postulated CMBR temperature in the early universe. Cosmic ray protons or atomic nuclei traveling through space at speeds approaching the speed of light would have been attenuated by a high temperature radiation field [16,17].

There remains something disquieting about this model. It contains a huge array of variables that can be changed pretty much at will. So flexible is it that some claim the model can be stretched to fit any observation [18].

Nowadays, it sometimes appears that the Big Bang model for the origin of the Universe is accepted as established fact, rather than simply another theory – albeit one with a multitude of ardent supporters and which seems to explain so much so satisfactorily. However, problems do remain and many have been addressed in the past by allowing additions to the basic theory – a privilege not afforded to rival theories [19].

The temperature of intergalactic space was predicted by Guillaume, Eddington, Regener, Nernst, Herzberg, Finlay-Freundlich and Max Born based on a universe in dynamical equilibrium without expansion. They predicted the 2.7 degree K background temperature prior to and better than models based on the Big Bang [20].

1.2 A New Cosmological Horizon

After Hubble's discovery of the redshift distance relation, Fritz Zwicky did not agree that the redshift should be interpreted as a Doppler expansion of the universe and he suggested that a static universe would still be viable if an alternative explanation of redshift by a mechanism causing photons to lose energy as it traveled through space could be developed [21]. This is the 'tired-light' concept developed in Chap. 7.

A static universe also has to describe a process for creation of Hydrogen since in an ancient or eternal universe there would no longer be star formation when the universe ran out of Hydrogen [22]. Matter creation has been addressed in the quasi-steady state model [23-28] but not by a formalism considered sufficient by most cosmologists.

The two main pillars of the Big Bang are the Doppler redshift and the Cosmic Microwave background Radiation (CMBR) which are formally addressed in Chap. 7. Here it is sufficient to make the challenge that Hubble discovered a cosmological redshift not a Doppler expansion of a Big Bang singularity. The COBE and WMAP satellites have found the CMBR to be a perfect blackbody radiation; but there is no reason to so stringently consider it solely as a relic of a Big Bang singularity.

Experimental science began in earnest when Galileo demonstrated that heavier objects do not fall faster than lighter objects in opposition to the logical reason at the time. It seems that now we have come full circle to a time when not only are there questions that science cannot answer, but that science draws seriously wrong conclusions from the data acquired. This book describes one such error in terms of the formulation of the Big Bang theory and represents one of the first applications of noetic principles to correct scientific error.

All theory formation has a metaphysical component; in addition to rigid adherence by its followers, this is another reason why some people call science a religion - whether a theory is formed before data is acquired to design an experiment or after in an attempt to model the world around us, intuition and experience play a significant role in determining the conclusions drawn.

> *A truth that represents a permanent and final grasp of some limited aspect of nature. Most people would say this is incompatible with the expectation that our theories will be falsified. I adhere to the expectation that our theories will be falsified, and look for the immutable truth only in those theories that have already been falsified. Newtonian mechanics...is an example of the most certain and permanent truth man has ever achieved. Its only failing is its scope; it does not cover everything* [29].

'The inherent tendency of an object to move toward its natural place depends on its composition. Heavy bodies, composed mainly of earth and water, are endowed with the property of gravity, a centripetal tendency to move toward the center of the universe. Light bodies, composed mainly of air and fire, are endowed with the property of levity'- Aristotle believed that the speed of a falling body is proportional to its weight. Aristotle's physics was qualitative rather than quantitative. Indeed, he believed that quantitative physics was impossible. More than 2,000 years ago, Aristotle concluded that heavier objects fall faster than lighter objects. Aristotle also surmised that the rate at which an object falls toward Earth when dropped is directly proportional to its mass, i.e., an object with twice the mass of another falls twice as fast. Reality, it was held, could be understood by pure reason - hence easily-disprovable logical errors like Aristotle's claim that a heavy object falls faster than a light one persisted for millennia. It is easy to see how scientific philosophers of antiquity embraced Aristotle's viewpoint as its apparent elegance was so intellectually appealing.

One can only speculate why a majority of scientists hold a preference for a more Darwinian-Naturalistic cosmogony rather than a deistic one. One reason might be that although it has been over four hundred years since Galileo was forced to recant heliocentrism and spend the last years of his life under house arrest by order of the Roman Inquisition, some residual resentment for the narrow mindedness with which the theocracy hindered the advance of science before the Renaissance. Another could be that while ~95% of the general population believe in some form of God only ~25 to 30% of scientists do so because it appears that scientists feel capable of finding explanations without resorting to putative deities.

Science etymologically from the Latin *scio*-'to know', is by definition supposed to be an unbiased search for truth; but this has never been the case. Human nature interjects popular or ones personal myopia. In early times the theological bias that everything in 'God's universe must be perfect – perfect spheres for example kept discovery of the heliocentric universe at bay for thousands of years. Such bias is human nature. We must confess a similar bias; but we do not profess a theistic cosmology solely for alignment with out belief system. As we hope to demonstrate in these chapters; it is the explanatory power of the anthropic cosmology that prospers the underlying predilections. We also believe that the human condition can be overcome or superceded by a second Galilean revolution – one that completes the tools of human epistemology [30].

References

[1] Van Flandern, T. (2002) The Top 30 Problems with the Big Bang, Apeiron, Vol. 9, No. 2.
[2] Arp, H.C., Keys, C.R. & Rudnicki, K (1993) Progress in New Cosmologies: Beyond the Big Bang, New York: Plenum.
[3] Kimble, S.W. (1978) Absolute Truth, Ensign, Sept 3, Salt Lake City: The Church of Jesus Christ of Latter-day Saints.
[4] Misner, C.W. (1974) Cosmology and theology, in W. Yourgrau & A.D. Breck (eds.) Cosmology, History, Theology, New York: Plenum.
[5] Smoot, G. & Davidson, K. (1993) Wrinkles in Time, quotation by Kiyoshi Shiraishi: *It's impossible that the Big Bang is wrong*, p. 69, New York: W. Morrow & Co.
[6] Eddington, Sir Arthur (1926) Internal Constitution of the Stars, Cambridge: Cambridge University Press, reprinted (1988).
[7] Smolin, L. (2006) The Trouble With Physics: The Rise of String Theory, the Fall of a Science, and What Comes Next, New York: Houghton Mifflin.
[8] Friedmann, Alexander (1922) *Über die Krümmung des Raumes, Zeitschrift für Physik* A 10: 377–386.
[9] Friedmann, A. (1924*) Über die Möglichkeit einer Welt mit konstanter negativer Krümmung des Raumes, Zeitschrift für Physik A* **21**: 326–332.
[10] Lazarides, G. (1998) Introduction to cosmology, Proceedings Corfu Summer Institute on Elementary Particle Physics; or arXiv:hep-ph/9904502v2 (2002).
[11] Lerner, E.J. (1991) The Big Bang Never Happened, New York: Random House.
[12] Steinhardt, P.J. & Turok, N. (2002) The cyclic universe: an informal introduction, arXiv:astro-ph/0204479v1; Phys. Rev. D 65:12.
[13] Ostriker, J.P. & Steinhardt, P. (2001) The quintessential universe, Scientific American, vol. 284, no. 1, January, pp. 46–53.
[14] Galianni, P., Burbidge, E.M., Arp, H., Junkkarinen, V., Burbidge, G. & Zibetti S. (2004) The discovery of a high redshift x-Ray emitting QSO very close to the nucleus of NGC 7319, arXiv:astro-ph/0409215v1.
[15] Jimenez-Bailon, E., Krongold, Y., Bianchi, S., Matt, G., Santos-Lleo, E., Piconcelli, E. & Schartel, N. (2008) Detection of blueshifted emission and absorption and a relativistic Iron line in the X-ray spectrum of ESO 323-G077, arXiv: astro-ph-0810.0513v1.
[16] Mitchell, W.C. (2002) Bye Bye Big Bang: Hello Reality, Carson

City: Cosmic Sense Books.
[17] Ashmore, L. (2006) Big Bang Blasted, www.booksurge.com.
[18] New Scientist Editorial (2005) Big bang doubts fuel cosmology boom, 02 July, issue 2506.
[19] Moore, M. & Dunning-Davies, J., Reflections and Thoughts on Tired Light, http://arXiv.org/ftp/arxiv/papers/0707/0707.3351.pdf.
[20] Assis, A.K.T. & Neves, M.C.D. (1995) History of the 2.7K temperature prior to Penzias and Wilson, Aperion, 2: 3; 79.
[21] Zwicky, F. (1929) On the red shift of spectral lines through interstellar space, Proceed. Nat. Acad. Sci., 15:10; 773–779.
[22] Hoyle, F., Burbidge, G., & Narlikar, J.V. (1993) A quasi-steady state cosmological model with creation of matter, Astro. J, 410: 437–457.
[23] Hoyle, F., Burbidge, G. & Narlikar, J.V. (1994) Astrophysical deductions from the quasi-steady state cosmology. Mon. Not. R. Astron. Soc., 267, 1007-1019.
[24] Hoyle, F., Burbidge, G. & Narlikar, J.V. (1994) Astrophysical deductions from the quasi-steady state: Erratum, Mon. Not. R. Astron. Soc., 269, 1152.
[25] Hoyle, F., Burbidge, G. & Narlikar, J.V. Further astrophysical quantities expected in a quasi-steady state Universe, Astronomy and Astrophysics, vol. 289, no. 3, pp. 729–739.
[26] Hoyle F, Burbidge G & Narlikar J, V. The basic theory underlying the quasi-steady state cosmological model. Proc. R. Soc. A 448, 191.
[27] Edward L. Wright (1994) Comments on the Quasi-Steady-State Cosmology. astro-ph/9410070.
[28] Hoyle, F., Burbidge, G. & Narlikar, J.V (1994) Note on a Comment by Edward L. Wright, astro-ph/9412045.
[29] Feynman, R.P. (1994) Character of Physical Law, New York: Random House, p. 115.
[30] Amoroso, R.L. & Pribram, K.H. (2010) The Complementarity of Mind and Body: Realizing the Dream of Descartes, Einstein and Eccles, Cambridge: MIT Press.

Chapter 2

Extending the Standard Model: Towards the Ultimate Evolution of String Theory

> *...maybe we should not try to quantize gravity. Is it possible that gravity is not quantized and all the rest of the world is?...Now the postulate defining quantum mechanical behavior is that there is an amplitude for different processes. It cannot be that a particle which is described by an amplitude, such as an electron, has an interaction which is not described by an amplitude but by a probability...it seems that it should be impossible to destroy the quantum nature of fields. In spite of these arguments, we should like to keep an open mind. It is still possible that quantum theory does not absolutely guarantee that gravity has to be quantized* [1] - R.P. Feynman.

String Theory, recast as M-Theory after the 1995 superstring revolution has remained highly controversial because until now direct methods for empirical tests of its parameters have remained elusive. There are purported to be 10^{100} candidates for the unique string background sought. One impetus for string theory in XD was the work of Kaluza and Klein showing that gravity and EM could be integrated by introducing a 5th dimension. We believe as Feynman wonders, that a search for a quantum gravity is not the way to orient the quest for a Theory of Everything (TOE). M-Theory has made great strides in developing, tinkering with and finding rich associations between the infinite parameters of string theory. Because string theory is still aligned with a Big Bang cosmology researchers seek one unique compactification from which the standard model of particle physics will drop out. Here because HAM cosmology is a complex self-organized continuous-state system we profess that all dimensionalities of compactification occur by what we call the HAM mantra: 'a continuous-state spin-exchange parallel transport deficit angle dimensional reduction compactification process' – a concatenation that is merely a fancy way of trying to elucidate SUSY breaking parameters.

2.1 Pre-Ambulatory Hoopla

Logically, since not long ago it appeared we lived in a Newtonian world, one might now assume we live in a Quantum world. By this logic should it not follow that after the unitary field is discovered one might surmise the universe is a form of unitarity as some monistic Eastern philosophies suggest. We believe this is not the case either, and suggest that the multiverse is a continuous-state Kantian antinomy [2] between the three. We realize this appears strange; but our evidence so far in examining the applications we have been able to develop based on such a view, i.e. surmounting uncertainty or universal quantum computing for example seem to indicate such a view should be embraced rather than ignored.

It appears logical in particle physics that supersymmetry (SUSY) is a symmetry that relates elementary particles of one spin to another particle that differs by a half unit of spin known as superpartners. In other words, in a supersymmetric theory, for every type of boson there exists a corresponding type of Fermion, and vice-versa. As of 2009 there is no direct evidence that supersymmetry is a symmetry of nature, a situation that physicists hope will change when the supercollider comes online late in 2009 if earlier initialization problems are overcome. We surmise the experimental protocol outlined in Chap. 9 provides a much simpler and direct low energy avenue for plumbing the HD domain making string theory readily testable. Other currently perceived 'outrageous' postulates of the continuous-state HAM cosmology are:

- There is no quantum gravity (not the regime of unification)
- There are no superpartners (sparticles)
- There is no Higg's Mechanism (New type wormhole instead)
- Photon mass anisotropy, oscillating duality, $m_\gamma \neq 0 \leftrightarrow m_\gamma = 0$
- Anthropic evolution drives self-organization within the Einstein H_R

The first three constructs, while intellectually appealing for some decades are remnants of Gauge Theory, although enormously successful, is only an approximation soon to be shown unable to continue to sustain such predictions. It is obvious that the standard model of particle physics-cosmology is incomplete (recent discovery of neutrino mass for example) justifying the alternative considerations presented here [3,4].

By current considerations supersymmetries are generated by objects transforming under a spin-statistics theorem, where spin-1 Bosonic fields commute while spin-½ Fermionic fields anticommute according to the tenets of the Copenhagen interpretation of quantum theory. By current thinking in order to combine the two fields a super-Lie-algebra is needed that doubles the number of fundamental particles (superpartners often called sparticles). The Higgs mechanism has been a primary motivation for SUSY because it entails inherent Boson-Fermion renormalization /symmetry breaking that can be formalized in XD. We believe the renormalization paradox is indicative of the immanent need for new physics in the same manner that the Raleigh-Jeans Law was indicative of the immanent appearance of quantum mechanics and that the infinities should not be considered a 'plague' but indicative of a lower order cut-off of the unitary field requiring a new set of transformations (see Chap. 5) beyond the current Lorentz/Poincairé to reveal their place in nature.

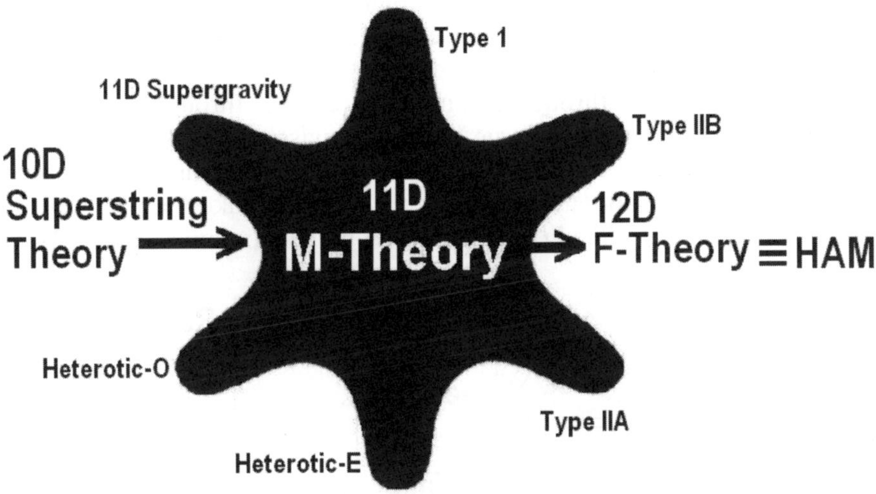

Figure 2.1 Various types of string theory combined to form M-Theory and a 12D form of F-Theory utilized by HAM cosmology to introduce the anthropic principle driving complex self-organization in the static Einstein 3-sphere, H_R.

Recent M-Theory modeling has been able to resolve the hierarchy problem [5,6] yielding insight into the disparity between the weak and gravitational interaction scales. The maximum number of super-symmetries is 32 (curiously the same as the number of crystal forms) suggesting the maximum number of dimensions is 11 [7]; but we implement a 12th XD in order to introduce the anthropic action principle driving the evolution of the complex self-organized continuous-state structural-phenomenology of the static-temporal Einstein-Hubble 3-sphere, H_R. Interestingly Smolin, architect of Loop Quantum Gravity (LQG) postulates that LQG does not require sparticles either and that LQG may turn out to be a component of M-Theory. Our problem with LQG is that it is limited to 4D; but its other interesting feature is that it is said to provide a background independent vacuum [8-13].

As stated the myopic Higgs mechanism requirement is an artifact of Gauge Theory being an approximation waiting for new physics. If no Higgs or superpartners, what then? The worm-hole-like topological dynamics of HD branes. Starting from the perspective of the Dirac spinorial spherical rotation of the electron requiring – $360^{\circ} - 720^{\circ}$; where the additional 2π rotation is indicative of rotation through HD topology before returning to the 3D point of origin. This 'Klein bottle' raising-lowering effect is amenable to a Wheeler wormhole concept where 'charge is topology'. According to Wheeler lines of force in a wormhole can thread through a handle and emerge through each mouth to give the appearance of charge in an otherwise charge free spacetime [14-16]. Wheeler originally failed; we believe because his approach was only 4D. An HD elaboration of this concept could take the place of the Higgs mechanism replacing sparticles with brane topologies (Fig. 2.2).

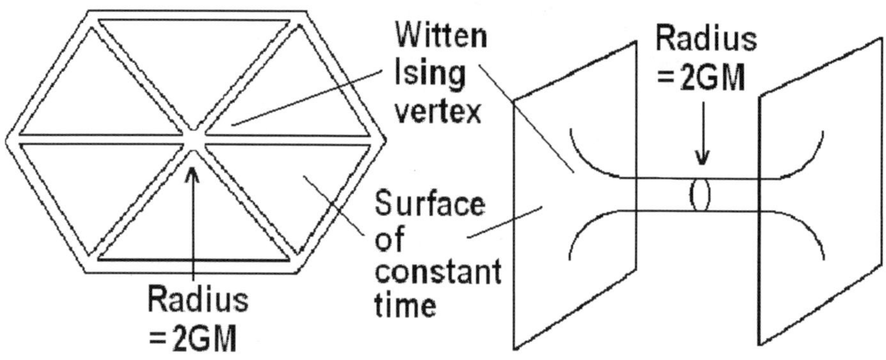

Figure 2.2 Wormhole modeling for M-Theory. When the continuous-state topological dynamics are fully understood it is postulated a form of Ising model wormhole lattice-gas structure of brane tension-coupling will be able to replace sparticles and the Higgs mechanism. If topology is charge as Wheeler postulated the standing-wave hysteresis loop in the HAM continuous-state least-unit transformation will enable such developments. Figure redrawn from [17].

Table 2.1 HAM COSMOLOGY SUSY PARAMETERS	
General Cosmological Parameters	**Symmetry Breaking Parameters**
• 12D HAM Multiverse • Anthropic Guidance Principle • Holographic Principle • Non-Compactified K-K Model • Cosmological Least-Unit • Large Scale XDs • Cosmos - A Self-Organized Complex System • Scale Invariance • Conformal Invariance • New HD Form of Absolute Space • Teleological Action – Evolution • Super Quantum Potential • Finite Photon Mass Anisotropy • Dark Energy as Λ Multiverse • 12 Dimensions • Unitary Wheeler Geon • Wheeler Wormholes • Continuous-State • Fine Tuning • Unitary Field - Quadrupole Photon - Graviton • Noeon – Topological exchange parameter of the Unitary Field • Holographic Multiiverse – Closed & finite in *t*, open and infinite in atemporal XD	• Continuous-State • Continuous Compactification • Dimensional Reduction • Spin Exchange • Parallel Transport • Deficit Angle • Bianchi Identities • Regge Calculus – Oscillating Spacetime Curvature ± 0 • Spin-Spin Coupling of Spacetime Topology • T-Duality / Mirror Symmetry • Ising Model Rotation of Riemann Sphere, $0 \leftrightarrow \infty$ • Topological Switching • Coordinate Leapfrogging • Dirac Spherical Rotation • Dirac Style Annihilation / Recreation of mass / Topology • Dynamic / Static Casimir Coordinate Leapfrogging • Strings / Branes • Standing Wave Symmetry • Rotations – circular, spherical, cylindrical, chiral, hyperspherical • Unique Vacuum Symmetry
DUAL STATIC - DYNAMIC PARAMETERS	
No Higgs Mechanism - Alternative	No Superpartners – Alternative
Arrow of Time From Large XDs	Oscillating String Tension, $\hbar + T_0$
Oscillating Cosmological Constant, $\Lambda \pm 0$	Unitary Formalism, $F_N = E / R$
$\hbar$ Not Fundamental, New Stoney Basis	Complex Self-Organization
Structural-Phenomenology	Future-Past Transaction

Table 2.1 SUSY parameters required by HAM cosmology for the continuous-state topological dynamics of dimensional reduction compactification process.

Although we propose radical changes for M-Theory we still consider it to be the best hope for a Theory of Everything (TOE); but it needs to take a lesson of some sort from LQG, revamp the concept of the Higgs mechanism and the origin of the fundamental parameters of particle physics forming sparticles such as the tension-coupling effects in the associated HD continuous-state topology. It is a 'torque' of some form in the energy dynamics of the spacetime least-unit [18,19] hysteresis loop in the SUSY breaking parameters of the 'continuous-state spin-exchange parallel transport deficit angle dimensional reduction process' that could be developed to fill this conceptual disparity if the Large Hadron Supercollider (LHC) fails to find a Higgs or neutralinos (lightest sparticles). M-Theory could still easily remain on track with SUSY requirements by incorporating the SUSY breaking parameters inherent in the alternatives presented by HAM cosmology.

String theory is currently aligned with the Copenhagen interpretation of quantum theory and Big Bang cosmology which have led to the quest for a putative 'quantum gravity'. In HAM cosmology none of these ideas form the correct basis for string theory and need to be replaced by new considerations that include a new cosmological perspective and an HD completed form of the de Broglie-Bohm-Vigier causal stochastic interpretation of quantum theory [20,21] compatible with the Transactional Interpretation of quantum theory [22] because it entails a mirror symmetry compatible with the dual Calabi-Yau 3-form and the associated SUSY breaking parameters being developed in string theory.

A snippet should be given regarding the TOE search. Recently some well known scientists like Hawking and Dyson have suggested that a TOE is impossible according to Gödel's incompleteness theorem [23] which simplistically states that nothing can be described in terms of itself because by definition that would be too limited a view; any complete description must come from outside the boundaries of the principle to be fully understood. If accepted this would appear to be a challenging philosophical conundrum; but from the HAM perspective it turns out to have a simple answer allowing one not to know everything and still have a TOE. The TOE is essentially about unifying the four fundamental forces and having essentially complete theories of particle physics with a connection between quantum theory and gravitations in a proper cosmological context. An anthropic cosmology by supposition is a complex self-organized system with the properties associated with such systems such as incursion, hierarchy and an inherent external action principle driving its self-organization. By applying Kant's antinomies [2] the Hubble sphere is closed and finite temporally, but open, infinite and causally separated in eternity such that in a Multiverse there is room for an infinite number of nested Hubble spheres each with their own fine-tuned laws of physics [24]. So a TOE is sufficiently developed for parameters within our Einstein-Hubble sphere as we begin peeking into the holographic Multiverse beyond it - compatible with Gödel's theorem.

2.2 Ultimate Evolution of M-Theory

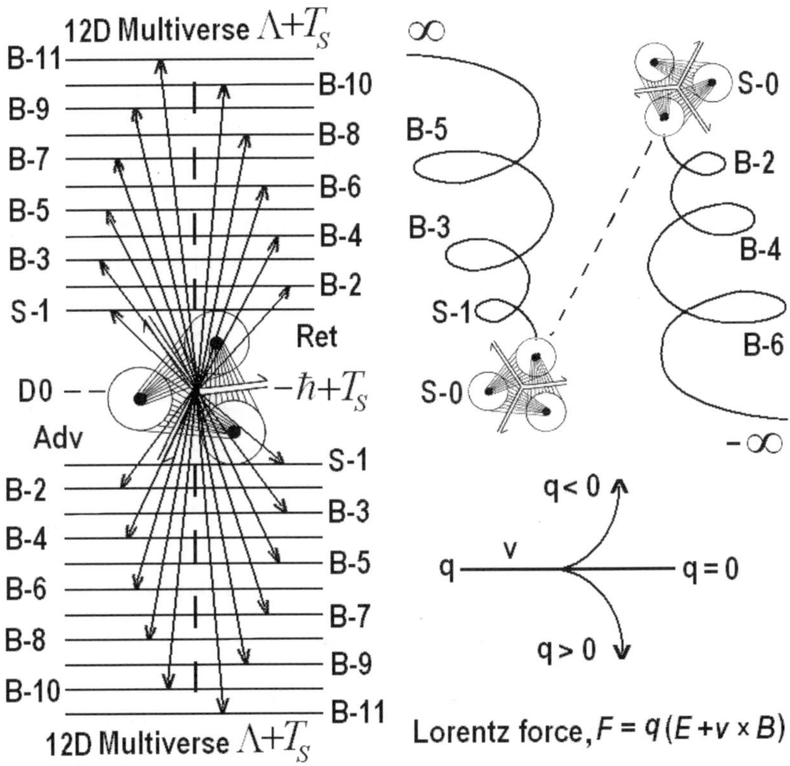

Figure 2.3 Conceptualized string (S) and brane (B) couplings in Advanced-Retarded spacetime arising from a least-cosmological unit, D0, S-0. a) String-brane duality couplings from 0 to 12D for odd-even Fermi-Bose topologies. b) Ising model spin-glass rotations which may be driven by an internal Lorentz-like force of the anthropic principle or external resonances for vacuum engineering.

Every Calabi-Yau manifold with mirror symmetry or T-duality admits a hierarchical family of supersymmetric toroidal 3-cycles as shown conceptually in Fig. 2.3. Figure 2.3a shows possible duality couplings and 2.3b is meant to illustrate the compactification-boost hierarchy as modeled by a Genus-1 helicoid 'parking garage' structure (Fig. 12.7). It is currently unknown whether the attempt to formalize this continuous-state structure should follow a Kaluza-Klein spin tower, logarithmic or golden ratio spiral, cyclotron resonance hierarchy, genus-1 helicoid 'parking-garage' or some other HD structure. We currently find the Genus-1 helicoid the most intellectually appealing because of its ability to incorporate Kahler manifolds compatible with M-Theory parameters listed in Table 2.1. Also of note is that the heterotic SO(32) Bosonic string introduces a tachyon which we do not consider anomalous but part of the internal field coupling of a Lorentz vacuum contraction. Type IIA & Type-IIB open/closed strings are cast in odd/even string/brane dimensionality which we postulate is an inherent part of the Ising model rotation of the Riemann sphere for 'parking-garage' helicoid raising-lowering indices of the continuous-state dimensional reduction compactification process. See Chap. 3. These complex constructs can only be adequately worked out with a move away from a Big Bang cosmology and limits imposed by Copenhagen-Gauge approximations.

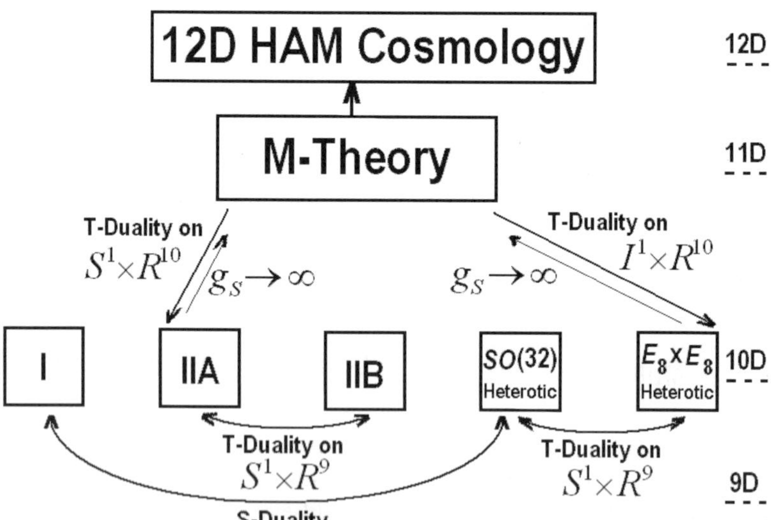

Figure 2.4 Mirror/duality transformations relating the 5 superstring theories to each other and the anthropic principle of HAM cosmology. Adapted from [25].

It is well known that it is possible to have supersymmetry in alternate dimensions. Because the properties of spinors change dramatically with dimensionality; each dimension has its own characteristics. In d dimensions, the size of spinors is roughly $2^{d/2}$ or $2^{(d-1)/2}$. Since the maximum number of supersymmetries is 32, the largest number of dimensions a supersymmetric theory can have is 11D. It is possible to have multiple supersymmetries and also have supersymmetric XDs.

If we accept the postulate of M-Theory that matter resides on the 3-brane along with the associated boundary conditions underlying the spinor elements of matter; along with duality/mirror symmetry this takes care of 6D. HAM cosmology is cast in 12D – 3 more for time and the final 3 for 'piloting' or the anthropic teleology for the continuous-state evolution of spacetime. HAM cosmology is built on the premises of extended EM theory [26-30], a covariant polarized Dirac vacuum [21,31,32] with photon mass anisotropy [33,34] giving the photon an internal motion coupling it to the vacuum. Since photons are not fermions its brane dynamics is different (simpler). Further we posit the photon as a periodic temporal 'pinch' of the continuous coherent unitary field – a geon or 12D ocean of light [35]. This could be one of the greatest contributions of this volume when properly understood.

2.3 String/Brane Dynamics

The purpose of this section is to illustrate the richness of string/brane transformations and to review the myriad fundamental component bases of the transformations not to necessarily demonstrate any particular action. In general this will include:

- String/brane action in 0 to 9D
- Linear, circular, cylindrical, spherical, chiral and hyperspherical rotations boosts transformations and compactifications.
- SUSY breaking
- Mirror symmetry, T-Duality
- Open – closed string-brane transformations
- String/brane tension-coupling dynamics
- Mass/energy/gravitation deficit angle parallel transport
- Annihilation/creation dynamics
- Teleological/anthropic action driving or piloting complex systems.

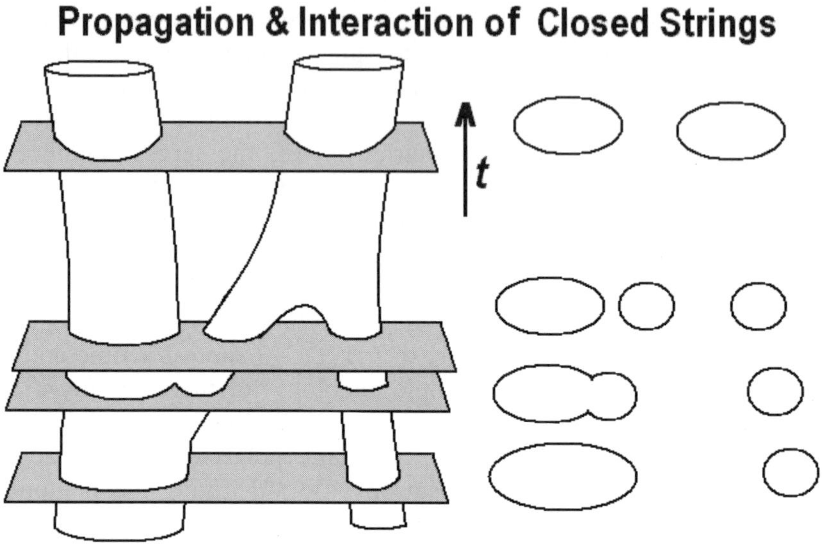

Figure 2.5 2D & 1D 'Pants diagram' for the topology of string interactions.

The five superstring models of M-Theory are:

- Type-I strings having one supersymmetry in 10D. Type-I strings are unoriented and open or closed while all the other types are oriented closed strings.
- Type-IIA & IIB string theories contain two 10D supersymmetries which differ in that IIA theory is non-chiral or parity conserving and IIB theory is chiral or parity violating.
- Heterotic strings, so named because they are left-moving and right-moving, are a hybrid of a type-I and Bosonic strings. There are two kinds of heterotic strings, the $E_8 \times E_8$ and SO(32) string.

In type-II string theories closed strings are free to move through the 10D bulk of spacetime, but the ends of open strings attach to D-branes. In type-IIA their dimensionality is odd – 1,3,5,7 and even in type IIB – 0,2,4,6. See Fig. 2.3. Through different gauge symmetry conditions various types of strings or branes are related by S-duality which relates the strong coupling limit of one type to the weak coupling limit of another type. T-duality relates strings/branes compactified on a circle of radius R, to strings/branes compactified on a circle of radius $1/R$.

Figure 2.6 Five possible open-closed string interactions which are forms of topological transforms.

Following work by Sundrum [36] for 5D General Relativity where the Einstein action is $\ni \partial_\mu$ or $\partial_5 Gr^0_{MN}(x) \to 0$ for large XD fluctuations

$$ds^2 \ni Gr_{55}\left(dx^5\right)^2 = Gr_{55}R^2 d\theta^2 \Rightarrow Gr^{(0)}_{55}(x) \equiv \text{dynamical XD radius.}$$

Randall and Sundrum [37] have found an HD method to solve the hierarchy problem by utilizing 3-branes with opposite tensions, $\pm\sigma$ residing at the orbifold fixed points which together with a finely tuned cosmological constant from sources for 5D gravity for a spacetime with a single S_1/Z_2 XD orbifold [38-40]. These 3-branes with opposite tensions residing at the orbifold fixed points along with their model of a finely tuned cosmological constant serve as sources for 5D gravity.

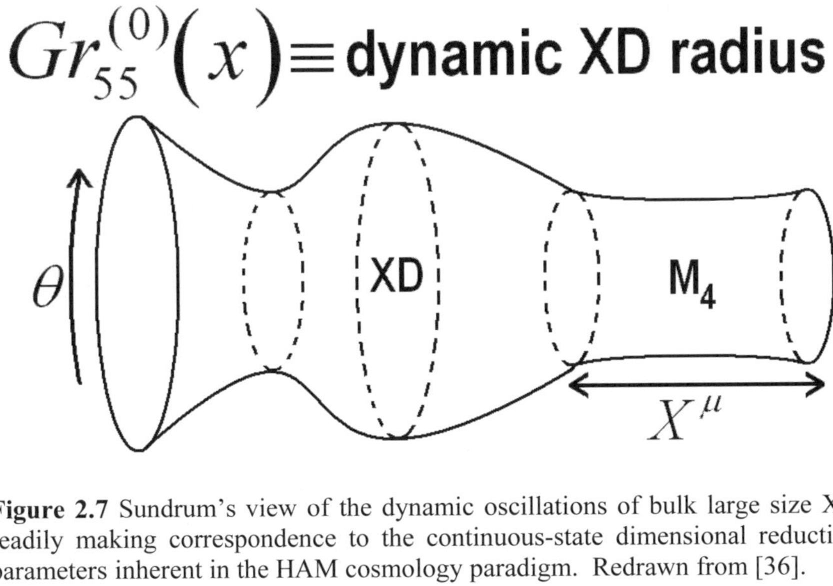

Figure 2.7 Sundrum's view of the dynamic oscillations of bulk large size XD readily making correspondence to the continuous-state dimensional reduction parameters inherent in the HAM cosmology paradigm. Redrawn from [36].

2.4 New Horizons Beyond the Standard Model

We did not choose to model our new cosmology after the steady-state or quasi-steady-state models of Bondi, Gold, Hoyle and Narlikar because they are set in expanding universe models. However in our continuous-state model continuous matter creation must still be addressed, which we do briefly at the end of Chap. 7. It is interesting that the matter creation requirement can be satisfied with 'one atom per cubic meter each 10 billion years, or about 1 atom per 10 cubic kilometers per year' [41].

The Standard Model of particle physics provides a good correlation with experimental data, but there are phenomena not accurately described by the theory which physicists hope might be resolved by experiments performed with the large hadron supercollider:

- The hierarchy problem
- The missing mass problem (dark matter and dark energy)
- The cosmological constant problem
- The strong CP problem
- Higgs Boson
- Non-zero neutrino mass

In the mid 1970's a postulate was bandied about the Stanford University physics dept. by a young graduate student (name forgotten) that went like this: '*if one assumes matter is a vector gluon, the leading light-cone singularity is modulated by a phase of the quark-gluon field*' which is not considered a sufficient descriptor. String theory postulates that matter is coupled vibration modes of string topology. The other element required is a mass producing Higgs spinor/twistor mechanism or an alternative like the one suggested here with wormhole dynamics as the basis. One might be willing to loosely accept that these elements provide an adequate conceptual framework for describing matter. The Stanford student's mantra adds an essential consideration especially if one embraces the de Broglie-Bohm model of relativistic quantum field theory where the wave-particle duality is not an either-or statistical complement but a physically real piloted matter-wave simultaneity considered highly relevant to our continuous-state self-organized postulate introduced in Chaps. 3 and 4. The electron is probably the simplest Fermion and the photon the simplest Boson. If string tension-coupling considerations are introduced into the Dirac spinor spherical rotation parameters one might begin to rigorously formalize the electron in terms of string theory. This is especially interesting in terms of the fact that high energy photons will undergo electron-positron pair production.

The highly symmetric supersymmetry of the Cramer-HAM vacuum might make it an easy task to design scattering experiments (when taking the additional protocols introduced in Chaps. 9 and 11 into account) to reveal the electrons ultimate fundamental structure. We predict that these high energy collisions produce a momentary superluminal boost (Lorentz transformation) to the photon's ($m_\gamma \neq 0$) internal motion, which because of the 'perfect' symmetry conditions allows an electron-positron pair (of equal total energy) creation. Already Chatterjee & Banerjee [42] have shown that angular momentum is conserved in the 5th dimension. We delve into this a bit more in Chap. 7 relative to the black-body cavity-QED exciplex model. Unraveling the fundamental structure of the photon may be simpler because the photon does not have the additional domain walls the electron has to keep it from unraveling when brought to rest. But as Einstein said 'every Tom, Dick and Harry believes they comprehend the photon nowadays; but they are sadly mistaken'. We take a stab at extending the understanding of the nature of the photon in Chap. 5 in the context of the arrow of time and putative nature of a photon-graviton complex as an sub-element of the unitary field.

Icosahedral symmetry has been used to illustrate how higher-lower dimensionality might enfold the vacuum state through a higher-dimensional polyhedron that Coxeter [43] described as Polytope 2,21 consisting of 27 points evenly distributed over the surface of a 5D sphere embedded in a 6D space that may have relevance to the study of stringy vacuum geometry [44].

Currently few physicists have reason to suspect that gravity should not be quantized. Geometrodynamics is a classical theory. Physicists have been busy quantizing or trying to quantize all classical domains. Because of the move from Newtonian mechanics to quantum mechanics most physicists have decided that we live in a quantum universe. Physicists might suppose that this includes the idea that the unified field is not a similar cosmological condition but just the integration of all forces and fields from within this quantum cosmology. Here we suggest that the multiverse in the reductionist sense is ultimately unitary, not in a monistic sense but rather one of a continuous-state complementarity between classical quantum and unitarity. Much of our motivation arose by noetic insight from Plato 'no matter how great one's intelligence or how broad one's knowledge base, noetic or transcendent insight is greater because it comes from beyond the individuals abilities' [24,45].

References

[1] Feynman, R.P. (1971) Lectures on Gravitation, Pasadena: California Inst of Technology.
[2] Kant, E. (1787) *Kritik der Reinen Vernunft*, Critique of Pure Reason, (1996) W. Pluhar (trans.), Indianapolis: Hackett.
[3] Gonzalez-Garcia, M.C. & Nir, Y. (2003) A review of evidence of neutrino masses and the implications, Revs. Mod. Phys. 75, pp.345–402.
[4] Pontecorvo, B. (1983) Pages in the Development of Neutrino Physics, *Usp.Fiz.Nauk*,141, 675; English Ed. Sov. Phys. Usp. 26, 1087.
[5] Sundrum, R. (2005) SSI lecture notes part 2, www.slac.stanford.edu/econf/C0507252/lec_notes_list.htm.
[6] Randall, L. (2005) Warped Passages: Unraveling the Mysteries of the Universe's Hidden Dimensions, New York: Harper-Collins.
[7] Schwarz, J.H. (1996) Anomaly-free supersymmetric models in six dimensions Physics Letters B, 371: 3–4; 223–230.
[8] Ling, Y. & Smolin, L. (1999) Supersymmetric spin networks and quantum supergravity, arXiv.org/abs/hep-th/9904016.
[9] Konopka, T., Markopoulou, F. & Smolin, L. (2006) Quantum graphity, arXiv.org/abs/hep-th/0611197.
[10] Smolin, L.(1992) The future of spin networks, in quantum gravity and cosmology, J. Perez-Mercader et al. (eds.) Singapore: World Scientific; or gr-qc/9702030.
[11] Barrett, L. & Crane, L. (1998) Relativistic spin networks and quantum gravity, J. Math. Phys. 39, 3296–3302; or gr-qc/9709028.
[12] Smolin, L. (2000) A candidate for a background independent formulation of M theory, ArXiv: hep-th/9903166.
[13] Smolin, L. (2006) The case for background independence, in D. Rickles, et. al. (eds.) The Structural Foundations of Quantum Gravity.
[14] Lemos, N.A. & Monerat, G.A. (2003) A quantum cosmological model with static and dynamic wormholes, General Relativity and Gravitation, Vol. 35, No. 3.
[15] Wheeler, J.A. (1955) Phys. Rev. 97, 511.
[16] Hawking, S.W. (1988) Wormholes in spacetime, Phys. Rev. D 4: 37; 904–910.
[17] Hawking, S.W. (1994) Wormholes in dimensions 1-4, in P. Nath & S. Reucroft (eds.) Proceedings of the 1st Int. Symposium on Particles, Strings & Cosmology, New Jersey: World Scientific.

[18] Amoroso, R.L. & Amoroso, P.J. (2004) The Fundamental Limit and Origin of Complexity in Biological Systems: A New Model for the Origin of Life, in D.M. Dubois (ed.) Computing Anticipatory Systems, AIP Conf. Proceedings V. 718, pp. 144–159, Melville: Am. Inst. Phys.
[19] Amoroso, R.L. (2006) The Holographic Conscious Multiverse, in R.L. Amoroso, B. Lehnert & J-P Vigier (eds.) Beyond the Standard Model: The Search for Unity in Physics, Orinda: Noetic Press.
[20] Bohm, D. & Vigier, J-P (1954) Model of the causal interpretation of quantum theory in terms of a fluid with irregular fluctuations, Phys. Rev. 96:1; 208–217.
[21] Vigier, J-P (2000) Selected papers, in S. Jeffers, B. Lehnert, N. Abramson, & L. Chebotarev (eds.) Jean-Pierre Vigier and the Stochastic Interpretation of Quantum Mechanics, Montreal: Aperion.
[22] Cramer, J. (1985) The transactional interpretation of quantum mechanics, Rev. Mod. Phys. 58, 647–687.
[23] Smullyan, R.M. (1992) Gödel's Incompleteness Theorem, Oxford: Oxford Univ. Press.
[24] Amoroso, R.L. & Pribram, K.H. (2009) The Complementarity of Mind and Body: Realizing the Dream of Descartes, Einstein and Eccles, Cambridge: MIT University Press.
[25] Szabo, R.J. (2002) Busstepp lectures on string theory: An introduction to string theory and d-brane dynamics, arXiv: hep-th/0207142v1.
[26] Lehnert, B. (2002) New developments in electromagnetic field theory, in R.L. Amoroso, G. Hunter, M. Kafatos & J-P Vigier (eds.) Gravitation & Cosmology: From the Hubble Radius to the Planck Scale, Dordrecht: Kluwer Academic.
[27] Lehnert, B. (1998) Electromagnetic theory with space-charges in vacuo, in G. Hunter, S. Jeffers & J-P Vigier (eds.) Causality and Locality in Modern Physics, Dordrecht: Kluwer Academic.
[28] Proca, A. (1936) *Compt. Rend.* 202, 1420.
[29] Evans, M. & Vigier, J-P (1994) The Enigmatic Photon Vols. 1-5, Dordrecht: Kluwer Academic.
[30] Roy, S. & Lehnert, B. (1998) Extended Electromagnetic Theory; Space Charge in Vacuo and the Rest Mass of the Photon, Singapore: World Scientific.
[31] Dirac, P.A.M. (1952) Nature (London) 169, 702.
[32] Cufaro, N., Petroni, C. & Vigier, J-P (1983) Dirac's ether in relativistic quantum mechanics, Found. Phys. 13; 253–286.

[33] Vigier, J-P (1990) Evidence for nonzero mass photons associated with a vacuum-induced dissipative redshift mechanism, IEEE Trans Plasma Sci, 18:1;64-72.
[34] Amoroso, R.L., Kafatos, M. & Ecimovic, P. (1998) The origins of cosmological redshift in spin exchange between Planck scale vacuum compactification and nonzero rest mass photon anisotropy, in G. Hunter, S. Jeffers & J-P Vigier (eds.) Causality & Locality in Modern Physics, Dordrecht: Kluwer Academic.
[35] Wheeler, J.A. (1955) Geons, Physical Review, 97:2, 511-536.
[36] Sundrum, R. (2005) SSI lecture notes part 2, www.slac.stanford.edu/econf/C0507252/lec_notes_list.htm.
[37] Randall, L. & Sundrum, R. (1999) A large mass hierarchy from a small extra dimension, Phys. Rev. Let. 83, 3370-3373; arxiv/hep-ph/9905221.
[38] Arkani-Hamed, N., Dimopoulos, S., Kaloper, N. & Sundrum, R. (2000) A small cosmological constant from a large extra dimension, arxiv:hep-th/0001197v2.
[39] Rizzoa, T.G. & Wells, J.D. (1999) Electroweak precision measurements and collider probes of the standard model with large extra dimensions, arXiv:hep-ph/9906234v1.
[40] Randall, L. & Schwartz, M.D. (2002) Unification and the hierarchy from AdS5, Phys. Rev. Let. 88, 081801; Arxiv/hep-th/0108115.
[41] Silk, J. (1989) The Big Bang, New York: W.H. Freeman & Co.
[42] Chatterjee, S. & Banerjee, A. (2004) C-field cosmology in higher dimensions, General Relativity & Gravitation, 36:2; 303-313.
[43] Coxeter, H.S.M. (1940) The polytope 2,21, whose twenty-seven vertices correspond to the lines on the general cubic surface, American Journal of Mathematics, v62, pp. 457-486.
[44] Saniga, M. (2000) Twenty-seven lines on a cubic surface and heterotic string spacetimes, arXiv:physics/0012033v1.
[45] Mitchell, E. (1974) Psychic Exploration; a Challenge for Science, New York: Putnam.

Chapter 3

Fundamental Parameters for a Continuous-State Holographic Anthropic Multiverse

It is sensible and prudent...to think about alternatives to the standard model, because the evidence is not all that abundant...and we do know that the standard cosmological model is pointing to another surprise...because (it) *traces back to a singularity* - P.J.E Peebles [1].

Although popular, Big Bang cosmology still contains critical untested assumptions and unresolved logical conflicts. Recent observational and theoretical insights suggest it has become feasible to consider developing a new standard model of cosmology. Parameters for developing such a Continuous-State Holographic Anthropic Multiverse (HAM) cosmology are developed herein. The new HAM cosmology is based primarily on a fundamental least cosmological unit tiling the spacetime backcloth of its 12D superspace that makes correspondence with the SUSY parameters of M-Theory, introduces the origin of complexity in self-organization and refines the role and nature of the observer in physical theory.

3.1 Introduction to the Cosmological Issues

We have recently entered one of the periodic transitional phases in the evolution of fundamental theories of physics, giving sufficient pause to reinterpret the general body of empirical data. Recent refinements in observation of cosmic blackbody radiation [2] and various programs of theoretical modeling [3,4] suggest that it might be reasonable to explore replacing the naturalistic Big Bang cosmology. A Continuous-State Holographic Anthropic Multiverse (HAM) based on alternative interpretations of the observational data is introduced in preliminary form. We begin by re-examining the main pillars of the Big Bang, briefly review alternate interpretations, and then introduce some of the alternative general parameters for HAM cosmology.

Reviewing the historical development of physical theory illustrates the fact that two general models, one unitary and the other dualistic, have evolved simultaneously in the scientific literature:

- Unitary Model. Naturalistic, Darwinian, Newtonian; a classically oriented model aligned with current interpretations of the standard models - i.e. Big Bang Cosmology, Bohr's phenomenological interpretation (Copenhagen) of Quantum Theory, standard Maxwellian electromagnetism (EM) and Einstein's General theory of Relativity. Many unanswered questions like the breakdown of Maxwell's equations at singularities remain.

- Dualistic Model. Includes all conventional wisdom pertaining to the above model plus extended theory like the de Broglie, Bohm, Vigier, models of quantum theory implying a covariant polarizable Dirac vacuum with additional parameters and interactions like a massive photon, m_γ and where Maxwell's equations do not cutoff at the vacuum. Best evidences are the Casimir, Zeeman & Aharanov-Bohm effects. Offers plausible explanations for unanswered questions like the Proca equation for EM theory. The model also allows room for teleological causalities.

Only in the context of the dualistic parallels of *extended theory* can a HAM cosmology be viably presented. The concept of a covariant polarizable Dirac vacuum introduces an additional causal order not deemed acceptable in physical theory because it was considered unreasonable that spacetime could contain such an ordered periodicity or significant additional symmetry. As discussed below a dual causality and additional vacuum symmetry invites extension of the Wheeler/Feynman [5] radiation law beyond Cramer's [6] transactional interpretation of quantum theory to string/brane topological dynamics of spacetime topology itself where an 'eternal' *present state* [7] is comprised of a continuous *future-past advanced-retarded* HD standing-wave [8,9].

The HAM is intended as the next evolutionary step in the progression of modern cosmological modeling stemming from Einstein's 1917 proposal of a Static Universe (ESU) [10] and the banner 1948 development of both the Steady-State Universe (SSU) of Bondi, Gold [11] & Hoyle [12] and the Big Bang by Alpher, Bethe & Gamow [13]. Although HAM cosmology could be considered a form of ESU or SSU modeling, it is sufficiently different to require a proliferation of nomenclature. For example the HAM has neither inflation or expansion; and the HAM is not confined to the limits of the 3(4)D+ N_c Einstein/Minkowski/Riemann/Hubble sphere, H_R of the current standard Big Bang and SSU models.

The HAM paradigm introduces a revolutionary structural change in the universe. The Hubble sphere, H_R represents only an observational limit not the physical limit as in Big Bang cosmology. Fundamental HAM space is a complementarity of a new absolute 12D space and our observed E_3/M_4 relational spacetime. HAM cosmology has HD holographic-like properties entailing a Multiverse of a potentially infinite number of nested *relational* Hubble-type domains, each with different fine-tuned laws of physics and complete causal separation from our 3D Euclidean, 4D Minkowski, E_3/M_4 realm [14]. The additional compact subspace dimensions, N_c [15,16] hypothesized as compactified in the initial Big Bang event are not a subspace of our E_3/M_4 domain, rather in HAM cosmology E_3/M_4 is the subspace of the 12D superspace.

'Our' whole *relational* Hubble sphere, H_R is a subspace of an absolute 12D hyperspace without dimensionality as now defined. Additional dimensions are not compact, but 'open' and of infinite size [17,18], undergoing a process of 'continuous compactification and dimensional reduction' for the benefit of the Earthly observer as the complex HD 'standing wave' of the present is continuously created and recreated by future-past advanced-retarded SUSY breaking dynamics. The idea of dimensionality in HAM cosmology is a tricky business on first bite. Under the umbrella of a 12D atemporal, timeless or eternal absolute space of infinite size dimensionality (the footstool of the Multiverse relative to our H_R) a domain of spacetime drops out. The properties of this spacetime are solely for the benefit of the observer imbedded in it. Spacetime dimensionality is a scaled continuum; SUSY properties are for large scale dimensionality > 4D and symmetry breaking < 4D compactifies to 0D where an Ising flip of the Riemann 3-sphere (least unit [19]) begins the process over again. As will be clarified in Chap. 5 this is what gives rise to the arrow of time and why the additional dimensions are not directly observed. It is the basis of observed reality.

3.2 Clarification of Pertinent Cosmological Nomenclature

Since the terms Holographic, Anthropic and Multiverse have many disparate uses it seems best to clarify these key terms before we begin to earnestly delineate the properties of HAM cosmology. This discussion is not exhaustive or even very detailed; it is only to provide an introduction and to distinguish our view from the others. Our viewpoint is neither the popular nor politically correct view. But this is the way it often is in the history of science, leapfrogging from one pole to the other.

MULTIVERSE COSMOLOGY

Generally Multiverse, sometimes called meta-universe or megaverse is the hypothetical set of multiple possible universes (including our Einstein-Hubble universe) that together comprise all of reality. The different universes within the multiverse are sometimes called parallel universes. The structure of the Multiverse, the nature of each constituent universe and the connection between them depends on the particular multiverse hypothesis being considered by the theory. But the term universe is supposed to represent the entirety of all existence; however with usages like 'Mr. Tompkins Universe' or the universe of the ant, one has become accustomed to the idea of many universes at least in the common vernacular. Interestingly the term Multiverse was first coined in 1895 by American psychologist William James. In scientific circles many disparate definitions of the Multiverse exist such as parallel universes, Bubble universes, alternate realities as in Everett's Many Worlds interpretation of quantum theory containing every possibility, or the 11D extension of string theory known as M-theory where our universe and others are purported to be created by collisions between membranes in an 11D space. An attempt to clarify and classify the various forms of possible Multiverse cosmologies into four types has been presented [20] but the attempt seems to have caused some controversy of its own. Unfortunately we contribute to the diversity of multiverse forms. For HAM cosmology the multiverse is an ensemble of holographically embedded Hubble domains each causally separated from each other and each with their own fine-tuned laws of physics. Each Hubble domain is a self-organized complex system and as such operates with the principles and dynamics attributed to such systems such as incursion and scale-invariance [8,9].

ANTHROPIC PRINCIPLE

The term 'anthropic principle' was first used at a 1973 symposium in honor of the 500th birthday of Copernicus by astrophysicist Brandon Carter. The anthropic principle refers to the assertion that scientists need to take into account the fact of the existence of life when developing their theories. This stems from the observation that the physical constants of nature all seem to be fine-tuned in a significantly balanced manner that promotes the existence of complex living systems [21]. If the four fundamental forces or fine-structure constant differed very much there would be no stars or chemical elements and then of course life as we know it could not exist.

Carter defined the two forms of the Anthropic Principle currently in use, the 'weak' anthropic principle referring to the idea of privileged spacetime locations in the universe, and the 'strong' form of the anthropic principle which has addressed the values of the fundamental constants of nature. Barrow and Tipler [22,23] in a detailed work formed different definitions of the weak and strong anthropic principles. They also argue extensively that it is highly probable that human life is the only intelligent life in the Milky Way galaxy. We strongly disagree. We believe that intelligent life is the rule, not the exception and that this is what the anthropic principle is all about. This is suggested twice in Carl Sagan's Hollywood film *Contact*, if not: "*it would be an awful waste of space*".

Weinberg suggests the Anthropic Principle could be utilized by cosmologists opposed to theism as a 'turning point' in science by applying it to the string landscape to "*... explain how the constants of nature that we observe can take values suitable for life without being fine-tuned by a benevolent creator*" [24]. Interesting that the same principle can be used for opposite purposes. Weinberg's view is opposite to the anthropic views presented here. We propose that teleological or eutaxiological bases are tantamount to the essence of anthropic cosmology itself suggesting that the anthropic principle entails an additional action principle driving or guiding cosmological evolution in opposition to the postulate of random Darwinian or naturalistic evolution of Big Band cosmologies. This new action is believed to be synonymous with the action of the unitary field which historically has also been equated with chi, ki, prana, the *élan vital* or spirit of God [8,9].

String theory predicts a universe with a virtually infinite number of possibilities (10^{1000}) for a unique string background or vacuum. This set of vacua has been called the Multiverse, anthropic landscape or string landscape. Susskind suggests this possibility for a large number of vacua strengthens anthropic reasoning: "*only universes whose properties are such as to allow observers to exist are observed, while a possibly much larger set of universes lacking such properties go unnoticed*" [25].

HOLOGRAPHIC PRINCIPLE

The holographic principle initially developed as a property of quantum gravity theories by 't Hooft [26] and Susskind [25,27] explained the information paradox of black holes in terms of string theory. Susskind states the principle as a description of a volume of space encoded on a boundary of the region, usually a light-like boundary such as the Schwarzschild gravitational horizon for a black hole. For black holes the holographic principle states that the information of all objects falling in is entirely contained in surface fluctuations of the black holes event horizon.

In a more speculative manner, it has been suggested that the multiverse is a two-dimensional information structure like a 2-brane hologram [26,28]. To create a theory of holographic cosmology is considered a challenge because of expansion of the cosmological horizon in the Big Bang model where a finite area expands over time. We don't see that this would necessarily be a problem as the structure would still remain invariant under the transformation. However there is no expansion or inflation in our model where we take the 'world as a hologram' idea fairly literally. We don't consider the 2D information surface of 't Hooft and Susskind [26,28] sufficient for a cosmology; for information of a black hole event horizon perhaps [29]. But for our HAM model we postulate that the hologram is an HD hypersurface of spacetime itself with the fundamental Gabor 'logons' being the dynamic system of cosmological least-units tiling it. This can only be elucidated by solving some of the fundamental issues of string theory such as: Does gravity require 6D, or Schöenflies Theorem [30] which states that there can be no torsion in the plane; and does this preclude the lower limit suggested by 't Hooft and Susskind for information on the surface of a holographic 2-brane. Of course their purpose was concerning black holes and not a complete multiverse cosmology.

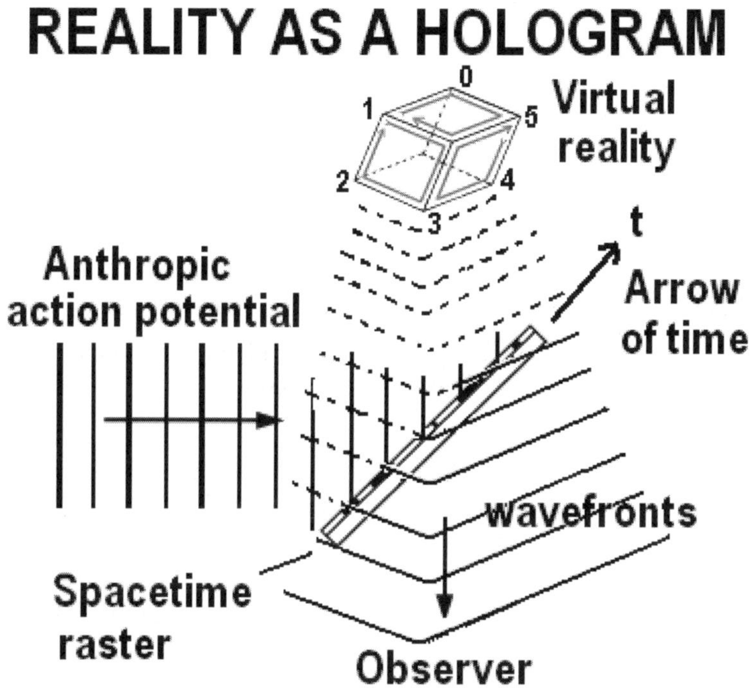

Figure 3.1 Reality modeled as an HD holographic construct with the 'laser' being the anthropic unitary action principle 'piloting' its continuous evolution.

A hologram is formed by simultaneously exposing a photographic plate to laser light from a reference beam and laser light reflected from the object employed. The two beams are out of phase which creates interference; it is the interference pattern that is recorded on the 2D photographic plate. A 3D image of the object is reconstructed by illuminating the hologram (plate) with coherent light. Eventually one might suppose an HD holographic multiverse cosmology could be fully developed to rigorously describe reality, matter and living systems [9] by the continuous-state action of the anthropic unitary field. See Fig. 3.2.

For HAM cosmology as we shall see in more detail in other chapters, a hyperdimensional holographic multiverse has room for an infinite number of nested Hubble spheres all in causal separation and out of phase with each other (invisible). This allows each Hubble sphere to have its own fine-tuned laws of physics.

3.3 Parallel Interpretations of Cosmological Data

TABLE 3.1

PARAMETER	BIG BANG	HAM COSMOLOGY
REDSHIFT	Doppler recession of an inflationary expanding universe. ($m_\gamma = 0$)	'Tired light' phenomena, non-zero mass photon ($m_\gamma \neq 0$) couples to vacuum dissipating energy.
CMBR	$2.75^\circ K$ blackbody remnant of initial hot cataclysmic explosion ~ 15 billion years ago.	Result of continuous-state blackbody emission by spacetime cavity-QED electrodynamics inherent in a continuous compactification D reduction process.
OLBER'S PARADOX	Expansion of the universe accounts for dissipation of luminosity.	Lifetime of stars insufficient to illuminate heavens; absorption by vacuum coupling and dispersion by interstellar media.
MATTER	Matter creation at initial Bigbang. Missing dark matter or dark energy required to explain galactic rotation etc.	Dark energy - balances the gravitational potential or Λ by multiverse matter. Results in flat spacetime. Spontaneous creation of matter; black hole evaporation removes evolved material.

3.4 Euclidean/Minkowski Geometry as Basis for Observed Reality

- The Euclidean line is assumed to be the real line [14] because it is what is observed. Logical reasons from supersymmetry and supergravity suggest there are a number of additional unobserved dimensions [15] leaving the issue of dimensionality as an open question. Euclidean space in classical Newtonian terms is a continuous 3D absolute space with time an independent parameter.
- Einstein's theories of relativity provided a discrete 3(4)D transmutable relational spacetime manifold. The debate between absolute space or substantivism and relational space still continues. Utilizing the standard definition of a straight line as the intersection of two rigid planes, measurements could be taken to observe whether the angles of a triangle add up to 180°; but settling the question definitively would require astronomical scale measurements where it appears physically impossible to apply the concept of a rigid body or to define a straight line in terms of a light ray by stellar parallax because of the effects of general relativity. Therefore all physics knows with certainty at the present time is that observed space is approximately Euclidean as is Minkowski space [14,31].
- According to the proof of Schöenflies theorem [30] there can be no torsion or topological knots in a plane. Therefore there can be no torsion in a 2D reality; thus the real line must be at least 3D Euclidean where the standard Pythagorean line element is:
$$ds^2 = dx_1^2 + dx_2^2 + dx_3^2 \tag{3.1}$$
- This assumption that the Euclidean line is the real line is intuitive. Currently there is no known method of empirical proof; and since the Euclidean line is what the Human mind apprehends it remains the formal basis for all scientific fact [14,22]. But this assumption remains profoundly problematical with issues stemming from both the foundations of mathematics and the nature of physical theory itself concerning the fundamental basis for sets, discreteness versus continuity, geometry and topology, and the relationship of real numbers to rational numbers for example [14].
- In general, the class of theories unifying gauge and gravitational fields by utilizing XD is called Kaluza-Klein theories. In these theories spontaneous symmetry breaking by coordinate transformation in 5D is a product of the standard 4D transformation and a local U(1) gauge group arising in basic form in a general relativistic

framework of five dimensions described according to the Einstein-Hilbert action

$$A = \int d^5 x \sqrt{g} R \,. \tag{3.2}$$

Where in Eq. (3.2) instead of postulating a 5D Minkowski space, $\hat{M}^5$ as the ground state, the ground state is taken to be the product $M^4 \times S^1$ where the circle S^1 is a U(1) group of rotations [15]. In conventional supersymmetry models the radius of circle S^1 is considered to be microscopically small on the order of the Planck scale, $(10^{-33} cm,\ 10^{-43} s)$, very short and very fast, used to explain why these XD are not observed. This will be discussed in more detail below where Planck's constant is recalculated utilizing the Larmour radius of hydrogen as it relates to non-compactified Kaluza-Klein theory [32].

An SU(3) × SU(2) × U(1) gauge symmetry group can be used to describe all known particle interactions. Following Witten [15], the minimum number of dimensions of a manifold with this symmetry is 7D. In this SU(3) × SU(2) × U(1) symmetry group gauge fields arise in the gravitational field as components of more than 4D. This yields a dimensionality for our reality of at least four non-compact and seven compact spacetime dimensions, $M^4 \times S^7 = 11D$, which Witten [15] calls a remarkable numerical coincidence since this 11D maximum for supergravity is the minimum for SU(3) × SU(2) × U(1) symmetry which also for symmetry reasons observed in nature is in practicality the largest group one could obtain from Kaluza-Klein theories in 7XD.

This gauge group for gravitational field components is insufficient to describe nature; for a complete theory quarks and leptons plus a Higgs or alternative type mechanism triggering symmetry breaking must be added to the Kaluza-Klein framework. In attempting to complete the theory, the gauge coupling constants are determined by calculating the Einstein action over the compact dimensions. This scales at a high power of $1/(M_p R)$, where M_p is the Planck length and R the radius of the XDs showing that R must actually be in the $10^{-33} cm$ range for these standard model gauge theories. If one adds the Lagrangian of a cosmological constant, Λ Witten finds one can form a reasonable theory [15].

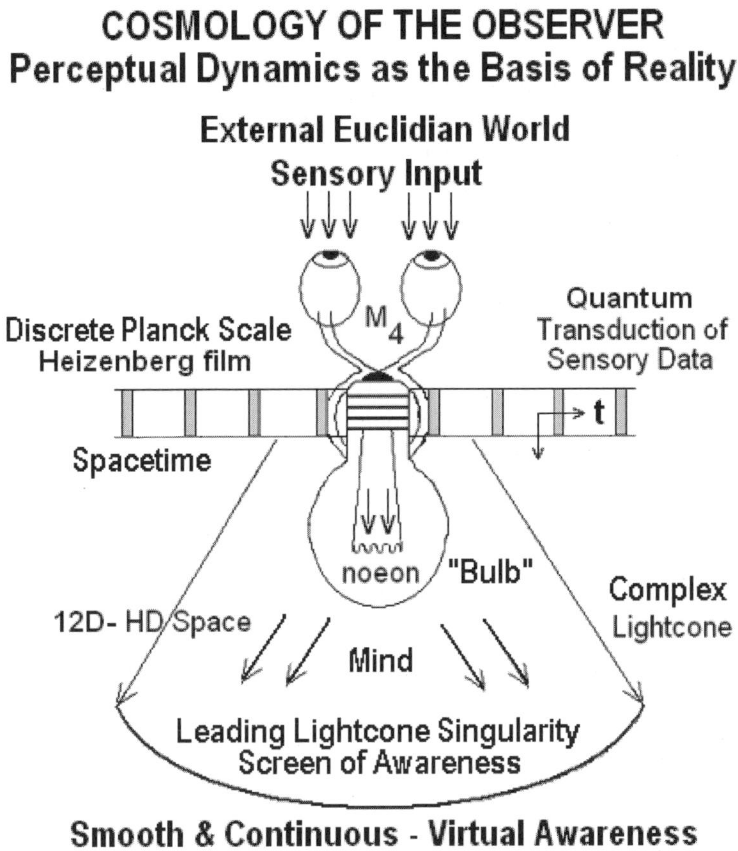

Figure 3.2 In an anthropic cosmology with a teleological action principle spacetime parameters can be driven in manner able to drive the evolution of self-organized complex systems.

Although only introduced in a preliminary form here, a different view is required by noetic theory because the Einstein gauge is both classical and incomplete. Noetic cosmology like any new theory must however bear correspondence to the established Einstein gauge. The existing derivation of Planck's constant represents classical mathematical limits only and are not actual physical limits in HAM cosmology. Since the Higg's mechanism also arises from the Einstein gauge it must also be called into question and be replaced by another mechanism when the noncompactified form of Kaluza-Klein theory is utilized [32].

3.5 Philosophy of Space in HAM Cosmology - Origin of Structure

Although the concept of Absolute Space (AS) as defined by Newton is discarded in contemporary physics, a deeper more fundamental form of AS nevertheless seems to exist and is a required foundation for HAM Cosmology. The HAM reintroduces a complementary AS that is non-Newtonian because Newtonian AS, once considered the basis of 'our space', first of all is only a form of Euclidean space without sufficient degrees of freedom to incorporate Quantum or Relativity theory. HAM Absolute Space is different, but similar enough that Newton deserves credit for realizing the importance of AS. Secondly the relational space of the Einstein universe contains insufficient symmetry parameters to describe the additional causal properties of a supralocal Multiverse. The Absolute Space (AS) proposed by the HAM (defined in postulate 3.1) represents the ground of all existence and 'resides' beyond the observed Hubble universe or even the infinite number of other possible supralocal nested Hubble-type spheres (with varied laws of physics) [9], Chap. 4. The ultimate nature of HAM Absolute Space remains ineffable at the moment, but empirical tests are being prepared [33,34]. See Chaps. 9,11. In the meantime we can deduce some Absolute Space properties to steer empirical investigations to higher order properties these deductions suggest.

Postulate 3.1: *Space is the most fundamental 'form or substance' of existence; and the origin of all structure. The demarcation and translation of which constitutes the basis of all energy or phenomenology. Space takes two forms in HAM cosmology, Absolute Space and the temporal relational subspaces that arise from it. A basis for energy (space geometry) is a fundamental form of information which signifies the cosmological foundation of causality. This postulate also connotes the most rudimentary basis of structural-phenomenology.*

The complementarity between the new concept of AS in HAM Cosmology and the contemporary relational space suggested by Einstein's theories of relativity can be simplistically represented as a 'virtual reality' by interpreting HAM AS as a fundamental background space of the related space fields referred to by Einstein's quote below.

Time is a complex process only just beginning to be addressed by physicists [35]. One can say that all forms of time [35-37] represent various types of motion and in that sense time can be discounted as a concept (i.e. - not absolutely fundamental). Then geometric translation or field propagation becomes more fundamental. Thus space (whatever it is) is the most fundamental concept of the universe. Space with boundary conditions or energy is fundamental to all forms of matter.

3.6 Space: Relational Versus Absolute

The conceptual disparity regarding the fundamental nature of space arises in terms of correspondence between the Newtonian worldview of a continuous Absolute Space in opposition to the current Einsteinian view of discreteness of the spacetime manifold. This debate about the nature of space has continued at least since Aristotle. Einstein in his last published statement regarding the nature of space and time said:

> The victory over the concept of absolute space or over that of the inertial system became possible only because the concept of the material object was gradually replaced as the fundamental concept of physics by that of the field...The whole of physical reality could perhaps be represented as a field whose components depend on four space-time parameters. If the laws of this field are in general covariant, then the introduction of an independent (absolute) space is no longer necessary. That which constitutes the spatial character of reality is then simply the four-dimensionality of the field. There is then no 'empty space', that is, there is no space without a field [38].

Einstein's view is a form of the *relational theory* of space introduced initially by Leibniz and Huygens [39,40]. Relationalism is in opposition to *substantivism* which gives space the ontological status of an independent reality as a kind of *substance* [39]; the Newtonian concept of absolute space being the prime example.

Finding the founding fathers of quantum theory credible in their declaration that the standard model is incapable of describing biological systems; means awareness can only be defined adequately by extending <u>all</u> the standard models since they are so intertwined. This means that:

- The standard cosmological model - the Big Bang is insufficient.
- The standard mechanistic model of biological naturalism is

inadequate.
- The standard Turing model of computation is inadequate.
- The standard model of gravitation is insufficient.
- The standard Copenhagen phenomenological model of quantum theory is inadequate.
- The standard model of EM is inadequate.
- The standard cognitive model of neuroscience is also insufficient.

This criticism does not mean these seven models are wrong; only that they go part way. The focus here is primarily on the cosmological model as it is the root of the problem. The required parameters of the post Big Bang universe will be stated axiomatically for simplicity. The domain of the Big Bang is defined in terms of the Hubble radius for the large-scale structure of the universe and the Planck scale for the microscopic. The large-scale observational limit according to Big Bang philosophy is caused by the Doppler effect on light propagation due to the recessional velocity of expansion of the universe. This observational limit occurs where light becomes attenuated by the redshift.

The Hubble radius, H_R remains an observational limit in Continuous-State Anthropic Multiverse (HAM) cosmology also but is not caused by the Doppler effect. It is due to a minute non-zero rest mass for the photon [3,41]. As a photon propagates it couples to the polarized Dirac vacuum and loses energy also attenuating to zero observability; but if one were able to travel to the Hubble limit observation would extend for another Hubble radius ad infinitum. Thus a critical difference in interpretation of redshift – a physical limit for the Big Bang and an observational illusion in HAM cosmology.

Einstein by the introduction of special and general relativity replaced the absolute 3D Newtonian continuum with a discrete 3(4)D relational spacetime manifold. This space can still be interpreted as a potential Big Bang space terminating at the impenetrable Planck backcloth of stochastic foam. Noetic cosmology changes the interpretation of this limit. The Planck barrier is a virtual mathematical barrier to Fermions as the present recedes into the past.

The HAM [41] is a Multiverse with the potential for an infinite number of nested Hubble spheres in causal separation and thus with their own laws of physics [42]. In the Big Bang the XDs laid down at the beginning of time are curled up at the Planck scale as a compactified subspace. In the Noetic HAM cosmology the opposite is true. A new form of HD Absolute Space projects a periodic 11(12)D space. The standard observed relational Einstein reality, 3(4)D M^4, is a subspace of the 11(12)D space projected from this new AS. An extension of the Wheeler-Feynman absorber theory of radiation [5] is utilized to define an eternal *present* as a standing wave of the *future-past* that is 'covered' at each level of scale by a HD Wheeler Geon [43] or ball of light. This HD Noetic light field filling the immensity of subspace is the unified field that acts as gravitation, the vital force, and light of the mind. As will be derived below this action principle can be described by a simple fundamental Noetic equation $F_{(N)}$ = E / R [41,44,45] (Chap. 4). This complex least unit explains the utility of the 12D space. All this will be discussed in detail in ensuing sections.

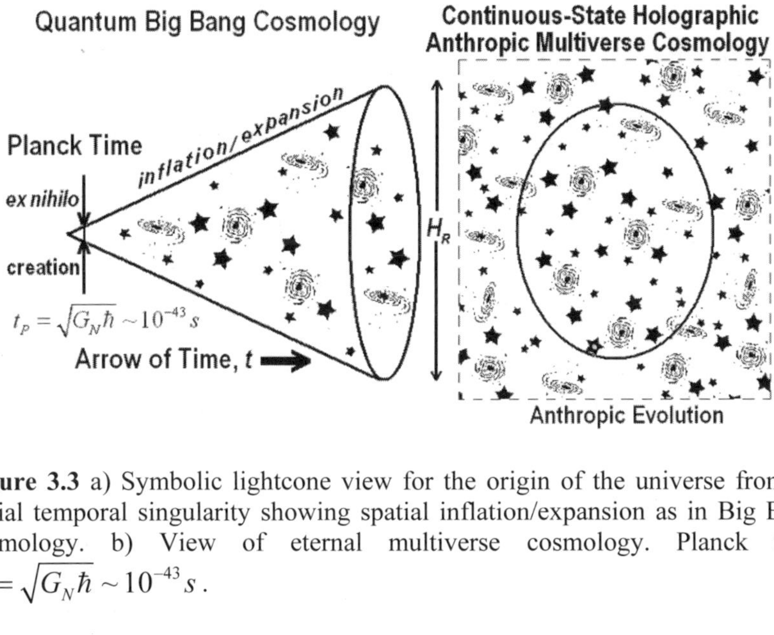

Figure 3.3 a) Symbolic lightcone view for the origin of the universe from an initial temporal singularity showing spatial inflation/expansion as in Big Bang cosmology. b) View of eternal multiverse cosmology. Planck time $t_P = \sqrt{G_N \hbar} \sim 10^{-43} s$.

The world lines of relational space are virtual extensions created and recreated harmonically by the torsion of the continuous compactification process. Therefore instead of a rigid impenetrable Planck barrier covered by a stochastic foam of particle creation and annihilation, HAM cosmology has a periodic ordered spacetime with a complex hyperstructure that is closed and finite in time for fermions, but open and infinite atemporally for bosons. In the HAM model, stochasticity, i.e.

zero-point string or brane dynamics, arises in the wake of unitary graviton propagation guiding the dynamics of the continuous-state.

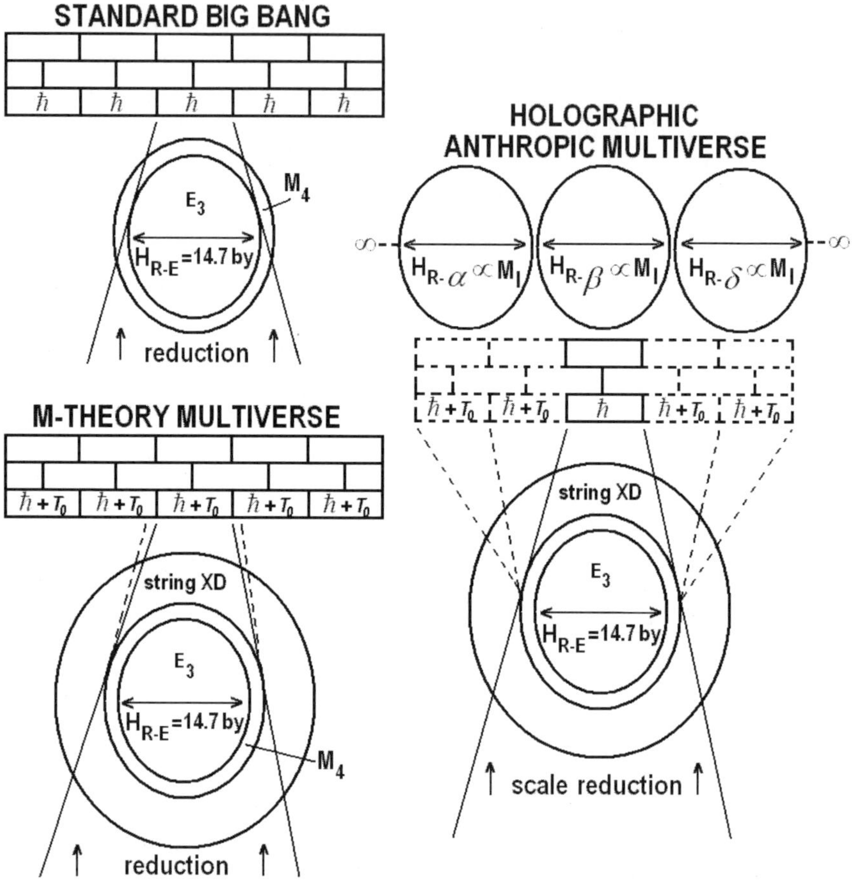

Figure 3.4 Contrast of the three main cosmological models illustrating dimensionality relative to the Planck constant, $\hbar$ and dimensionality. $H_{R\text{-}E}$ is the Hubble radius for an Earth observer, E_3 & M_4 Euclidian and Minkowski space.

The Noetic graviton, is a quadrupole photon complex confined to the spacetime metric like quarkonium [4]. The Planck singularity $(10^{-33}\,cm)$, $10^{-43}\,s$ is *virtual*, a geometric orientation that arises as the present recedes into the past [41]. The Big Bang is said to originate from an initial singularity; this is only an observational illusion in the HAM model where the arrow of time arises from continuous-state spin-exchange dimensional reduction compactification by an uncoupling recoupling process during deficit angle production during parallel transport around the close-packed cosmological least-units tiling spacetime. That mantra is a lot to swallow and is addressed more adequately in Chap. 5.

In Fig. 3.4 H_R is an observational limit, not because of temporal Doppler expansion of the universe as postulated in Big Bang cosmology but because of infinitesimal photon mass, m_γ [3,46]. See Chap. 13 for an interesting additional reason. Because Gauge Theory is only an approximation, the Planck constant, $\hbar$ is not a fundamental ultra-microscopic singularity and is reformulated in HAM cosmology [47]. Its zero point oscillates from the usual $\hbar$ to $\hbar + T_0$ which has an upper limit of the Larmour radius of the hydrogen atom, see Chap. 4. This is because the new singularity is the string vertex [48] (see Fig. 3.6).

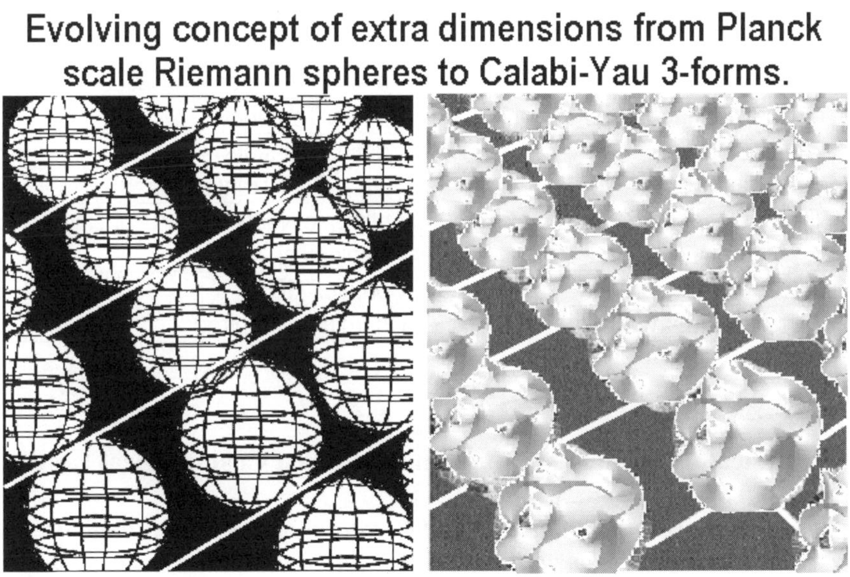

Figure 3.5. a) XDs as originally considered to be microscopically curled up at each spacetime point. b) More complex view 30 years later in terms of Calabi-Yau 3-form topology. With the application of HAM string tension, $\bar{\lambda} + T_0$ the 3-forms may stretch to infinite size.

Because our observed Euclidean temporal reality is a virtual subspace of an HD eternal realm, the arrow of time, the propagation of which 'creates' our observed reality, time is a complex system. The Hubble radius, H_R is still an observational cosmological limit, but instead of being indicative of a Doppler expansion of spacetime arises as part of the complex self-organized structural-phenomenology of the continuous-state dynamics. The energy of the photon with infinitesimal mass, m_γ anisotropy [3,46] attenuates to zero such that an observer traveling to H_R would be able to see out to an additional H_R.

According to Cramer: The transactional interpretation of quantum mechanics is a nonlocal relativistically invariant alternative to the Copenhagen interpretation. It requires a 'handshake' between retarded, (ψ) and advanced waves, (ψ^*) for a quantum event which he calls a 'transaction' in which energy, momentum, angular momentum, and other conserved quantities are transferred as a standing wave [6].

Table 3.2 PROCESSES FORMING THE STANDING WAVE PRESENT

Dimensional reduction	Continuous Compactification
Spin exchange	Parallel Transport
Deficit Angle	Continuous-state
Holographic Principle	Future-Past Transaction
Super Quantum Potential	Anthropic Noetic Action Principle - Teleological
Advanced / Retarded	Complex HD
Mirror symmetry/duality	Coordinate leapfrog

3.7 Physical Cosmology of Fundamental Least Cosmological Unit

Theories avoiding completely the notion of the continuum are, of course possible in principle. But the attempt is not so simple as you seem to believe. The interesting question is if on logical grounds (simplicity) a plausible choice of axioms is possible. Of course, the concept of time (continuum) could not enter such a theory - Albert Einstein, 1952.

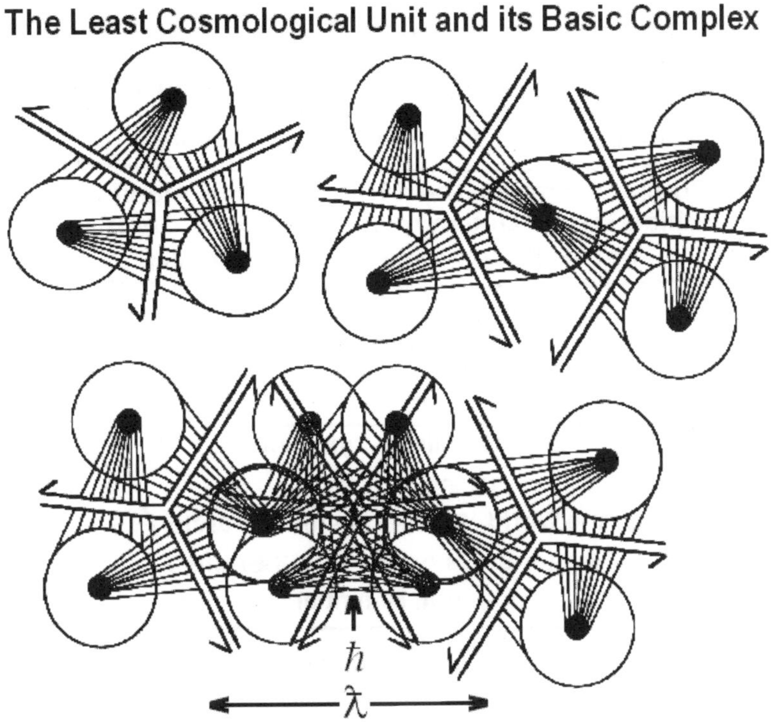

Figure 3.6 a) Conceptualization of the triune nature of an isolated least-unit not existing in nature. b) Coupling of two isolated least-units along an x coordinate. c) The central portion denoted by $\hbar$ represents the realization of one virtual Euclidean point which oscillates harmonically to ~ the Larmour radius of a hydrogen atom here denoted as $\bar{\lambda}$ which represents the new *Stoney* representation of $\hbar$ plus string tension, T_s. This model can be considered a Cramer transaction.

In Fig. 3.6 an advanced-retarded future-past transaction is represented as an instant of the eternal present. The Planck constant, $\hbar$ still exists in HAM cosmology but represents a virtual lower limit of the SUSY topology as constituents of the least-unit complex are compactified as they recede into the past in preparation for the next HD Cramer type transaction cycle. The *Stoney* representation, $\bar{\lambda}$ represents the 'open' future orientation of this portion of the continuous-state cycle. This complex structure is only touched upon in Chaps. 2-4; its main delineation is presented in Chap. 5.

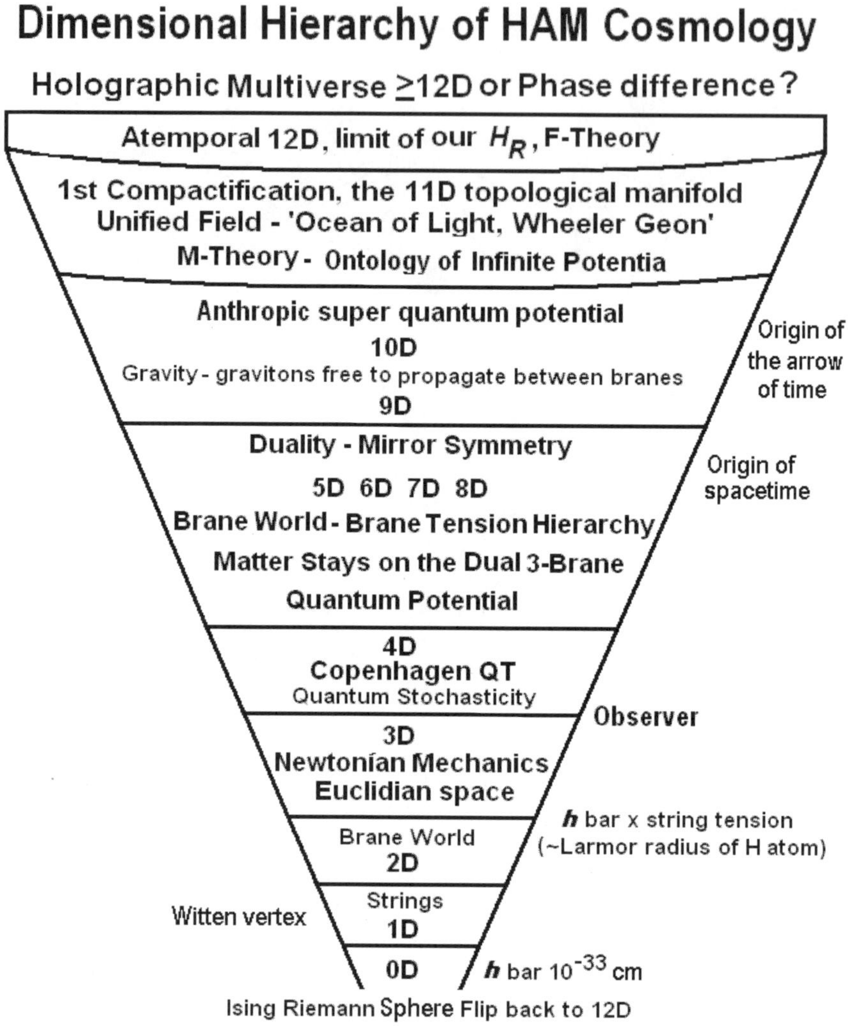

Figure 3.7 The 12D HAM cosmology hierarchy underlying the continuous-state compactification process that produces our observed virtual reality.

Conventional 11D M-Theory searches for one unique compactification within which to formalize a completed string theory model. This line of reasoning is a product of Big Bang cosmology and the Copenhagen interpretation of quantum theory suggesting nucleons were created around the time of the original singularity and a particular compactification was produced forming the basis of our reality. HAM cosmology does not have these constraints. Compactification is a self-organized continuous-state process through all possible compactification formats. This gives the cosmology unique characteristics; especially the anthropic action principle driving its self-organization.

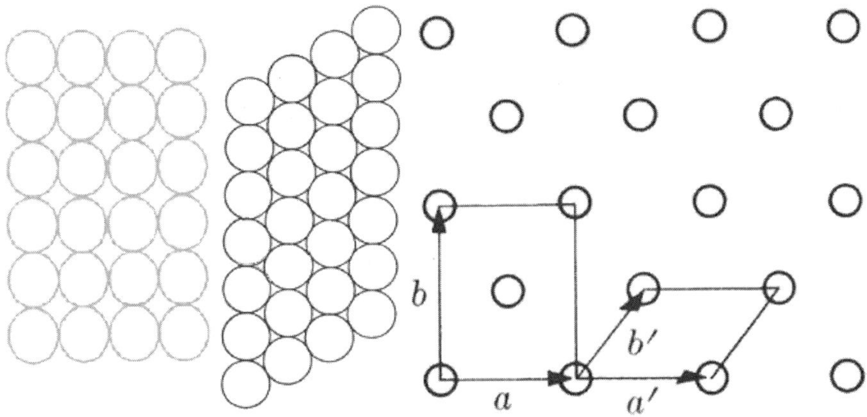

Figure 3.8 Conceptualization of close-packed least cosmological units and how Euclidian/Minkowski geometry might naturally emerge from its domain walls.

> *Time and space are modes by which we think and not conditions in which we live.* - Albert Einstein, 1941.

Awareness is introduced as a fundamental physical quantity [8,9]. The context for defining awareness is an advanced form of Einstein's model of a static universe, called the Continuous-State Anthropic Multiverse (HAM). The new cosmology is based on principles of the Wheeler-Feynman absorber theory of radiation extended to the topology of a periodic 12D spacetime. The fundamental *least-unit of awareness* is shown to be a scale invariant complex cosmological system. Time arises naturally as a 'beat frequency' in the translating boundary conditions of a spin exchange 'continuous state' dimensional reduction compactification process. A new set of Noetic transformations beyond the Galilean and Poincairé-Lorentz are called for to show how the macroscopic nature of awareness arises from microscopic action principles inherent in the Dirac polarized vacuum. The inherent topology of the Noetic transformations are derived by coupling superluminal Lorentz boosts with noncompactified Kaluza-Klein theory in the context of an energy dependent spacetime metric, $\hat{M}_4$.

The standard model for a living system, biological mechanism, presumes that life can be completely described by parameters of chemistry and physics. In general this biological naturalism is described by quantum theory which deals with the mechanics of atomic and related systems. Quantum theory is described formally by the Schrödinger equation which takes myriad forms, but simply equation (3.3)

$$ih(\partial\psi / \partial t) = H\psi \tag{3.3}$$

describes the action of a particle on a manifold. But the founding fathers of quantum theory said the standardized Copenhagen interpretation was incapable of describing biological systems. Therefore the bulk of this paper is devoted to developing the proper cosmological framework for introducing a fundamental definition of awareness.

3.8 Holographic Anthropic Multiverse Cosmology (HAM)

WHAT IS THE HOLOGRAPHIC ANTHROPIC MULTIVERSE?

- There Is No Big Bang (Temporal Singularity), Expansion or Inflation.
- Redshift Is Non-Doppler Due To Periodic Photon Mass (Tired Light)
- CMBR Is Cavity-QED Blackbody (BB) Radiation.
- Thus CMBR Is Emission & Redshift Absorption For BB Equilibrium.
- The HAM Is Closed & Finite In Time / Open And Infinite Eternally.
- This Relates To The Holographic Principle – A Multiverse With Potential For An Infinite Number Of Nested Hubble Spheres (H_R) Each With Their Own Laws of Physics.
- Dark Energy Arises From The Rest Of The Multiverse Beyond H_R.
- Cosmological Constant Is Based On This Horizon - Fluctuating Near Zero As -, 0, +.
- The HAM Is Not Static, Steady-State Or inflationary, But A Continuous-State (CS).

- The CS Provides A Standing-Wave Present From Future-Past Elements In A Spin Exchange Dimensional Reduction Compactification Process- i.e. The XD Can Range From Infinitely Large > 4D And to Planck Scale < 3D.
- 3D Reality Is A 'Pocket Space' Or Temporal Subspace Of A New Form of Absolute Space 12D Eternity.
- A Conscious Universe Requires An Additional Teleological Or Anthropic (Noetic) Action Principle Guiding Evolution And Governing Hierarchical Complex Self-Organization.

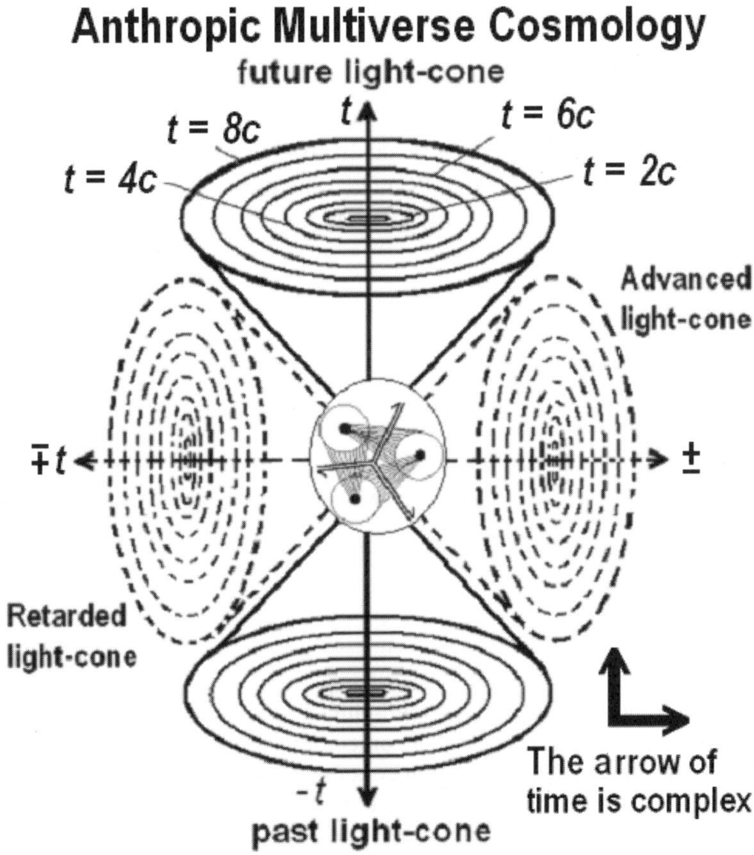

Figure 3.9 Conceptualization of the topological backcloth of continuous-state Holographic Anthropic Cosmology. It could also be considered to model a Cramer transaction of the Hubble 3-sphere.

§1: The Holographic Principle[1] is extended

§2: By the anthropic principle of the physical basis of intelligence and Einstein's mass-energy relation, $E = mc^2$, a tenet of HAM cosmology is that the mass-energy of the Earth is equivalent to the mass-energy of all life over its ~ 5 billion year history.

§2A: This mass-energy relation determines the observational limit in cosmology equated with the Hubble radius, H_R.

§2B: Through parameters of string tension/coupling and the fine-structure constant $c \equiv c$ as empirically measured rather than of infinite velocity.

§2A and §2B are fine-tuned variables, different in each multiversal nested Hubble 3-sphere. An associated principle relates to conformal scale invariance. At the microscopic limit the Anthropic field equation balances the oscillation of the Planck constant; $\hbar$ at the cosmological scale, the Anthropic field equation balances the fluctuation of the cosmological constant, Λ.

3.9 Overview of the Formalism for Noetic Cosmology

Noetic Cosmology is cast in a 12D harmonic superspace $S_N = S_0 + S_1 + S_2$ in the context of an extended Wheeler/Feynman absorber theory [5] where standard Minkowski space M_4 is a 'standing wave' of the future-past [6]. This takes the general form

---

[1] The usual rendition of the Holographic Principle attributed to 't Hooft, Susskind and Wheeler [26,28,29] is a conjecture about quantum gravity theories, claiming that all of the information contained in a volume of space can be represented by information living in the boundary of that region. In other words, if you have an empty sphere, all of the events within can be explained by the arrangement of information on the surface of the sphere. The theory also suggests that the entire universe can be seen as a 2D information structure 'painted' on a boundary surface, and that the three observed dimensions are illusory. String theorists (11D M-theory) claim the holographic principle could form the theory of everything (TOE).

$$R^{S_{N0}}_{symM_4} = \frac{1}{2}\left[R^{S_{N1}}_{retC_4} + R^{S_{N2}}_{advC_4}\right] \quad (3.4)$$

or simplistically stated the 12D noetic superspace S_N represents a complex Minkowski metric $M_4 + C_8$ (or $\pm C_4$). S_N thus combines the standard M_4 four *real* dimensions (D) plus 8 imaginary D representing a *retarded* and *advanced* complex hyperspace topology which adapts the complex ($M_4 + C_8$) Minkowski metric from the standard stationary form to a periodic form. $S_0 = M_4$ represents the noetic 3(4)D 'standing wave' Minkowski 'present' spacetime; $S_1 = -C_{4(ret)}$ represents the past component and $S_2 = +C_{4(adv)}$ represents the future for complex correspondence to the standard 4 real dimensions utilizing 8 imaginary dimensions. The 8 imaginary dimensions, while not manifest generally (locally) on the Euclidean real line, are nevertheless 'physical' in the HAM and can be represented by complex coordinates

$$X = \pm(x + i\xi), Y = \pm(y + i\eta), Z = \pm(z + i\zeta) \text{ and } t = \pm(t + i\tau) \quad (3.5)$$

designating correspondence to real and retarded/advanced continuous spacetime transformations. For symmetry reasons the standard Minkowski line element metric $ds^2 = g_{ij}dx^i dx^j$ is expanded into periodic *retarded* and *advanced* topological elements fundamental to relational space 'extension' giving Noetic Superspace S_N its continuous state dimensional reduction standing wave periodicity. This is illustrated conceptually in Fig. 3.10 below.

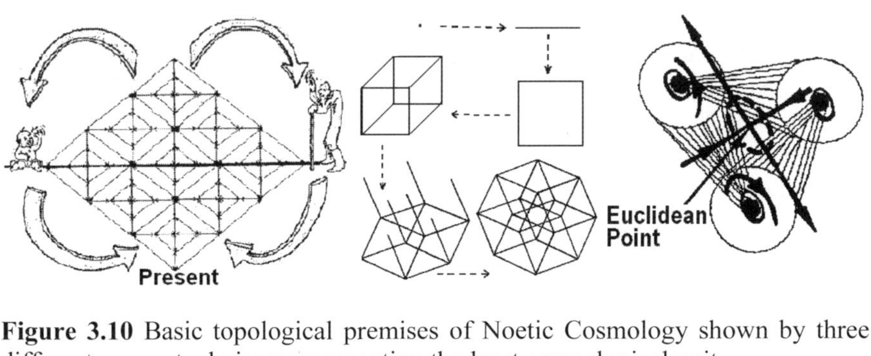

Figure 3.10 Basic topological premises of Noetic Cosmology shown by three different conceptual views representing the least cosmological unit.

Figure 3.10 reveals some of the main parameters of HAM cosmology; in 3.10a) the baby and old man represent the *relational* periodic basis of spacetime by applying extended Wheeler/Feynman absorber theory where the present is a standing wave of the future/past. 3.10b) The 11(12)D harmonic superspace translates in a continuous state dimensional reduction compactification process. A 12D HAM provides enough degrees of freedom so that two complex imaginary $\pm 3(4)$D spacetime packages can topologically transform into a "standing wave" present, i.e. the present has a future-past basis by extending Wheeler-Feynman radiation law to include the continuous state transformation of the topology of spacetime dynamics itself. 3.10c) A 3-torus illustrating a virtual standing wave 'creation' of a discrete virtual Euclidean point; a different conceptual view of figure 1a and 1b. The three 3(4)D ($S_0 = M_4$, $S_1 = -C_{4(ret)}$ and $S_2 = +C_{4(adv)}$) spacetime packages surround a virtual Planck scale singularity, (in the form of a 3-torus $\left[\sqrt{(x^2+y^2)}-R\right]^2 + z^2 = r^2$) the continuous propagation of which 'create and recreate' periodically the 'standing wave' Euclidean real line illustrating the virtual basis of relational Einsteinian reality as a subspace of absolute HD HAM space. This Noetic 'least unit' represents a Wheeler/Feynman future/past periodicity and a continuous cycling of *classical* $\rightarrow$ *quantum stochasticity* $\rightarrow$ *fundamental unitary* ($R_C \rightarrow R_Q \rightarrow R_U$) in the D reduction compactification $D_s \rightarrow D_t \rightarrow D_E$ transformation process [9].

The Kaluza-Klein model utilized is set in a noncompactified D = 12 harmonic Noetic Superspace S_N since it is the foundation of a conscious universe. For symmetry reasons shown in the text this superspace is comprised of an 11D hypersurface in a 12D universe, giving it theoretical correspondence to 10D superstring theory and 11D supergravity and providing a context to solve the disparity between them. The general appeal of the Kaluza-Klein model is that physics seems simplified in HD, especially integration of the electromagnetic (EM) and gravitational field.

Periodic Noetic superspace S_N entails a continuous state of dimensional reduction that operates under transformations beyond the Poincairé / Lorentz where spatial dimensions D_S through superluminal boosts are transformed into temporal dimensions D_t and further in terms of a non-compactified Kaluza-Klein model [32] into energy dimensions D_E by $D_S \rightarrow D_t \rightarrow D_E$. This requires the properties of an energy dependent spacetime metric first developed by Einstein where standard Minkowski space, M_4 is a topologically invariant homeomorphic manifold of an energy dependent spacetime metric $\hat{M}_4$

$$f : M_4 \rightarrow \hat{M}_4 . \tag{3.6}$$

According to the principle of relativity a spacetime region which is a 'perfect vacuum' (no matter and no fields) must be isotropic and covariant in the Lorentz sense [28]. The deformed region $\hat{M}_4$ of S_N and the symmetry of S_N itself reduces to the Einstein relativistic metric and is assumed compatible with a covariant polarized Dirac vacuum.

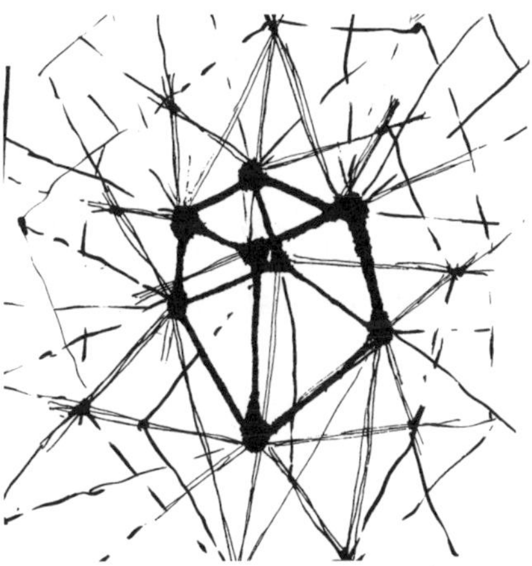

Figure 3.11 Two additional conceptual views of Fig. 3.10. Fig. 3.11a) conceptualizes the relational nature of Minkowski space emerging from the polarized vacuum 3.11b) represents a snapshot in time. The central hypersphere represents the atemporal hidden HD covering the standing wave present. The larger peripheral tubes represent open orientation toward the future; and the narrower coupled tube forming a square represents a phase of recessional compactification toward the past, the final phase of which would end up like that of Fig. 3.10c – a virtual Planck scale singularity. This figure hints at why the Planck constant needs to be recalculated. Related to the past – the resultant of measurement, the Planck constant apples as usual. In the *eternal now*, the Planck constant takes the form of the Larmour atomic radius and is an unbounded component of the unitary field in the future orientation.

3.10 Transformation of Space into Time

It is well known that Superluminal Lorentz Transformations (SLT) change real quantities into imaginary ones. Following Cole [49] and Rauscher [50] we illustrate the transformation of complex spatial dimensions into temporal dimensions by orthogonal superluminal boosts (SLB). For example an SLB in the x direction with velocity $v_x \pm \infty$ the SLT is $x' = \pm t,\ y' = -iy,\ z' = -iz,\ t' = x$. In complex Minkowski space the coordinates are $z^u = x^u_{\text{Re}} + ix^u_{\text{Im}}$ where z is complex and x_{Re} and x_{Im} are real and the index u runs over 0,1,2,3. Using classical notation for simplicity

$$t = t_{\text{Re}} + it_{\text{Im}},\ x = x_{\text{Re}} + ix_{\text{Im}},\ y = y_{\text{Re}} + iy_{\text{Im}},\ z = z_{\text{Re}} + iz_{\text{Im}}. \tag{3.7}$$

To clarify the meaning of imaginary quantities in an SLT it is helpful to represent time as a 3D vector $t_x,\ t_y,\ t_z$; therefore time is defined as $t = t_x\hat{x} + t_y\hat{y} + t_z\hat{z}$ where

$$t_x = t_{x\,\text{Re}} + it_{x\,\text{Im}},\ t_y = t_{y\,\text{Re}} + it_{y\,\text{Im}},\ t_z = t_{z\,\text{Re}} + it_{z\,\text{Im}} \tag{3.8}$$

Finally for the SLB for velocity $v_x \pm \infty$ along x the transformations are

$$\begin{aligned} &x'_{\text{Re}} + ix'_{\text{Im}} = t_{x\,\text{Re}} + it_{x\,\text{Im}},\quad y'_{\text{Re}} + iy'_{\text{Im}} = y_{\text{Im}} - iy_{\text{Re}}, \\ &z'_{\text{Re}} + iz'_{\text{Im}} = z_{\text{Im}} - iz_{\text{Re}}\quad t'_{x\,\text{Re}} + it'_{x\,\text{Im}} = x_{\text{Re}} + ix_{\text{Im}}, \\ &t'_{y\,\text{Re}} + it'_{y\,\text{Im}} = t_{y\,\text{Im}} - it_{y\,\text{Re}},\quad t'_{z\,\text{Re}} + it'_{z\,\text{Im}} = t_{z\,\text{Im}} - it_{z\,\text{Re}} \end{aligned} \tag{3.9}$$

where the SLT in the x direction of M_4 spacetime transforms real components into imaginary and imaginary complex quantities into real quantities as one major property of the periodic nature of Noetic HAM spacetime [49-51].

3.11 Energy Dependent Spacetime Metric

Einstein originated the concept of an energy dependent spacetime for explaining temporal rate change in the presence of a gravitational field by generalizing the special relativistic line element

$$ds^2 = (1 + 2\phi / c^2)c^2 dt^2 - dx^2 - dy - dz^2 \qquad (3.10)$$

with the introduction of time curvature [1] where ϕ is the Newtonian gravitational potential. This utilizes the deformed Minkowski metric $\hat{M}_4$ (introduced above by Eq. 3.6) which is imbedded in the periodic HD Noetic space chosen axiomatically for HAM cosmology to take the form of a noncompactified Kaluza-Klein theory [32].

Kaluza's initial demonstration of gravity in 5D, ${}^5G_{AB} = 0$ with AB running 0,1,2,3,4 contained 4D General Relativity with an EM field ${}^4G_{\alpha\beta} = {}^4T_{\alpha\beta}^{EM}$, with α, β running 0,1,2,3 [15,16]. The currently less common non-compactified Kaluza-Klein model is utilized by Noetic Cosmology where also dependence on the extra D is required; this yields the same result for Einstein's equations ${}^5R_{AB} = 0$ except that the EM energy momentum tensor ${}^4T_{\alpha\beta}^{EM}$ is replaced by a general one ${}^4T_{\alpha\beta}$ instead [15,16]. We demonstrate the feasibility of an energy domain pervading HD spacetime with properties similar to Wheeler's Geon proposal discussed in section 3.12 below. In a generalized deformed spacetime metric $\hat{M}_4$, spacetime is fixed by the energy and has the metric

$$\eta(E) = diag.(a(E),\ -b(E),\ -c(E),\ -d(E)). \qquad (3.11)$$

3.12 The Wheeler Geon Concept Extended to Noetic Superspace

Wheeler [43] postulated a photonic mass of sufficient size to self cohere into a spherical ball of light. In Wheeler's notation the Geon is described by three equations. The first (3.11) is the wave equation, followed by two field equations the first (3.12) of which gives a mass distance relationship and the second (3.13) variation of the factor Q:

$$d^2 f / d\rho^{*2} + [1 - (l^* Q / \rho)^2 (1 - 2L / \rho)] f = 0 \qquad (3.12)$$

with circular frequency $c\Omega$ related to the dimensionless radial coordinate $\rho = \Omega r$ such that $d\rho^*$ is the abbreviation for $d\rho^* = Q^{-1}(1 - 2L / \rho)^{-1} d\rho$

$$dL / d\rho^* = (1/2Q)[f^2 + (df / d\rho^*)^2 + (l^* Qf / \rho)^2 (1 - 2L / \rho)] \quad (3.13)$$

$$dQ / d\rho^2 = (\rho - 2L)^{-1} [f^2 + (df / d\rho^*)^2] \quad (3.14)$$

L and f are mass and field factors respectively; Q is a scale correction factor. The factor l relates to a family of modes with distinct frequencies associated with the well-known completeness theorem of spherical harmonics. HD extended modes of l are key elements in propagation of the noetic field. Wheeler states that these equations permit change of distance scale without change of form [43] which is compatible with the Noetic action principle F_N = E/R derived below [44,45].

Postulate 3.2: *The Supralocal Hyper-Geon is the most fundamental energy or phenomenology of existence. This Energy arises from the ordering and translation of AS 'space' (i.e. information or change of entropy). This fundamental Geon energy, is the unified field, the primary quantum of action of all temporal existence; filling the immensity of space (nonlocally) controls the evolution of the large scale structure of the universe, the origin of life ('elan vital') of classical philosophy and finally is the root and 'light of consciousness'.*

3.13 The Hyper-Geon Domain of HAM Noetic Field Theory

As summarized in section 3.12 above Wheeler defined the Geon as a theoretical classical spacetime construct not yet observed in nature. A complex Hyperdimensional Geon is postulated to *cover* our observed 3(4)D relational spacetime and filters through each dimensional reduction like a waterfall as the de Broglie-Bohm pilot-wave quantum potential. This is described by a new set of Noetic transformations for HAM cosmology [35]; acting on all levels of scale from the Einstein/Hubble radius to the Planck scale. Because of its contact with the Multiverse it relates also to balancing the cosmological constant, Λ by the 'dark energy' responsible for the postulated missing dark matter that causes galaxy rotation to be like a solid disk rather than with a centripetal vortex with increasing speed with distance from the center. HAM cosmology postulates this missing energy to arise from the rest of the Multiverse. The Geon also forms the lower energy boundary of a projected 12D space making it synonymous with the unified field. This unitary Noetic field is the origin of the teleological anthropic action principle [52] guiding evolution. This coalesced region of nonlocal photon-gravitons – The hyper-geon superspace cover acts as:

- Gravitation (The graviton in HAM cosmology is a confined quadrupole photon $\hat{M}_4$ complex; thus teleological action of the unified field orders the large scale structure of the universe – which is a non-Darwinian guided evolution)
- Causal action of the quantum potential or pilot wave (An additional causal action principle pertinent to extended quantum theory and its completion)
- *Élan vital* or life force (The long sought vital principle required to legitimize dualism / interactionism)
- "light' of the mind (Bosonization of the Eccles psychon [52] as it couples to dendrons etc. to become the qualia of awareness [9]).

3.14 Conclusions

Scientific theory, whether popular or unpopular at any point in history, must ultimately be based on description of natural law, not creative fantasies of a scientist's imagination. Only by adequate determination of natural law can a theory successfully model reality. *"There is good reason for the taboo against the postulate of new physics to solve new problems, for in the silly limit one invents new physics for every new phenomena* [15]*"*. Not long ago cosmology was not considered to be a viable science; one saying went – 'first there is speculation, followed by speculation squared, then comes cosmology'. Is Cosmology becoming a mature science; mature enough that there is no room for surprises? We don't think so; and we have hinted at some of the surprises here.

A new model of the universe called the Holographic Anthropic Multiverse (HAM) provides a fundamental framework for introducing a scale-invariant complex cosmological system where life is the rule not the random exception because of the anthropic principle guiding its evolution. Many controversial principles stated emphatically; but Noetic cosmology is empirically testable so it will now be possible to settle many of these questions experimentally.

References

[1] Peebles, PJE (1993) Principles of physical cosmology, Princeton: Princeton University Press.
[2] de Bernardis, P. et al. (2000) A flat universe from high-resolution maps of the cosmic microwave background radiation, Nature, 404, 955–959.
[3] Amoroso, R.L., Kafatos, M. & Ecimovic, P. (1998) The origins of cosmological redshift in spin exchange between Planck scale vacuum compactification and nonzero rest mass photon anisotropy, in Hunter, G., S. Jeffers & J-P Vigier (eds.) Causality & Locality in Modern Physics, Dordrecht: Kluwer.
[4] Vigier, J-P & Amoroso, R.L. (2002) Can one unify gravity and electromagnetic fields? in R.L. Amoroso, G. Hunter, S. Jeffers & M. Kafatos, (eds.) Gravitation & Cosmology: From the Hubble Radius to the Planck Scale, Dordrecht: Kluwer Academic.
[5] Wheeler, J.A. & Feynman, R. (1945) Rev. Mod. Physics, 17, 157.
[6] Cramer, J.G. (1986) Reviews of Mod. Physics, 58:3, 647–687.
[7] Franck, G. (2000) Time and presence, in Science and The Primacy of Consciousness, R.L. Amoroso et al, (eds.) Orinda: Noetic Press.
[8] Amoroso, R.L & Amoroso, P.J. (2004) The fundamental limit and origin of complexity in biological systems: A new model for the origin of life, in D. Dubois, (ed.) Computing Anticipatory Systems, Proceedings of CASYS04, 6th Int, Conf., Liege, Belgium, 11-16 Aug, 2003, AIP Proceedings Vol. 718, pp. 144–159.
[9] Amoroso, R.L & Pribram, KH (2010) The Complementarity of Mind and Body: Realizing the Dream of Descartes, Einstein and Eccles, Cambridge: MIT University Press.
[10] Einstein, A. (1917) *Kosmologische Betrachtungen zur allgemeinen Relativitätstheorie*, Königlich Preußische Akademie der Wissenschaften (Berlin) Sitzungsbericht, pp. 142–152. Reprinted in facsimile as Doc. 43 in Kox et al. (ed.) 1996, Translation: Cosmological Considerations on the General Theory of Relativity. In: H. A. Lorentz et. al.(eds.) The Principle of Relativity, New York: Dover, 1952, pp.177–188.
[11] Bondi, H. & Gold, T. (1948) MNRAS, 108, 252.
[12] Hoyle, F. (1948) A New Model for the Expanding Universe, MNRAS, 108, 372–82.
[13] Alpher, R.A., Bethe, H. & Gamow, G. (1948) The origin of chemical elements, Physical Review, 73: 7; 803–804.

[14] Sen, R.N. (1999) Why is the Euclidian line the real line?, Found. Physics, 12:4;328–345.
[15] Witten, E. (1981) Search for a realistic Kaluza-Klein theory, Nuclear Physics B186, 412–428.
[16] Overduin, J.M. & Wesson, P.S. (1997) Kaluza-Klein gravity, Physics Reports, 283, pp. 303–378.
[17] Randall, L & Sundrum, R. (1999) An alternative to Compactification, Physical Review Letters, 83, 4690–4693.
[18] Randall, L. (2005) Warped Passages, New York: Harper-Collins.
[19] Einstein, A. (1952) letter to H. Stevens, from Stevens, H (1989) Size of a least unit, in M. Kafatos (ed.) Bell's Theorem, Quantum Theory and Conceptions of the Universe, Dordrecht: Kluwer Academic.
[20] Ellis, G.F.R., Kirchner, U. & Stoeger, W.R. (2004) Multiverses and physical cosmology, Monthly Notices of the Royal Astronomical Society, 347: 921–936, or arXiv:astro-ph/0305292v3.
[21] Dicke, R. H. (1957). Gravitation without a Principle of Equivalence. Reviews of Modern Physics 29: 363–376.
[22] Barrow, J.D. & Tipler, F.J. (1988) The Anthropic Principle, Oxford: Oxford Univ. Press.
[23] Tipler, F. (1998) The Physics of Immortality: Modern Cosmology, God and the Resurrection of the Dead New York: Doubleday.
[24] Weinberg, S. (2007) Living in the Multiiverse, in B. Carr (ed) Universe or multiverse?, Cambridge: Cambridge University Press, arXiv:hep-th/0511037v1 (2005).
[25] Susskind, L. (2008) The Black Hole War - My Battle with Stephen Hawking to Make the World Safe for Quantum Mechanics, Boston: Little, Brown & Co.
[26] 't Hooft, G. (2000) The holographic principle. hep-th/0003004.
[27] Susskind, L. (2003) Superstrings (Features: November 2003) Physics World 16 (11).
[28] Susskind, L. (1994) The World as a Hologram, arXiv:hep-th/9409089 v2.
[29] Bousso, R. (2002) The holographic principle, Rev Mod Phys 74: 825-874.doi:10.1103/Rev. Mod. Phys. or arXivhep-th/0203101.
[30] Brown, M. (1960), A proof of the generalized Schöenflies theorem. Bull. Amer. Math. Soc., vol. 66, pp. 74-76.
[31] Smart, J.J.C. (1972) Space, V. 7, 506-511, Encyc. Philosophy, New York: Macmillan.
[32] Darabi, F., Saikoand, W.N. & Wesson, P.S. (2000) Quantum

cosmology of 5D non-compactified Kaluza-Klein theory, Class. Quantum Grav. 17; 4357–4364, or arXiv:gr-qc/0005036v1.
[33] Chu, M,-Y, & Amoroso, R.L. (2008) Empirical mediation of the primary mechanism initiating protein conformation in prion propagation, in D. Dubois (ed.) Proceedings of CASYS07, IJCAS, vol. 22, pp. 292–313, Liege, Belgium.
[34] Amoroso, R.L. & Rauscher, E.A. (2008) Emergence of generalized F-theory 2-branes from SUSY spacetime parameters of the discrete incursive oscillator in D. Dubois (ed.) Proceedings of CASYS07, IJCAS, Vol. 22, pp. 283–291, Liege, Belgium.
[35] Amoroso, R.L. (2000) The parameters of temporal correspondence in a continuous state conscious universe, in L. Buccheri & M. Saniga (eds.) Studies on the Structure of Time: From Physics to Psycho(patho)logy, London: Plenum.
[36] Burns, J.E. (2002), Vacuum radiation, entropy and the arrow of time, in R.L. Amoroso, G. Hunter, S. Jeffers & M. Kafatos, (eds.) Gravitation & Cosmology: From the Hubble Radius to the Planck Scale, Dordrecht: Kluwer Academic.
[37] Zeh, H.-D. (1989) The Physical Basis of the Direction of Time, Springer-Verlag, New York.
[38] Misner, C.W., Thorne, K. & Wheeler, J.A. (1973) Gravitation, San Francisco: Freeman.
[39] Sklar, L. (1995) Philosophy and Spacetime Physics, Berkeley: Univ. of California Press.
[40] Reichenbach, H. (1957) Philosophy of Space and Time, NY: Dover.
[41] Amoroso, R.L. (2002) The continuous state universe, in R.L. Amoroso, G. Hunter, S. Jeffers & M. Kafatos, (eds.), Gravitation & Cosmology: From the Hubble Radius to the Planck Scale, Dordrecht: Kluwer Academic.
[42] Kafatos, M. Roy, S. & Amoroso, R. (2000) scaling in cosmology & the arrow of time, in Buccheri, di Gesu & Saniga, (eds.) Studies on Time, Dordrecht: Kluwer.
[43] Wheeler, J.A. (1955) Geons, Physical Review, 97:2, 511–536.
[44] Amoroso, R.L. (2002) The Physical Basis of Consciousness: A Fundamental Formalism, Part 1 Noesis, XXVI, Romanian Academy.
[45] Amoroso, R.L (2000) Derivation of the fundamental equation of consciousness, Part I, Boundary conditions, Noetic J 3:1, pp. 91–99.
[46] Amoroso, R.L. & Vigier, J.-P. , The origin of cosmological redshift and CMBR as absorption/emission equilibrium in cavity-QED blackbody

dynamics of the Dirac vacuum, In Amoroso, R.L., Hunter, G., Kafatos, M., Vigier, J.-P. (eds.) Gravitation & Cosmology: From the Hubble Radius to the Planck Scale, 2002, Dordrecht: Kluwer Academic.
[47] Amoroso, R.L. & Rauscher, E.A. (2008) Derivation of the string tension formalism, and super quantum potential as inherent parameters of a holographic conscious multiverse cosmology, in R.L. Amoroso, I. Dienes & Cs. Varga (eds.) Unified Theories, Oakland: The Noetic Press.
[48] Witten, E. (1996) Reflections on the fate of spacetime, Physics Today (April), pp. 24–30.
[49] Cole, E.A.B. (1977) Il Nuovo Cimento, 40:2, 171–180.
[50] Rauscher, E. (2002) Non-Abelian gauge groups for real and complex Maxwell's equations, in R.L. Amoroso, G. Hunter, S. Jeffers & M. Kafatos, (eds.), Gravitation & Cosmology: From the Hubble Radius to the Planck Scale, Dordrecht: Kluwer Academic.
[51] Rauscher, E.A. & Targ, R. (2001) The speed of thought: investigation of a complex space-time metric to describe psychic phenomena, J Sci Exp., 15:3; 331–354; or www.bibliotecapleyades.net/archivos_pdf/speedthought_rauscher_targ.pdf.
[52] Eccles, J.C. (1986) Do mental events cause neural events analogously to the probability fields of quantum mechanics?, Proc. Royal Soc. London B227, pp. 411–428.

Chapter 4

An Alternative Derivation of String Tension Determining a Unique Background Independent String Vacuum

The standard model is a non-abelian gauge theory of symmetry group U(1)×SU(2)×SU(3); and as one knows all gauge theories including General Relativity or Quantum Electrodynamics are only approximations anticipating additional physics. It appears that cosmology is the domino that must fall to revolutionize all the related fields of physical science as it tumbles through them in a domino effect. One important consequence is that Planck's constant, $\hbar$ will no longer be fundamental in the manner currently considered. Since our extension of Einstein's Static Universe Model utilizes an energy dependent spacetime metric, $\widehat{M}_4$ the original *Stoney* is seen as a more appropriate basis than $\hbar$ for extending the lower limit of physical cosmology and particle physics. In this scenario the new term modulating $\hbar$ is that of string tension, T_S. The various iterations of M-Theory are said to provide $10^{1,000}$ possibilities for delineating the fundamental string vacuum. With the addition of our version of the Anthropic Principle we find a putative way to derive T_S in a manner suggesting a unique background independent vacuum leading to a number of surprising results based on what is called the 'least cosmological unit' tiling spacetime as cast in a new Holographic Anthropic frame that includes a 12D regime of unitary absolute space.

4.1 Introductory Prolegomena

Absolute Space, in its own nature, without regard to any thing external, remains always similar and immovable - Isaac Newton [1]

Before the ultimate spinor, twistor, SUSY, unitary or some-such structural-phenomenology of matter can be rigorously defined, the dynamic brane topology of a unique background-independent string vacuum must be formalized because this is the context from which this structural-phenomenology arises. A critical clue is simplistically illustrated in the $360° / 720°$ Dirac spherical rotation of the electron which can be considered to be like two manifolds of a 4D Klein bottle forming a structure preserving map $f : M \rightarrow N$ representing a diffeomorphic topological imbedding such that $f^{-1} : N \rightarrow M$ allowing the electrons spinor structure to undergo continuous transformation between 3D and 4D. This process is probably some form of Calabi-Yau transform. The photon, quarks, electron and positron are considered to have the simplest of these complex structures with internal and external motion, couplings and tensions. Einstein has said 'that an aether or medium for the propagation of electromagnetic waves is not required because the photon provides its own medium'. This is partially true in a semi-classical approach, the mantle under which Einstein received his education and formulated his relativities. Even though he demonstrated that 4D spacetime is relational and not absolute, he didn't realize in his day how much further spacetime physics had to go, that a photon isn't an independent entity, that its propagation can't be sufficiently separated from reality itself to be its own medium because its complete description must be imbedded in a cosmological background. This has been a major theoretical challenge because current understanding suggests there are 10^{1000} possibilities for the unique string vacuum, assuming of course a form of M-Theory is the way to proceed. We embrace a version of string theory here called F-Theory, which may seem surprising when the reader finds that we reject quantum gravity, the Higgs Boson and superpartners outright. But it still remains all about SUSY principles and symmetry breaking albeit from a radical HD topological Holographic Anthropic Multiverse (HAM) perspective.

It is said that the usual formulation of String Theory has only one parameter, that of string tension, T_S. Large XD SUSY models [2] have an additional fine-tuning parameter, that of the bulk (an HD space within which our 4D realm exists as a subspace) cosmological constant [3,4]. Also if the universe is stringy Planck's constant, $\hbar$ is not fundamental, especially since Gauge Theory despite its phenomenal success, is only an approximation [5] which of course sets the stage for new physics like that proposed here. The correction applied to $\hbar$ is the string tension parameter, T_S [6]. These are cosmological components that the new cosmology brings into correspondence with current 4D theory and which provide the setting for developing an alternative derivation of the string tension formalism as inherent parameters of the HAM cosmology which putatively leads to a physically real unique background-independent string vacuum.

In HAM cosmology the observed 3(4)D, i.e. +,+,+,- signature temporal reality is a virtual subspace of an 11(12)D 'eternity' in correspondence with the tenets of a unique F-Theory incarnation of M-Theory [7]; 12D being the minimum number of dimensions (D) to signify causal separation from temporality. Succinctly HAM cosmology postulates an infinite number of nested Hubble spheres each with their own fine-tuned laws of physics; the universe is closed and finite in time but open and infinite in the holographic multiverse. This relates to our extended interpretation of the holographic principle [8-11] where this rest of the Multiverse is responsible for dark energy and the properties of the cosmological constant [12,13]. The HAM 'eternal present' [14] is a continuous dynamic instant, an HD standing-wave array of least cosmological units [15] of the 12D Superspace undergoing a Continuous-State Spin-Exchange Dimensional Reduction/ Compactification parallel transport process based on new SUSY extensions of the Wheeler-Feynman-Cramer transactional models of radiation and quantum theory [16,17]. This HAM dynamic entails an energy dependent spacetime metric, $\widehat{M}_4$ as 1st proposed by Einstein [18].

This means that HD properties of the 'continuous-state dimensional reduction/compactification standing-wave metric' entail a form of future-past advanced-retarded hysteresis loop [19] (formalized below). The energetics of this so-called hysteresis loop of the 12D least-unit structure reveal an inherent new action principle driving the evolution of HAM cosmology which allows cosmology itself (inside the H_R) to be postulated as a form of self-organized complex system. As is well known self-organized systems are driven by external action [20]. Since the HAM is covariant and scale-invariant these energetics also apply to self-organized autopoietic living systems [20-22] and the nature of the observer. This new teleological, noetic or anthropic action principle is shown to be associated with the unitary physical field acting as a form of 'super-quantum potential' as postulated by de Broglie and Bohm [23,24]. Using the context of this plethora of parameters, an alternate derivation of the string tension formalism, T_S is derived. As already mentioned it is suggested that this form of the string tension formalism includes a recalculating of Planck's constant and leads to a program for completing quantum theory. (See Chaps. 9 and 11.) HAM cosmology is empirically testable and an experimental protocol for vacuum engineering by manipulating the new energy dynamics is presented in Chaps. 9 and 11.

These fundamental least-units (section 4.7) entail a form of Incursive Oscillator (IO) [25-28] inherent in the continuous-state topology of HAM spacetime. Simulated application of the IO is shown to produce a natural emergence of generalized M-Theory 2-branes from the superspace backcloth. This result could be instrumental in solving the problem of deriving parameters of the fundamental string vacuum especially emergence of the new action principle driving the evolution of its self-organization, and achieve the ultimate goal of M-Theory that of investigating the fundamental stringy structure of matter.

M-Theory, is based essentially on one parameter, string tension, T_S

$$T_S = e / l = (2\pi\alpha')^{-1}; \tag{4.1}$$

where e is energy, l is length of the string and α the fine structure constant, $e^2 / \hbar c$ where this e is the electron charge. See Eqs. (6.13-6.15). It is well known that the gauge condition is an approximation [5] suggesting Planck's constant, $\hbar$ needs to be recalculated to satisfy the parameters of M-Theory [7]. Since HAM cosmology is aligned with an extension of Einstein's energy-dependent spacetime metric $\hat{M}_{3(4)}$, the Stoney e^2 / c, an electromagnetic precursor to Planck's constant, [29-32] is therefore the choice for studying the recalculation. The factor added to $\hbar$ is string tension T_S, where T_0 can increase the size of $\hbar$ to the Larmour radius of the hydrogen atom in the small scale and lead to infinite size additional dimensionality cosmologically [2,5,6,33]. Thus the fine-tuned Stoney, $\lambda\!\!\!^{-}$ and the cosmological constant, Λ adjust the microscopic and cosmological domain limits of H_R respectively. Equation (4.2) illustrates the initial historical basis for this distinction

$$l_P = \sqrt{\frac{\hbar}{mc} \cdot \frac{Gm}{c^2}} \quad or \quad l_S = \sqrt{\frac{e^2}{4\pi\varepsilon_0 mc^2} \cdot \frac{Gm}{c^2}}, \tag{4.2}$$

where l_P *and* l_S are the length of the Planck and Stoney respectively.

One example for rescaling Planck's constant comes from Wolf [34]

$$\Delta x \Delta \rho = h \to h_0 \pm \Delta h. \tag{4.3}$$

He then suggests that

$$\Delta h = \frac{hv^2}{c} \tau_0 L_0 \tag{4.4}$$

where τ_0 and L_0 are time uncertainty and a discrete spacetime correction respectively. Wolf is able to speculate that this Planck rescaling has application to Neutron stars, CMBR and black hole formation [34]. Our approach for a time, τ_0 and spacetime corrections, L_0 are different.

4.2 Scaling in Cosmology and the Continuous-State Postulate

Fundamental theories must ultimately not only account for the structure and evolution of the universe, and the physics of fundamental interactions but also lead to an understanding of why this particular universe follows the physics that it does. Such theories must lead to an understanding of the values of the fundamental constants themselves. Moreover, the understanding of the universe has to utilize experimental data from the present to deduce the state of the universe in distant regions of the past and also account for certain peculiarities or coincidences observed. The continuous-state postulate for the static HAM cosmology replaces the concept of expansion/inflation in Big Bang cosmologies. The prevalent cosmological view has been the Big Bang, inflationary evolutionary model. Although serious problems remain, e.g. the need to postulate undetectable cold, dark matter in amounts much larger than all observable matter put together, or the recent need to re-introduce the cosmological constant, Big Bang cosmology has nevertheless achieved impressive results [35].

Here we take a radically different approach than the usual evolutionary picture where the physics itself is assumed invariable. We study numerical relations among fundamental constants starting from relations first proposed by Weinberg [36], which are equivalent to those found by Dirac [37], and explore a new scaling hypothesis relating the speed of light c and the scale of the universe R. For simplicity we develop an axiomatic approach resulting in an *apparent* expanding universe, yielding the same successes as present Big Bang cosmology but without the need to postulate inflation, cold dark matter, or any of the artificialities of current theory. The "coincidences" of Dirac [37] and Eddington [38] concerning large numbers and ratios of fundamental constants are not explained, just accepted and in the process yield a fundamentally different view of the cosmos. The fundamental constants are fine-tuned and can be assumed to vary within each nested Hubble sphere. The assumption that redshift is Doppler has led to an *apparent* expansion of the universe. Here this 'energy' is internalized in the continuous-state process which reveals the nature of the arrow of time (Chap. 4) as perceived by an Earthly observer. A fundamentally different view of the cosmos arises in this cosmology that is closed and finite in time, open and infinite in eternity with room for an infinite number of nested Hubble spheres each with their own fine-tuned laws of physics.

4.3 Fine Tuning Implied by Astrophysical Observation

Numerous observations must be applied in any cosmological theory attempting to explain the observed structure of the universe:

- The universe is observationally flat, meaning the density of the universe is close to the so-called closure or critical density,

$$\rho_{crit} = 2\times10^{-29}\left(H_0 / 100\text{km s}^{-1}\ \text{Mpc}^{-1}\right)^2 \text{gr cm}^{-3} \tag{4.5}$$

  where H_o is the Hubble constant defined as the apparent rate of expansion with distance, $\dot{R}/R$ and where R is the scale of the universe. In Big Bang cosmology, this so-called 'constant' is actually a function of cosmic time, i.e. it is a variable. Its present-day value seems to be ~ 75 km s^{-1} Mpc^{-1}. The universe appears to be *close* to a flat, Euclidean, Einstein-de Sitter state as indicated from (3.3), and yet it is still not clear what the geometry of the universe is; *exactly* flat (which would be required by the inflationary scenario); open (yielding a forever-expanding, negatively curved space-time); or closed (yielding a maximum expansion and a positively curved space-time). The HAM model we discuss is an antinomy between open and closed [39], i.e. closed and finite in time, H_R and open and infinite in the eternal Multiverse.

- If one is to assume that the universe followed an inflationary period in the distant past, then the universe must be *exactly* flat to one part in 10^{50} near the time of Big Bang. This is the so-called *flatness problem*: This is such a remarkable requirement that the usual interpretation proposed in the early 1980's is that early on, the universe was in an inflationary state, washing out any departures from flatness on time scales of 10^{-35} sec. The inflationary model proposed by Guth and others [40,41] has been developed in various forms to account for the flatness of the universe and also is proposed to solve the *horizon problem*, or apparent homogeneity of the $2.73°K$ black body radiation seen by COBE [42]. The latter problem involves the observation that although the $2.73°K$ radiation was emitted ~ 10^5 years after the beginning, opposite sides of the sky at that time were out of causal contact, separated by ~10^7 light years. Other structures involving large-scale correlations in the universe exist such as very large structures in the distribution of matter [43]. These structures are progressively hierarchical to the scale of the universe itself.

- If the universe is indeed flat, observations indicate that baryons (and luminous matter) can only contribute at most ~0.05 of the closure density at present. We should ultimately be able to detect the other 90% or more of the matter required to give closure density, presumed to be in the form of cold dark matter [44]. Nevertheless, attempts to detect such exotic matter in the laboratory have, so far, failed. Moreover, the recent realization that the cosmological constant, Λ may have to be re-introduced [45] has also led to the probability of Λ itself varying and other similar notions [46]. Without though some direct laboratory verification or overwhelming requirements imposed by particle theory (neither of which presently exists), the nature of dark matter remains elusive. This is clearly a very unsatisfying situation for Big Bang cosmologists.

- As we saw, present-day approximate flatness yields to an *exact* flatness in the distant past (this was one of the main reasons why the inflationary scenario was introduced to begin with). The alternative is to accept fine-tuning in the universe. In fact, the flatness of the universe is not the only fine tuning. In considering other fundamental observed facts, the universe appears to be extremely fined tuned. It was Eddington [38,47] and Dirac [37] who noticed that certain cosmic "coincidences" occur in nature linking microscopic with macroscopic quantities [48,49]. A most unusual relationship is the ratio of the electric forces to gravitational forces (this ratio is presumably a constant in an expanding universe where the physics remains constant), or

$$e^2/Gm_e m_p \sim 10^{40} \tag{4.6}$$

while the ratio of the observable size of the universe to the size of an elementary particle, or

$$R/\left(e^2/m_e c^2\right) \sim 10^{40} \tag{4.7}$$

where in the latter relationship the numerator is changing as the universe expands because the scale of the universe R is constantly changing in an expanding universe.

Dirac formulated the so-called *Large Number Hypothesis* which simply states that the two ratios in (4.6) and (4.7) are in fact equal for all practical purposes and postulates that this is not a mere coincidence. Various attempts were made to account for the apparent equality: A possibility that constants such as the gravitational constant may be varying was proposed by Dirac [38] himself and others [50]. Other ratios such as the ratio of an elementary particle to the Planck length,

$$\frac{e^2/m_e c^2}{\left(\hbar G/c^3\right)^{1/2}} \sim 10^{20} \tag{4.8}$$

can also be constructed [51] yielding to the conclusion that fine tuning is prevalent in the multiverse. These relationships seem to be indicative of the existence of some deep, underlying symmetries involving the fundamental constants and linking microcosm to macrocosm. Physical theory has not, however, accounted for these in a self-consistent way, perhaps with the anticipated unification of all physical forces at the level integrating quantum theory and gravity.

- Evidence [52] has recently been found which seems to be consistent with a time-varying fine structure constant $\alpha = e^2 / (\hbar c)$. A varying speed of light theory (with $\hbar \alpha c$) has also been proposed by Albrecht and Magueijo [53]. These two theories correspond to different representations of a varying α in terms of varying dimensional constants. The minimal varying-*c* theory is of interest because it offers a means of solving the so-called cosmological problems: the horizon, flatness, cosmological constant, entropy and homogeneity problems. Barrow and Magueijo [52] tried to show that there exists a set of duality transformations between these two representations. On the other hand, recent observations of astrophysical events at high redshift [54,55] can be used to place severe limits on the variation of the speed of light itself ($\Delta c / c$), as well as the photon mass (m_γ).

- Several ideas such as Quantum-like correlations [56] and the Anthropic Principle [57], developed here, have been proposed to account for the above fine tuning properties of the universe.

4.4 Numerical Relations Coupled to the Concept of Scaling

The critical density of the universe, ρ_{crit} in (4.1) is defined as

$$\rho_{\rm crit} = \frac{3H_0^2}{8\pi\ G} \tag{4.9}$$

Let N_p be the number of nucleons in the universe, then

$$m_p = \frac{M}{N_p} = \frac{R\dot{R}^2}{2GN_p} \tag{4.10}$$

where m_p and M are the mass of the nucleon and mass of the universe, respectively. Weinberg [36] noticed that one can find a relationship linking the masses of elementary particles to the Hubble constant and other fundamental constants

$$m_\pi \sim \left(\frac{8\hbar^2 \mathrm{H}_0}{\mathrm{Gc}}\right)^{1/3} \tag{4.11a}$$

and, correspondingly,

$$m_e \sim \left(\frac{\hbar \mathrm{e}^2 \mathrm{H}_0}{(8\pi)^3 \mathrm{Gc}^2}\right)^{1/3} \tag{4.11b}$$

where, m_π and m_e are the pion and electron masses, respectively. These relations can be rewritten as

$$m_p \sim X_{p\pi}\left(\frac{8\hbar^2 (\dot{R}/H)}{\mathrm{Gc}}\right)^{1/3} \text{ with } = X_{p\pi} = \frac{m_p}{m_\pi} \text{ and} \tag{4.12a}$$

$$m_p \sim X_{pe}\left(\frac{\hbar \mathrm{e}^2 (\dot{R}/R)}{\mathrm{Gc}^2 (8\pi)^3}\right)^{1/3} \text{ with } X_{pe} = \frac{m_p}{m_e} \tag{4.12b}$$

From (4.10) and (4.11a) one easily gets

$$G^2\hbar^2 c^{-1} \sim X_{p\pi}{}^{-3} N_p{}^{-3} \frac{R^4 \dot{R}^5}{64} \tag{4.13}$$

We also have
$$m_p = X_{p*}\sqrt{\frac{\hbar c}{G}} \tag{4.14}$$

where $X_{p*} = m_p / m_*$, m_* being the Planck mass and the suffix * indicates Planck quantities. Combining (4.12) and (4.8), yields

$$\mathrm{cG}\hbar \sim 1/4\left(N_p{}^{-2} X_{p*}^{-2} R^2 \dot{R}^4\right) \tag{4.15}$$

Similarly, from (4.13) and (4.14)

$$\mathrm{c}{\sim}2^{2/3} N_p{}^{-1/3} X_{p*}^{-4/3} X_{p\pi} \dot{R} \tag{4.16}$$

The multiplier factor for $\dot{R}$ in (4.16) is equal to $2^{2/3} N_p^{-1/3} X_{p*}^{-4/3} X_{p\pi}$, and is ~1. Conversely, if we choose to set $2^{2/3} N_p^{-1/3} X_{p*}^{-4/3} X_{p\pi} = 1$, one gets the simple relationship linking the speed of light to $\dot{\mathrm{R}}$, $\mathrm{c} = \dot{\mathrm{R}}$ with $\mathrm{N_p} \sim 3.7 \times 10^{79}$, which is a good estimate of the number of particles in the current universe. The relationship $c = \dot{R}$ could be interpreted as the

Hubble Law $\dot{R} \sim c$, although we emphasize that this is just a relationship and in no way implies that an expansion is indeed taking place. Similar considerations apply if one chose to work with the relations applying to electrons.

If we start by assuming a heuristic relation

$$c \equiv \dot{R} \tag{4.17}$$

i.e. the speed of light is *identical* to the rate of change of the scale of the universe, we build an axiomatic approach equivalent to Hubble's Law that can be considered an alternative to the mysterious coincidences of Eddington and Dirac [37,38] which Weinberg called "so far unexplained... a real though mysterious significance."

It can be further shown that all lengths, such as the Planck length, l_*, the classical electron radius, r_e, etc., are also proportional to

$$R, l_*, r_e, \sim (...)R \tag{4.18}$$

For example,

$$l_* \sim \left(2^{-7/3} N_p^{-1/3} X_{p*}^{5/3} X_{p\pi}^{-2}\right) R \tag{4.19}$$

Similar relations can be found for r_e and r_p where r_e and r_p be the electron and proton radii. Combining (4.15) with (4.16) we obtain

$$G\hbar = \frac{R^2 \dot{R}^3}{4} N_p^{-2} X_{p*}^{-2} \sim 3.4 \times 10^{-122} R^2 \dot{R}^3 \tag{4.20}$$

A relationship linking the gravitational and Planck's constant to R and $\dot{R}$, and where the last relationship in (4.20) holds for the current values of $N_p^{-2} X_{p*}^{-2}$ in the universe.

Let us now set the following initial conditions, i.e.,

$$R \to l_* \tag{4.21}$$

$$\dot{R} \to \frac{l_*}{t_*} = c \tag{4.22}$$

where l_* and t_* are the Planck length and Planck time, respectively.

Then $N_p^{-2} X_{p*}^{-2} / 4 \to 1$ at those initial conditions, while for the present universe the value of this quantity is ~3.4×10^{-122}. The limit $N_p \to 1$ indicates that in this model "in the beginning" there was only one bubble-like object or a "cosmic egg" [58]. Moreover in a Big Bang sense, $R \to l_*$ and $N_p \to 1$ imply that $X_{p*} \to 1$ as well (similarly for all ratios of masses, X's), which in turn indicates that the masses of all particles were equal to each other at these initial conditions. Also, "in the beginning" $R/(e^2/m_e c^2) \sim (e^2/m_e c^2)/(\hbar G/c^3)^{1/2} \sim 1$, rather than the large values of 10^{40} and 10^{20} which these ratios are equal to, respectively, today. "In the beginning" all lengths were equal, all masses were equal and there was only one particle or cosmic egg. Today for Big Bang cosmology, these ratios are not unity, there is a very large number of particles in the universe and R is equal to $\sim 10^{28}$ cm. However, conformal scale-invariant relationships such as $c \equiv \dot{R}$; all lengths are proportional to each other, etc. still hold. Israelit & Rosen [58] proposed a cosmological model where the universe emerges also from a small bubble ('cosmic egg') at the bounce point of a de Sitter model filled with a cosmic substrate ('prematter').

The survey above of scaling and the large number hypothesis has so far been from the more usual perspective as might be applied to Big Bang perspective. For our HAM model this cosmic egg is not primordial but would represent the whole Hubble 3-sphere, H_R as a continuous-state and the 'prematter' would be the whole background of 'Heisenberg 'potentia'. It is still covariant and scale-invariant but looked at from a different point of view. The concept may seem difficult to grasp at first and one we remain challenged to convey clearly. Should our attempt be to map out its properties in terms of the holographic principle or from the standing-wave perspective? One thinks of a standing-wave as a 1D violin string or the surface of a 2D drum oscillating in 3D; but this is too simplistic and reveals little of the internal or scale-invariant structure. Certainly we could perform our delineation from the perspective of SUSY symmetry breaking which is what it is all about; but this is already considered in M-Theory and doesn't clarify the distinction because all forms of compactification of the theory are considered to be primordial whereas in HAM cosmology it is 'non-stop' which for us becomes key to understanding the fundamental nature of the arrow of time (Chap. 4).

In other words (and from either perspective), $c \equiv \dot{R}$ at the 'initial time' which would be Big Bang T_0 or continuous holographic lightcone T_0 when $N_p \to 1$ and all $X's \to 1$, and this relationship remains invariant even at the present universe or virtual reality of the observer (cf. (4.12) and (4.14)). The self-consistency is obtained by calculations for the value of N_p from (4.12) and (4.16). This relation is a type of a scaling law connecting the microcosm and the macrocosm in both cases. Now if irrespective (and it is even immaterial) of whether there is expansion/inflation of the universe or not, if R itself is changing from the Planck scale to the size of the observable universe, H_R then the fundamental constants like G, $\hbar$, and c also *all* are changing. The key new idea is that this change in R in the sense of the continuous-state model is the internal inherent conformal scale-invariance of the symmetry breaking of the dynamics of the dimensional reduction process. Note, however, that the actual variation or the initial value of c and other constants cannot be deduced from the usual observations: The relationship $c \equiv \dot{R}$ is not enough to tell us the actual variation or even over 'how long' it takes place. It is a scale-invariant relationship.

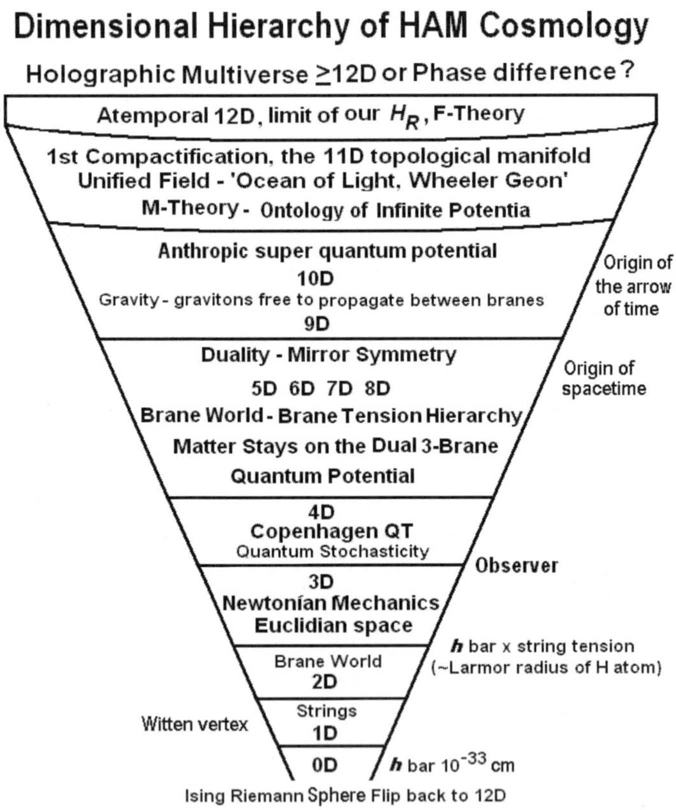

Figure 4.1 The continuous-state $12D \Leftrightarrow 0D$ Riemann topology Ising spin flip lattice-gas model dimensional hierarchy in HAM cosmology a form of $0 \leftrightarrow \infty$ advanced-retarded future-past standing-wave hyperstructure.

However we have devised a curious anthropic procedure to show how this can be achieved in Chap. 13. If we rewrite it as a scale-invariant relationship, $c(t_*)/c(t_0) = \dot{R}(t_*)/\dot{R}(t_0)$ where t_* and t_0 could conveniently be taken as the Planck time and present "age" of the universe, then this relationship is insufficient for the evolution of $\dot{R}$ or even values of t_* and t_0. Hence it cannot tell us how c itself is varying or even if it is varying. If we wanted to insist that *c is constant,* then all the other 'constants' like G and $\hbar$ *are really constant* as well. But if c is not constant, then all the other 'constants' are varying as well (or are different in other multiverse domains, Chap. 13). In both cases, however, the number of particles is changing (or appears to be changing), the ratios of masses are changing and the ratios of scales or lengths are also changing.

An arrow of time can therefore, be introduced. See Chap. 5. In this picture, invariant relationships hold and from unity, there is evolution into diversity. One cannot though conclude how the variations are taking place, over what timescales they are taking place or even how old the universe is. The universe could be 10^{10} years old or 5×10^{-44} sec (the Planck time) old, any time in between or timeless. *Time is strictly a parameter* that can be introduced in the scale-invariant relationships. It has no meaning by itself and is a virtual effect for the sake of the observer in HAM cosmology. The universe *appears to be evolving* as the number of particles and ratios are varying.

To summarize, the existence of horizons of knowledge in cosmology, indicate that as a horizon is approached, ambiguity as to a unique view of the universe sets in. It was precisely these circumstances that apply at the quantum level, requiring that complementary constructs be employed [59]. At the *initial time*, which could be conveniently taken as the Planck time, if we set the conditions like $c = \dot{R}$, as proposed in this chapter, we can axiomatize the numerical relations connecting the microcosm and the macrocosm or our Hubble sphere or any of the infinite other Hubble spheres comprising the holographic multiverse. One then has scale-invariant relationships, the range of which, $\lim \sum_{\lambda}^{\Lambda} H_R$ sets the stage to explore inter-dimensional multiverse relationships. During the *evolutionary* process of the multiverse, the fundamental constants may be changing or constant. In the former case, we don't know *how* or even over *what* timescales or domains they are changing. In the latter case, one gets the usual Big Bang evolutionary universe. This is a clear case where complementarity applies; and the dynamic which we wish to build into a HAM cosmology that is closed and finite in time and open and infinite in eternity with room for an infinite number of nested Hubble spheres each with their own fine tuned laws of physics.

In other words as N_p is changing from the initial value of 1 (unity) to the present large value of $\sim 10^{80}$ (diversity), more particles are created as R and all length scales as well as all masses are changing. This could be interpreted by an observer as an "expansion of the universe". An observer, who is inside the Hubble universe will perceive an "arrow of time" and an "evolving universe". But equivalently, as the "constants" change (in contrast to previous works, they would *all* have to be changing), or even if they are truly constant, there *appears* to be an evolution because of the hierarchical coupling and uncoupling of the continuous-state symmetry breaking process through all levels of scale. As $N_p \to 10^{80}$, the present number of the nucleons in the universe, the fundamental "constants" achieve their present values. This cosmic egg business can make it appear as if there is just one particle, a wave function of the universe, $H\Psi = 0$ with a complementary quantified hierarchical degrees of separation in the unitary holographic fabric.

To recapitulate, the arrow of time can be related to a kind of complementarily between two constructs, the fundamental "constants" *are truly constant*, on the one hand; and the fundamental "constants" *are changing*, on the other hand. In summarizing this section, we found that by adopting Weinberg's relationship (equivalent to Dirac's relationships (4.4) and (4.5) when the latter are equated to each other), we can obtain a relationship linking the speed of light c to the rate of change of scale of the universe, $\dot{R}$. In fact, the proportionality factor is ~ 1 if one substitutes for values of fundamental quantities like the present number of particles in the universe, etc. The next step assumes that the relationship linking c and $\dot{R}$ is an identity, i.e. $c \equiv \dot{R}$ (for example, at the Planck time, one observes that this relationship still holds if the ratios of all masses $\to 1$ and the number of particles also $\to 1$). As such, it is possible (but not necessary) to state that *all* the fundamental constants are changing and not just one of them as was assumed in prior works.

It is interesting that the possibility of the cosmological constant, Λ itself changing has been suggested [46]. As such, what we are suggesting here as a framework for the universe is a natural extension of previous ideas. Therefore, as N_p changes from an initial value of 1 to the present value of 10^{80} ($1 \rightarrow 10^{80}$), the universe would appear to be evolving to an observer inside it or an arrow of time is introduced. Finally, the outcomes of this prescription are not just that an arrow of time is introduced and the mysterious coincidences of Dirac and Eddington now can be understood as continuous-state scale-invariant relationships linking the microcosm to the macrocosm; but in addition, all scales are linked to each other in the continuous-state and what one calls, e.g. the *fundamental length,* etc. is purely a convention. In the same way, time itself is not as fundamental as the scale-invariant relationships linking the microcosm to the macrocosm but arises from it. In Chap. 5 we propose that the continuous-state symmetry breaking conditions of the dynamics of the scale-invariance hierarchy 'produces' an arrow of time for elucidating the world view of the conscious observer. These relations are as seen below as essential to the alternative derivation of the string tension formalism. Finally the other mind-boggling concatenation is that the action of 'the all' on 'the one'; this zeroth order 'cosmic egg' is like Mach's principle in that in some contexts the action of the infinite and minute are one in the same like being imbedded in a fractal where above is infinitely the same as infinitely below.

4.5 Physical Cosmology of the Close-Packed Fundamental Least Unit for an Energy Dependent Spacetime Metric

A stochastic zero-point field quantum foam of Planck units has often been suggested to tile the spacetime backcloth; but such a convention does not serve our purposes here, however the controversy between absolute-relational continuous-discrete remains. The idea of a cosmological fundamental least-unit could be drawn from the unit cells forming the periodic array of crystal structure [60-62]. The idea of a least cosmological unit is not entirely new and may be attributed to a question posed by Einstein in 1952 [15,63,64]. Stevens has suggested "Contact…between least-units…is taken to be 5, making the aggregation of least-units fourth dimensional. This leads to a cosmology in which our 3D physical space occurs as the surface of a hypersphere of close-packed least units" [15]. From this Stevens makes a preliminary calculation for the size of a least-unit as 1/3 the diameter of a nucleon [15].

What we are looking for in terms of a cosmological least-unit for

HAM cosmology is a periodic structure incorporating the structural-phenomenology of the unitary field from which all of cosmology and matter can be built up by conformal scale-invariance. In the context of an advanced form of Einstein's model of a static universe a new Continuous-State [65-67] cosmology based on principles of the Wheeler-Feynman absorber theory of radiation [16] is extended to the topology of a periodic 12D spacetime. The fundamental *least-unit* is shown to be a scale-invariant complex self-organized cosmological system. The translating boundary conditions of a spin exchange 'continuous-state' dimensional reduction compactification process are inherent in the Dirac polarized vacuum. The topology is derived by coupling superluminal Lorentz boosts with non-compactified Kaluza-Klein theory in the context of a 12D Complex energy dependent spacetime metric, $\hat{M}_4 \pm C_4$.

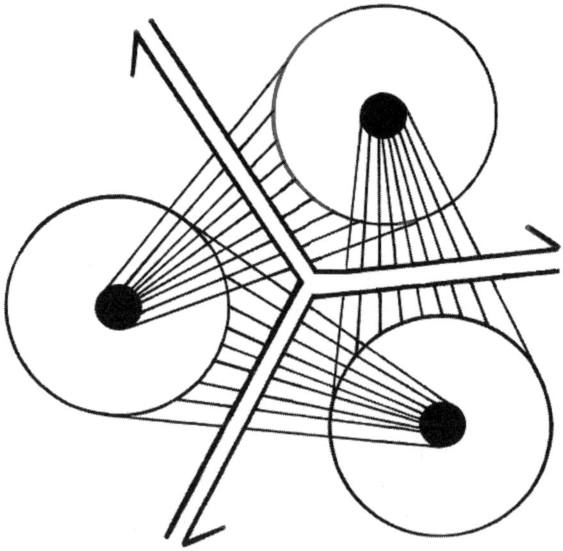

Figure 4.2 Conceptualization of an isolated HAM least cosmological unit which would not occur in nature showing the continuous-state static and dynamic Casimir boundary conditions around a central Witten Ising model string vertex. Compare Fig. 4.4.

Einstein originated the concept of an energy dependent spacetime for explaining temporal rate change in the presence of a gravitational field by generalizing the special relativistic line element

$$ds^2 = \left(1+2\phi/c^2\right)c^2dt^2 - dx^2 - dy^2 = dz^2 \tag{4.23}$$

with the introduction of time curvature where ϕ is the Newtonian gravitational potential. This utilizes the deformed Minkowski metric which is imbedded in the periodic HD Noetic space chosen axiomatically for HAM cosmology to take the form of a non-compactified Kaluza-Klein theory [68,69]. The feasibility of an energy domain pervading HD spacetime with properties similar to Wheeler's Geon [70] proposal is discussed below. In a generalized deformed spacetime metric, spacetime is fixed by the energy and has the metric

$$\eta(E) = diag(a(E), -b(E), -c(E), -d(E)). \tag{4.24}$$

There is no need to develop the toy model further at present as it sufficiently illustrates pertinent aspects of the noetic transformation that show how boundary conditions transform the dimensionality of space and time along with the energy covering of the unified field by

$$D_S \Leftrightarrow D_t \Leftrightarrow D_E \quad \text{or} \quad S \to t \to E \tag{4.25}$$

The symmetry breaking during dimensional reduction of the continuous compactification of noetic superspace is a harmonic oscillation between the future and past as space transforms to time transforms to energy.

- The ordered spin exchange structures alternate in a hysteresis cycle, the area of which represents the energy of the Noetic Field 'injected' into each spacetime point and piloting quantum dynamics. Various stressors may alter the geodesic pathway of this energy and interfere with the arrow of time. These changes are best described by catastrophe theory and provide a new basis for vacuum engineering.

4.6 The Formalism for Noetic HAM Cosmology

$$S_n = S_0 = S_1 + S_2 \tag{4.26}$$

in the context of an extended Wheeler/Feynman absorber theory [16] where standard Minkowski space M_4 is a 'standing wave' of the future-past. Taking the general HAM form

$$R^{S_{N0}}_{sym\hat{M}_4} = \frac{1}{2}\left[R^{S_{N1}}_{retC_4} + R^{S_{N2}}_{advC_4}\right] \tag{4.27}$$

or simplistically 12D noetic superspace S_N represents a complex Minkowski metric combining the standard 4 *real* dimensions (D) plus eight imaginary D representing *retarded* and *advanced* complex hyperspace topology which adapts the complex (+) Minkowski metric from the standard stationary form to a periodic form. *R* represents the noetic 3(4)D 'standing wave' Minkowski 'present' spacetime; *adv* represents the past component and *ret* represents the future for complex correspondence to the standard 4 real dimensions utilizing 8 imaginary dimensions, $\pm C^4$. The 8 complex imaginary dimensions, while not manifest (locally) on the Euclidean real line, are 'physical' [71] in HAM and can be represented by complex coordinates.

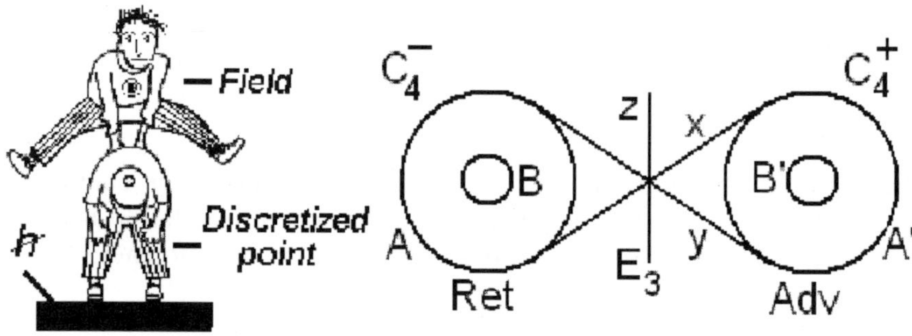

Figure 4.3 Conceptual illustration of the leap-frog see-saw dynamics of the field-particle duality of future-past advanced-retarded complex symmetry breaking parameters of HD spacetime topology. These are part of the critical elements of continuous-state dimensional reduction compactification dynamics.

In the Continuous-State there is a complementarity between field and discretization similar to the Dirac spherical rotation for the electron, only here it applies to the topology of spacetime itself and the standing wave Euclidian grid of perceived reality.

4.7 Transformation of Space into Time and String Tension

If nature is stringy, *h* is not a fundamental constant [5,6]. Natural units for the string won't have $\hbar = 1$, but $T_s = 1/\pi$. String tension, *h* and *c* can be combined to form a length, *L*. This means that $\hbar$ in string theory must be multiplied by T_0. New string theory suggests *L* can be the size of the Larmour radius of the Hydrogen atom. Randall [69] has suggested XD may extend up to infinite size!

The unified field governing gravitation and the super-quantum potential guides the action of translation along certain allowed pathways. For example if l, w or h is removed from a cube the object collapses to a plane. Removing a dimension from a plane causes compactification to a line and so on as applied to any hyperstructure. The released space is not initially empty. At the 1st stage of D reduction space transforms to time; at the 2nd stage into the energy that couples with the energy governing it as compactification is completed for that particular least unit.

We now introduce our preliminary formalism for a generalized complex 8D metric in special relativistic terms following Hansen and Newman [72]. The general relativistic formalism with gravitational fields present requires Riemannian (curved) geometry. We will utilize the invariant line element expressed in Einstein's special relativity theory. Hansen and Newman [71,72] have shown that the complex 8-space metric yields the proper solutions to Einstein's field equations only in the asymmetrically flat condition of Euclidean geometries for the case of weak gravitational fields. Thus, this formalism approximates, in general terms conditions described by special relativity. Einstein used a 3D geometric figure termed the light-cone to represent the usual four space Minkowski metric, M_4 in a 2D plane, based on the conic sections diagrams developed by the ancient Greeks. This geometric picture is formed by a figure with two axes, the ordinate is time, t and the abscissa is formed from the three dimensions of space as one axis $X = x, y, z$. The speed of light, c forms the sides of the two cones apex to apex (which represents 'now') with the t axis in the vertical direction. The purpose of this picture is to define the relationship between events in 4-space. For events connected by signals of $v < c$, events occur within the top of the light cone (forward time) or bottom (past). These are termed time-like signals. Event connections outside the light cone surface $c = c$, are connected by $v > c$ are called space-like signals and are not often addressed in standard physics.

In defining causality conditions for the usual 4-space, distance ds^2 is invariant and given as $ds^2 = g_{ab}dx^a dx^b$ where the indices a, b run 1 to 4. We use the metric signature (+,+,+,-) for the three spatial and one temporal component in metric g_{ab}. This metric is expressed as a sixteen element 4 x 4 matrix which represents a measure of the form and shape of space. This is the metric defined on (within) the light cone, connecting time-like events. This insures Einstein's postulate that $v \leq c$ for any given velocity of event connection. Rauscher [73] and Newman [72] construct a second intersecting light cone identifiable with the four imaginary dimensions. We express the complex 8-space metric as $M_4 \pm C^4$ because it represents the complexification of four spacetime dimensions. The complex space is expressed in terms of the complex 8-space variable Z^{μ}, where $Z^{\mu} = X^{\mu}_{\text{Re}} + iX^{\mu}_{IM}$, and $Z^{*\nu}$ is the complex conjugate of Z^{μ} so that $Z^{\nu} = X^{\nu}_{\text{Re}} - iX^{\nu}_{IM}$. We now form the complex eight space differential line element $dS^2 = \eta_{\mu\nu} dZ^{\mu} Z^{*\nu}$ where the indices run 1 to 8. The generalized complex metric in the previous equation is analogous to the usual Einstein 4-space metric in the above paragraph. In our formalism, we proceed by extending the usual 4D Minkowski space into a four complex dimensional spacetime. This new manifold (or space-time structure) is analytically expressed in the complexified 8-space.

Here X_{Re} is represented by $x_{\text{Re}}, y_{\text{Re}}, z_{\text{Re}}$ and t_{Re} i.e. the dimensions of our usual 4-space. Likewise, X_{Im} represent the four additional imaginary dimensions of $x_{\text{Im}}, y_{\text{Im}}, z_{\text{Im}}$, and t_{Im}. Hence, we represent the dimensions of our complex space as Z^{μ} or $x_{\text{Re}}, y_{\text{Re}}, z_{\text{Re}}, t_{\text{Re}}$, $x_{\text{Im}}, y_{\text{Im}}, z_{\text{Im}}$, and t_{Im}. These are all <u>real</u> quantities. It is the *i* before the x_{Im}, etc. that complexifies the space. Now we write the expression showing the separation of the real and imaginary parts of the differential form of the metric: $dZ^{\mu} dZ^{*\mu} = \left(dX^{\mu}_{\text{Re}}\right)^2 + \left(dX^{\mu}_{\text{Im}}\right)^2$. We can write in general for real and imaginary space and time components in the special relativistic formalism.

$$ds^2 = \left(dx^2_{\text{Re}} + dx^2_{\text{Im}}\right) + \left(dy^2_{\text{Re}} + dy^2_{\text{Im}}\right) + \left(dz^2_{\text{Re}} + dz^2_{\text{Im}}\right) - c^2\left(dt^2_{\text{Re}} + dt^2_{\text{Im}}\right) \tag{4.28}$$

As is well known Superluminal Lorentz Transformations (SLT) change real quantities into imaginary. Following Rauscher [71,73,74] we illustrate the transformation of complex spatial dimensions into temporal dimensions by orthogonal superluminal boosts (SLB). For example a 4D SLB in the x direction with velocity $v_x \pm \infty$ the SLT is

$x' = \pm t,\ y' = -iy,\ z' = -iz,\ t' = x$. In complex Minkowski space the coordinates are $z^u = x_{\mathrm{Re}}^u + ix_{\mathrm{Im}}^u$ where z is complex and x_{Re} and x_{Im} are real and the index u runs over 0,1,2,3. Using classical notation for simplicity

$$t = t_{\mathrm{Re}} + it_{\mathrm{Im}},\ x = x_{\mathrm{Re}} + ix_{\mathrm{Im}},\ y = y_{\mathrm{Re}} + iy_{\mathrm{Im}},\ z = z_{\mathrm{Re}} + iz_{\mathrm{Im}}. \quad (4.29)$$

To clarify the meaning of imaginary quantities in an SLT it is helpful to represent time as a 3D vector $t_x,\ t_y,\ t_z$; therefore time is defined as $t = t_x\hat{x} + t_y\hat{y} + t_z\hat{z}$ where

$$t_x = t_{x\mathrm{Re}} + it_{x\mathrm{Im}},\ t_y = t_{y\mathrm{Re}} + it_{y\mathrm{Im}},\ t_z = t_{z\mathrm{Re}} + it_{z\mathrm{Im}} \quad (4.30)$$

Finally for the SLB for velocity $v_x \pm \infty$ along x the transformations are

$$\begin{aligned} &x'_{\mathrm{Re}} + ix'_{\mathrm{Im}} = t_{x\mathrm{Re}} + it_{x\mathrm{Im}},\ y'_{\mathrm{Re}} + iy'_{\mathrm{Im}} = y_{\mathrm{Im}} - iy_{\mathrm{Re}}, \\ &z'_{\mathrm{Re}} + iz'_{\mathrm{Im}} = z_{\mathrm{Im}} - iz_{\mathrm{Re}},\ t'_{x\mathrm{Re}} + it'_{x\mathrm{Im}} = x_{\mathrm{Re}} + ix_{\mathrm{Im}}, \\ &t'_{y\mathrm{Re}} + it'_{y\mathrm{Im}} = t_{y\mathrm{Im}} - it_{y\mathrm{Re}},\ t'_{z\mathrm{Re}} + it'_{z\mathrm{Im}} = t_{z\mathrm{Im}} - it_{z\mathrm{Re}} \end{aligned} \quad (4.31)$$

Where an SLT in x of M_4 spacetime transforms real components into imaginary and imaginary complex quantities into real quantities as a major property of the periodic nature of Noetic HAM spacetime [65-67].

This is the first part of the HAM spacetime transformation. Not illustrated is the second set of SLB where these spatial dimensions boosted into temporal dimensions are boosted again into dimensions of energy $D_S \rightleftarrows D_t \rightleftarrows D_E$. One might consider that this arises from the historical consideration of Kaluza-Klein where energy is the 5th dimension. But in HAM cosmology this energy is the quantum potential in 4-space, the super-quantum potential in hyperspace and the unitary noetic field in 12D.

4.8 Alternative Derivation of String Tension in HAM Cosmology

Recently an alternative derivation of T_S has been discovered in the context of HAM cosmology [75]. It is interesting to note that both the Schrödinger equation and Einstein's equations for geometrodynamics (and Newton's) reduce to Newton's second Law of Motion. Newton's dimensionless second law of motion $F = ma$ is the starting point for deriving the noetic formalism. We pull this rabbit out of the hat to form the continuous-state cosmology from the $\dot{R} \equiv c$ hypothesis above. First by substituting Einstein's mass-energy relation $E = mc^2$ into Newton's second law we obtain: $F_N = E / c^2 a$ where F_N is to be the noetic force and E a form of self-organized energy. E is scale-invariant and covariant through all levels of scale in HAM cosmology beginning at the highest level in the supralocal Multiverse as a hyperdimensional Wheeler Geon, a *ball* of photons of sufficient size to gravitationally self cohere [70]. At the micro level the Geon becomes synonymous with the de Broglie-Bohm quantum potential and relates to balancing by the cosmological constant, Λ. This 'energy' rather than dark energy is responsible for 'phonograph record' form of galactic rotation. Cosmologically this is like an 'ocean of light', a super quantum potential synonymous with the unitary field. Next the derivation of the noetic equation is generalized for the holographic multiverse by taking an axiomatic approach, based in part on Eddington's large number hypothesis above, to cosmological scaling that suggests all lengths in the universe are scale-invariant [17,37,38,39].

Beginning with the heuristic relation $c \equiv \dot{R}$ or $\dot{R} = L / t = c$ where $\dot{R}$ represents the rate of change of scale in the universe. This corresponds to the putative Hubble relation for Doppler expansion of the universe where $H_0 = \dot{R} / R$ and $a = \dot{R} \times H_o$. By substituting $\dot{R}^2 / R$ for a in the original equation $F_N = E / c^2 a$, then for final substitution we have $F_N = E / c^2 \times \dot{R}^2 / R$. Since $c = \dot{R}$ the c^2 and $\dot{R}$ terms cancel leaving

$$F_N = E / R \tag{4.32}$$

which takes the same form as Eq. (4.1) for T_S. Don't be fooled by the apparent simplicity of this equation; its expansion to dynamic-static Casimir boundary condition of brane topology, photon propagation, the graviton, and black body cavity QED for redshift/CMBR equilibrium etc. become sufficiently complex to tax the imagination. Its beauty is in its simplicity especially as it describes the action of the unitary field and the arrow of time (Chap. 5) in describing the nature of our 3D virtual reality.

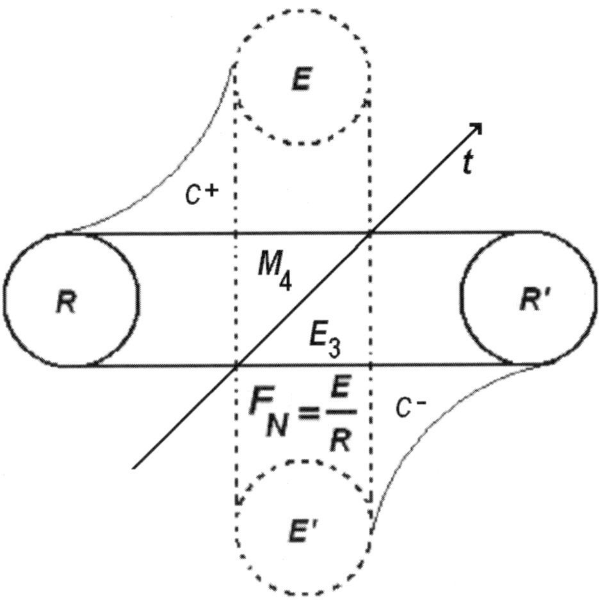

Figure 4.4 Geometric representation of the unitary Noetic Field equation, $F_N = E/R$ for an isolated least-unit not occurring in nature. Solid lines represent extension, dotted lines represent field. Where $F_{(N)}$ is the anthropic force of the unitary field driving self-organization of the structural-phenomenology, E equals the hysteresis loop energy of the hypervolume, R is the scale-invariant radius of the action and the semicircles tension, T_0.

Note that R is a complex relativistic rotational length with standing wave properties. It is scale-invariant and becomes associated with the radii of various HD hyperspheres in the continuous-state compactification process. Any temporal slice or cross section would be considered a Cavity-QED hysteresis loop suggesting pertinent localized volumes from which energy ranges and limits can be calculated. It should be emphatically noted again and again that Hubble discovered a cosmological redshift not a Doppler expansion of the universe. HAM cosmology provides an alternative interpretation for redshift suggesting the possibility of profound new applications. HAM cosmology contains the same energy of motion perceived as expansion or inflation but operationally its action is internalized as an inherent component of the relativistic properties of the continuous-state dimensional reduction compactification SUSY symmetry breaking process.

4.9 Parameters of the Spacetime Incursive Oscillator (IO)

This model is empirically testable and amenable to computer simulation through application of the incursive harmonic oscillator associated with complex systems, which we profess HAM cosmology is, i.e. the Hubble radius as acted upon by the anthropic action principle. That said, motion of a one dimensional classical harmonic oscillator as well known is given by $q = A\sin(\omega t + \varphi)$ and $p = m\omega A\cos(\omega t + \varphi)$ where A is the amplitude and φ is the phase constant for fixed energy $E = m\omega^2 A^2 / 2$. For state $|n\rangle$, with $n = 0,1,2,...,\infty$ and with Hamiltonian $E_n = (n + 1/2)\hbar\omega$, $\langle n|q^2|n\rangle = \hbar / 2m\omega \langle n|(a^\dagger a + aa^\dagger)|n\rangle = E_n / m\omega^2$ becomes the quantum harmonic oscillator becomes and $\langle n|p^2|n\rangle =$ $1/2(m\hbar\omega)\langle n|a^\dagger a + aa^\dagger = mE_n$ where a and $a^\dagger$ are the annihilation and creation operators, $q = \sqrt{\hbar / 2m\omega}(a^\dagger + a)$ and $p = i\sqrt{m\hbar\omega / 2}(a^\dagger a)$. For the 3D harmonic oscillator each equation is the same with energies $E_x = (n_x + 1/2)\hbar\omega_x$, $E_y = (n_y + 1/2)\hbar\omega_y$ and $E_z =$ $(n_z + 1/2)\hbar\omega_z$ [76,77].

In Dubois' notation (developer of the IO) the classical 1D harmonic oscillator according to Newton's second law in coordinates *t* and *x(t)* for a mass *m* in a potential $U(x) = 1/2(kx^2)$ takes the differential form

$$\frac{d^2x}{dt^2} + \omega^2 x = 0 \quad \text{where} \quad \omega = \sqrt{k / m} \tag{4.33}$$

which can be separated into the coupled equations [6-9]

$$\frac{dx(t)}{dt} - v(t) = 0 \qquad and \qquad \frac{dv(t)}{dt} + \omega^2 x = 0. \tag{4.34}$$

From incursive discretization, Dubois creates two solutions $x(t + \Delta t)$ $v(t + \Delta t)$ providing a structural bifurcation of the system which together produce Hyperincursion. The effect of increasing the time interval discretizes the trajectory as in Fig. 4.6. This represents a background independent discretization of spacetime [25-28].

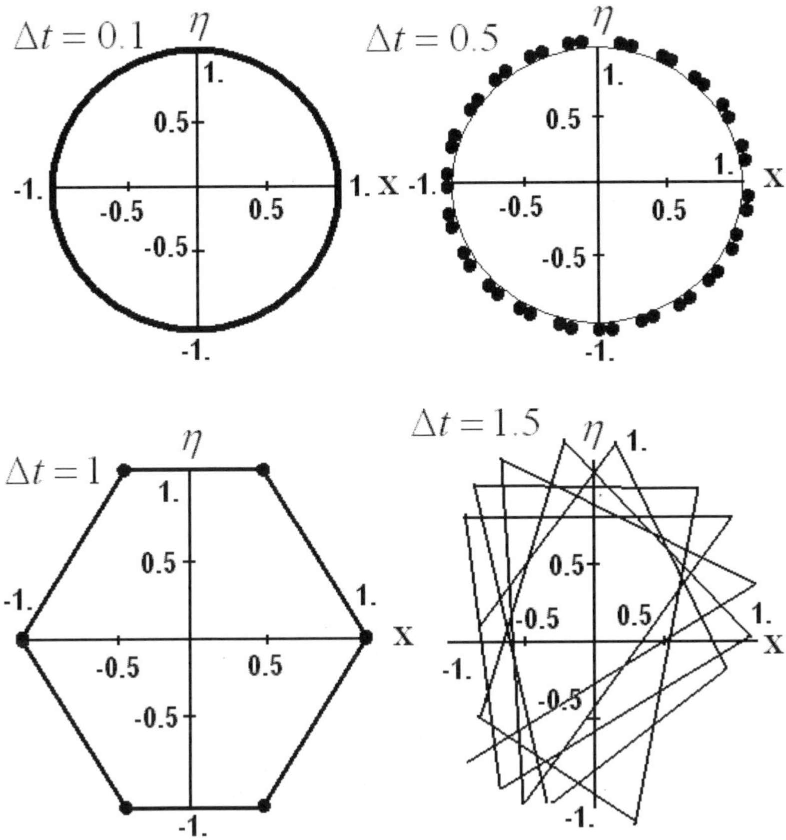

Figure 4.5 Numerical simulation of the phase space trajectory of the Dubois *superposed incursive oscillator* based on coordinates and velocities $x_n = 1/2[x_n(1) + x_n(2)]$ $v_n = 1/2[v_n(1) + v_n(2)]$ is shown in the figure for values of $\Delta\tau = \omega t$ equal to 0.1, 0.5, 1.0 and 1.5. Initial conditions are $\chi_0 = 1, \eta_0 = 0$ and $\tau_0 = 0$ with total simulation time $\tau = \omega t = 8\pi$. Figure adapted from [25-28].

Each mode of the field of a quantum harmonic oscillator is associated with the cavity-QED dynamics, hexagon lattices in Figs. 4.1 and 4.2, of spacetime topology as it undergoes its continuous transitions. E is the state of energy for n photons. For $n = 0$ the oscillator is in the ground state, but a finite energy $1/2\hbar\omega$ of the ground state, called the zero-point energy, is still present in the region of the cavity. According to equation (4.31) of the quantum harmonic oscillator the field energy of the photons undergo periodic annihilation and recreation in the periodic spacetime [78]

$$E_n = (n + \frac{1}{2})\hbar\omega . \tag{4.35}$$

The simulation is meant to demonstrate generally how the inherent periodic holophote action, flashing metaphorically like a light house beacon, injects HD geon energy into each virtual moment of the present during the continuous transformation of the Cavity-QED topology of the 12D superspace of the noetic least-unit [79] to produce the natural emergence of F-Theory 2-branes [7,80]. As an example, we illustrate one of a number of possible models of how, at the semi-classical limit from the stochastic background of the vacuum zero-point field, this energy is harmonically injected into every point and atom in spacetime by a mechanism like a 'chaotic gun' [79,81,82]. This action and the geometric-topology of the polarized vacuum is putatively suggested to generate F-branes.

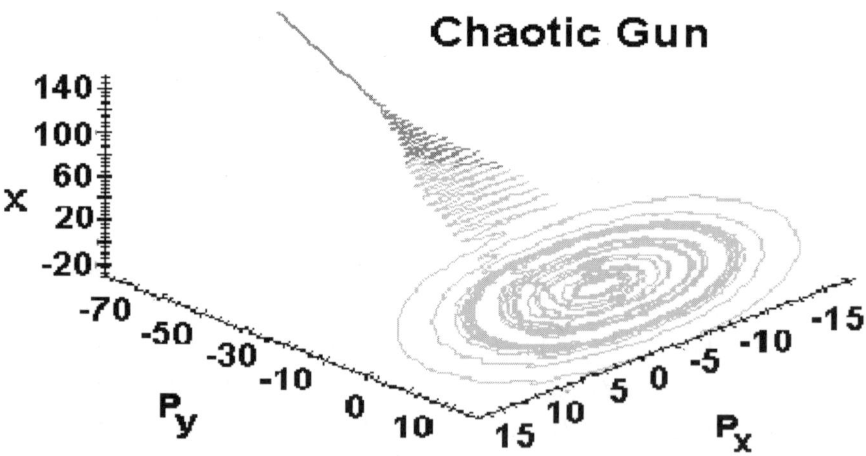

Figure 4.6 Ciubotariu's Chaotic Gun is another way of modelling energy injection into spacetime points at the quantum level. Figure adapted from N. Ciubotariu [81,82].

Possible quantum model for entry of the new quantum of action. A 3D rendering of the phase space where Bosons of the Noetic Field (noeons) are injected into each spacetime point (least unit) and every atom by a periodic 'gun effect' of the continuous holophote action of the continuous state dimensional reduction inherent in the topology of Noetic space.

Ciubotariu's equations combine Maxwell's equations and relativistic equations of motion for the phase space where the additional Ω terms represent the cyclotron frequency of the *chaotic gun* effect. The noeon *Bosons* mediating the life force field are *emitted* from the spacetime cavities only in certain preferred directions allowed by the parallel transport conditions of dimensional reduction and compactification. This effect occurs in Noetic HAM Cosmology because in the energy dependent spacetime metric, just as the periodicity of wave and particle moments occur in the propagation of a photon, so does charge or energy arise in periodic moments in the hysteresis looping of the Noetic least unit. Because Wheeler showed in 1962 that 'charge is topology' [83].

Using equations for a spacetime chaotic gun developed by the Ciubotarius [81,82] the nonlinear dynamics of the model for injecting a charged noeon, defined as the quanta of the noetic unified field, into a spacetime cavity can putatively occur as follows:

$$\dot{X} = \frac{\mathrm{d}X}{\mathrm{d}T} = \frac{1}{\gamma} P_x = \frac{1}{(1 + P_x^2 + P_y^2)^{1/2}} P_x, \tag{4.36}$$

$$\dot{P}_x = \frac{\mathrm{d}P_x}{\mathrm{d}T} = \Omega_c[\beta \cos(X - T) + 1]P_y, \tag{4.37}$$

$$\dot{P}_y = \frac{\mathrm{d}P_y}{\mathrm{d}T} = -\Omega_c[\beta \cos(X - T) + 1]P_x + H\cos(X - T). \tag{4.38}$$

Equations (4.34) - (4.36) illustrate a possible quantum model for entry of the new noetic action principle into the 3D phase space P_X, P_Y, X where unitary bosons of the Noetic field (noeons) are injected into each point or least-unit QED cavity in spacetime and every atom by a periodic 'gun-like effect' of the continuous holophote action. This process occurs in the context of continuous state spin-exchange dimensional reduction compactification inherent in the topology of Noetic Superspace which acts like a hysteresis loop [21,65-67]. Ciubotariu's equations combine Maxwell's equations and relativistic equations of motion for the phase space P_X, P_Y, X. The Ω terms represent the cyclotron resonant frequency of the chaotic gun effect. Infusion of the noeon Boson field, which mediates the action of self-organization and evolution, into spacetime cavities only occurs in certain preferred directions allowed by the symmetry conditions of what is called parallel transport [84] in the dimensional reduction compactification spin-exchange process [21,65-67].

The holophote effect appears in the Noetic cosmology because in its energy dependent spacetime metric $\hat{M}_4$, just as a periodicity of wave and particle moments occur in photon propagation through space, so does charge or energy arise in periodic moments of the Noetic least-unit transformation. Because as Wheeler demonstrated [83] 'charge is topology'. According to Wheeler lines of force in a wormhole can thread through a handle and emerge through each mouth to give the appearance of charge in an otherwise charge-free spacetime [83].

4.10 Emergence of 2-Branes from Inherent Spacetime Oscillations

In this section we create a toy model for emergence of generalized F-Theory 2-branes from spacetime parameters of discrete supersymmetric incursive oscillations. If our Hubble sphere is a self-organized complex system principles associated with complex systems such as hierarchy, conformal scale invariance, recursion/incursion and anticipation should be inherent parameters in the cosmological model, and discovered as principles of nature and thus be revealed in the laws of physics. The most important principle associated with complex systems is that they are driven by an external evolutionary force. We postulate that this action principle is the Anthropic Principle.

The evolutionary search for the fundamental background independent string vacuum has been cast recently in a Twelve Dimensional (12D) form of M-Theory called F-Theory [7]. Generally String Theory has remained aligned with naturalistic Big Bang Cosmology not perceived as compatible with a covariant Dirac polarized vacuum essential for extended electromagnetic theory and finite photon mass, m_γ. Photon mass has been discounted by physicists because it is believed that this would violate gauge theory which has been highly successful. Firstly gauge theory is only an approximation suggesting more theory is anticipated and secondly gauge theory describes a finite regime in the same way Newtonian mechanics did before the discovery of quantum theory, so this myopic criticism does not even apply. Most critically, if gauge theory is not fundamental, Planck's constant is not fundamental either and needs to be recalculated. This is where string tension comes in as an additional parameter added to the Planck constant. A recently formulated highly symmetric continuous-state cosmology called the holographic anthropic multiverse (HAM) utilizes a 12D energy dependent standing wave superspace based on extensions of the Wheeler-Feynman-Cramer transactional model providing a context where scale-invariant least cosmological units of the Superspace act as a complex self-organized system. These fundamental least-unit entail a form of incursive oscillator inherent in the continuous-state topology of HAM spacetime. Simulated application of the Incursive Oscillator (IO) is shown to produce a natural emergence of generalized F-Theory 2-branes from the superspace backcloth potentially bringing the IO program into closer alignment with mainstream physical cosmology which could be instrumental in solving the problem of deriving parameters of the fundamental string vacuum, especially emergence of a new action principle driving the evolution of its self-organization.

4.11 Summary of Noetic Spacetime Parameters

The periodic symmetry of HAM cosmology contains an inherent beat frequency during the continuous state dimensional reduction spin-exchange compactification topological transformation which introduces energy by the holophote action of the Noetic Force F_N energy through every spacetime point into every atom during the process of dimensional transformation as $D_s \rightarrow D_t \rightarrow D_E$ [65-67] and as $R_U \rightarrow R_Q \rightarrow R_C$ where spatial dimensions, D_s continuously transform into temporal dimensions, D_t and into energy, D_E in a cyclical process of unitarity, R_U to quantum, R_Q to classical, R_C; a relativistic process representing an additional set of Noetic transformations: Galilean$\rightarrow$ Lorentz-Poincairé$\rightarrow$ Noetic. A deficit angle occurring in the parallel transport [84] around the noetic least-unit leads to a new model for the arrow of time, offering an explanation for why the XD are not considered sub-Planckian in HAM cosmology but still unobserved.

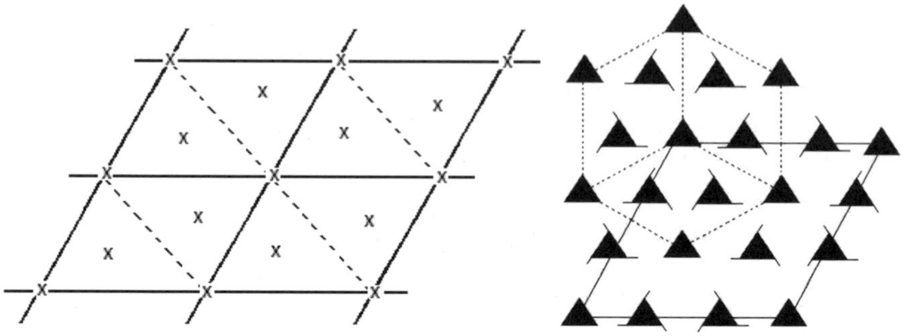

Figure 4.7 a) A triangular spacetime lattice in the complex plane for production of a torus T_0 with Z_3 symmetry utilized in the study of compactification in string theory. b) Elaboration of how a hexagon lattice in a) arises from the spin structure of the spacetime fabric. Solid triangles become fixed coordinates, while 'propellers' or screws have left/right handed spin axes representing field parameters for 'bumps and holes' in the Dirac sea. These spacetime structures in conjunction with Fig. 3.2 putatively support the basis for F-brane emergence from the future-past standing-wave hysteresis of the spacetime least-unit continuous-state structure. Figures redrawn from [60-62,85].

From generalized examples of spacetime topology possible conditions for string propagation are illustrated for the noetic stringy vacuum, considered a form of the covariant Dirac polarized vacuum [86] so that $S_N = S_0 + S_1 + S_2 \to S_{12} \to \hat{M}_4 \times K_8 \to \hat{M}_4 \times \pm C_4$ [21,65-67]. The 12D Noetic Superspace S_N is triune, comprised of the standing-wave Minkowski present $\hat{M}_4$ and two complexified future-past elements $\pm C_4$, where for the intermediate subset $\hat{M}_4 \times K_6$ the $\hat{M}_4$ is a 4D energy dependent Minkowski space and K_6 a compactified 6D torus. A realistic example is given below. First points $z \approx z+1 \approx z+e^{2i\pi/3}$ admitting Z_3 symmetry are identified in the complex z plane and three tori T_i, $i = 1,2,3$ are obtained whose product is a torus of six real dimension, three of which are complex [86], on which string propagation is considered. From the well-known symmetry groups rotations can generate discrete symmetry elements accompanied by various translation components τ parallel to a spin axis A such that n translations τ equal an integral number p of lattice translations t along the axis

$$n\tau = pt \tag{4.39}$$

where n and p are integers. When $p/n < 1/2$, the screw is right-handed, when $p/n \leq 1/2$, left-handed and when $p/n = 1/2$ it is zero [60-62]. A translation t' normal to axis A of a screw produces a translation equivalent to A' as well as nonequivalent but equal screw operations about parallel axes B and C along the perpendicular bisector of AA' at a distance $t' \cot \alpha / 2$ from AA'. These screw operations accumulate along axes B and C making screw axes parallel to A [60-62]. The resulting sets of symmetry elements are repeated by the lattice translations to constitute infinite sets of parallel axes as extrapolated from Fig. 4.7b into Fig. 4.8.

4.12 Simplistic Computer Simulated Production of the 2-Brane

From the proof of Schöenflies theorem [87] there can be no topological knots in a plane. Therefore there can be no topological torsion, and thus no Einstein type geometrodynamics, in a 2D reality. Information 2d According to tenets of M-Theory 'matter remains on the 2-brane and gravity is free to pass between branes'. A simulated creation of an F-Theory 2-brane from the Dirac polarized vacuum [85] is demonstrated utilizing the Autodesk Chaos Software [88].

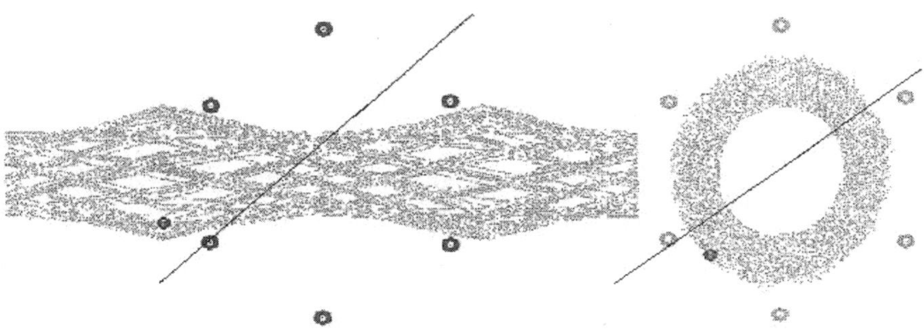

Figure 4.8 Two views of one form of computer simulated production of a 2-brane from parameters of the hexagonal geometry (Fig. 4.7) of the putative covariant Dirac polarized vacuum. Hysteresis loop harmonic oscillation of the future-past dynamics produces branes by incursive resonance.

The software simulation of 2-brane emergence from the geometry of spacetime least-units is achieved by applying a harmonic oscillator generated by the energy of the Noetic Action Principle. The oblique lines in each figure are insertion angles and the two tiny points are holophote injection points of the unitary noeon energy.

Table 4.I Spacetime Harmonic Oscillator Parameters
As Utilized in running the Autodesk Chaos Software

PARAMETER	VALUE USED	POSSIBLE RANGE
Charge	3	±500
Magnetic Capture Radius	5	0 to 20
Magnetic Field Radius	11	1 to 60
Pull Towards Center	27	±500
Frequency	33	2 to 10,000
Friction (String Tension)	1.37	0 to 500

4.13 Conclusions

The approach presented is a work in progress, but its initial success suggests that more comprehensive calculations and simulations can add further rigor to the results and with more sophistication be used to study more complex brane dynamics and the structure of matter. Further if the theory indeed reveals a sound physical basis, a demonstration of the production of F-Theory 2-branes from more specific vacuum parameters of complexified HD space could shed light on determining the actual physical vacuum sought for M-Theory. Simulations with sufficient complexity could be developed to aid in determining the actual spin structures and geometric topology of actual matter which is one of the ultimate goals of string theory.

We took a fairly simple and straightforward approach in this alternative derivation of the string tension formalism and concentrated only on the seminal idea and the $D_S \rightleftarrows D_t \rightleftarrows D_E$ relativistic dimensional boosting which has been ignored as a SUSY breaking parameter in M-Theory. We think it could be insightful to finding the actual physical cosmology of our local reality. The idea of a continuous-state cosmology as opposed to Big Bang evolution is probably hard to swallow for many astrophysicists / cosmologists at present so we kept it somewhat superficial so we could sit back and see how long it takes some insightful stringy postdocs to run with it…

References

[1] Newton, I. (1687) *Philosophiæ Naturalis Principia Mathematica*, (1999) The Principia: Mathematical Principles of Natural Philosophy, Berkeley: U. of California Press.
[2] Rizzoa, T.G. & Wells, J.D. (1999) Electroweak precision measurements and collider probes of the standard model with large extra dimensions, arXiv:hep-ph/9906234v1, hep-ph/9906234.
[3] Arkani-Hamed, N., Dimopoulos, S., Kaloper, N. & Sundrum, R. (2000) A small cosmological constant from a large extra dimension, Phys Lett. B, 480: 1-2; 193-199.
[4] Mukohyama, S. & Randall, L. (2004) A dynamical approach to the cosmological constant, http://arXiv.org/abs/hep-th/0306108.
[5] Grotz, K., & Klapdor, H.V. (1990) The Weak Interaction in Nuclear, Particle and Astrophysics, New York: Adam Hilger.
[6] Hatfield, B. (1992) Quantum Field Theory of Point Particles and Strings, Reading: Addison-Wesley.
[7] Kaku, M. (1999) Introduction to Superstrings and M-Theory, New York: Springer-Verlag.
[8] Bousso, R. (2002) The holographic principle, Rev Mod Physics 74: 825-874, arXiv:hep-th/0203101.
[9] Susskind, L. (1995) The world as a hologram, J. Math. Phys. 36, 6377, hep-th/9409089.
[10] Susskind, L. & Witten, E. (1998) The holographic bound in Anti-de Sitter space, hep-th/9805114.
[11] 't Hooft, G. (2000) The holographic principle, hep-th/0003004.
[12] Cooperstock, F.I. & Tieu, S. (2005) General relativity resolves galactic rotation without exotic dark matter, arXiv:astro-ph/0507619.
[13] Korzyński, M. (2005) Singular disk of matter in the Cooperstock-Tieu galaxy model, arXiv:astro-ph/0508377v2 29.
[14] Franck, G. (2000) Time and presence, in Science and The Primacy of Consciousness, R.L. Amoroso et al, (eds.) Orinda: Noetic Press.
[15] Stevens, H.H. (1989) Size of a least-unit, in M. Kafatos (ed.) Bell's Theorem, Quantum Theory and Conceptions of the Universe, Dordrecht: Kluwer Academic.

[16] Wheeler, J.A. & Feynman, R.P. (1945) Rev. Mod. Phys., 17, 157.
[17] Cramer, J. (1986) The Transactional Interpretation of Quantum Mechanics, Rev. Mod. Phys. 58, 647-687.
[18] Cardone, F., Francaviglia, M. & Mignani, R. (1999) Energy-dependent phenomenological metrics and 5Dl Einstein equations, Found. Phys. Lett., 12:3;281-289, arXiv:hep-th/0505149v1 (2005).
[19] Sides, S.W., Rikvold, P.A. & Novotny, M.A. (1998) Hysteresis loop areas in kinetic Ising models: Effects of the switching mechanism, arXiv:cond-mat/9710244.
[20] Amoroso, R.L. & Pribram, K.H. (2009) The Complementarity of Mind and Body: Realizing the Dream of Descartes, Einstein & Eccles, Cambridge: MIT Univ Press.
[21] Amoroso, R.L. (2004) The Fundamental Limit and Origin of Complexity in Biological Systems: A New Model for the Origin of Life, in D.M. Dubois (ed.) Computing Anticipatory Systems, AIP Conf. Proceedings Vol. 718, pp. 144-159, Melville: American Inst. of Physics.
[22] Varela, F.G., Maturana, H.R. & Uribe, R. (1974) Autopoiesis: The organization of living systems, its characterization and a model, BioSystems, 5, 187-196.
[23] de Broglie, L. (1926) *Compt. Rend.* 183, 447.
[24] Bohm, D. (1963) Quantum Theory, p. 353, Englewood Cliffs: Prentice-Hall.
[25] Dubois, D.M. (2001) Theory of incursive synchronization and application to the anticipation of delayed linear and nonlinear systems, in D.M. Dubois (ed.), Computing Anticipatory Systems: CASYS 2001-5th Intl Conference, Am Inst of Physics: AIP Conf. Proc. 627, pp. 182-195.
[26] Antippa, A.F. & Dubois, D.M. (2008) The synchronous hyper-incursive discrete harmonic oscillator, Proc. CASYS07, preprint.
[27] Dubois, D.M. (2008) The quantum potential and pulsating wave packet in the harmonic oscillator, proceedings of CASYS07, preprint.
[28] Antippa, A.F. & Dubois, D.M. (2004) Anticipation, orbital stability and energy conservation in discrete harmonic oscillators, in D.M. Dubois (ed.) Computing Anticipatory Systems, AIP Conf. Proceedings Vol. 718, pp.3-44, Melville: American Inst. of Physics.
[29] Stoney, G.J. (1881) On the physical units of nature, Phil. Mag. 11, 381-391.
[30] Einstein, A. (1909) *Phys. Zeitsch.* 6, 185.
[31] Schrödinger, I. (1922) *Über eine bemerkenswerke eigenshaft der quantenbahnen eines einzelnen elektrons*, Z. Phys. 12, 13-23.

[32] Eddington, A.E. (1935) New Pathways in Science, Cambridge, Cambridge Univ. Press.
[33] Randall, L. (2005) Warped Passages, New York: Harper-Collins.
[34] Wolf, C. (2002) Anomalous statistics and the rescaling of Planck's constant, Apeiron, Vol. 9, No. 3.
[35] Silk J. (1989) The Big Bang, New York : W. H. Freeman.
[36] Weinberg, S. (1972) Gravitation and Cosmology: Principles and Applications of the General Theory of Relativity, Wiley, New York
[37] Dirac, P.A.M. (1937) Nature, 139, 323.
[38] Eddington, A.S. (1931) M.N.R.A.S., 91, 412.
[39] Kant, I. (1878) The Critique of Pure Reason, JMD Meiklejohn, (Trans.) London, G. Bell & Sons, Pdf - http://books.google.com/books.
[40] Guth, A. (1981) Phys. Rev. D., 23, 347.
[41] Schaefer, E.B. (1999) Severe Limits on Variations of the Speed of Light with Frequency, Phys. Rev. Lett. 82, 4964-4966.
[42] Smoot, G.F. (1996) in Examining the Big Bang and Diffuse Background Radiations, ed. M. Kafatos & Y. Kondo, 31, Kluwer Academic Publishers, Dordrecht.
[43] Geller, M.J., Huchra, J. (1989) Science, 246, 897.
[44] Novikov, I. (1996) in Examining the Big Bang and diffuse background radiations, in M. Kafatos & Y.Kondo, (eds.) Dordrecht: Kluwer Academic Publishers.
[45] Peebles, P.J.E. (1998) Is Cosmology Solved?, paper presented at The Nature of the Universe Debate: Cosmology Solved?, Smithsonian Institution.
[46] Glanz, J. (1998) Science, 282, 2156.
[47] Eddington, A.S. (1939) The Philosophy of Physical Science, Cambridge: Cambridge University Press.
[48] Kafatos, M. (1989) Horizons of knowledge in cosmology, in Bell's Theorem, Quantum Theory and Conceptions of the Universe, M. Kafatos (ed.) Dordrecht: Kluwer Academic Publishers.
[49] Kafatos, M. Roy, S. & Amoroso, R. (2000) Scaling in Cosmology & the Arrow of Time, in Buccheri, di Gesu & Saniga, (eds.) Studies on Time, Kluwer.
[50] Dyson, F.J. (1972) in Aspects of Quantum Theory, A. Salam & E.P. Wigner (eds) Cambridge: Cambridge University Press.
[51] Harrison, E.R. (1981) Cosmology: The Science of the Universe, Cambridge: Cambridge University Press.

[52] Barrow J.D. & Magueijo, J. (1998) Varying theories and solutions to the cosmological problems, Astro-ph/ 9811072.
[53] Albrecht A. & Magueijo J. (1998) A time-varying speed of light as a solution to cosmological puzzles (preprint).
[54] Amelio-Camella G., et al. (1998) Nature, 393, 763.
[55] Schaefer, E.B. (1998) Severe limits on variations of the speed of light with frequency, Yale University (preprint).
[56] Roy S. & Kafatos M. (1998) Bell-type Correlations and Large Scale Structure of the Universe (preprint).
[57] Barrow, J.D. & Tipler F.J. (1986) The Anthropic Cosmological Principle, Oxford: Clarendon Press.
[58] Israelit M. & Rosen N. (1989) A singularity-free cosmological model in general relativity ApJ, 342:25; 627-634.
[59] Bohr, N. (1961) Atomic Theory and the Description of Nature, Cambridge: Cambridge University Press.
[60] Buerger, M.J. (1977) Elementary Crystallography, Cambridge: MIT University Press.
[61] Buerger, M.J. (1971) Introduction to Crystal Geometry, New York: McGraw-Hill.
[62] Buerger, M.J. (1991) Crystal symmetry, in RG Lerner & GL Trigg (eds.) Encyclopedia of Physics, 2nd Ed. , 221-229, New York: VCH Publishers.
[63] Stevens, H.H. (1978) Spec. Sci. Tech. 1; 99-107.b Speculations in Science and Technology.
[64] Stevens, H.H. (1980) Spec. Sci. Tech.3; 587-593. Speculations in Science and Technology.
[65] Amoroso, R.L. (2002) Developing the cosmology of a continuous state universe, in R.L. Amoroso, G. Hunter, M. Kafatos & J-P Vigier (eds.) Gravitation & Cosmology: From the Hubble Radius to the Planck Scale, Dordrecht: Kluwer Academic.
[66] Amoroso, R.L. (2005) Paradigm for a continuous-state holographic conscious multiverse, R.L. Amoroso & B. Lehnert (eds.) Oakland: Noetic Press.
[67] Amoroso, R.L. (2010) Defining a context for the cosmology of awareness, in R.L. Amoroso & K.H. Pribram (eds.) The Complementarity of Mind and Body: Realizing the Dream of Descartes, Einstein and Eccles, Cambridge: MIT Univ. Press.
[68] Darabi, F, Sajko, W.N. & Wesson, P.S. (2000) Quantum cosmology of 5D non-compactified Kaluza-Klein theory, Classical Quantum

Grav. 17; 4357-4364.
[69] Randall, L. & Sundrum, R. (1999) An Alternative to Compactification, Physical Review Letters, 83, 4690-4693.
[70] Wheeler, J.A. (1955) Geons, Phys. Rev., 97:2, 511-536.
[71] Rauscher, E. A. (1983) Electromagnetic Phenomena in Complex Geometries and Nonlinear Phenomena, Non–Hertzian Waves and Magnetic Monopoles, Millbrae: Tesla Books; (2009) 2nd edition, Oakland: The Noetic Press, and references therein.
[72] Hansen, R.O. & Newman, E.T. (1995) A complex Minkowski space approach to twistors,. Gen. Rel. Grav. 6, 361-385.
[73] Rauscher, E.A. (2002) Non-Abelian Gauge Groups for Real and Complex Amended Maxwell's Equations, , in R.L. Amoroso, G. Hunter, M. Kafatos & J.-P. Vigier (eds.) Gravitation & Cosmology: From the Hubble Radius to the Planck Scale, Dordrecht: Kluwer Academic.
[74] Ramon, C. & Rauscher, E.A. (1980) Superluminal transformations in complex Minkowski space, Found. Physics, 10:7/8; 661-9.
[75] Amoroso, R.L. & Rauscher, E.A. (2008) Derivation of the string tension formalism, and super quantum potential as inherent parameters of a holographic conscious multiverse cosmology, in R.L. Amoroso, I. Dienes & Cs. Varga (eds.) Unified Theories, Oakland: The Noetic Press.
[76] Bohm, D. (1963) Quantum Theory, pg. 353, Englewood Cliffs: Prentice-Hall.
[77] Messiah, A. (1999) p. 438, 444, New York: Dover.
[78] Loudon, R. (1994) The Quantum Theory of Light, Oxford: Clarendon Press.
[79] Amoroso, R.L. (2004) The Fundamental Limit and Origin of Complexity in Biological Systems: A New Model for the Origin of Life, in D.M. Dubois (ed.) Computing Anticipatory Systems, AIP Conf. Proceedings Vol. 718, pp. 144-159, Melville: American Inst. of Physics.
[80] Amoroso, R.L. & Rauscher, E.A. (2009) Emergence of Generalized F-Theory 2-Branes from SUSY Spacetime Parameters of the Discrete Incursive Oscillator, in D. Dubois (ed.) IJCAS, Univ. Liege Belgium, Preprint.
[81] Ciubotariu, C. & Ciubotariu, C. (2002) A chaotic-stochastic model of an atom, in R.L. Amoroso, G. Hunter, M. Kafatos & J-P Vigier (eds.), Gravitation and Cosmology: From the Hubble Radius to the Planck Scale, Dordrecht: Kluwer Academic.

[82] Argyris, J. & Ciubotariu, C. (1999) A new physical effect modeled by an Ikeda map depending on a monotonically time-varying parameter, Int. J. Bif. Chaos, 9:1111-1120.
[83] Wheeler, J.A. (1977) Gravitational and Electromagnetic wave flux compared and contrasted, Phys. Rev. D, 16:12, 3384-3389.
[84] Misner, C.W., Thorne, K. & Wheeler, J.A. (1973) Gravitation, San Francisco: Freeman.
[85] Green, M.B., Schwarz, J.H. & Witten, E. (1987) Superstring Theory V-2, Cambridge: Cambridge Univ. Press.
[86] Pettroni, N.C. & Vigier, J.-P. (1983) Dirac's aether in relativistic quantum mechanics, Found. Physics, 13:2, 253-285.
[87] Hocking, J.G. & Young, G.S. (1988) Topology, New York: Dover.
[88] Gordon, J., Rucker, R. & Walker, J. (1990) James Gleick's Chaos: The Software, Version 1.0, Sausalito: Autodesk, Inc. www.mathcs.sjsu.edu/faculty/rucker/chaos.htm.

Chapter 5

Formalizing the Ultimate Geometry of Reality: Dimensionality, Awareness and Arrow of Time

> *...the right hand side includes all that cannot be described so far in the Unified Field Theory, of course, not for a fleeting moment, have I had any doubt that such a formulation is just a temporary answer, undertaken to give General Relativity some closed expression. This formulation has been in essence nothing more than the theory of the gravitational field which has been separated in a somewhat artificial manner from the unified field of a yet unknown nature*- A. Einstein [1].
>
> *In fact this new feature of natural philosophy means a radical revision of our attitude as regards physical reality* - Niels Bohr [2].

Utilizing the natural projection of fundamental parameters inherent in a Holographic Anthropic Multiverse (HAM) cosmology we introduce a delineation of dimensionality, awareness and the arrow of time as they might arise relative to a temporal subspace of an absolute timeless HD space – a regime of the unitary field. Temporal asymmetry - the observed arrow of time has remained one of the most paradoxical problems in physics, recently considered more fundamental than quantum theory. The noetic approach delineated here assumes a 'supra-local' domain more fundamental than the sub-quantum hidden variable regime proposed by de Broglie and Bohm and introduces a whole new cosmology defining time and its origin. Current thinking suggests that there are five arrows of time, four physical and one psychological; in this chapter it is suggested that all arrows of time are a function of the phenomenology of the observer which calls for a reformulation of the basis of physical theory. Incorporating this assumption into scientific epistemology requires a 'continuous-state anthropic multiverse' entailing an additional set of transformations beyond the Galilean, Lorentz-Poincaré groups of transformations and a description of the correspondences reducing to the current standard models of particle physics and quantum theory.

5.1 Introduction

This work attempts to demonstrate the utility of an emerging new cosmological framework by increasing our understanding of the arrow of time in a putative new context pertaining to the nature of reality itself and role of the observer which was irrelevant in Newtonian Mechanics; this mentality has continued to persist among physicists where time is looked at with a certain amount of abhorrence because it seems to demand addressing the nature of the observer in a fundamental way including addressing the nature of consciousness or awareness itself [3].

Although resistance to Big Bang cosmology has remained since its inception, only recently have such criticisms become more acceptable. We assume cosmology to be a multiverse, a form of covariant scale-invariant self-organized complex system (Chap. 3) within which the spacetime backcloth is a dynamical process of the inherent self-organization. We consider this fundamental backcloth to be a form of Dirac covariant polarized vacuum [4,5]. The dynamical surface of this vacuum is purported to be a zero point field. In this context the arrow of time is defined as a locus of events conjoined with the continuous transformation of this spacetime. At the fundamental level an event is generally defined as an interaction of reversible trajectories in the dynamics of a physical system; because the dynamical equations of physics are time-reversible no preferred direction of time is considered relevant. A new event requires a preferred direction which produces a change in the initial conditions of the dynamical trajectories because such transformations of the dynamical trajectories of a system consist of shifts in the spatial and momentum coordinates within the limits of the uncertainty principle $<\delta x^2>^{1/2}<\delta p_x^2>^{1/2}=\hbar/2$ [6]. At this fundamental level the vacuum structure is stochastic and quantum fluctuations of vacuum radiation [7] randomize the momentum distribution of particle coordinates and by the uncertainty principle there is a continuous fundamental ambiguity relating to any direction for an arrow of time. This is in contrast to the concept of Heisenberg 'potentia' where all properties of a system remain unresolved before occurrence of an event. Thus an arrow of time can be defined as a cumulative resultant of oriented new events producing a hierarchical long-range order in the dynamics. From this starting point a new continuous-state HD offshoot is developed.

5.2 Current Philosophy of Temporal Science

In Newtonian clockwork mechanics awareness was irrelevant or nonexistent. The advent of quantum mechanics introduced a troublesome observer involved measurement. Although highly successful quantum mechanics is deemed incomplete - unable to describe biological systems suggesting that extended theory is required. The additional theory must describe the fundamental nature of time and fully include the observer.

Traditionally five arrows of time [8-10] have been described:

- ELECTROMAGNETIC: It is generally observed, which leads to the belief that all electromagnetic waves propagate into the future only. However this is not necessarily true in interpretations of the transactional / absorber theory of radiation [11,12]; and represents a component of the illusion represented in Fig. 5.2 [3,13,14]. This is consistent with Maxwell's equations which are symmetrical in time.
- PERCEIVED EXPANSION OF THE UNIVERSE: Distinguishes between a past and future in the evolution of matter in the universe. In HAM cosmology this arrow of time has no fundamental significance because it is an observational illusion proposed by an incorrect interpretation of astrophysical data [3,13,14].
- THERMODYNAMIC: The observation of temporal asymmetry in thermodynamic processes represents the most important arrow of time because it provides all of our phenomenological experience, and the existence of biological activity. Irreversible processes move toward a thermodynamic equilibrium of maximum entropy.
- KAON DECAY: Nuclear reactions may occur in either direction with one exception occurring between elementary particles that are not part of ordinary matter - Neutral Kaons. There are three kinds of K meson but only the neutral Kaon exhibits a temporal asymmetry. Because neutral Kaon decay asymmetry is a spacetime property not associated with ordinary matter its description can be formalized into evidence of the supralocality of the HAM [3,13,14] by illustrating the variation

of decay paths in terms of gravitational coupling to spatial and temporal nonlocal spacetime spin-exchange dynamics.

- THE OBSERVER: The subjective flow of time that reveals to our moment by moment experience that all actions flow from the present into what we define as the past. The contrast between the four arrows of time defined by physical laws and the subjective arrow of time has often led to the belief that time is an illusion.

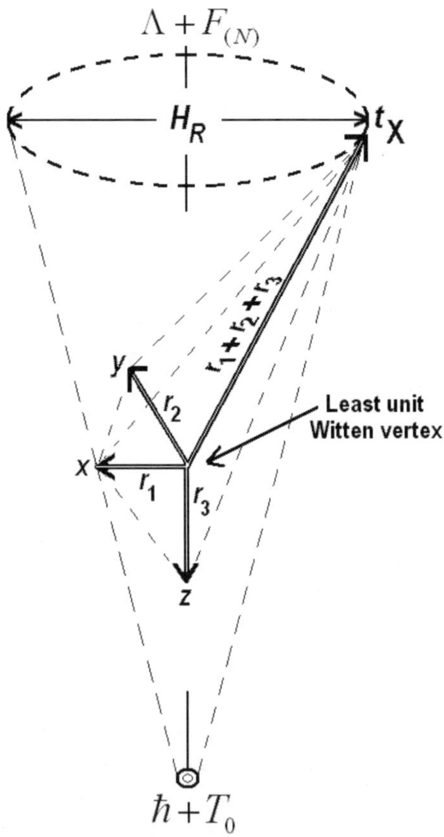

Figure 5.1 Arrow of time as vector sum of continuous-state dimensional reduction deficit angle spin-exchange compactification of spacetime least-units.

Currently physicists describe: 1) Physical time and 2) Psychological time separately, both of which are incompletely understood. A major premise of this work is that all five arrows of time are an illusion related to the phenomenology of the observer's awareness. But because mind in HAM cosmology is completely physical; all the arrows of time are likewise actually 'physical' and can be investigated with new experimental methods leading to discovery of the teleology of the noetic field [15]. The unitary nonlocal noetic field couples classical dynamics and general relativistic effects in a complementarity through the pathways of neural dynamics [3]. This is directly responsible for the perceived arrow of time because this matter-spacetime medium is what 'we' are made of and 'imbedded' in. The noetic timeless domain is the entry point of awareness coupling eternal time through special relativistic dynamic transformations independent of classical gravitation to temporality.

5.3 Complementarity of Physical Time and Observer Time

Einstein remarked that 'if one could ride on a photon, one could circumnavigate the universe without the passage of time'. We postulate that the view from sitting in the saddle of that photon is of an ocean of light or universal Wheeler geon [16] equated with an HD regime of the unitary field. This view is shielded from the observer by the arrow of time which forms the domain walls of our virtual reality. See Chaps. 3 and 4. Virtual pretty much in the sense depicted in the Hollywood trilogy, The Matrix. This view is readily metaphored by Plato's 'analogy of the cave' from antiquity or by the more contemporary view of an observer seated in a movie theater. See Figs. 5.2 and 5.3. Discrete frames of film passing over the projector lens appear smooth on the screen because the motion of the film at a few *cm/sec* is too fast for the eye.

We may extend this to a holographic model in which the observer is imbedded in the cosmology itself and made out of the same materials as the surrounding matter. Because of the relativistic effects the observer does not notice the virtual nature of the reality. See Fig. 5.3. The anthropic action of unitary noeons (exchange unit of the unitary field) is the 'light' of the projector bulb or laser for the spacetime hologram. This is an extension of the Holographic Principle [17,18] to its penultimate form. The nonlocal nature of quantum theory and the EPR principle give the best indicia of this kind of fundamental basis for reality.

As discussed elsewhere anthropic reasons for the 1st person – 3rd person barrier and why we are not easily able to perceive the unitary 'ocean of light' relates to the nature of the arrow of time [3,19]. In HAM cosmology the additional dimensions are not invisible because they are curled up at the Planck scale but because of a spin-exchange deficit-angle mechanism to be developed further below.

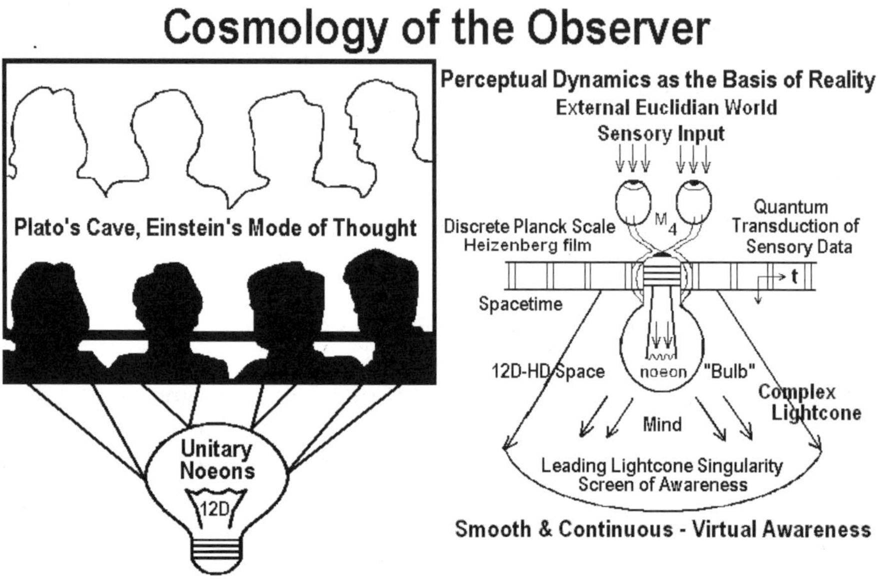

Figure 5.2 a) Plato said 'reality is as if we are dwellers in a darkened cave chained up facing a wall viewing events as shadows by the light of a fire projected from behind and never knowing the true nature of existence'. If we were released to turn toward the light at first we would be blinded by its brightness still having our perception clouded. b) Movie theater model of perceived 4D virtual holographic reality. Discrete Planck scale least-units propagate the arrow of time. Because of relativistic effects reality appears smooth on the 'screen'.

Now that some cosmological properties are reviewed it is easier to show the relationship of physical time to conscious time. All arrows of time reduce to the spacetime topology of the Dirac polarized vacuum [10,19]. From within the microscopic action of the complex hierarchical cosmology of the least-unit of awareness, macrophysical phenomena, which include thermodynamic processes, appear asymmetric because of a complementarity of Cramer-like standing-wave boundary conditions related to human awareness and other physical conditions. There is no preferred temporal direction in the microphysical laws of physics. When this atemporality is reduced to the temporal domain (when it becomes a subspace) many parameters are subtracted out through the symmetry breaking of the spin exchange compactification dimensional reduction process occurring at the speed of light. But this microscopic annihilation governed by teleological causality produces an orthogonal summation creating the macroscopia of perception. The velocity c of the reduction / compactification receding from the eternal present has a discrete microscopic beat frequency perceived macroscopically as continuous.

We introduce the suggestion by Franck [20] that an eternal now occupies the center of awareness and all points in spacetime. We assume that awareness, a fundamental physical principle like the concept of 'charge' [3,19,21], is associated with the 'least unit' in HAM cosmology (Chap. 3). The least cosmological unit is governed by a new Noetic Transform for the self-organized anthropic action guiding the evolution of information from the Planck scale continuously boosting it through M_4 into the 12D Hyper-Geon domain of the unitary field as an Ising model rotation of the Riemann sphere modeled after extended Wheeler-Feynman-Cramer future/past standing-wave parameters [11,12].

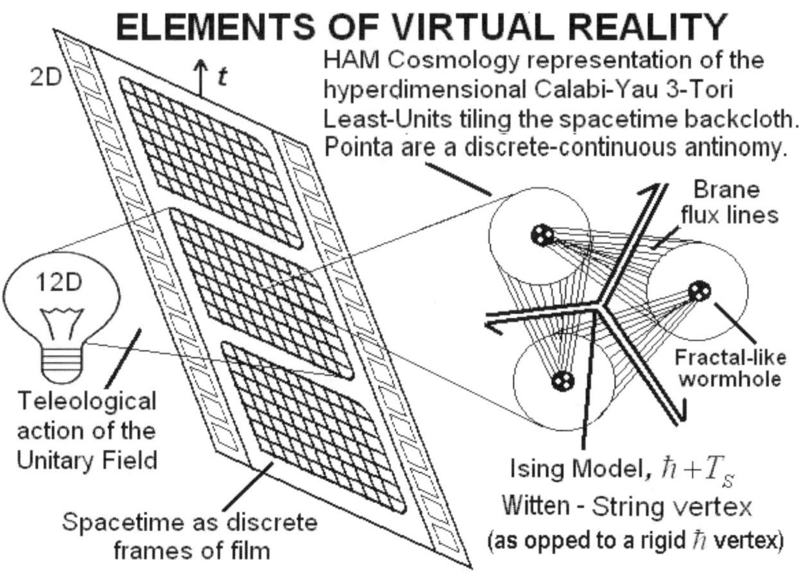

Figure 5.3 Cosmology of holographic reality. Anthropic action of the unitary field pilots the continuous-state evolution of spacetime and quantum dynamics through modulation of the supersymmetric topology of fundamental close-packed Calabi-Yau least-units as a finite regime tiling the spacetime backcloth.

If we utilize the metaphor of a movie theater (Figs. 5.2 and 5.3) to describe the structural-phenomenology of the mind / body and apply Huygens' principle of wave train addition in a manner similar to how sunlight shines through discrete raindrops summating into the smooth image of a rainbow, we can begin to understand the human psychosphere [3,14,19,22]. The psychosphere is the standing wave light-cone surface of human awareness impinged by qualia [19]. It is not confined to the brain; but occupies the total boundary conditions of the human mind-body that extends from the Euclidean brain occupying M_4 to the limits of the HD Noetic Geon. There is a complementarity between these two domains of the human psychosphere. Fermi-Dirac statistic describe the temporal dynamics in the M_4 brain / body region and Einstein-Bose statistics describe the atemporal HD domain applicable to the holophote action of the Noetic hyper-geon. This is the view of Franck's 'eternal now' [20]. The two domains are mediated by the noeon of the unitary noetic field.

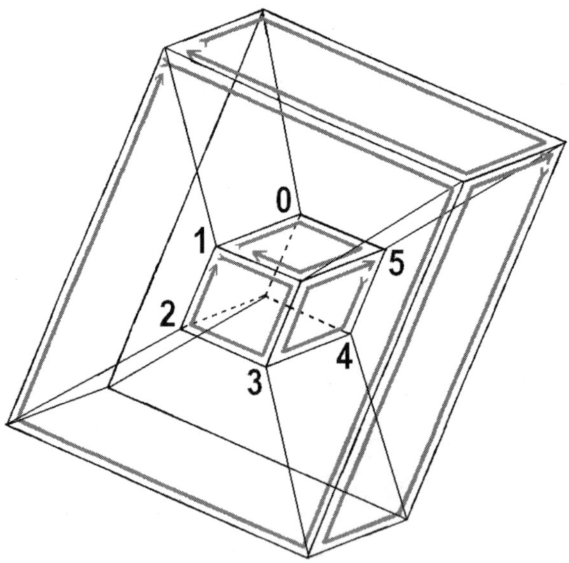

Figure 5.4 Parallel transport by the Bianchi identities through a 3D Witten Ising model vertex. As the 4D hypercube transforms the small central cube becomes the larger outer cube. In the 12D standing-wave model of HAM cosmology these small central *x,y,z* cubes continuously rotate in and out through three complex, $x = x_{Re} + ix_{Im}$, $y = y_{Re} + iy_{Im}$, $z = z_{Re} + iz_{Im}$ orientations simultaneously. This provides part of the basis of our Euclidean virtual reality.

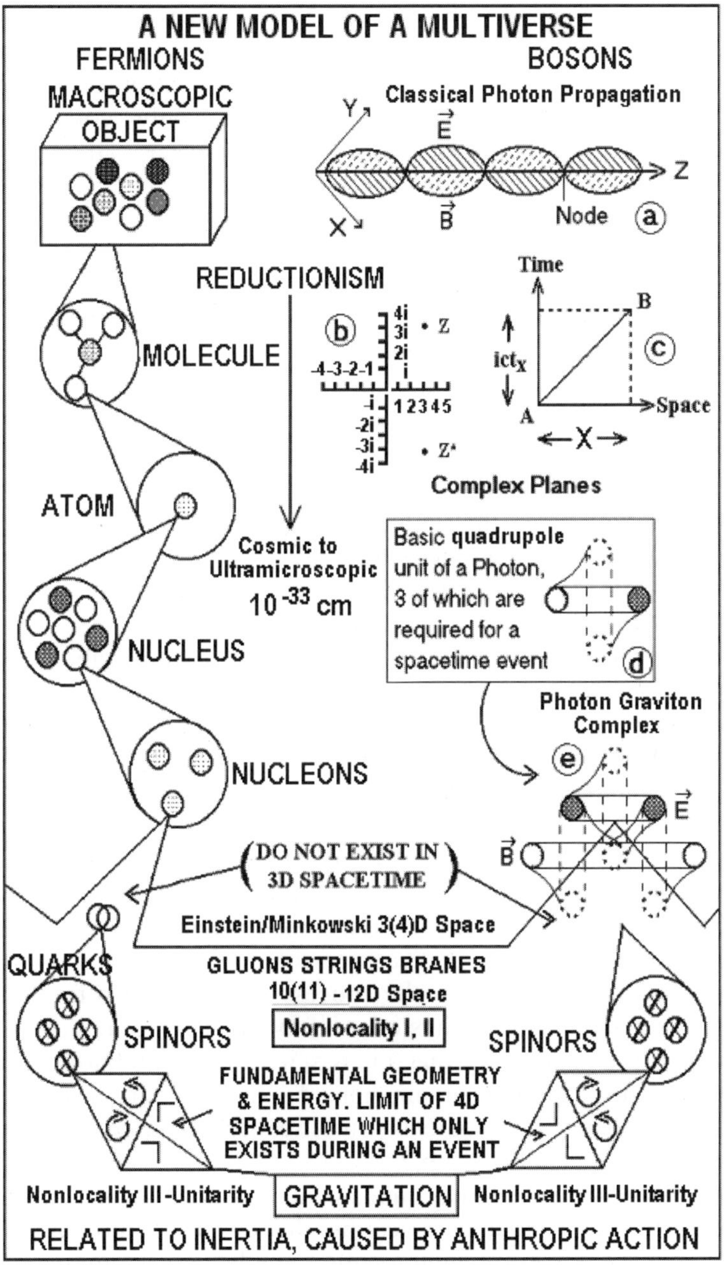

Figure 5.5 Bose-Fermi reductionist model of matter from macroscopic wave-particles to the photon-graviton noeon-quadruplex of the unitary field.

Clock time appears absolute in the Newtonian sense or mutable for relativistic observers. The perception of time is coupled to dynamical processes often associated with entropic flow; but entropic time does not correlate with clock time. And clock time only correlates with psychological time for certain states of consciousness. If a process is harmonic or occurs at a microscopic level where the laws of physics are temporally symmetric all conception of time can be lost; thus the nature of time has maintained itself as a dilemma. As Einstein said 'The distinction between past, present, and future is an illusion'.

Einstein's notion of eternal time in a static universe is extended to develop a framework for correspondences that unify all aspects of time within the new HAM cosmological model [14,15]. Within the HAM framework human existence is composed of a two-fold complementarity; a Cartesian type body/mind dualism comprised of a *res extensa*, and *res cogitans* [3]. The former component associated with the body (obeying Fermi-Dirac statistics); the latter component (obeying Bose-Einstein statistics) is imbedded in an 'eternal' 11(12)D HD space that mediates the *élan vital* [23] through action of a teleological anthropic principal synonymous with the unitary noetic field [3,19].

Time, which is not considered fundamental, is inexorably connected with space in the evolution of the entity spacetime, and within which 'We' and the properties of all the matter we perceive is imbedded. It is the properties of this matter, that the awareness of the observer is imbedded in, with which we define the measuring rods of duration and extension that are the fundamental basis of all physical science [1]. To understand the nature or foundations of time we must be able to comprehend more fully the nature of space and the *potentia* from which it arises. As Einstein obviated the absolute space of Newtonian mechanics; we must now obviate the classical basis of measurement used up to this point in the history of empiricism, and in the process obviate and reformulate the whole fundamental basis of physics itself by inclusion of the true basis of the observer. It is mandatory that a new scientific methodology is devised to investigate the 'absolute noumenon of existence' that resides behind the 'facade of phenomenological virtual reality'. If scientists are finally ready to open this Pandora's box?

It has been suggested that fundamental awareness is comprised of three base states imbedded in spacetime that evanesce into our perceived reality in a manner metaphorically similar to a movie theater [3,10,14,19,24-32]. This concept was described in antiquity by Plato's 'analogy of the cave' as shown in Fig. 5.2. More recently Einstein said: "Time and space are modes by which we think and not conditions in which we exist." [33]. Awareness arises as a complementarity of three fundamental interacting base states

$$|\Psi_M\rangle = |B|\psi_b\rangle + |\psi_e\rangle + |\psi_a\rangle \tag{5.1}$$

where $|B|\psi_b\rangle$ represents the Fermi brain, $|\psi_e\rangle$ an individuals components of eternal elemental intelligence [34] and $|\psi_a\rangle$ the teleological anthropic action principle [3,10,14,19,24-32]. Within the current adamant vogue of cognitive theory this represents an extremely unpopular position which is also quite complicated. We have made our preliminary case elsewhere [3,10,14,19,24-32] especially in the companion mind-body volume [3]. Suffice it to say here that within the bounds of an individual psychosphere the above triune base states produce a superradiance that becomes awareness. This occurs through the holophote injection of the anthropic action principle into the unique hyperdimensional 'standing wave' domain of an individual coupled to their local eternal spacetime present.

The physics of time (thermodynamic processes, kaon decay etc.) seems independent of psychological time. But in an *anthropic multiverse*, all arrows of time are interrelated and arise from one central process in the hierarchy of unitary translation of the close-packed noetic least cosmological unit tiling the spacetime backcloth. An understanding can be garnered by explaining the amplification of microscopic phenomena by processes inherent in fundamental awareness. Observation synonymous with measurement is the obverse of the process of awareness. William James stated that 'there is no splitting of experience into consciousness and what the consciousness is of'. So between experience *A* and experience *B* there is no gap; no collapse of the wave function is observed in thought processes. If one attempts to bring a photon to rest it destructs. This observed reduction of the wave function in the external world has confused conceptions of what occurs in the mind where there is no collapse and as in the photon analogy there cannot be. With large-scale XD the 'continuous-state spin-exchange compactification dimensional reduction process' occurring at the speed of light suggests why large XD are not readily observed. The deficit angle arising in the parallel transport of the continuous-state topology subtracts out one half of the HD standing wave parameters during the compactification process between the gaps in the film so to speak (Figs. 5.2 to 5.4) leaving us with a limited view which is the geometric origin of time.

According to the Copenhagen interpretation all quantum measurements are associated with reduction of the wave function, a thermodynamically irreversible process. Only the final observed component of the ensemble is considered to be *real* [35] by

$$\sum_i c_i \psi_i \to \psi_i \,. \tag{5.2}$$

This action directly creates boundary conditions separating the fundamental reversible aspects of microscopic natural law into the perceptual macroscopia and an additional HD physical realm not perceived by neurophysiology. Noetic cosmology proposes that this temporal asymmetry is completely observer related and the ensuing boundary conditions delete essentially half of the systems information cosmology. Bohr stated from the beginning that the Copenhagen interpretation did not describe biological systems; therefore a full physical description must utilize extended de Broglie/Bohm ontological forms of quantum theory without state reduction and therefore loss of systems information. The big question then is what is the utility of the unobserved parameters of this cosmology?
Here is where the main utility of the Noetic least unit transform enters in.

The complementary superluminal boosting of the 'standing wave' eternal present undergoing the continuous cycling spatial, *s* temporal, *t* and energy, *E* dimensions through unitary, *U* quantum, *Q* and classical, *C* states

$$D_s \to D_t \to D_E \; : \; R_U \to R_Q \to R_C \tag{5.3}$$

that produces and maintains the perceptual macroscopic amplification of microscopic phenomena. The Noetic boosts reduce the flux of all physical fields at the domain wall boundary conditions by absolute parallelism of the Bianchi identities, $\partial \circ \partial = 0$ where the boundary of a boundary equals zero [36] facilitating this whole cosmological process. We begin with the description of the electromagnetic field. Following Kafatos and his collaborators [37] suggesting the importance of $\dot{R} \equiv C$ for universal continuous-state boundary conditions (see Chap. 4) which are also relevant to the velocity required for the observers mind to escape microphysics and become coupled to a virtual macroscopia for EM by

$$\vec{c} = \frac{2\vec{E} \times \vec{B}}{\vec{E}^2 + \vec{B}^2} \tag{5.4}$$

where, according to Wheeler [38,39], velocity $\vec{c} = \vec{n} \tanh \alpha$ and the numerator is the Poynting flux and the denominator the energy density. This boost equation describes the reduction of the EM field to mutual parallelism which according to the Bianchi identity describes how the boundary of a boundary equals zero [36] (see Fig. 5.10). Allowing half the universe to cancel out of awareness into the resultant standing wave covering. (see Figs. 5.7 to 5.9). The covering is piloted by the de Broglie-Bohm wave-particle energy. An application of Huygens' principle of wave addition might produce the smooth feel of the evanescence of reality we observe while we are *surfing* as it were on the face of the discrete elements of atemporal microphysics!

5.4 The Vacuum Origin of Thermodynamics and Entropy

Temporal asymmetry is a fundamental problem because the microscopic laws of physics are time reversible. The macroscopic arrow of time arises from translation of the complex boundary conditions of the observer, which ultimately is a property of the unified field. Although this is a perceptual phenomenology it is still physical. The most fundamental basis, more fundamental than for quantum interactions of matter is the unified electromagnetic-gravitational arrow of time; from which the thermodynamic and all other arrows arise. The continuous-state dimensional reduction compactification process within the topological structure of the polarized Dirac vacuum has a *beat frequency* associated with the inherent *Jacob's ladder-holophote* of least unit translation.

Entropy increase in thermodynamic systems can be accounted for by vacuum radiation; and this interaction of vacuum radiation with matter is time-reversible. Therefore whether entropy increase in thermodynamic systems can be considered to produce an arrow of time depends on what controls the vacuum photons. Both cases are consistent with quantum mechanics. Position and momentum perturbation on particles by vacuum zero-point radiation is limited by uncertainty to

$$\langle \delta x^2 \rangle^{1/2} \langle \delta p_x^2 \rangle^{1/2} \tag{5.5}$$

where the first root mean square value is position and the second momentum respectively, Burns [6,7]. According to Zeh, [35]

$$< \delta x^2 >^{1/2} = (\hbar t / m)^{1/2}, \tag{5.6}$$

(where m is particle mass), can be obtained both from classical SED and the stochastic interpretation of quantum mechanics. Substituting the result into the uncertainty principle yields a fractional change in momentum coordinates, $< \delta p_x^2 >^{1/2} / p$, p is the total momentum, $2^{-3/2} (\hbar / Et)^{1/2}$, E is the kinetic energy. As vacuum radiation interacts with particles, momentum is exchanged. When an initial fractional change $< \delta p_x^2 >^{1/2}$ in momentum is amplified by the lever arm of molecular interaction,

$$< \delta p_x^2 >^{1/2} / p \geq 1 \tag{5.7}$$

it becomes greater than one in only a few collision times [6,7,35]. Therefore the momentum distribution of a collection of interacting particles is randomized in that time, and the action of vacuum radiation on matter can account for entropy increase in thermodynamic systems; i.e. it can be related to the atemporal / temporal microscopic / macroscopic cosmology of fundamental awareness.

Dynamical interactions occurring at the molecular level are time-reversible, but thermodynamic processes associated with entropy increase, like diffusion and heat flow, only proceed unitarily in time. Entropy increase appears to be only a macroscopic phenomenon, appearing when a coarse-grained average is taken of microscopic processes. No averaging of time-reversible processes has been shown to account for temporally irreversible phenomena [35]. The reduced or temporal subspace nature of human perception filters out half of the microscopic action by the continuous dimensional reduction process. This action occurs at the speed of light explaining perspective – narrowing of the railroad tracks into the distance; which would not occur for a HD atemporal observer like 'God'.

In the standard model (utilizing only the positive set of Maxwell's equations) electromagnetic waves emanate from a source to infinity only, and do not converge from infinity to a source. Collapse of the wave function is a one-way process [40,41]. Burns [6,7] has shown that entropy increase in thermodynamic systems is produced by the interaction of vacuum radiation with matter. This interaction is time reversible. Whether an arrow of time is ultimately involved in entropy increase depends on how vacuum radiation is produced. In Noetic cosmology which utilizes an extension of the Wheeler / Feynman absorber theory of radiation EM waves from infinity do converge with the standing wave source. There are extended quantum domains without collapse of the wave function where noncomputable ontological superpositions occur; and vacuum radiation is governed by teleological cosmological action principles inherent in the HD vacuum topology [42,43].

The exchange particle of the Noetic Unified Field, the noeon, follows preferred paths within the continuous spin-exchange dimensional reduction compactification process. It should be noted that 'exchange particle of the unitary field' is a bit of a misnomer as the exchange is energyless and ontological – a form of topological switching; we will deal with that conundrum later. It is reminiscent of a quantized traveling arc or *Jacob's ladder* where the 'charge' enters with a harmonic *holophote action* at the bottom (Planck scale) and travels to the HD region where it is released or reabsorbed cyclically as the *eternal present* [20] remains a continuous-state of the future-past HD topology. This again is the movie theater metaphor where discrete frames of film pass over the projector bulb (Planck scale holophote noeon emission into every point and atom in spacetime) propagating *up* the Jacob's ladder (psychosphere light cone surface) to the *screen* (smooth continuous raster of awareness) as qualia[1].

5.5 Peripheral Physical Properties Related to the Observer

Twelve D is the minimum number required to describe eternity. By eternity we mean a continuous-state topological manifold able to completely transform out of contact and be causally free and independent of the temporal reality of the observer. This is a property of observed Euclidean/Minkowski space being a standing wave subspace of the 12D HAM Absolute Space. The rigorous description of this property requires a new set of anthropic transformations beyond the standard Lorentz/ Poincaré transformations. Planck's constant is also reformulated and quantum theory to be completed.

---

[1] Qualia- In philosophy of mind 'The quality of the feel' of a moment of awareness, the sensation of 'redness for example. We take this much further here in that the duality of the reality of the observer is like a 'Qualia of the Multiverse'; part of the cosmologies inherent conformal scale-invariance [3].

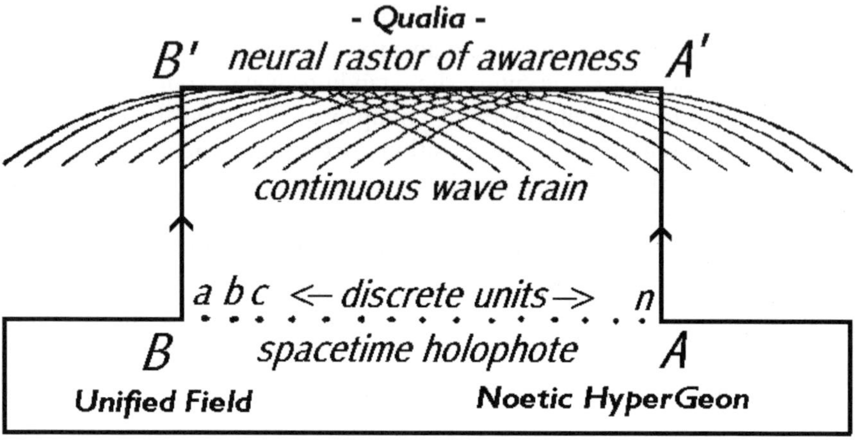

Figure 5.6 *Movie theater* view of the light-cone boundary. All *D* suppressed except one extended spatial element *B' – A'*. Noeons (exchange particles of the Noetic Unified Field) propagate within the discrete Planck scale backcloth of the polarized Dirac vacuum, not in free space, but confined to the metric of the HD fabric like quarks.

Unitary Noeons also represent both the life principle, *élan vital* and light of consciousness [3,19]. They propagate with an inherent beat frequency along preferred paths of the Jacob's ladder holophote by the Noetic spacetime transform of HAM cosmology. The smoothness of reality is the leading edge of the lightcone kept in phase by a Huygens'-like principle of wave train addition of the oscillating Planck scale holographic least-units conceptually illustrated in Figs. 5.2b, 5.3 and 5.6.

5.6 Introduction to Spin Exchange Compactification Dynamics and the Permutation of Dimensions in the Noetic Transformation

Photon mass is not continuously maintained in HAM cosmology but occurs only during a period of internal motion (angular momentum) when the centrum of the wave - the particulate moment, couples to the vacuum; so the photon in propagation cycles harmonically from mass to masslessness as a property of the future-past symmetry of its wave-particle duality. This is a new property of photon propagation introduced by the continuous-state parameters of HAM cosmology.

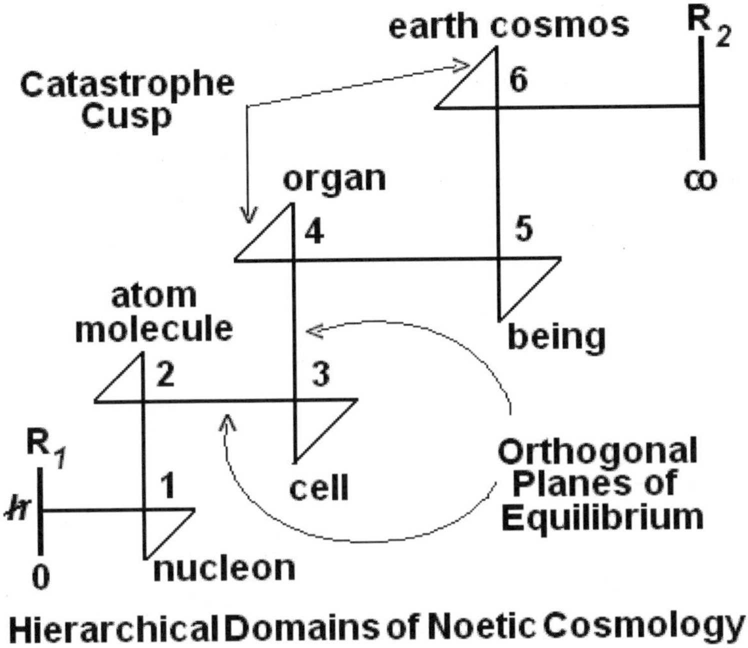

Figure 5.7 Basic conceptualization of the covariant scale-invariant hierarchical structure and function of HAM cosmology from microscopic to cosmic.

Photon mass anisotropy is a major feature of the HAM model. It is indicative of the ubiquitous occurrence of the properties of spherical rotation discovered by Dirac initially attributed only to the spin of the electron where it 720° instead of 360° to return to the origin. The HAM spacetime Cavity-QED paradigm is based on the fundamental premise that the energetic interplay of the fundamental forces of nature, mass, inertia, gravitation and spacetime is based on a unified symmetry of internal spin-spin coupling and spin exchange compactification with a 'super quantum potential' [44] ultimately being the anthropic unitary action and control principle of the evolution of spacetime which within the Einstein Hubble 3-sphere is considered a complex self-organized system which gives it the known properties of such systems [19]. Spin exchange symmetry breaking through the interplay of a unique topological control package orders the compactification process providing a template from which superstring or twistor theory could be clarified if the tenets of Chap. 4 are applied (assuming they are correct of course).

One purpose of compactification dynamics is to allow the Einstein 3-sphere of temporal reality to stochastically 'surf' as it were on the superstructure of an HD eternity creating our virtual reality and the perceived arrow of time allowing nonlocal interactions not possible in a Newtonian absolute space. Stated another way, the domain of quantum uncertainty stochastically separates the classical regime from the unitary regime revealing why large XD can be relativistically unobservable.

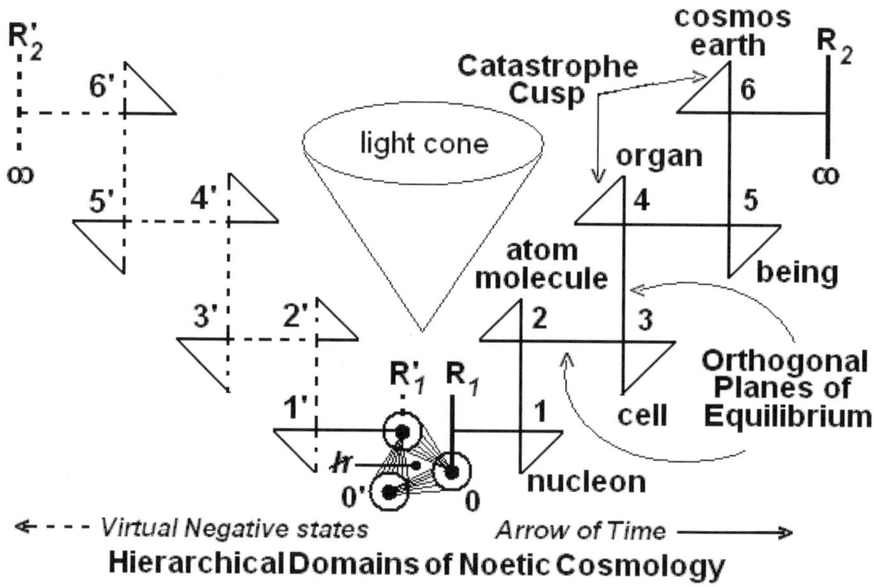

Figure 5.8 Further conceptualization of the advanced-retarded future-past mirror symmetry/duality of the scale-invariance and function of the standing-wave properties of HAM dimensionality from a 0D least-unit to 12D limit of the Hubble 3-sphere setting the stage for application to HAM model arrow of time.

By parallel transport of the topological boundary conditions of the continuous-state dimensional reduction compactification process the deficit-angle produced in the hysteresis loop of the standing-wave eternal present allows half of the parameters to drop out during the 'leap-frogging' of coordinate fixing and re-fixing as the awareness of the observer relativistically couples uncouples and re-couples as a baton passing in a relay race to observed reality. This seemingly complicated process creates the arrow of time and also reveals why the XD are not perceived even though they are large in scale during the retarded portion of the process. Only certain pathways for parallel transport by spin exchange dimensional reduction (D down scaling) and superluminal boosting (D up scaling) are allowed by the Wheeler-Feynman symmetry breaking relations in the continuous maintenance of the standing wave present.

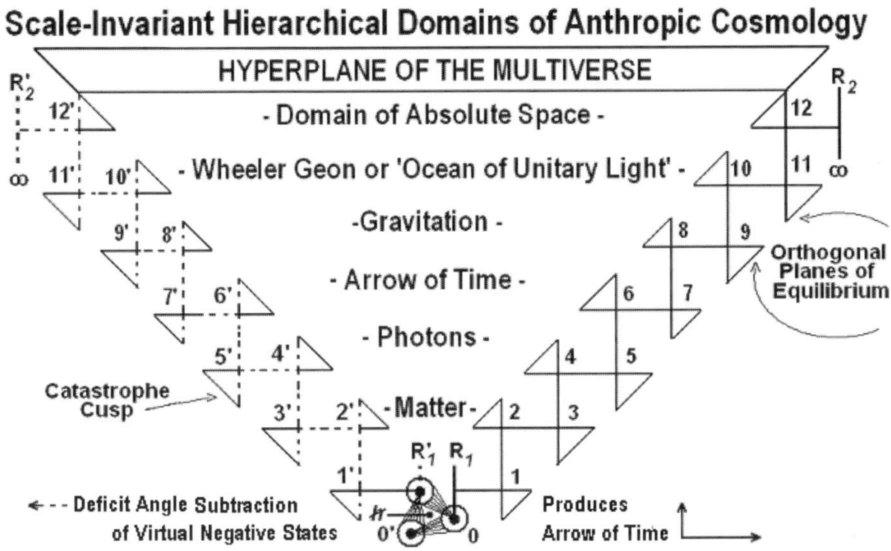

Figure 5.9 By deficit angle parallel transport during the continuous-state spin-exchange dimensional reduction compactification process the arrow of time emerges naturally by subtraction of the advanced portion of the standing-wave topological elements of spacetime relative to the quantum state of the observer.

It is useful to further clarify the utility of parallel transport begun in association with Fig. 5.4 above in terms of the Regge equations [36] relation to the Bianchi identity 'of a boundary of a boundary' being equal to zero $(\partial \circ \partial \equiv 0)$ [36,45,46]. Figure 5.10 shows the three counter-propagating circular permutations of the face plane of a tetrahedron representing parallel transport which creates a deficit angle [47] allowing uncoupling from Euclidean reality. Allowed pathways and orientations restricted by the symmetry breaking conditions allow boosting of the information or energy associated with one domain to transform by topologically switched parametric up-down conversion into another regime.

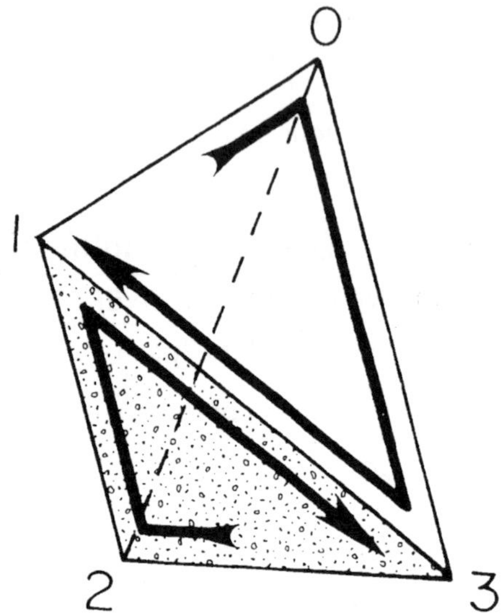

Figure 5.10 Bianchi identities of absolute parallelism for a tetrahedron.

Ordering vertices as shown in Fig. 5.10 induces an orientation on the tetrahedrons two dimensional boundary, which consists of four oriented triangles by $\partial(0123) = (012) - (013) + (023) - (123)$. This in turn induces an orientation on the edges of the one dimensional boundaries $\partial(012) =$ (01) - (02) + (12). Summing the dimensional boundaries cancels them in pairs [(01) - (01) = 0]. This is the Bianchi identity $\partial \circ \partial = 0$ described by the Regge equations for parallel transport where the boundary of a boundary is zero. Or suggesting the tetrahedron is edgeless because the 1D boundary of the 2D boundary of the 2D region is zero [45-47].

5.7 Dirac Spherical Rotation Inherent to the Transformation of the Fundamental Least-Unit

Typically the Dirac dual (2π) spinor rotation applies to the observation that an electron undergoes 720° of rotation (not the usual 360°) before returning to the initial orientation. Traditional thinking has assumed this to be some property of matter. But the discovery of the complex structure of spacetime has shown that this is not a property fundamental to the electron;

but rather to the superspace the electron is imbedded in and part of. Dirac spherical rotation as it is also called, is more fundamentally a primary property of space than it is matter. This is revealed in the complex hierarchical structure of the least unit discussed in the paper.

The Dirac String Trick

Take a square and tie the four corners to another larger square by loose string as shown in the figure below (alternatively, tie the initial square to the four corners of the room). Now rotate the small square by 360° about a vertical axis, that is, in a horizontal plane. The strings will become somewhat tangled, and it is not possible to untangle them without rotating the square. If we rotate through another 360°, for a total of 720°; it is now possible to untangle the string without further rotation of the square by simply allowing enough space for the strings to be looped over the top of the square! You won't believe it unless you check it out for yourself. It is advisable for your experiments to use bulldog clips to attach the ribbons to the squares, so that it can be undone easily if it gets too tangled. A similar idea works for a rotation through 720° about any axis.

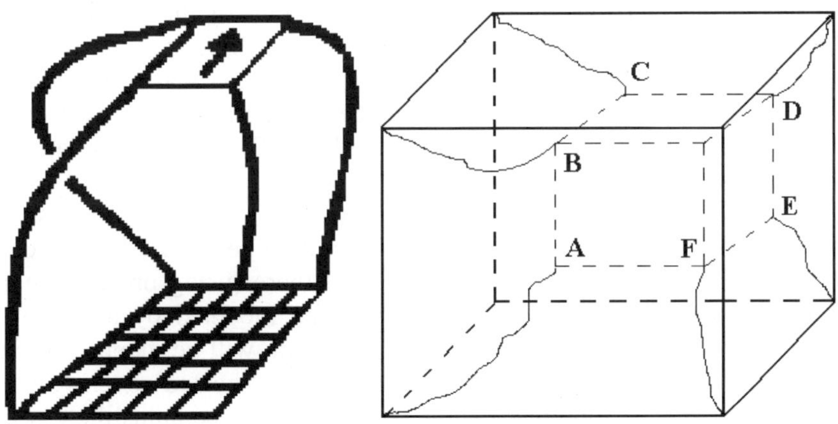

Figure 5.11 Two forms of demonstrating the Dirac string trick to illustrate how spin ½ particles like the electron must undergo 720° of rotation instead of the usual expected 360° to return to the starting point. Figure adapted from [36].

Another version of the Dirac string trick is called the Philippine wine dance. A glass of water held in the hand can be rotated continuously through 720° without spilling any water. These geometrical demonstrations are related to the physical fact that an electron has spin ½. A particle with spin 1/2 is something like a ball attached to its surroundings with string. Its amplitude changes under a 360° (2π) rotation and is restored by rotation of 720° (4π). The formal description of such complex phenomena typically requires sophisticated mathematics (algebra, group theory, topology, quaternions...) since they are not part of everyday experience.

According to Kauffman [48] features of certain spin networks can be viewed as particles with similarities to Bosons and Fermions of the standard model of particle physics by looking at topological elements of the Artin braid group [49-51] that could be used as the basis for introducing quantum numbers. The focus of Kauffman et al. is the manipulation of braid forms, not specific correlations to actual physics, but the work establishes a useful basis for physical implications in future works especially for twist words for fermions that can be matched to quantum numbers such as weak isospin, hypercharge, baryon number or lepton number for example [48]. Kauffman says 'The spinor rotation does not contain a twist of a knot (A knot is the closure of specific braids). What occurs in the Dirac 'knot trick' is that a certain kind of belt twist can model the fact that the first homotopy group of the rotation group SO(3) is Z/2Z. This can be easily visualized (Fig. 5.11) giving an understanding of how the phase change in a fermion wave-function can occur as the result of a rotation in 3-space which can be represented via SU(2) on the quantum wave function with SU(2) appearing relative to SO(3) as its double covering space' [52].

The twist as demonstrated with a belt (Fig. 5.11) happens in 3-space. But this topology is not directly associated with a geometric linking of an electron with its surroundings. We only get there (using present theory) by noting that any quantum process must be modeled by a family of unitary transformations. And then a 360° rotation will be mapped up into SU(2) and end up on the second sheet of the two sheeted covering space SU(2)---SO(3). The topology of this covering space contains the essence of the Dirac belt trick (Fig. 5.11). But the belt trick itself is part of something occurring in 4D, namely the quaternions. See [53] for a discussion of this relationship. A topological theory of the electron where the Dirac 'belt trick' rotation property is connected directly with the physical properties of a particle is currently stymied by the standard Copenhagen quantum theory because there is no 'physical particle' only the result of measurements of an electron wave function which gives only statistical parameters of the wave function.

But as well known this is not true in the de Broglie-Bohm-Vigier (DBV) causal stochastic interpretation of quantum theory where wave and particle are physically real and may both exist simultaneously [54]. But DBV has not been completed. We believe when this explanatory gap has been filled it will show that there is no Quantum Gravity. The quantum regime ends with Copenhagen and the unification of quantum theory and gravity will be shown to occur at the level of unitarity. Our view here is initially more conceptual, we think that a certain rotation point of the belt where the twist occurs in 3D becomes like a Klein bottle that can only be untwisted by rotation through 4D where it is not intertwined. We encourage the reader to perform the little trick with belts or strings from Fig. 5.11. When the electron is rotated 360° the 3D observer sees the twist that in that perspective cannot be untied except by another 360° rotation that occurs in 4D.

In string/brane theory there is a putative Kaluza-Klein spin tower compactification gradient of T-duality/Mirror symmetry for a pair of Calabi-Yau 3-forms or Kahler manifolds where the raising and lowering of the dimensionality with the string/brane tension-coupling parameters passes through Fermi-odd and Bose-even spin symmetries relating the branes to each other. Our postulate is that the rotation of the electron is indicative of a topological process that might be conformally scale-invariant through this whole convoluted hierarchy of dimensionality...We assume, the Dirac $360\text{-}720^{\circ}$ spherical spinor rotation of the electron contains a 'pinch or twist' in the midst of the transformation assumed to be indicative of a 4D topological background component of the rotation.

Is there a braid-form that might be scaleable to even higher dimensions; a form that might require the mirror/dual symmetry conditions purported to occur in string-brane topologies to perform the pinch and unpinch? The Dirac spherical rotation concept appears to be indicative of a covariant scale invariant cosmological principle applying to the entire dimensional nature of reality itself not just the electron. This cosmological twist then would occur as the Copenhagen regime separating Newton classical mechanics from unitarity. We have done our best to introduce an empirical protocol to falsify this prediction (low energy methods without accelerator that can also test string theory).

5.8 Preparing the Noetic Spacetime Transformation

Noetic HAM cosmology implies that so-called 'real space' is a relational standing wave subspace of an absolute HD space, where a continuous-state Dirac type spin-exchange dimensional reduction compactification process is central to the scale invariant periodic Ising model topological structure. It is useful to initiate the description by introducing a toy model of the lower *D* space and build it up toward the actual HD space.

Maintaining the extended Wheeler-Feynman-Cramer property of the present as a function of the advanced-retarded future-past (Figs. 5.1, 5.8, 5.9 and 5.13-5.16). We begin by describing a discrete Einstein type point in the relational spacetime manifold. Since points are defined as singularities where dimensionality breaks down, a dimensionless 0D point cannot be topologically 'covered'. This property will be shown to be a valuable criteria as a 'hole' for oriented orthogonal superluminal boosts in the noetic transformation. This also contrasts the nature of continuity (Absolute space) with discreteness (relational space); points are not absolute because the universe turns out not to be a Newtonian continuum.

The 1D Case

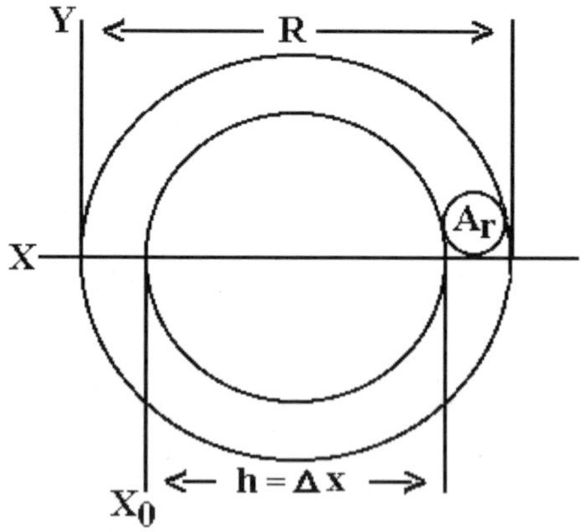

Figure 5.12 The 2-torus appearing as a donut slice acts as a covering of an infinitesimal 1D topological least unit, the line $h = \Delta x$. Any point of, $h = 0$ is dimensionless and cannot be covered (or confined). But the line, $h = \Delta x$ acting as a transient 1D unit of extension, may be covered by a 2-torus as shown. One additional dimension is required to cover the next lower D space (3D).

Therefore we begin the construction of dimensionality with the 1D scalar case. Assuming an arbitrary, discrete, infinitesimal, oriented least unit $h = \Delta x$ as in Fig. 5.12; an entourage of additional HD's are required to 'cover' or confine each subspace level. Usually the entourage has one more D than its subspace. The least unit, h on coordinate x can be covered by a 2-torus when the orthogonal generating circle A, of radius r is located at distance $R > h_{\Delta x}$ from x_0 and not on h, is rotated through dimension y into an HD plane x, y. Thus a 2D flat torus covers the least unit $h_{\Delta x}$ with an x, y plane. The rotation through y (of growing importance later) may occur in counterpropagating directions. Finally the 1D case utilizes a $\pm 2D$ covering for the $h = \Delta x$ unit of extension which may 'wink' in and out of existence since it is a complementarity of 0D and 1D. This is supposed to be similar to a virtual particles lifetime.

The 2D Case

Covering the least unit of a plane, $h = \Delta x, \Delta y$ uses a method similar to the 1D case except that two modes of covering are allowed:

- Type 1. Energy–Time. An intermediate covering of region h by a $\pm 2D$ flat torus in the plane x, y as in the 1D case which leaves room for access of a 3rd energy or time coordinate utilizing either the spin exchange dimensional reduction process or superluminal boosts into HD; also allowing action of a quantum potential or anthropic action.
- Type 2. Spatial. Region $h = \Delta x, \Delta y$ is completely covered by a 3-torus. This occurs by rotating a generating circle orthogonal to x, y through the z direction. This covering represents the lower limit of standard M_4 space with the addition of time, t.

There is no need to develop the toy model further at present as it sufficiently illustrates pertinent aspects of the noetic transformation that show how boundary conditions transform the dimensionality of space and time along with the energy covering of the unified field by $D_S \rightarrow D_t \rightarrow D_e$. The unified field governing gravitation, and the quantum potential guides the action of translation along certain allowed pathways. For example if either l, w or h is removed from a cube the object collapses to a plane. Removing a dimension from the plane causes compactification to a line and so on. The released space is not initially empty. At the first stage of D reduction space transforms into time; and at the second stage into the energy that couples with the energy governing it as compactification is completed for that particular spacetime unit.

5.9 Developing the Line Element for Noetic Superspace

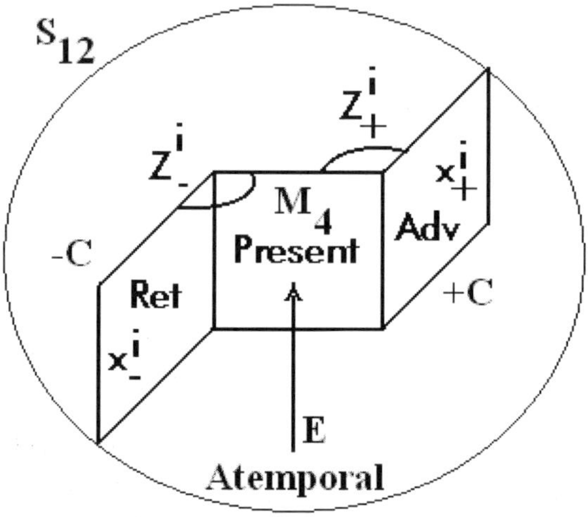

Figure 5.13 A 2D representation of the three 4D spacetime packages making up the 12D periodic noetic superspace of post Big Bang cosmology. M_4 is the Euclidean based Minkowski / Riemann standing-wave present with two HD complex spacetime packages $\pm C$ representing the four retarded and four advanced dimensions respectively which puts certain constraints on the description of the noetic line element.

The real parameters for the line element in standard Einstein-Minkowski space, M_4 is

$$dS_0^2 = dx_1^2 + dx_2^2 + dx_3^2 - dt^2 \tag{5.8}$$

to which noetic superspace must make correspondence. We begin by developing the associated future-past and advanced-retarded 8D complex dual spaces following work initiated by Amoroso [3,19], Rauscher [55,56], Cole [57] and Hansen & Newman [58] on complex Minkowski space, $\hat{M}^4 \pm C^4$. For 12-space variable, Z^μ where $Z^\mu = \pm X^\mu_{\text{Re}} + iX^\mu_{\text{Im}}$ and $Z^{*\nu}$ is the complex conjugate of Z^μ so that $Z^\nu = \pm X^\nu_{\text{Re}} - iX^\nu_{\text{Im}}$ with the 12-space differential line element as

$$dS^2 = \pm\eta_{\mu\nu} dZ^\mu dZ^{*\nu} \tag{5.9}$$

with indices running 1 to 12 where $\pm X_{\text{Re}}$ is $\pm x_{\text{Re}}, \pm y_{\text{Re}}, \pm z_{\text{Re}}, \pm t_{\text{Re}}$, the usual 4D metric plus addition of Wheeler-Feynman-Cramer future-past conditions [11,12]. We must then finally introduce the additional complex noetic relations, $\pm N_8$ to include the advanced-retarded Dirac annihilation-creation ladder operators to complete the parameters required for the complex noetic transformation

$$W^{jk} = \pm[Z^j_{\text{Re}} + iZ^k_{\text{Im}}],\ \pm[\overline{Z}^j_{ret} + \overline{Z}^k_{adv}] \tag{5.10}$$

again with indices j, $k = 1$ to 12. Then for complex advanced space, $+N_4$ we have $Z^{jk}_{adv} = X^{jk}_{\text{Re}(adv)} + iX^{jk}_{\text{Im}(adv)}$, $\overline{X}^{jk}_{\text{Re}(adv)} + \overline{X}^{jk}_{\text{Im}(adv)}$ with $j,k = 1$ to 12. For complex retarded space, $-N_4$ the relation is $Z^{jk}_{ret} = X^{jk}_{\text{Re}(ret)} + iX^{jk}_{\text{Im}(ret)}$, $\overline{X}^{jk}_{\text{Re}(ret)} + \overline{X}^{jk}_{\text{Im}(ret)}$ with indices running $k = 1$ to 12. Then the noetic line element is

$$\Delta W^2 = \pm\left(\eta_{\mu\nu} dZ^\mu dZ^{*\nu} + \eta_{jk} dZ^{jk}_{adv} Z^{jk}_{ret}\right) \tag{5.11}$$

5.10 Formalizing the Noetic Group of Transformations

We postulate an additional set of transformations beyond the Lorentz-Poincaré called the 'noetic group' with another causal relationship distinct from the strong causality of the standard model allowing spatially separated systems to exchange information without orthodox collapse of the wave function. This occurs through a nonlocal coupling of unitary field effects which produce a geodesic deviation mediated by intentionality or anthropic teleological control depending on the segment of the scale-invariant regime being acted upon. The dynamics of particle and fields are described by various groups of transformations; the Galilean group describes Newtonian mechanics, and the Lorentz transformations describe modern relativistic and quantum theories. This action is outside the current limits described by the Galilean, Lorentz and Poincaré groups of transformations. This additional noetic transformation of a normally null path is allowed in extended electromagnetic theory by nonzero restmass photon anisotropy [59] without violating gauge theory [60]. The correspondences in physical theory, for example the reduction of quantum mechanics to classical mechanics or the recovery of thermodynamics from its successor statistical mechanics will also apply in relation to the mind and the nature of time. A correspondence between a complementary stable and unstable causality is shown to reduce to the null path of the standard model.

Current thinking for an 'energy regime' extends only to the indicia provided by the proposal of primal Kaluza-Klein theory of energy as a 5th dimension [61], Eastern philosophical dogma of 'consciousness as the monistic ground of all being' [62] or the Judeo-Christian doctrine 'the spirit of God fills the immensity of space' [34] which physicists like Einstein and Schrödinger equated with the unified field. Behind the facade of reality lies an atemporal hyper-geon [16] or ubiquitous unitary regime that is likened unto an 'ocean of light'. Any usual EPR state is a parametric down-conversion [63] of simultaneity or bi-local entanglement of this holographic state [17,18] into the fabric of the spacetime view of the Euclidean observer. Our immediate question is if one is to parametric up-convert [63] such an entangled EPR state between two locally separated observers what should the description of the transformation to that state entail? The Lorentz transform [64,65] adds a relativistic warp factor to the Galilean transform for both x and t coordinates. In the HAM model do the rotating ends of the standing-wave strings have opposing ends with a velocity gradient or range from $v < c \rightarrow v \geq c$ which could be ignored for the purposed here if our interest remained only in delineating the final state; but the coupling/uncoupling process to this state's intermediates would require elucidation of the dynamics of the complex topological gradient.

Another major issue is that resonance modes for each intermediate coordinate boost individually entail a description of orthogonal planes of equal phase; but the final result is the imbedding of the observer's awareness in the surface of a hypersphere of information or charge equal to the area/volume of the topology where the mutually exclusive orthogonality of the intermediates is returned to parallelism allowing or producing independence from the initial Euclidean plane as illustrated in Fig. 5.24. We have two major problems at the moment keeping us from rigorously formalizing the noetic transformation;

- Where to draw the line in the sand. We have a 12D holographic model, but at this point following the Kaluza-Klein logic of 'energy as a 5th dimension [61]; how many dimension does the transform requires to both contain and transform the topology in order to ontologically exchange the information. So 'today' we make some best guesses and introduce a preliminary noetic transform based on postulates of M-theory and noetic philosophy. If we start with the premise of string theory that 'matter resides on the 3-brane and gravity is free to pass between branes', then by including the premises of duality/mirror symmetry along with HAM cosmologies embrace of Wheeler-Feynman future-past symmetry conditions [11,12] we seem to end up with a local-nonlocal 6D QED spacetime cavity.
- Secondly, if the use of the transform happened to be the exchange of mental information between two separated subjects with the usual x', x Lorentz coordinate separation which could be locally mapped; until the special class quantum computer is built there is no known manner of finding the correct holographic hyperplane resonance mode to couple the two systems to be entangled by the transform. Even if one logically assumes because of the holographic principle that the information to be transferred is ubiquitous there is still the same problem in initializing the receiver.

Physical understanding of coordinate transformation laws began with the Galilean transform for correlation between two Newtonian coordinate systems X', X with velocity, $v < c$ with time absolute and independent of the motion of the different observers

$$x' = x - vt, \;\; y' = y, \;\; z' = z, \;\; t' = t. \tag{5.12}$$

With the advent of quantum theory and special relativity the need for the Lorentz/Poincaré group of transformations arose for velocities, $v \leq c$ and time becoming a new concept of spacetime

$$x' = \gamma(x - vt), \;\; y' = y, \;\; z' = z, \;\; \gamma\left(t - v_x / c^2\right) \tag{5.13}$$

with the 'warp factor' $\gamma = 1/\sqrt{1 - v^2 / c^2}$. Now the need has arisen to fully integrate not only the role but also an inherent imbedding of the very existence of the observer as an essential element of the anthropic Multiiverse; for which a new set of 'noetic' transformations is required. How to proceed has been fraught with challenging conceptual dilemmas like elucidating the proper cosmological framework and the restrictions imposed by extended EM, quantum and M-theories. Another challenge is reflected in what the nature or basis of the final state the transform should be like, where it is and what happened the information; because the observer has not necessarily traveled anywhere in Euclidean space as in the former transformation laws. We consider the Galilean-Lorentz-Poincaré transforms to reflect a virtual quantum reality of 'parametric down-converted' states; and what the new noetic transform requires is a description of a 'parametric-upconverted' state [63] that entangles the two observers in an HD regime with time again becoming independent of the final state. This is different than Newtonian temporal independence in that the Galilean conditions have no relevant quantum entanglement; and the Lorentz-Poincaré basis, although EPR entangled is lacking the overt simultaneity between the two observers that would be considered a violation of Copenhagen causality and the uncertainty principle. Also instead of the focus being for quanta in motion along a manifold the interest here lies with the information field itself and therefore must address conditions of relativistic quantum field theory with static de Broglie waves for all coordinates, *x,y,z* and *x',y',z'* simultaneously.

With this in mind we might begin outlining the Noetic transform for a coordinate regions *x',y',z'* and *x,y,z* with each axis having their own warp factors, α, β, γ respectively as

$$
\begin{aligned}
&x' = \alpha(x - vt),\; y' = \beta(y - vt),\; z' = \gamma(z - vt), \\
&t_x' = \alpha\left(t_x - v_x / c^2\right),\; t_y' = \beta\left(t_y - v_y / c^2\right),\; t_z' = \gamma\left(t_z - v_z / c^2\right),
\end{aligned}
\tag{5.14}
$$

with γ for example the usual Lorentz warp term

$$
\gamma' \equiv 1 \Big/ \sqrt{\left(\frac{v_x^2}{c^2} - 1\right)}
\tag{5.15}
$$

But this is far too simplistic. In order to add Dirac spherical rotation as an element of the transformation for rotational parameters into HD one needs to apply additional superluminal Lorentz boost conditions and the XD supersymmetry conditions of noetic 12-space which is comprised of an energy dependent Minkowski spacetime present, $\widehat{M}_4$ derived from an extended HD Cramer type [12] transaction; model where this 'eternal' present is a virtual standing-wave of Wheeler-Feynman advanced-retarded future-past elements [11]. It is well known that superluminal Lorentz boosts may transform spatial dimensions into temporal dimensions [55,56]. The Noetic transform requires a double boost; the former and a second boost which transforms the complex spatial dimensions into the original 'Kaluza-Klein' concept of energy dimensions [61] equated here with the unitary field. For simplicity we initially consider just the initial superluminal Lorentz transformation (boost), $v_x = \infty$ only along the positive x direction where the space and time vectors in a real Minkowski space, $\widehat{M}_4$ cyclically transform as [55]

$$x' = +t, \; y' = -iy, \; z' = iz, \; t' = x \tag{5.16}$$

for real and imaginary parts separately, where x, y, z, t are real quantities for one frame, and x', y', z', t' are the real quantities in the second frame. For the initial 6D representation in complex Minkowski space, M_6 the above superluminal boost $(v_x = +\infty)$ becomes [55,56]

$$\begin{aligned} &x'_{\mathrm{Re}} + ix'_{\mathrm{Im}} = t_{x,\mathrm{Re}} + it_{x,\mathrm{Im}}, \quad y'_{\mathrm{Re}} + iy'_{\mathrm{Im}} = y_{\mathrm{Im}} - iy_{\mathrm{Re}}, \\ &z'_{\mathrm{Re}} + iz'_{\mathrm{Im}} = z_{\mathrm{Im}} - iz_{\mathrm{Re}}; \quad t'_{x,\mathrm{Re}} + it'_{x,\mathrm{Im}} = x_{\mathrm{Re}} + ix_{\mathrm{Im}}, \\ &t'_{y,\mathrm{Re}} + it'_{y,\mathrm{Im}} = t_{y,\mathrm{Im}} - it_{y,\mathrm{Re}}, \quad t'_{z,\mathrm{Re}} + it'_{z,\mathrm{Im}} = t_{z,\mathrm{Im}} - it_{z,\mathrm{Re}}. \end{aligned} \tag{5.17}$$

The points of interest are that a superluminal Lorentz boost cyclically transmutes spatial dimensionality into temporal dimensionality while also preserving the magnitude of the line element but not the sign:

$$-x'^{\mu} x'^{\nu} = x^{\mu} x^{\nu}, \tag{5.18}$$

where indices μ and ν run over 1,2,3,4 representing 1 as the time vector and 2,3,4 as spatial vectors with signature (+++-). Situation (5.16) must be carried out additionally for the y and z coordinates which we will discuss later on.

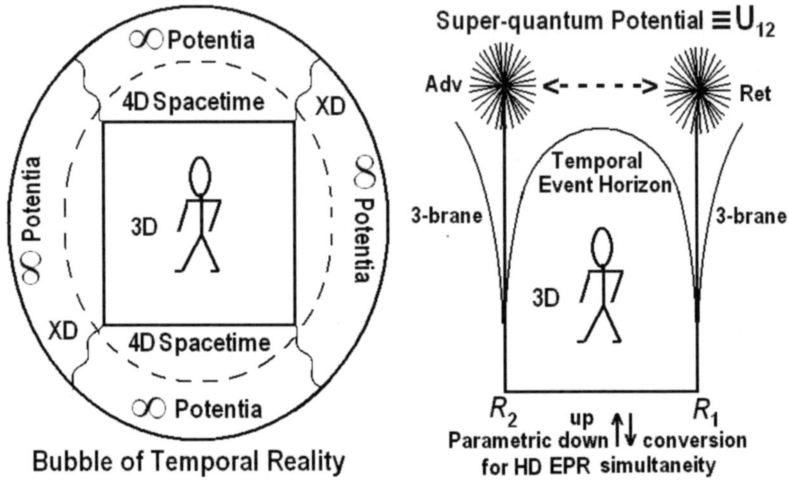

Figure 5.14 Observed reality is a temporal subspace of a 12D superspace potentia. The noetic transform is meant to upconvert and entangle the 3D observer with complete HD EPR correlations. 5.14a) and 5.14b) can be considered top views and side views respectively.

There are still a few things left to do; a second Lorentz boost to convert the first boosted $t_{x,\mathrm{Re}} + it_{x,\mathrm{Im}}$ dimensions to 'energy' dimensions is required to complete this component of the transformation such that $s \to t \to E$ which isn't so terrible if we remember that the original Kaluza-Klein model considered energy as the 5th dimension [61] or that in the usual signature +,+,+,- spatial dimensions could be considered as 'realized' or cut-offs of a topological field whereas the temporal dimension is a 'field' in flux. Of course here we consider this 'energy' regime as an indicator of a de Broglie-Bohm super-quantum potential [44] or in still higher dimensionality as the anthropic action principle. This additional HD domain is an essential part of the hysteresis loop of least-unit propagation as an inherent element of the 12D continuous-state. See Chaps. 3, 4 and 9.

§5.1 First boosted complex dimension $x'_{\mathrm{Re}} + ix'_{\mathrm{Im}} \rightarrow t_{x,\mathrm{Re}} + it_{x,\mathrm{Im}}$ is second boosted to an 8D hypersurface $E_{w,\mathrm{Re}} + iE_{w,\mathrm{Im}}$ where E_w is an HD Kaluza-Klein-like hypercube energy field coordinate. When the secondary boost is performed simultaneously along all 6 positive and negative axes including the temporal dimensions the temporal dimensions cancel and the 'attachment' of the observer couples to hypersphere coordinates.

Before elaborating on §5.1 we introduce another aspect of the transform. We have mentioned that in the noetic transformation it is necessary to double boost all three coordinates, *x, y, z* simultaneously such that $s \rightarrow t \rightarrow E$; this includes advanced-retarded future-past Cramer-like standing-wave parameters such that the double boost includes coordinates, $\pm x, y, z$ which could be considered action on a dual/mirror symmetry Calabi-Yau 3-form. To accomplish this task for a boost singling initially an arbitrary direction with velocity, $\vec{v}$ it is necessary to decompose the spatial vectors, $\vec{r}$ into perpendicular, $\vec{r}_\perp$ and parallel, $\vec{r}_\parallel$ components to the velocity vector, $\vec{v}$; then one may 'warp' only the $\vec{r}_\parallel$ component of $\vec{v}$ by the γ factor

$$t' = \gamma\left(t - \frac{\vec{r}\cdot\vec{v}}{c^2}\right) \qquad (5.19)$$

$$\vec{r}\,' = \vec{r}_\perp + \gamma\left(\vec{r}_\parallel - \vec{v}t\right)$$

where the γ factor then becomes

$$\gamma \equiv \frac{1}{\sqrt{1 - \vec{v}\cdot\vec{v}/c^2}} + \vec{r}\,' = \vec{r} + \left(\frac{\gamma - 1}{v^2}\left(\vec{r}\cdot\vec{v}\right) - \gamma t\right)\vec{v}\,. \qquad (5.20)$$

This allows the perpendicular components, $\vec{r}_\perp$ to remain stationery or coupled to the original local position of the observer. This separation of vectors allows an easier description for the implementation of § 5.1 for an Ising model rotation of $\vec{r}_\parallel$ by a $0 \rightarrow \infty$ rotation of the Riemann sphere while the $\vec{r}_\perp$ components remain coupled to the stationery Euclidean regime.

The two-stage triple coordinate boost of the noetic transformation rotates the Euclidean space through two sets of three mutually orthogonal complex planes (i.e. the future-past advanced-retarded coordinates). Each coordinate is simultaneously orthorotated to HD in a process that violates the Copenhagen regime quantum uncertainty principle and the usual associated causal conditions. This is required to uncouple the observer from the Euclidean perspective in order to recouple ontologically to the HD perspective.

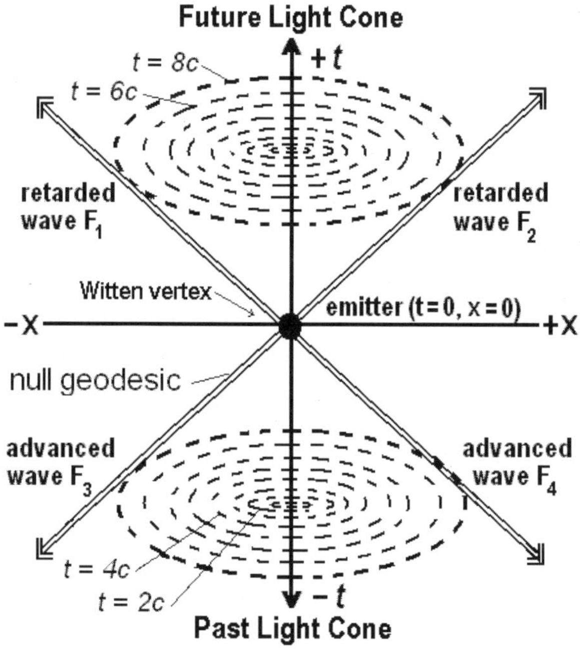

Figure 5.15 Adaptation of a complex Minkowski light-cone showing advanced-retarded future-past Cramer wavefront transactions with a central Witten Ising lattice string vertex able to undergo symmetry transformations.

Four solutions emerge: two retarded (F_1 and F_2) connecting processes in the forward light cone and two advanced, (F_3 and F_4) connecting processes in the backward slight cone [12]. These four solutions are

$$\begin{aligned} F_1 &= F_0 e^{-i(-kx-\omega t)}, \quad F_2 = F_0 e^{i(kx-\omega t)}; \\ F_3 &= F_0 e^{i(-kx+\omega t)}, \quad F_4 = e^{i(kx+\omega t)} \end{aligned} \tag{5.21}$$

with F_1 for a wave moving in the (-x, + t) direction, F_2 is for a (+x, +t) moving wave, F_3 is for a (-x, -t) moving wave, and F_4 is a (+x, -t) moving wave. F_1 and F_4 are complex conjugates of each other and F_2 and F_3, are complex conjugates, so that $F_1^+ = F_4$ and $F_2^+ = F_3$. Then the usual solutions to Maxwell's equations are retarded plane wave solutions.

In Fig. 5.16 the twelve points labeled C_4 symbolize a conceptualization of the twelve dimensions comprising a fundamental least unit. The complex plane is suppressed for simplicity. Counterpropagating, complex, future-past, 'hyper-Geon' elements act in concert to 'create' instantaneous harmonic elements of localized Euclidean 3-sphere extension. They are 'standing wave' relational spacetime extensions R(t) of the absolute 12D hyperspace that form the fundamental basis of observational reality representing a metric framework for events and interactions. Extension is mediated by the noumenal action principle of the unified field by $F_n = E_n / R_{(t)}$, where E_n is energy of the unified field (see Chap. 4).

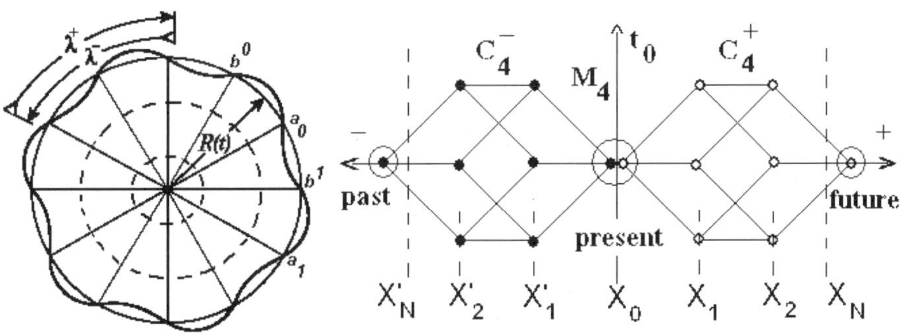

Figure 5.16 Another conceptual view of the symmetry of a least-unit in Noetic Superspace. a) 2D standing wave. b) 12D relationship depicted as points. The 12 $\pm C_4$ points represent future-past potenia for a single M_4 point, X_0 the cyclic continuous iteration of which becomes the locus of points for the arrow of time t_0. The larger center circle represents a Minkowski, M_4 present comprised of the smaller circles at each end representing future/past components that comprise it.

We begin discussion of the actual operation of the new transform by introducing the concept of planes or surfaces of constant phase which we hope to eventually correlate with the equilibrium regions on the genus-1 helicoid parking garage in Fig. 11.7. But the starting configuration is shown in Fig. 5.17 below where k is the propagation vector for a plane wave along the z axis and the magnitude of k is the wavenumber

$$\psi \propto e^{i\phi}, \ \phi = k \cdot r - \omega t \qquad (5.22)$$

where $\omega = \omega(k)$ is the dispersion relation and $\lambda = 2\pi / k$ the wavelength. The positions of r at time, t where the phase, ϕ has a fixed value defines the planes of equal phase perpendicular to the propagation vector, k [66].

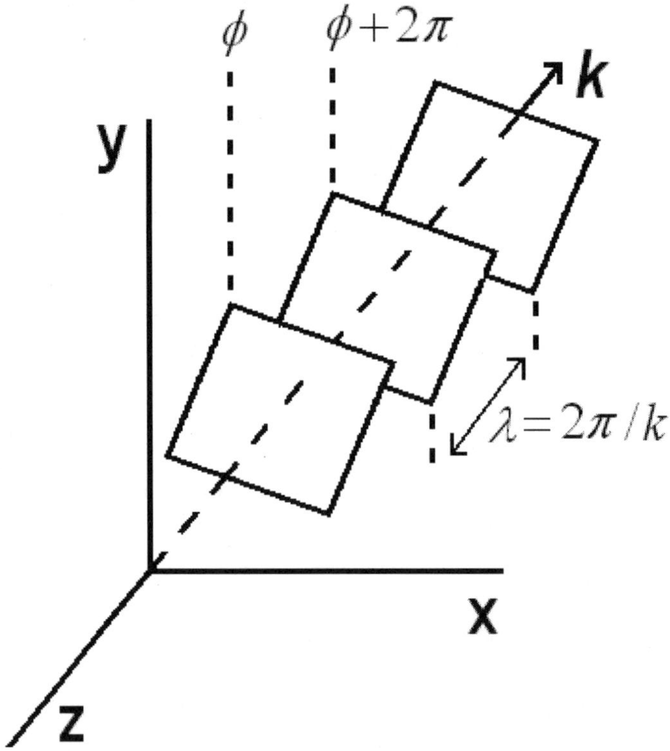

Figure 5.17 Planes or surfaces of constant phase, ϕ along the z axis for the plane wave propagation vector k.

Next we want to associate these planes of equal phase with spherical sectors (Fig. 5.18 below) of the close-packed Riemann sphere least-units tiling the spacetime backcloth and then apply this to the harmonic oscillator properties of the future-past advanced-retarded standing-wave properties of a present instant. A spherical sector is generally formed by rotating a section of a circle about diameter where the volume of the spherical sector would be $V = (D^2 h)/2$ which can be considered to be a torus similar to the energy levels of a harmonic oscillator.

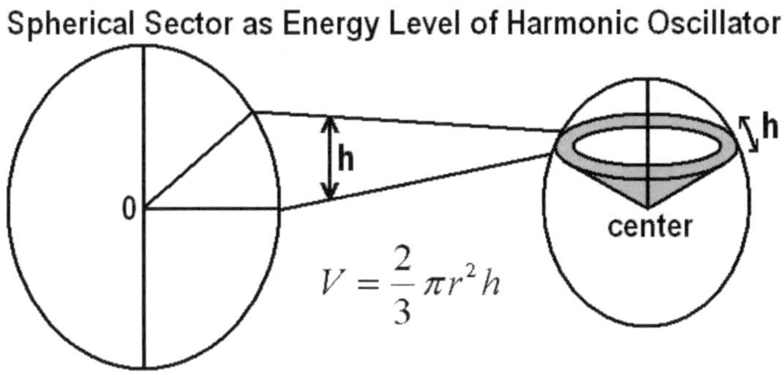

Figure 5.18 Spherical sector, a volume formed by rotating a section of a circle around a diameter.

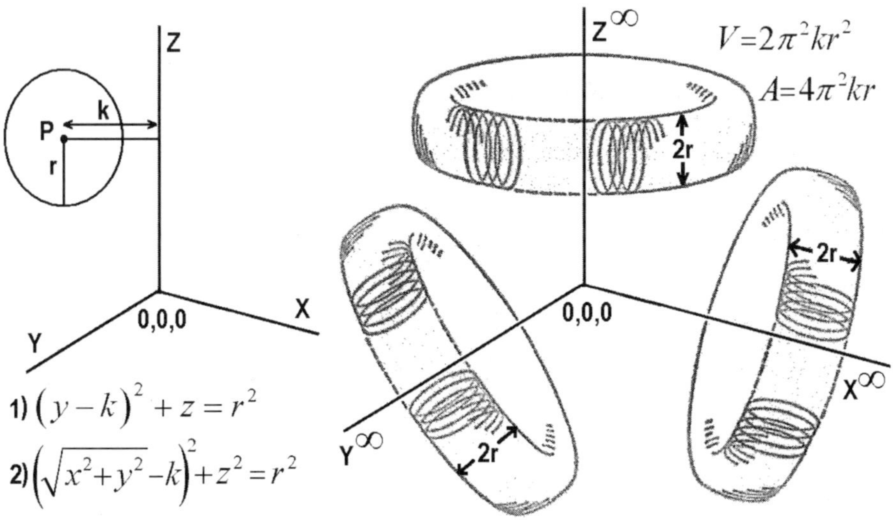

Figure 5.19 a) Generator of a toroidal surface formed by rotating a circle, P of radius, r at distance, k from the rotation axis, z in the plane of the circle. b) Spherical sectors from Fig. 5.16 modeled to form tori on the *x,y,z* axes to represent regions of equal phase.

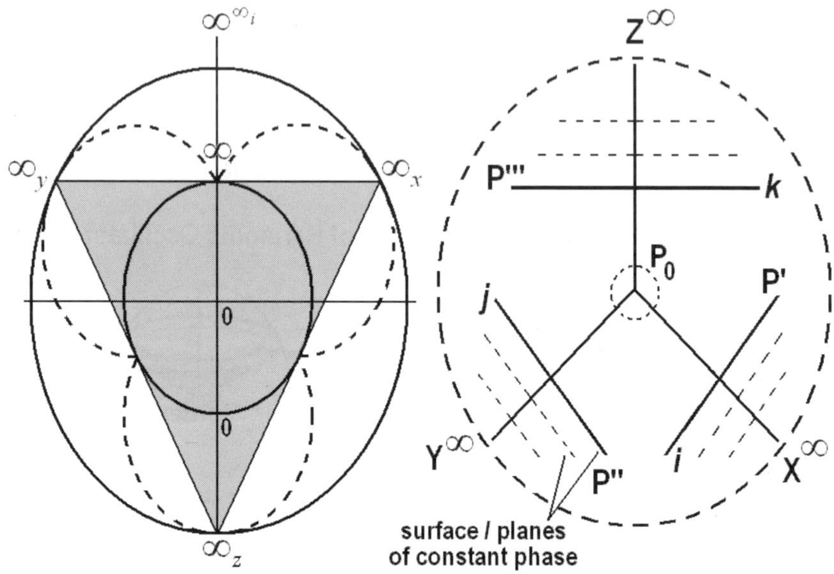

Figure 5.20 a) Riemann sphere representation of subelements of toroidal phase. b) Combining symbolism of Figs. 5.16 and 5.17 to form hierarchical model of toroidal planes of equal phase.

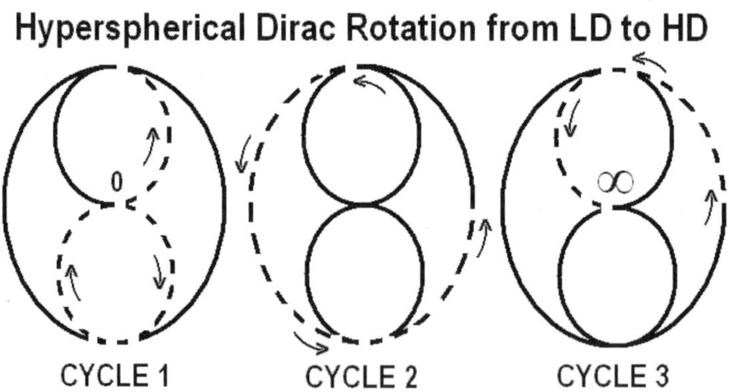

Figure 5.21 Dirac hyperspherical rotation of the Riemann sphere from Euclidean 3-space to hyperspace during the 360-720° rotation process.

These additional figures above and below are meant to help illustrate some additional conceptual geometrical and topological components needed to develop the noetic transform.

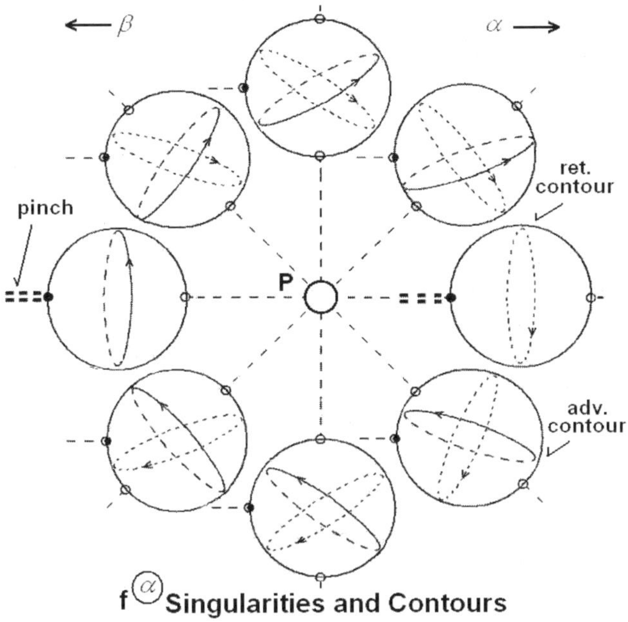

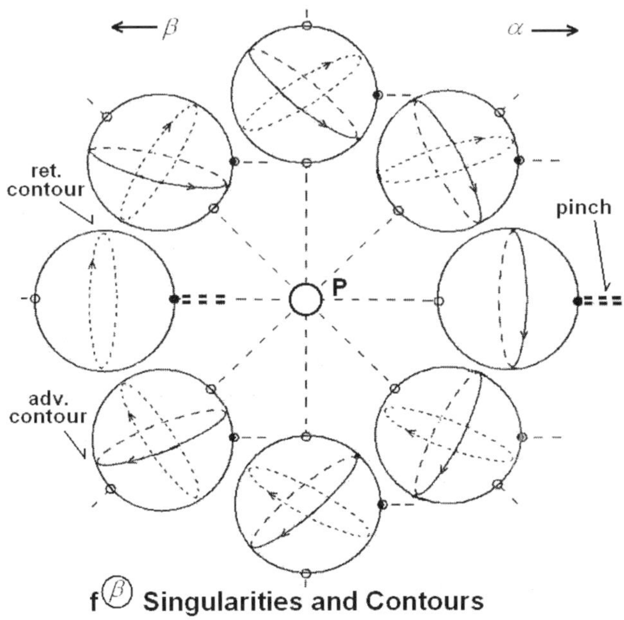

Figure 5.22 Singularities and advanced-retarded phase contours for the Dirac rotation to reflect duality/Mirror symmetry. Figure adapted from [67].

Referring back earlier in the section to equations 5.19 and 5.20 these later figures are meant to graphically illustrate the action of the basic element of the noetic transform where unlike the Galilean and Lorentz/Poincaré transformations the action is both stationary and along all axes simultaneously. Figure 5.21 is simplistic rendition of the contours and pinches in Fig. 5.22. This is the Dirac spherical rotation originally for the spinor rotation of the electron through 720° which HAM cosmology suggests is also a conformal scale-invariant parameter of the universe itself. It is easy to see the similarity of Fig. 5.23 to Fig. 5.21; the additional factor is that the topology of the inner core rotates to the outer skin in a continuous-state manner. Figures 5.18-5.20 are meant to illustrate (in the context of the other figures here) the additional requirements for boosting and transforming the coordinates, $s \to t \to E$.

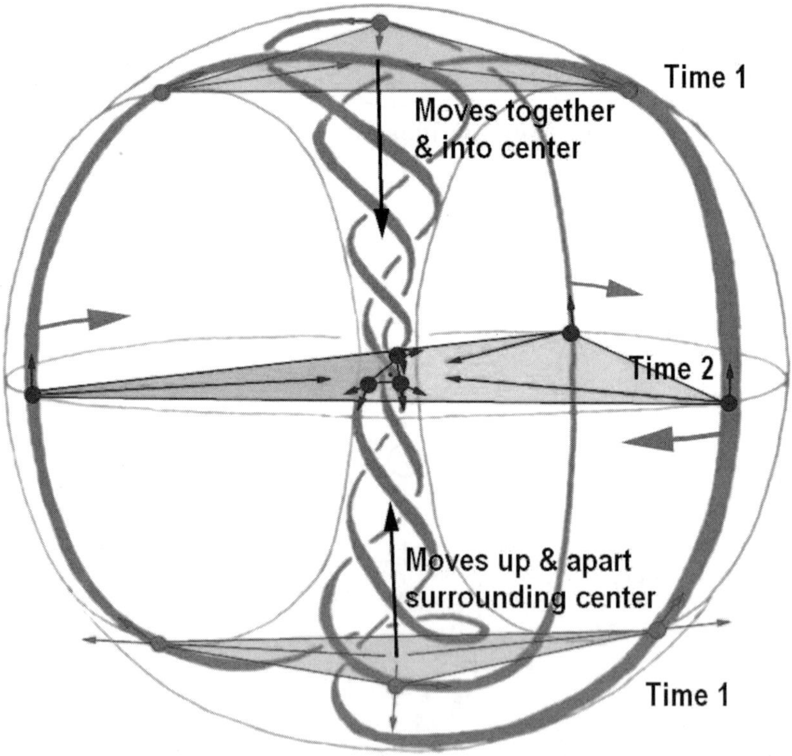

Figure 5.23 Symbolic of how transformation of inner Euclidean coordinated rotate continuously to become external HD elements. Figure adapted from [68].

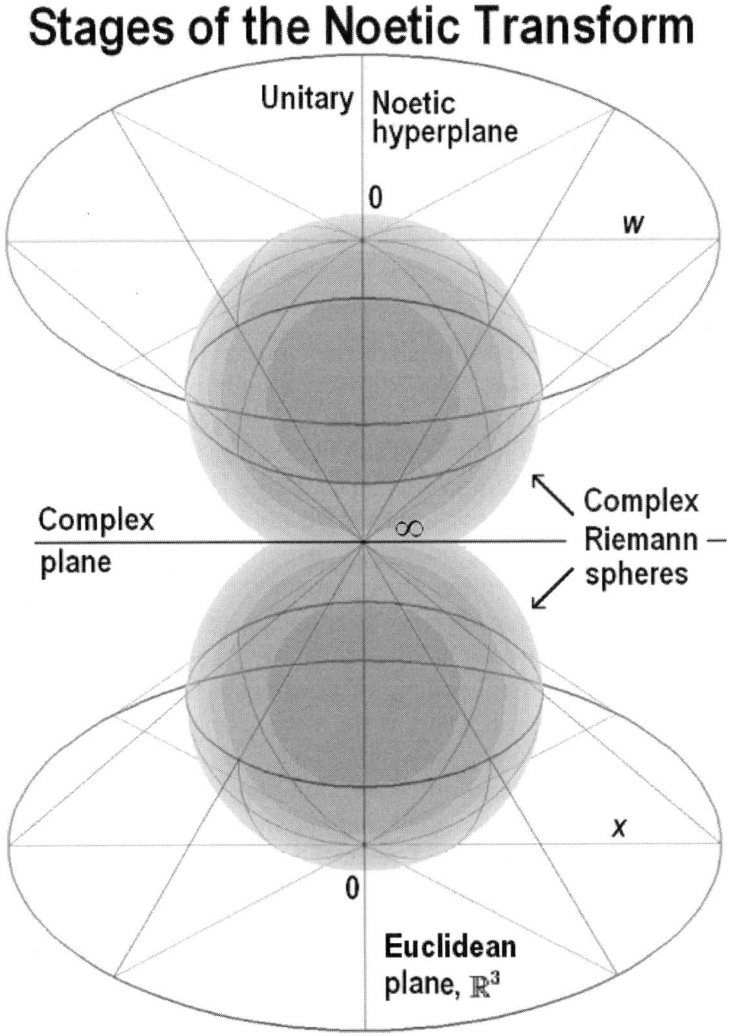

Figure 5.24 Conceptualization of the basic rotational element of the static noetic transform which performs parametric up-down conversions between Euclidean space and the unitary HD hyperplane by a form of Dirac spherical rotation that does not return to the original position after a 720° rotation but instead uncouples the energy-information content of the internal hysteresis loop and recouples the awareness of the observer or exchanges information with the HD hyperplanes.

Figure 5.24 above represents a final conceptualization as far as our thinking enables us to go at the time of this writing. Simply for the stationery Euclidean observer at point, 0 the first set of rotations and boosts complexify the space to an intermediate domain. Finally a second set of boosts either recouples the observer to the HD unitary regime or if operated in reverse allows the HD noetic information to enter into the observers awareness in the Euclidean ground state.

5.11 Final Remarks

We were not able to finish formalizing the noetic transform at the time of this writing; however one may know with relative precision the local coordinates for an observer in 3-space but those details are somewhat irrelevant for the critical details of the noetic transformation because one wants to access information in HD. If the HD regime is actually holographic and therefore the information ubiquitous, that is the same as being nowhere because one does not currently know how to address an infinite number of possibilities in a manner that automatically finds the one. In 3-space the unique signifier of an object is simply position; but in a holographic HD what does position mean in Einstein's relativistic sense of riding a photon and being everywhere at once. How does one plumb a point in that infinite ocean? This is the way revelation, telepathy or *déjà vu* works. But it is currently no trivial manner to mechanically create the simultaneity by parametric up-conversion that is so easily created by parametric down-conversion. Each person can be uniquely identified by DNA. Is there some unique HD resonance mode applicably oriented to the local position of an observer by some form of unique resonance? In theology a persons 'elemental intelligence is coeternal with God' and 'each person is given a bound (boundary conditions for the soul) or they cannot exist' [34]. But where in the wide holographic multiverse is this? How could it be accessed? MacKinnon [69] has given an idea that may prove helpful; he says a stationery de Broglie wave packet in some specific location is a result of a focus of the de Broglie waves of all observers in the whole universe. At our current level of understanding we believe only research using the special class of quantum computer postulated to model the mind-body interface as a naturally occurring form dubbed the 'conscious quantum computer' would fulfill the requirements for a test platform to discover whatever the unique parameters of this resonance mode is. A person skilled in meditation or a 'psychic' receives information spontaneously, but what is the coupling process.

Secondly what conditions does string theory impose on the process. String theory states that 'matter resides on the 3-brane and gravity is free to pass between'. With duality/mirror symmetry this would take care of 6D. There are three time dimensions in HAM cosmology leaving three for the quantum or anthropic potential for a total of 12D. HAM cosmology equates gravitons with the unitary field of anthropic information, but this doesn't yet help us either. Where do we draw the line? Einstein said 'if one could ride a photon, one could circumnavigate the universe without the passage of time'. Time is removed in the resultant of the noetic transform. What is a field? In 3-space a field has a coordinate representation; but how can this be described in a holographic arena of infinite potentia where 'something' is everywhere and nowhere? In this domain of ubiquity the railroad tracks do not recede because of the atemporality; this is like viewing the inside and outside rotations of Fig. 5.23 simultaneously. So for the moment we can only go as far as intermediate Fig. 5.24 above illustrates: stationery simultaneous coordinate boosts from E^3 to W^8.

We are 'flatlanders' [70] to the complete nature of the multiverse. This is the great separator of the observers awareness from the infinite *potentia* of noumenal reality of which the temporal reality of everyday existence is only a limited subspace. This entails a subtractive process that produces the arrow of time (see Chap. 4). One could say a stroboscopic beat frequency of the holophote action of the unitary field subtracts out the additional HD parameters because this mirror symmetric wave-particle duality of the topology of reality is a form of HD harmonic oscillator with nodes of constructive and destructive interference. Plato's cave again.

We are comprised of the 'matter', which is actually a resonant array of domain walls comprised mostly of empty space, projected from this fundamental absolute absolute space or potentia; and our consciousness is likewise coupled to that limited lower part of it, and imbedded in this same material. It is generally known that the standard models of quantum theory and cosmology do not include consciousness or give an adequate description of the nature of time; suggesting that the elucidation of these ideas must come from extended theoretical insights; perhaps like those offered here. Human perception is indicative of a flow of time - from past, to present to future in accordance with the 2nd law of thermodynamics by appropriate changes in entropy of the system observed as it undergoes evolution. Thus the observed temporal order seems related to entropic order; and these dynamics constitute how we perceive 'action' or translation in this particular dimension, the dimension of time. It is a stretch; but probably the whole thermodynamic-entropic process is ultimately just a razzmatazz for the myopic virtual reality of the observer. Believe it or not.

Is there a force or inherent 'action' in the cosmology of mind that couples awareness to the entropic activity observed in the perceived external reality that we equate with the flow of time? Do we 'ingest' time parameters through an '*axis mundi*' into our psyche during transit through the cosmological sea as a whale ingests plankton while swimming in the ocean? If it is true, that the entire perception of time is a creation of a normative human ontology through an innate or habituated philosophic tension [71] that drives the orientation of our mind, a decoupling from the flux of this noetic field would allow a reorientation of our 'psychosphere' as suggested by Plato [10] and the possibility to re-tune the perception of our psyche to additional or alternative parameters of entropic 'action', time or atemporality. This is a key point in understanding the nature of time. Philosophical tension [71] that couples consciousness to the 'physicality' the entity is imbedded in with specific fine-tuned laws for our Hubble sphere in a holographic multiverse with the potential for an infinite number of nested Hubble spheres. See Chap. 13. The action of intentionality modulates the spin alignment angular momentum of the noetic field. More in the future…

References

[1] Einstein, A. (1950) The Meaning of Relativity, Princeton: Princeton Univ. Press.
[2] Bohr, N. (1935) Can quantum mechanical description of physical reality be considered complete? Physical Rev, 48; 696-702.
[3] Amoroso, R.L & Pribram, KH (2009) The Complementarity of Mind and Body: Realizing the Dream of Descartes, Einstein and Eccles, Cambridge: MIT University Press.
[4] Dirac, P.A.M. (1952) Nature (London) 169, 702.
[5] Cufaro, N., Petroni, C. & Vigier, J-P (1983) Dirac's Ether in Relativistic Quantum Mechanics, Found. Phys., 13, 253-286.
[6] Burns, J.E. (1998) Entropy and vacuum radiation, Found. Phys. Lett. 28:7; 1191-1207.

[7] Burns, J.E. (2002) Vacuum radiation, entropy and the arrow of time, in R.L. Amoroso, G. Hunter, S. Jeffers & M. Kafatos, (eds.) Gravitation & Cosmology: From the Hubble Radius to the Planck Scale, Dordrecht: Kluwer Academic.
[8] Morris, R. (1984) Time's Arrows, New York: Simon & Schuster..
[9] Davies, P. (1974) Physics of Time Asymmetry, Berkeley: University of California Press.
[10] Amoroso, R.L. (2000) The parameters of temporal correspondence in a continuous-state conscious universe, in L. Buccheri & M. Saniga (eds.) Studies on the Structure of Time: From Physics to Psycho(patho)logy, London: Plenum.
[11] Wheeler, J.A. & Feynman, R. (1945) Rev. Mod. Physics, 17, 157;
[12] Cramer, J.G. (1986) Transactional interpretation of quantum mechanics, Rev. of Modern Physics, 58:3; 647-687.
[13] Amoroso, R.L. (2002) The continuous state universe, in R.L. Amoroso, G. Hunter, S. Jeffers & M. Kafatos, (eds.) Gravitation & Cosmology: From the Hubble Radius to the Planck Scale, Dordrecht: Kluwer Academic.
[14] Amoroso, R.L. (2005) Paradigm for the cosmology of a continuous-state holographic conscious multiverse, in R.L. Amoroso, B. Lehnert & J-P Vigier (eds.) Beyond the Standard Model: Searching for Unity in Physics, Oakland: The Noetic Press.
[15] Chu, M-Y & Amoroso, R.L. (2008) Empirical mediation of the primary mechanism initiating protein conformation in prion propagation, in D. Dubois (ed.) Proceedings of CASYS07, IJCAS, vol. 22, pp. 292-313, Liege, Belgium.
[16] Wheeler, J.A. (1955) Geons, Physical Review, 97:2, 511-536.
[17] Susskind, L. (1994) The World as a Hologram, arXiv:hep-th/9409089 v2.
[18] Bousso, R. (2002) The holographic principle, Reviews Modern Physics 74:825-874; arXiv:hep-th/0203101.
[19] Amoroso, R.L & Amoroso, P.J. (2004) The fundamental limit and origin of complexity in biological systems: A new model for the origin of life, in D. Dubois, (ed.) Computing Anticipatory Systems, Proceedings of CASYS04, 6th Int, Conf., Liege, Belgium, 11-16 Aug, 2003, AIP Proceedings Vol. 718, pp. 144-159.
[20] Franck, G. (2000) Time and presence, in Science and The Primacy of Consciousness, R.L. Amoroso et al, (eds.) Orinda: Noetic Press.
[21] Chalmers, D. (1996) The Conscious Mind, Oxford: Oxford:

University Press.
[22] Amoroso, R.L. & Martin, B. (1995) Modeling the Heisenberg matrix: Quantum coherence and thought at the holoscape manifold and deeper complementarity, in J. King & K. Pribram (eds.) Scale in Conscious Experience, Mahwah: Lawrence Earlbaum.
[23] Bergson, H. (1911) Creative Evolution, A. Mitchell (trans.) New York: Henry Holt & Co.
[24] Amoroso, R.L. (1999) An Introduction to Noetic Field Theory: The Quantization of Mind, Noetic Journal, Vol 2, No. 1, pp. 28-37.
[25] Amoroso, R.L. (1997) Consciousness, a radical definition, Noetic Journal, Vol.1 No.1, p.19-27.
[26] Amoroso, R.L. (1997) A Brief Introduction to Noetic Field Theory: The Quantization of Mind, pp. 297-302, In Brain & Consciousness, L. Rakic, G. Kostopoulos, D. Rakovic & D. Koruga (eds.) ECPD: Belgrade.
[27] Amoroso, R.L. (1997), Engineering a conscious quantum computer, In M. Gams, et al (eds.) Mind Versus Computer, IOS Press.
[28] Amoroso, R.L. (1996) The production of Fröhlich and Bose-Einstein coherent states in in vitro paracrystaline oligomers using laser phase control interferometry, Bioelectrochemistry & Bioenergetics, 40: 39-42.
[29] Amoroso, R.L. (1995) The extracellular containment of natural intelligence, Informatica, 19: 585-90.
[30] Amoroso, R.L. (2002) The Physical Basis of Consciousness: A Fundamental Formalism, Part 1 Noesis, XXVI, Romanian Academy.
[31] Amoroso, R.L. (2000) Consciousness, a radical definition: Substance dualism solves the hard problem, In Amoroso, R.L., Antunes, R., Coelho, C., Farias, M., Leite, A., & Soares, P. (eds.) Science and the Primacy of Consciousness, Orinda: The Noetic Press.
[32] Amoroso, R.L (2000) Derivation of the fundamental equation of consciousness, Part I, Boundary conditions, Noetic J, 3:1, pp. 91-99.
[33] Einstein, A. (1954) Ideas and Opinions, New York: Crown.
[34] Smith, J. (1989) The Doctrine and Covenants, Salt Lake City: The Church of Jesus Christ od Latter-day Saints.
[35] Zeh, H-D (1989) The Physical Basis of the Direction of the Arrow of Time, New York: Springer.
[36] Misner, C.W., Thorne, K.S. & Wheeler, J.A.(1973) Gravitation, San Francisco: W.H. Freeman.
[37] Kafatos, M., Roy, S. & Amoroso, R.L. (2000) Scaling in cosmology and the arrow of time, In R. Buccheri & M. Saniga, (eds.) Studies on the Structure of Time: From Physics to Psych(patho)logy, Dordrecht: Kluwer.

[38] Wheeler, J.A. (1977) Gravitational and Electromagnetic wave flux compared and contrasted, Phys. Rev. D, 16:12, 3384-3389.
[39] Wheeler, J.A. (1962) Geometrodynamics, N.Y.:Academic Press.
[40] Prigogine, I. (1997) From Poincaré's divergences to quantum mechanics with broken time symmetry, *Zeitschrift für Naturforschung* 52a, 37-47.
[41] Petrosky, T. & Rosenberg, M. (1997) Microscopic non-equilibrium structure and dynamical model of entropy flow, Foundations of Physics 27: 2; 239-259.
[42] Puthoff, H.E. (1989) Source of vacuum electromagnetic zero-point energy, Phys. Rev. A 40(9), 4857-4862.
[43] Puthoff, H.E. (1991) Reply to "Comment on 'Source of vacuum electromagnetic zero-point energy'", Phys. Rev. A 44(5), 3385-3386.
[44] Holland, P.R. (1995) The Quantum Theory of Motion: An Account of the de Broglie-Bohm Causal Interpretation of Quantum Mechanics, Cambridge: Cambridge Univ. Press.
[45] Cartan, E. (1966) The Theory of Spinors, New York: Dover.
[46] Cartan, E. (1979) Elie Cartan - Albert Einstein Letters on Absolute Parallelism 1929-1932, R. Debever (ed.) Princeton: Princeton Univ. Press.
[47] Schutz, B.F. (1980) Geometrical Methods of Mathematical Physics, Cambridge: Cambridge Univ. Press.
[48] Bilson-Thompson, S., Hackett, J. & Kauffman, L.H. (2009) Particle topology, braids and braided belts, arXiv:0903.1376v1.
[49] Bilson-Thompson, S.O. A topological model of composite preons, arXiv:hep-th/0503213.
[50] Bilson-Thompson, S.O., Markopoulou, F. & Smolin, L. (2007) Quantum gravity and the standard model, Class. Quant. Gravity 24; 3975-3993; arXiv:hep-th/0603022.
[51] Magnus, W (1974) Braid groups: A survey, In Lecture Notes in Mathematics, vol. 372, pp. 463-487; or (1974) Proceedings of the 2nd Intl Conf. on the Theory of Groups, Canberra, Australia, 1973, New York: Springer.
[52] Personal communication, Lou H. Kaufman (2009).
[53] Kaufman, L.H. (2001) Knots and Physics, 3rd ed., Singapore, World Scientific.
[54] Vigier, J-P (2000) selected papers, in S. Jeffers, B. Lehnert, N. Abramson, & L. Chebotarev (eds.) Jean-Pierre Vigier and the Stochastic Interpretation of Quantum Mechanics, Montreal: Aperion.

[55] Rauscher, E.A. (1993) Electromagnetic Phenomena in Complex Geometries and Nonlinear Phenomena, Non-Hertzian Waves and Magnetic Monopoles, Millbrae: Tesla Books.
[56] Rauscher, E.A. (2002) Non-Abelian guage groups for real & complex Maxwell's equations, in R.L. Amoroso, G. Hunter, S. Jeffers & M. Kafatos, (eds.) Gravitation & Cosmology: From the Hubble Radius to the Planck Scale, Dordrecht: Kluwer.
[57] Cole, E.A.B. (1977) Il Nuovo Cimento, 40:2, 171-180.
[58] Hansen, R.O. & Newman, E.T. (1975) Gen. Rel. and Grav. 6, 216.
[59] Amoroso, R.L., Kafatos, M. & Ecimovic, P. (1998) The origin of cosmological redshift in spin exchange vacuum compactification and nonzero rest mass photon anisotrophy, in G. Hunter, S. Jeffers & J-P Vigier (eds.) Causality and Locality in Modern Physics, Dordrecht: Kluwer.
[60] Atchison, I. & Hey, A. (1982). Gauge Theories in Particle Physics, Bristol: Adam Hilger.
[61] Overduin, J.M. & Wesson, P.S. (1997) Kaluza-Klein gravity, Physics Reports, 283, pp. 303-378.
[62] Goswami, A. (1995) The Self-Aware Universe, New York: Putnam.
[63] Marshall, T. (2002) Engineering the vacuum, in R.L. Amoroso, G. Hunter, S. Jeffers & M. Kafatos, (eds.), Gravitation & Cosmology: From the Hubble Radius to the Planck Scale, pp. 459-468, Dordrecht: Kluwer.
[64] Lorentz, H.A. (1899) Simplified theory of electrical and optical phenomena in moving systems, Proc. Acad. Sci. Amsterdam I, 427-443.
[65] Lorentz, H.A. (1904) Electromagnetic phenomena in a system moving with any velocity less than that of light, Proc. Acad. Science Amsterdam IV, 669–678.
[66] Peebles, P.J.E. (1992) Quantum Mechanics, Princeton: Princeton Univ, Press.
[67] Penrose, R, (1982) Real-space topology for advanced/retarded twistor functions, Twistor Newsletter, No. 14, pp. 16-18.
[68] The Meru Foundation, Stan & Lavanah Tenen, www.meru.org/.
[69] MacKinnon, L. (1979) The de Broglie wave packet for a simple stationery state, Foundations of Physics, 9:9-10; 787-791.
[70] Abbot, E.A. (1983) Flatland, New York: Harper Perennial.
[71] Drãgãnescu, M.C. (1991) The Philosophical tension and the cosmic feeling, Bucharest: The Romanian Academy.

Chapter 6

Integration of Gravity & Electromagnetism in Terms of a Dirac Polarized Vacuum

Conventionally Maxwell's equations describe transverse elements described as 'EM' waves; but by utilizing the Einstein/de Broglie relations one can derive additional degrees of freedom so that Maxwell's equations are not 'cut off' at the vacuum. Therefore one must employ the $\mu\nu$ fields in addition to the standard EM suggesting also that the photon is piloted. The two sets of coordinates for the EM or $\mu\nu$ fields are mutually exclusive and generally considered to be independent of each other. In this work a method is developed for integrating them in terms of a Dirac covariant polarized vacuum and extended theoretical perspectives.

6.1 Introduction to Fixing the G/EM Framework

The integration of Gravity and electromagnetism (EM) has been one of the holy grails of physics for the last century. In this chapter Gravity and EM are unified in terms of the covariant density distribution of a real average covariant Dirac vacuum built with extended random elements filling flat space-time. Although the Newton and Coulomb potentials have similar forms, the two theories have developed separately leaving their unification an unsolved problem throughout the history of Modern Science. In the past most attempts at unification have been within a frame associating electromagnetism with new geometrical properties of spacetime [1-3]. The approach of this integration is different. Following Puthoff and others [4-7], both fields are represented by four-vector field densities, A_μ; where one considers both types of phenomena as different types of motions within the same real physical zero-point field in a flat spacetime, i.e. as two different vacuum types of collective perturbations carried by a single vacuum field moving in such a space. Our hope is that since this approach suggests new types of experimentation and new interpretations of unexplained effects it could, if confirmed, help to disentangle the present theoretical discussion.

The basis of this model is as follows:

A) The first basis is observational. The universe apparently does not change with distance [8-10] (as it would for Big-Bang type theories). This leads to the possibility of a non-Doppler redshift [11] (which suggests a non-zero photon mass, $m_\gamma \neq 0$) with the velocity of light isotropic in an absolute inertial frame, I_0 , in time.

B) The second basis is that our essential instrument for distance observation (i.e. electromagnetic waves) is more complex than initially thought. De Broglie and Einstein demonstrated that $E = h\nu = \mathrm{mc}^2$, with $m = m_0 . \left(1 - v^2/c^2\right)^{-1/2}$ so that individual massive photon's could be considered as piloted by real non zero-mass Maxwellian waves allowing the electromagnetic field to be represented by a vector density, A_μ. As shown by the Aharonov-Bohm effect, this implies that the EM field is not completely represented by the $\mu\nu$ fields [12].

Maxwell's equations [3] conventionally describe Transverse elements denoted as 'EM' waves; by utilizing the Einstein / de Broglie relation one may derive additional degrees of freedom such that Maxwell's equations are not 'cut off' at the vacuum, but lead to Longitudinal wave components and non-zero electric conductivity of the vacuum. Thus our distinct need for the utility of the $\mu\nu$ fields instead of just the standard 'EM'. This also suggests that the photon is 'piloted'. One must 'fix' the coordinates of either the EM field or the $\mu\nu$ field we have chosen the latter. It should be noted that while c is constant in the rest frame and the velocity of massive photons would be frequency dependent; there is no contradiction because as Dirac himself stated according to coordinate law the pilot wave and the photon decouples [13]. The two sets of coordinates, EM or $\mu\nu$ are mutually exclusive and would generally be independent. In this work a method is developed for integrating them.

It is well known that the usual form of Maxwell's equations in vacuum (describing zero mass photons) possess infinite families of boundary free exact solutions with Longitudinal electric or magnetic fields; this is the usual $\mu\nu$ theory where $B^{(3)} = 0$ and photon mass, $m_\gamma = 0$. This is also true for the vector potential in the Lorentz gauge according to the equation, $A_\mu = 0$. But of interest to the task here, for massive photons there is only one family and one set of boundary conditions!

C) The third basis has its theoretical origin in the introduction by Dirac et al. of a real covariant chaotic physical aether which fills space-time, carries real physical observable wave-like and particle like (soliton-like) perturbations or local extended elements, whose four momenta and angular momenta are statistically and evenly distributed on specific hyperbolic surfaces, at each given point, in all given inertial frames. This vacuum distribution thus appears, as invariant isotropic chaotic and undetectable (except in specific physical cases) for all inertial observers. The form taken by an aether within Relativity Theory carrying both particles and waves is now described in terms of collective motions on the top of a real essentially stochastic covariant background. Such an aether theoretically justifies the statistical productions of Quantum Mechanics (in its causal stochastic interpretation) and SED theory, and has a direct experimental justification in the Casimir effect. This implies a background friction (associated with absolute local conservation of total momentum and angular momentum) and collective motions which provide a new interpretation of the observed cosmological red-shift [11,14] and yields new possibilities to interpret (also in terms of local frictions) the anomalous red-shifts observed by Arp, Tifft and other astronomers [15].

From these bases, section 6.3, describes the gravitational results of General Relativity in Maxwellian terms. Section 6.4 develops a possible unification model of both theories. Section 6.5 briefly discusses possible consequences of the preceding attempt. This aether is locally defined by a particular real Poincaré frame, I_O, in which (measured with real physical instruments) the velocity of light is identical in all directions at all observable frequencies. All observers tied to other frames passing through local inertial motions will see (measure) different space-time properties (associated with their velocity and orientations) defined by the corresponding Poincaré transformations.[1] Local variations of physical properties of the aether correspond to local transitions relating differential inertial frames at neighboring points.

6.2 Flat Spacetime and a Real Physical Aether

This model depends on the existence of a real physical vacuum (or zero point field) built with extended wave-like individual elements [16,17] centered on points in an external flat space-time, where such elements can overlap and interact (i.e. carry) collective motions corresponding to excess (electromagnetic 'bumps') or defects (gravitational 'holes') in the average density of the local aether elements. The model could be described as a gas of extended elements within flat space-time. These elements can interact locally (i.e. carry collective motions) and the gas' local scalar density thus carries waves (and solitons) associated with excess (electromagnetic) or defects (gravitational) in density, with respect to the average local vacuum density. One thus defines field variables associated with these two possible (excess or defect) local density variations. The vector fields, for example, in this paper, represent localized excess or density defects with respect to the local vacuum density. This model thus implies:

- A description of real physical vacuum properties in terms of real extended vacuum elements average behavior.
- A description of the behavior of its collective excess (above average) associated with recently observed electromagnetic effects.
- A description of the behavior of its collective defects (below average) associated with observed gravitational effects.

Introducing these new concepts into Maxwell's equations and the description of gravitational fields along the same lines (in terms of vector fields, A_μ) suggests a new type of unification of both theories. Instead of looking for a common geometrization of gravity and light (i.e. their

---

[1] To quote Kholmetsky "In order to pass from one arbitrary inertial frame I_1 to another one I_2 it is necessary to carry out the transformation from I_1 to the absolute frames I_0 and then from I_0 to I_2" [18].

unification within a unique form of extended space-time geometry) one could assume the following from Newton and Lorentz :

A) The evolution of extended (fields) and of localized (sources) in terms of 1) vacuum (aether) 2) gravitational fields, 3) the electromagnetic field, reflects the time evolution (motions) and interactions of perturbations of a real material substance moving in a 3-dimensional flat space. This means that all three field and particle sub-elements are localized at given points, at each instant, in this 3-space and move continuously (i.e. locally transform) according to causal laws[2].

This assumption (distinction of space and fields) is now supported by the existence of a special particular experimental inertial cosmological frame I_0 in which

- the 2.7°K microwave radiation frame is isotropic and non rotating.
- The average distribution of different types of galaxies (spiral, elliptical, QSO's) is isotropic not changing with distance [15].
- The observable anisotropy of the velocity of light propagation in different directions and around massive objects reflects the real motions of real fields described with respect to the I_0 frame in any real inertial Poincaré frame by covariant (local) four-vector scalar chaotic average density $\rho(x_\mu)$ around each absolute space-time point x_μ in I_0 i.e. by average four-vectors $A_\mu^0(x_\alpha)$ where the $({}_0)$ denotes average measures taken in I_0.[3]

B) That all real physical observations rest on:

- The utilization of real physical apparatus based on electromagnetic fields and gravitational material with charged (or uncharged) particles.

---

[2] As a consequence of the failure of the geometrical unification program Einstein was still obliged in 1954 to consider the electromagnetic field as filling curved space-time, but never reached a final satisfying model.

[3] This implies 1) the existence of a basic high density of sub-elements in vacuum, 2) the existence of small density variations above (for light) and below (for gravity) the average density with the possibility of propagating density variation on the top of such a vacuum model as initially suggested by Dirac.

- Where observers are also built with the same materials, i.e. influenced by the said fields and particles.

In other terms all observers (and their observations, inertial or not) are an integral part of fields and particles since they are part of the same overall real field and particle distribution. This fact determines their relation with all real phenomena. A physical theory should explicitly provide (within its context) a definition of the means whereby the quantities with which the theory is built and can be measured. The properties of light rays and massive particles are thus sufficient to provide the means of making basic measurements. Since real clocks and rods are the real instruments utilized in physics, we shall thus first define, for an individual inertial observer, the behavior of such instruments with respect to each other: since this determines, for every inertial observer possessing them, the behavior, with respect to I_0, of the material fields around him.

As a consequence of the covariant distribution character observed in I_0, the very small resistance to motion and assumed non-zero photon rest mass, real spin of possible extended vacuum sub-elements and their internal possible motions (and associated local interactions) one can describe the four-momenta and angular momenta of all extended sub-elements passing through a small four-volume with a constant average density on a hyperboloid, Σ_0. The four-momenta and angular momenta of extended elements are distributed at each point $P(x_\mu)$ with constant density $\rho(x_\mu)$ on space-like hyperboloids.

C) Following an idea of Noether the local analysis of moving fields and extended particles at each point by real observers tied to this point, is defined by local clocks and rods which move with the corresponding element. It is thus locally performed at each point of coordinates $x_\mu(\tau)$ which follows a world-line L. To this point are attached local (in I_0) internal variables $b^{(\lambda)}$, which describe its neighborhoods physical properties and thus depend on τ. The evolution is given by $x_\mu(\dot{x}_\mu)$, $b^\lambda(\dot{b}^\lambda)$, where $\cdot$ denotes the proper time derivative with respect to τ when x_μ describes a world-line L. A scalar Lagrangian thus represents the evolution of the real physical medium in I_0, which depends on a local

Lagrangian, L and is thus given by Poisson brackets. This description on I_0 is assumed to correspond to local space-time translations and four dimensional rotations which are determined by a Lagrangian L invariant under the local group of Poincaré transformations (i.e. the inhomogeneous Lorentz group). They contain [8,9]:

1) The operators P_μ of infinitesimal translations of X_μ only and can be described by $P_\mu \cdot X_\lambda = g_{\mu\lambda}$.

2) The operators $M_{\mu\nu}$ of infinitesimal four rotations in I_0 which act simultaneously on X_μ and on the internal variables. We have at X_μ:

$$M_{\mu\nu} x_\lambda = x_\mu g_{\nu\lambda} - x_\nu g_{\mu\lambda}. \tag{6.1}$$

Their action on internal local variables depends on their choice.

3) A choice of L leads to the momenta

$$G_\mu = \frac{\partial L}{\partial \dot{x}_\mu} \quad \textit{and} \quad \beta^{(\lambda)} = \frac{\partial L}{\partial \dot{b}^{(\lambda)}} \tag{6.2}$$

yielding a constant impulsion vector

$$G_\lambda P_\mu x_\lambda = G_\lambda g_{\mu\lambda} = G_\mu : \tag{6.3}$$

and the total angular momentum:

$$M_{\mu\nu} = G_\lambda M_{\mu\nu} x_\lambda + \beta^{(\lambda)} M_{\mu\nu} b^{(\lambda)},$$

so that

$$M_{\mu\nu} = x_\mu G_\nu - x_\nu G_\mu + S_{\mu\nu}, \tag{6.4}$$

with

$$S_{\mu\nu} = \beta^{(\lambda)} M_{\mu\nu} \beta^{(\lambda)}.$$

These quantities satisfy the Inhomogeneous Lorentz group commutation relations $[P_\mu, P_\lambda] = 0$

$$[M_{\mu\nu}, P_\alpha] = g_{\alpha\beta} P_\nu - g_{\alpha\nu} P_\mu \tag{6.5}$$

i.e. Poisson Group Relations :

$$[G_\mu, G_\nu] = 0 \ , \ [M_{\mu\nu}, G_\alpha] = g_{\alpha\beta} G_\nu - g_{\alpha\nu} G_\mu \tag{6.6}$$

$$[M_{\mu\nu}, M_{\alpha\beta}] = g_{\mu\alpha} M_{\nu\beta} + g_{\nu\beta} M_{\mu\alpha} - g_{\mu\beta} M_{\nu\alpha} - g_{\nu\alpha} M_{\mu\beta}.$$

With these quantities one can also define local conservation laws for free elements i.e.

$$\begin{aligned} \dot{G}_\mu &= 0 \\ \dot{M}_{\mu\nu} &= 0 \\ \dot{S}_{\mu\nu} &= G_\mu \dot{x}_\nu - G_\mu \dot{x}_\mu \end{aligned} \tag{6.7}$$

and introduce a constant local mass term M_0 with $G_\mu G_\mu = -M_0^2 \cdot c^2$.

4) An associated center of gravity y_μ is defined by the introduction of the four-vector

$$R_\mu = \left(\frac{1}{(M_0^2 c^2)} \right) \cdot S_{\mu\nu} \cdot G_\nu \tag{6.8}$$

associated with x_μ i.e.

$$y_\mu = x_\mu - R_\mu; \tag{6.9}$$

which implies that locally extended real media in I_0 are described by pairs of points as first suggested by Yukawa.

5) An inertial mass (usually not constant) μ_0 defined by

$$-M_0 c^2 = G_\mu \cdot \dot{x}_\mu \tag{6.10}$$

can also be attributed to x_μ: M_0 being located at y_μ since one has:

$$\dot{y}_\mu = \dot{x}_\mu - \dot{R}_\mu = \dot{x}_\mu - \frac{1}{M_0^2 c^2} (G_\mu \cdot \dot{x}_\nu - G_\nu \dot{x}_\mu) G_\nu = \frac{\mu_0}{M_0^2} \cdot G_\mu \tag{6.11}$$

so that the motion of y_μ is locally rectilinear and y_μ has a proper time Θ, (with $d\lambda / d\Theta = M_0 / \mu_0'$) and we have :

$$y'_\mu = \dot{y}_\mu \cdot \frac{d\tau}{d\Theta} = G_\mu / M_0 = \text{constant}$$

and

$$\mu_{\mu\nu} = R_\mu G_\nu - R_\nu G_\mu + S_{\mu\nu}, \tag{6.12}$$

with respect to the center of gravity. Local instantaneous four rotations are described by :

- A specific *beigrössen* 4-frame b_μ^ξ (ξ=1,2,3,0) with

  $\dot{x}_\mu = b_\mu^4 = \frac{ic}{6}\varepsilon_{\mu\nu\alpha\beta} \cdot \varepsilon^{rst} b_\nu^r b_\alpha^s b_\beta^t$, $b_\mu^\xi = (i/2)\varepsilon_{\mu\nu\alpha\beta}\dot{x}_\nu S_{\alpha\beta}$ and

  $S_{\alpha\beta} = I \cdot \dot{b}_\alpha^\xi \cdot b_\beta^\xi$.
- A specific four-frame a_μ^ξ centered on y_μ with $M_{\alpha\beta} = K \cdot \dot{a}_\alpha^\xi \cdot a_\beta^\xi$ for a_μ^4 along y'_μ and $a_\mu^3 = (i/2M_0 c) \cdot \varepsilon_{\mu\nu\alpha\beta} G_\nu \mu_{\alpha\beta}$.

This set of relations must be completed by relations which will define the interactions between the extended elements i.e. the propagation in the aether of collective motions corresponding to observed gravitational and electromagnetic phenomena. Before the introduction of such interactions one must recall that such proposals have already been made in the past.

We only mention here:

- Weyssenhof's proposal [16] $S_{\alpha\beta}\dot{x}_\beta = 0$ extensively discussed in the literature.
- Nakano's proposal [19] $S_{\alpha\beta}\dot{x}_\beta = I \cdot \ddot{x}_\alpha$.
- Roscoe's proposal with photon mass [20].

6.3 General Relativity Represented as a Polarizable Vacuum

Since all observed effects of gravity in distant space rest on light observation (including γ and radio EM waves coming through space from distant sources) a simple model endows the polarizable vacuum with properties that might account for all the phenomena in terms of distortions. This initial proposal of Wilson and Dicke has been recently revived with astonishing success by Puthoff [4] and Krogh [21]. We first summarize their model and will complete it with a supplementary mass term in electro-magnetism.

One starts from the idea that in flat space the electric field moves in a real vacuum medium with a point varying dielectric constant *K*: so that this *D* field satisfies the vacuum equation:

$$D = K \cdot \varepsilon_0 \cdot E. \tag{6.13}$$

This corresponds to a variable fine structure constant

$$\alpha = \frac{e^2}{4\pi\varepsilon_0 \hbar c} \cdot \left(\frac{\mu(K)/\mu_0}{K} \right)^{1/2} : \tag{6.14}$$

so that the vacuum has permittivity and permeability constants given by

$$\varepsilon_0 \to \varepsilon = K \cdot \varepsilon_0 \quad \text{and} \quad \mu_0 \to \mu = K \cdot \mu_0, \tag{6.15}$$

and an impedance $(\mu/\varepsilon)^{1/2} = (\mu_0/\varepsilon_0)^{1/2}$ to satisfy Eötvos-type experiments. The local velocity of light for a given frequency ν varies like $V_\nu = c/K$ i.e like $1/(\mu\varepsilon)^{1/2}$. The corresponding principle of equivalence implies that the self energy of a system changes when K changes; so that a flat-space energy E_0 in flat space changes into

$$E = E_0 \cdot (K)^{-1/2}; \tag{6.16}$$

and one has

$$m = m_0 \cdot K^{3/2}. \tag{6.17}$$

As a consequence the condition $\mathrm{E} = \hbar \cdot \omega$ becomes

$$\omega = \omega_0 (K)^{-1/2} \tag{6.18}$$

along with the time and length variations Δt *and* Δr given by the relations:

$$\Delta t = \Delta t_0 (K)^{1/2} \quad \text{and} \quad \Delta r = \Delta r_0 (K)^{-1/2}. \tag{6.19}$$

These relations are evidently equivalent to a local curvature of space. Indeed a dx_0 length rod shrinks to $d_x = d_{x_0} \cdot (K)^{-1/2}$ and would measure dx_0, where the rod remains rigid, is now expressed in terms of dx-length rod as $dx_0 = (K)^{1/2} dx$.

Using the same argument for dt and dt_0 we find that one can write:

$$dS^2 = c^2 dt_0^2 - (dx_0^2 + dy_0^2 + dz_0^2) \tag{6.20}$$

which transforms into

$$dS^2 = \frac{1}{K} c^2 dt^2 - K(dx^2 + dy^2 + dz^2): \tag{a}$$

i.e. (6.21)

$$dS^2 = g_{ij} ... dx^i dx^j, \tag{b}$$

with $g_{00} = 1/K, \; g_{11} = g_{22} = g_{33} = -K \; \; and \; \; g_{ij} = 0 \; \; for \; \; i \neq j.$

In the case of a spherically symmetric mass distribution one writes

$$\begin{cases} K = e^{2G\cdot M/rc^2} \\ K = 1 + 2\dfrac{G\cdot M}{rc^2} + \dfrac{1}{2}\left(\dfrac{2GM}{rc^2}\right)^2 + \ldots \end{cases} \tag{6.22}$$

where G is the gravitational constant, M the mass and r the distance from its origin located at the center of mass. Puthoff [4] has recently shown that this model accounts (sometimes with better precision) for all known experimental tests of General Relativity in a simple way i.e. one can describe

- The gravitational redshift given by $\omega = \omega_0 /(K)^{1/2}$ (so that $\Delta\omega/\omega \cong (GM/R^2c^2)h$ has a 1/100 precision).
- The bending of light rays by the sun and stars.
- The advance of the Perihelion of Mercury.

He has also shown that one can derive the form of (6.22) from a general Lagrangian with a variable, K leaving aside vacuum interaction in I_0 :

$$\begin{aligned} L = &-\left[\frac{m_0c^2}{K^{1/2}}\left(1-\left(\frac{v}{(c/K)}\right)^2\right)^{1/2} + q\cdot\phi - q\cdot\vec{A}\vec{V}\right]\delta^3(r-\bar{r}) \\ &-\frac{1}{2}\left(B^2/(K\cdot\mu_0) - K(\varepsilon_0 E^2)\right) - \frac{\lambda}{K^2}\left[(\nabla K)^2 - \frac{1}{(c/K^2)}\left(\frac{\partial K}{\partial t}\right)^2\right] \end{aligned} \tag{6.23}$$

This association of gravitational theory with electromagnetic theory based on the introduction of a variable dielectric vacuum constant K has recently been made more explicit by Krogh [21]. Noting that:

a) Electromagnetic theory implies the effects of electromagnetic vector four-potential vectors A_μ on the phases S of quantum mechanical waves so that one has

$$\Delta S = \frac{q}{h}\int \phi dt - \frac{q}{hc}\int \vec{A}\cdot d\vec{S} \tag{6.24}$$

for charged particles moving under the influence of the four-vector, A_μ.

b) If $m_\gamma \neq 0$ (m_γ is the mass term introduced into Maxwell's equation) the force on charged particles takes the form

$$F = q\left(E + \frac{V \times B}{c}\right) + q \cdot V \tag{6.25}$$

where the first term is the usual transverse Poynting force on currents and the second a longitudinal force along currents (resulting from non zero photon mass) recently observed by Graneau [22] and Saumont [23].
c) One can describe gravity with a four-vector density A_μ^g so that the gravitational (Newton) and electromagnetic (Coulomb) potentials have the same form, but different coupling constants. This suggests that both wave fields and singularities are just different aspects of the same fundamental field.

6.4 Maxwell's Equations Extended

This discussion opens the possibility to test new types of extensions of Maxwell's equations in the laboratory. Since this has already been attempted some results (derived within the frame of the model) are given here:

a) From a non-zero vacuum conductivity coefficient $\sigma \neq 0$ [24,25] we have in vacuum div E =0 with curl $H = \sigma E + \varepsilon_0 \chi_0 \partial E / \partial t$ and div $H = 0$ with curl $E = -\mu_0 \chi_m \partial H / \partial t$.

b) From an associated non-zero photon mass term ($m_\gamma \neq 0$) (with $A_\mu A_\mu \rightarrow 0$ where A_μ denotes the total four-potential density in Dirac's aether model. This introduces a non-zero fourth component of the current $J_\mu = \sigma E, j_0$ (where $j_0 \neq 0$) into the vacuum corresponding to a real detectable space. With present technology this implies that the present vacuum really carries space-charge currents [25] (so that the divergence of the electric field is different from zero in Vacuo) and the corresponding existence of a displacement current (i.e. a curl of the magnetic field) and its associated current density[4].

6.4.1 *The Infinitesimal Mass of Photons*

Unifying massive spin 1 photons piloted by electromagnetic waves built with massive extended sub-elements has been developed in a series of books by Evans, Vigier et al. [24] The model implies the introduction of spin and mass with an associated energyless magnetic field component $B^{(3)}$ in the direction of propagation and a small electrical conductivity in the Dirac vacuum also implying a new 'tired light' mechanism [11,14,24]. Corresponding equations will be given below.
In the absolute inertial frame I_0 all massive particles are governed by a gravitational potential four-vector $\phi_g, \vec{A}_g / c$, associated with a small mass m_g which can be decomposed into transverse, longitudinal and gradient potentials.

We can thus associate the relations

$$\square\, \phi = -\frac{\rho}{\varepsilon_0} + \mu\phi \text{ and } \square\, \vec{A} = -\vec{d}_0 / \varepsilon_0 c + \mu\vec{A} \qquad (6.26)$$

which represent the electromagnetic field in vacuum in any inertial frame, Σ_0 the relations:

$$\square\, \phi_g = 4\pi G m \rho_\mu + \mu_g \cdot \phi_g \text{ and } \vec{A}_g = 4\pi \cdot G \cdot \vec{j}_m + \mu_g \vec{A}_g, \qquad (6.27)$$

which represent the gravitational field in the same vacuum; where ρ_μ refers to mass density, j_m to mass current and μ and μ_g to EM and gravitational mass (both very small $\cong 10^{-65}$ grams) and $\rho \cdot c_0$ in the $\square$ terms ($\square = \nabla^2 - (1/c_0^2)\partial / \partial t^2$) represents the corresponding wave velocities (which except in I_0 depend on the directions in flat space-time) so that one has:

$$c_0 = c \cdot e^{2\phi_g / c^2}; \qquad (6.28)$$

---

[4] Such attempts have been recently published in a book by Lehnert & Roy [25] so we shall only present a summary of some results and assumptions.

where c is the value in the absence of a gravitational potential A_μ^g. In this model, one assumes, with Sakharov [5-7], that the gravitational field corresponds to local depressions in the immensely positive energy of the zero-point field; and gravitational fields represent regions of diminished energy (i.e. that their momentum gravity corresponds to holes in vacuum energy or local defects of vacuum elements). Their effective momentum is thus opposite and corresponding gravitational forces are attractive.

Such an association also suggests that although measuring devices (observations) in local inertial Poincaré frames are altered by gravitational potentials (they are part of the same real physical background in this model). There is no effect on the geometry of flat space and time. For any given real inertial local Poincaré frame, Σ_0 real space is Euclidean and one uses Poincaré transformations between Σ_0 and I_0 to describe real motions which include consequences of gravitational potentials. For example a reduction of the velocity of quantum mechanical waves, including light, is taken as a fundamental effect of gravitational potentials. Clocks are slowed and measuring rods shrink in such potentials by a factor e^{ϕ_g / c^2}.

6.4.2 *Divergence of the Electromagnetic Field*

A non-vanishing divergence of the electric field given below, can be added to Maxwell's equations which results in space-charge distribution. A current density arises in vacuo and longitudinal electric non-transverse electromagnetic terms (i.e. magnetic field components) appears (like $B^{(3)}$) in the direction of propagation.

Both sets of assumptions were anticipated by de Broglie and Dirac. They imply that the real zero-point (vacuum) electromagnetic distribution

- is not completely defined by $F_{\mu\nu}$ but by a four-vector field distribution given by a four-vector density, A_μ associated with a de Broglie-Proca equation i.e.

$$\Box A_\mu(x_\alpha) = -\frac{m_\gamma^2 c^2}{\hbar^2} A_\mu(x_\alpha) \tag{6.29}$$

and its complex conjugated equation.

- the A_μ field potential equation also contains a gradient term so one has in vacuum:

$$A_\mu = A_\mu^T + A_\mu^L + \lambda \partial_\mu S \tag{6.30}$$

with $A_\mu A^* \to 0$ and a small electrical conductivity in vacuo.

6.5 Possible New Consequences of the Model

Since such models evidently imply new testable properties of electromagnetic and gravitational phenomena we shall conclude this work with a brief discussion of the points where it differs from the usual interpretations and implies new possible experimental tests.

If one considers gravitational and electromagnetic phenomena as reflecting different behaviors of the same real physical field i.e. as different collective behavior, propagating within a real medium (the aether) one must start with a description of some of its properties.

We thus assume that this aether is built (i.e. describable) by a chaotic distribution $\rho(x_\mu)$ of small extended structures represented by four-vectors $A_\mu(x_\alpha)$ round each absolute point in I_0. This implies

- the existence of a basic local high density of extended sub-elements in vacuum
- the existence of small density variations $\delta\rho(x_\mu)A_\alpha(x\mu)$ above $\delta\rho > 0$ for light and below $(\delta\rho < 0)$ for gravity density at x_μ.
- the possibility to propagate such field variations within the vacuum as first suggested by Dirac [13].

One can have internal variations: i.e. motions within these sub-elements characterized by internal motions associated with the internal behavior of average points (i.e. internal center of mass, centers of charge, internal rotations) and external motions associated with the stochastic behavior, within the aether, of individual sub-elements. As well known the latter can be analyzed at each point in terms of average drift and osmotic motions and A_μ distribution. It implies the introduction of non-linear terms.

To describe individual non-dispersive sub-elements within I_0, where the scalar density is locally constant and the average A_μ equal to zero, one introduces at its central point $Y_\mu(\theta)$ a space-like radial four-vector $A_\mu = r_\mu \exp(iS/\hbar)$ (with $r_\mu r^\mu = \mathrm{a}^2 =$ constant) which rotates around Y_μ with a frequency $\nu = m_\gamma c^2 / h$. At both extremities of a diameter we shall locate two opposite electric charges e^+ and e^- (so that the sub-element behaves like a dipole). The opposite charges attract and rotate around Y_μ with a velocity $\cong c$. The +e and –e electromagnetic pointlike charges correspond to opposite rotations (i.e $\pm\ \hbar/2$) and A_μ rotates around an axis perpendicular to A_μ located at Y_μ, and parallel to the individual sub-element's four momentum $\partial_\mu S$.

Assuming electric charge distributions correspond to $\delta m > 0$ and gravitation to $\delta m < 0$ one can describe such sub-elements as holes ($\delta m <$ 0) around a point 0 around which rotate two point-like charges rotating in opposite directions as shown in Figure 6.1 below.

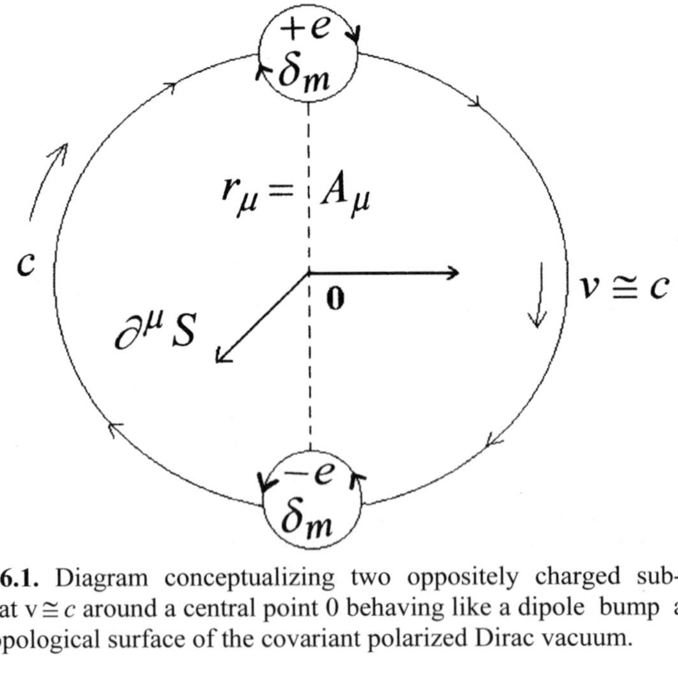

Figure 6.1. Diagram conceptualizing two oppositely charged sub-elements rotating at $\mathrm{v} \cong c$ around a central point 0 behaving like a dipole bump and hole on the topological surface of the covariant polarized Dirac vacuum.

These charges themselves rotate with a velocity c at a distance $r_\mu = A_\mu$ (with $r_\mu r_\mu$ = Const.). From 0 one can describe this by the equation

$$\square A_\mu - \frac{m_\gamma^2 c^2}{\hbar^2} \cdot A_\mu = \frac{\left[[](\mathrm{A}_\alpha^* \mathrm{A}_\alpha) \right]^{1/2}}{(A_\alpha^* A_\alpha)^{1/2}} \cdot A_\mu \tag{6.31}$$

with $A_\mu = r_\mu \cdot \exp[iS(x_\alpha)/\hbar]$ along with the orbit equations for e^+ and e^- we get the force equation

$$m \cdot \omega^2 \cdot r = e^2 / 4\pi r^2 \tag{6.32}$$

and the angular momentum equation:

$$m_\gamma \cdot r^2 \cdot \omega = \hbar / 2 \tag{6.33}$$

Eliminating the mass term between (31) and (33) this yields

$$\hbar\omega = e^2 / 2r \tag{6.34}$$

where $e^2/2r$ is the electrostatic energy of the rotating pair. We then introduce a soliton-type solution

$$A_\mu^0 = \frac{\sin \cdot K \cdot r}{K \cdot r} \cdot \exp[i(\cot - K_0 x)] \tag{6.35}$$

where

$$K = mc/\hbar, \quad \omega = mc^2/\hbar \quad and \quad K_0 = mv/\hbar \tag{6.36}$$

satisfies the relation (31) with $r = ((x - vt)^2 \cdot (1 - v^2/c^2)^{-1} + y^2 + z^2)^{1/2}$ i.e.

$$\square A_\mu^0 = 0 : \tag{6.37}$$

so that one can add to A_μ^0 a linear wave, A_μ (satisfying $\square A_\mu = (m_\gamma^2 c^2/\hbar^2) A_\mu$) which describes the new average paths of the extended wave elements and piloted solitons. Within this model the question of the interactions of a moving body (considered as excess or defect of field density, above or below the aether's neighboring average density) with a real aether appears immediately[5].

As well known, as time went by, observations established the existence of unexplained behavior of light and some new astronomical phenomena which led to discovery of the Theory of Relativity.

In this work we shall follow a different line of interpretation and assume that if one considers particles, and fields, as perturbations within a real medium filling flat space time, then the observed deviations of Newton's law reflect the interactions of the associated perturbations (i.e. observed particles and fields) with the perturbed average background medium in flat space-time. In other terms we shall present the argument (already presented by Ghosh et al. [26]) that the small deviations of Newton's laws reflect all known consequences of General Relativity.

The result from real causal interactions between the perturbed local background aether and its apparently independent moving collective perturbations imply absolute total local momentum and angular momentum conservation resulting from the preceding description of vacuum elements as extended rigid structures.

6.6 Extending Newton's Model with Inertia and Vacuum Drag

Starting from an aether built with moving small extended structures with an average real distribution isotropic in an inertial frame I_0 (i.e. examining the effects in a given inertial frame I centered on a point Y_μ of the real vacuum distribution on a test particle moving with absolute velocity V^0 and angular momentum $\omega^0_{\alpha\beta}$) one can evaluate more precisely, the collective interactions carried by this aether between two extended neighboring regions centered on points A and B with two centers of mass situated at X_A and X_B.

Starting with $\delta\rho < 0$, i.e. for gravitational effects, it appears immediately

[5] According to Newton massive bodies move in the vacuum, with constant directional velocities, i.e. no directional acceleration, without any apparent relative friction or drag term. This is not true for accelerated forces (the equality of inertial and gravitational masses are a mystery) and apparent absolute motions proposed by Newton were later contested by Mach.

a) if one assumes the gravitational potential is spherical in the rest frame I_B of its source *B*,
b) that the motion of *A* undergoes a velocity dependent inertial induction with respect to *A* i.e. a friction depending on the velocity *v* of *A* with respect to *B*
c) that this motion is also submitted to an acceleration dependent inertial with respect to I_B i.e. also an acceleration depending on its acceleration a measured in I_B.
d) possible terms depending on higher order time derivations which we will neglect in the present analysis we can write (6.19) the force on *A* due to *B* in I_B in the form $F = F_S + F_v + F_a$ where

$$F = -G \cdot \frac{m_A \cdot m_B}{r^2} - G' \frac{m_A \cdot m_B}{c^2 r^2} \cdot v^2 \cdot f(\theta)\hat{U}_r - G'' \cdot \frac{m_A \cdot m_B}{c^2 r} \cdot af(\phi)\hat{U}_r \quad (6.38)$$

The terms *G, G', G"* are scalars possibly dependent on *v*. The terms m_A and m_B are the gravitational masses in I_B, $\hat{U}$, is the unit vector along *r*. $f(\theta)$ and $f(\phi)$ must have the same form i.e. 1/2 cos ϕ or cos $\phi|\cos\phi|$. If we also accept the preceding velocity dependent analysis for contracting rods and retarded clocks then we should write $G = G'$ in (38) and take $f(\theta) = \cos\theta\,|\cos\phi|$ as done by Ghosh [26]. Moreover, if we compare the form given by Weber to the repulsion of two electric charges of the same sign:

$$F_{AB}^{e} = \frac{e_A \cdot e_B}{4\pi\varepsilon r^2} \cdot \left[1 - \frac{1}{c^2}\left(\frac{dr}{dt}\right)^2 + \frac{2r}{c^2} \cdot \frac{d^2 r}{dt^2}\right] \quad (6.39)$$

corresponding to electromagnetism, with the recent form given by Assis [27,28] to attracting interacting masses m_A and m_B i.e.

$$F_{AB}^{g} = -G \cdot \frac{m_A . m_B}{r^2}\left[1 + \frac{6}{c^2}\left[r \cdot \frac{d^2 r}{dt^2} - \frac{1}{2}\left(\frac{dr}{dt}\right)^2\right]\right] \quad (6.40)$$

we see they have exactly the same form; the difference of their coefficients being compatible (within our interpretation) since they correspond to opposite variations of the average vacuum density. Their interpretation in terms of $\delta\rho > 0$ (for electromagnetism) and $\delta\rho < 0$ (for gravitation) also explains (at last qualitatively) why extended depressions repel or attract when they rotate through parallel or antiparallel directions and only attract when $\delta\rho < 0$. This also explains why a reduction of attraction between two masses has been observed when one puts another mass between them (the LAGEOS satellite). In this model this similarity is indeed comparable to similar behaviors of vortices for gravitation and Tsunamis for electromagnetism on an ocean surface.

If one assumes the absolute local conservation of four-momentum and angular momentum in regions containing the preceding aether carrying its associated collective electromagnetic and gravitational motions one can evaluate the effects of their interactions. With a real physical aether there is no such thing as free electromagnetic or gravitational phenomena. Drag theories (described as inertial induction) are always present and responsible for Casimir type effects in the microscopic domain. Real consequence of the aether appear, at various levels, in the macroscopic and cosmological domains… as has already been suggested in the literature and tested in laboratory or astronomical phenomena. We only mention here:

1) Possible consequences of modifying and testing the Newton and Coulomb forces.
2) The redshift and variable velocity of electromagnetic waves results from the rotational inertial drag of extended photons moving in vacuum: an effect already observed in light traversing around the earth [28].
3) The possible measurable existence of the redshift of transverse gravitational waves… possible in the near future.
4) Observational redshift variations of light emitted by Pioneer close to the solar limb, i.e. also of photons grazing a massive object [28].
5) The observed anisotropy of the Hubble constant in various directions in the sky [28] associated with various galactic densities.
6) Observed torques on rotating spheres in the vicinity of large massive bodies. This also appears in some experiments, i.e.:
   a) Secular retardation of the earth's rotation.
   b) Earth-moon rotation in the solar system etc.
7) Apparent evolution with time of angular momentum in the solar-planetary system.
8) Different variation of redshift of light traveling up and down in the Earth's gravitational field…Which also supports existence of photon mass.

6.7 Relativistic Maxwell's Equations in Complex Form

We will now outline the relativistic formalism which gives a more comprehensive explanation of the complexification scheme. Such issues as the Higgs (soliton) monopole depend on considering Lorentz invariance and relativistic causality constraints. We will also relate the complexification of Maxwell's equations to models of nonlocality. We examine, for example, the manner in which advanced potentials may explain the remote connectedness which is indicated by the Clauser test of Bell's theorem. Similar arguments apply to Young's double slit experiment. The collective coherent phenomena of superconductivity is also explainable by considering the relativistic field theoretic approach in which wave equations are solved in the complex Minkowski space (such as the Dirac equation).

6.7.1 *Relativistic conditions on Maxwell's equations in complex geometries and the invariance of the line element*

This section introduces the relativistic form of Maxwell's equations. The fields $\underline{E}$ and $\underline{B}$ are defined in terms of ($\underline{A},\phi$), the four-vector potential; and the relativistic form of $\underline{E}$ and $\underline{B}$ is presented in terms of the tensor field, $F_{\mu\nu}$ (where indices μ and ν run 1 to 4). We then complexity $F_{\mu\nu}$ and determine the expression for the four-vector potential $A_{\mu} = \left(\underline{A}_j, \phi\right)$ in terms of $F_{\mu\nu}$ (index j runs 1 to 3). Discussion of line element invariance is given in terms of $F_{\mu\nu}$.

In section 6.8 we describe the complex form of A_{μ} fields and through the formalism in this section we can relate this to the complex forms of $\underline{E}$ and $\underline{B}$. We utilize Weyl's action principle to demonstrate the validity of the use of the complex form of $F_{\mu\nu}$. Weyl relates the gravitational potential, $G_{\mu\nu}$, to the EM 'geometerizing' potential A_{μ}, or geometrical vector, using the principle of stationary action for all variations $\delta G_{\mu\nu}$ and δA_{μ} . The quantity A_{μ}, or vector potential, which we identify with A^{μ}, is related to $F_{\mu\nu}$, the EM force field, by a set of gauge invariant relations. The EM force $F_{\mu\nu}$ is independent of the gauge system. The curl of A_{μ} has the important property

$$F_{\mu\nu} = \frac{\partial A_{\mu}}{\partial x_{\nu}} - \frac{\partial A_{\nu}}{\partial x_{\mu}} \tag{6.41}$$

where $F_{\mu\nu}$ is antisymmetric or $F_{\mu\nu} = -F_{\nu\mu}$, and changing $\sim A_{\mu}$ to $A'_{\mu} = A_{\mu} + \partial\phi / \partial x_{\mu}$ is a typical gauge transformation where the intrinsic state of the world remains unchanged.

We define the four-vector potential as A_{μ}, which can be written in terms of the three-vector, $\underline{A}_j$ and ϕ, where ϕ is the fourth or temporal component of the field. The indices μ, ν run 1 to 4 and *j* runs 1 to 3.

Then we can write Maxwell's equations in compact notation in their usual tensor form in terms of $F_{\mu\nu}$, (for $c = 1$);

$$F_{\mu\nu} = \begin{pmatrix} 0 & -B_z & B_y & E_x \\ B_z & 0 & -B_x & B_y \\ -B_y & B_x & 0 & E_x \\ -E_x & -E_y & -E_z & 0 \end{pmatrix} \tag{6.42}$$

then the equations $\nabla \times E = -(1/c)(\partial \underline{B} / \partial t)$ and $\nabla \cdot \underline{B} = 0$ can be written

$$\text{as} \frac{\partial F_{\mu\nu}}{\partial x} + \frac{\partial F_{\mu\nu}}{\partial y} + \frac{\partial F_{\mu\nu}}{\partial z} = 0 \tag{6.43}$$

or $\nabla \times F_{\mu\nu} = 0$ for $x^1 = x, \; x^2 = y, \; x^3 = z,$ and $x^4 = t$.

To complexity the elements of $F_{\mu\nu}$ we can take conditions,

$$\text{For } (F_{41}, F_{42}, F_{43}) = i\underline{E} \text{ and } (F_{23}, F_{32}, F_{12}) = B,$$
$$\text{or } (E_x, E_y, E_z) = iE \text{ and } (B_x, B_y, B_z) = \underline{B}. \tag{6.44}$$

We can write the complex conjugate of the electric and magnetic fields in terms of the complex conjugate of $\underline{F}$ or $F^*_{\mu\nu} = -F^{\mu\nu}$. There is a useful theorem stating [29] $\nabla_{123} \times \underline{F} = \nabla^4 \cdot F^*$ or $(\nabla_{xyz} \times F = \nabla^t \cdot F^*)$.

Then for $\left(F^{23*}, F^{31*}, F^{12*}\right) = i\underline{E}$ and $\left(F^{41*}, F^{42*}, F^{43*}\right) = -B$ we obtain $\partial F^{\mu\nu*} / \partial x^{\nu} = 0$ or $\nabla \cdot \underline{F}^{*} = 0$ which gives the same symmetry between real and imaginary components as ours and in Inomata's formalism [30].

The expressions for the other two Maxwell equations $\nabla \cdot \underline{E} = 4\pi\rho$ and $\nabla \times \underline{B} = \frac{1}{c}\frac{\partial E}{\partial t} + J_e$ can be obtained by introducing the concept of the vector potential in the Lorentz theory as first noticed by Minkowski [31]; we have the four-vector forms $\left(\phi_1, \phi_2, \phi_3\right) = \underline{A}$ and $\phi_4 = i\phi$, then $\underline{B} = \nabla \times \underline{A}$ and $\underline{E} = -\nabla\phi - \frac{1}{c}\frac{\partial \underline{A}}{\partial t}$. Then we have $F_{\mu\nu} = \frac{\partial A_{\nu}}{\partial x^{\mu}} - \frac{\partial A_{\mu}}{\partial x^{\nu}}$ or $\underline{F} = \nabla \times \underline{A}$ for the vector and scalar potentials $A = \left(A_1, A_2, A_3, \phi\right)$. If A is a solution to $F = \nabla \times \underline{A}$ then $\phi_{\mu}' + \frac{\partial \phi}{\partial x^{\mu}}$ also is (by gauge invariance) and $\nabla \cdot \underline{A} + \frac{1}{c}\frac{\partial \phi_4}{\partial t} = 0$. We term the fourth component ϕ or ϕ_4 interchangeably. Then from Lorentz theory we have the 4D form as $\frac{\partial A^{\mu}}{\partial x^{\mu}} = 0$ or $\nabla \cdot \underline{A} = 0$. We can now write the equations for $\nabla \cdot \underline{F} = 4\pi\rho$ and $\nabla \cdot \underline{B} = \frac{1}{c}\frac{\partial E}{\partial t} + J_e$ as

$$\frac{\partial F^{\mu\nu}}{\partial x^{\nu}} = s^{\mu} \quad \text{or} \quad \nabla \cdot \underline{F} = S. \tag{6.45}$$

The most general covariant group of transformations of the EM field equations (more general than the Lorentz group) is formed by affine transformations which transform the equation of the light cone, $s^2 = 0$ into itself. (The properties of the spacetime manifold are defined in terms of the constraints of the line element, which relate to the gravitational potential, $g_{\mu\nu}$. We also form an analogy of the metric space invariant to the EM source vector, s_{μ} [32].)

This group contains the Lorentz transformations as well as inversion with respect to a 4D sphere, or hyperboloid in real coordinates. Frank [33] discusses the Weyl theory and gives a proof that the Lorentz group together with the group of ordinary affine transformations is the only group in which Maxwell's equations are covariant [33]. Recall that an affine transformation acts as $x^{\mu} = \alpha^{\mu}_{\nu} x^{\nu}$ with an inverse $x^{\nu} = \alpha^{-\iota}_{\mu} x'^{\mu}$. The affine group contains all linear transformations and the group of affine transformations transforms $s^2 = 0$ on the light cone into itself.

In the Weyl geometry, if we have from before, $\underline{F} = \nabla \times \phi$ and

$$\nabla \cdot \underline{F} = \frac{1}{\sqrt{g}} \frac{\partial \sqrt{g} F^{\mu}}{\partial x^{\mu}} \tag{6.46a}$$

and

$$\nabla^{\mu} \cdot F = \frac{1}{\sqrt{g}} \frac{\partial \sqrt{g} F^{\nu}}{\partial x^{\nu}} \tag{6.46b}$$

with the signature (+,+,+,-). Then using the theorem in W. Pauli [34],

$$\nabla_{\mu} \cdot \nabla \times \underline{F} = \nabla_{\mu} \nabla \cdot \underline{F} - \Box F_{\mu} \tag{6.47}$$

and from before, $\nabla \cdot F = S$ and since $\nabla \cdot \phi = 0$ and then $\partial \phi^{\mu} / \partial x^{\mu} = 0$ and we have from

$$\nabla_{\mu} \cdot \nabla \times \underline{A} = \nabla_{\mu} \nabla \cdot \underline{A} - \Box \phi_{\mu} = S_{\mu} \tag{6.48}$$

or

$$\Box A_{\mu} = -S_{\mu} \tag{6.49}$$

for our potential equation, where $\Box$ is the D'Alembertian operator, and

$$\Box = \partial^{\mu} \partial_{\mu} = \eta^{\mu\nu} \partial_{\nu} \partial_{\mu} = \frac{\partial^2}{\partial x^2} + \frac{\partial^2}{\partial y^2} + \frac{\partial^2}{\partial z^2} - \frac{1}{c^2} \frac{\partial^2}{\partial t^2} = \nabla^2_{R^3} - \frac{\partial^2}{\partial t^2} \tag{6.50}$$

The important aspect of this consideration [35] is our ability to relate the EM potential to a corresponding spacetime metric interval s or s^2. Hence we can construct the invariant relations for our fields in terms of our Lorentz invariance four space conditions. We can also relate the introduction of a complex spacetime to the complex expansion of the electric and magnetic fields in this section and demonstrate their self-consistency. We will look at this in more detail at the end of this section where we consider a generalized affine connection. We can relate the EM potential, A_{μ} and ϕ_4 to $g_{\mu\nu}$ as $\sqrt{g}$ and also to the square root of the invariant, or s.

The key to the relationship of complex $F_{\mu\nu}$ and complex spacetime is the analogy between ϕ and $g_{\mu\nu}$. We can relate the EM scalar potential into the interval of time as in Eq. (6.49), $\square A_{\mu} = \phi = -S_{\mu}$ and we make the analogy of A_{μ} to $g_{\mu\nu}$ which is tied to the invariance conditions on s^2. Both potentials are then related to spacetime or spacetime interval separation. Note that in the $\square A_{\mu} = -s$ equation we have a $\sqrt{g}$ factor in order to form the invariant. In the equation for s^2, the invariant is found directly as $s^2 = g_{\mu\nu} x^{\mu} x^{\nu}$. We will write a set of invariant relations for the case of complex $\underline{E}$ and $\underline{B}$ fields at the end of this section. We can relate this then to the de Sitter algebras and the complex Minkowski metric.

Note that we associate the E_x component of $F_{\mu\nu}$ or $F_{41} = E_x$ with ϕ_4' as follows:

$$F_{41} = E_x = \phi_4' \frac{e}{r^2} \tag{6.51}$$

in which $4\pi e$ is associated with electric charge on the electron. This approximation is made in the absence of a gravitational field. Maxwell's equations are intended to apply to the case in which no field of force is acting on the system or in the special system of Galilean coordinates $A^{\mu} = \left(A_x, A_y, A_z, \phi\right)$, where $A^{j} = \left(A_x, A_y, A_z\right)$ is the vector potential and ϕ is the scalar potential and A^{μ} is the covariant form. Also, for the contravariant form, we have $A_{\mu} = \left(-A_x - A_y - A_z, \phi\right)$. And in empty space we have

$$\square A_{\mu} = 0 \tag{6.52}$$

In non-empty space then

$$\square A^{\mu} = J^{\mu} \tag{6.53}$$

or we can write this as

$$\nabla^2 A^{\mu} - \mu\varepsilon \frac{\partial^2 A^{\mu}}{\partial t^2} = -J^{\mu} \tag{6.54}$$

which is true only approximately in the assumption of flat space for Galilean coordinates. This is the condition which demands that we consider the weak Weyl limit of the gravitational field.

The invariant integral, $\mathcal{J}$ for $F^{\mu\nu}$ is given by

$$\mathcal{I} = \frac{1}{4}\int F^{\mu\nu} F_{\mu\nu} \sqrt{g}\, d\tau \tag{6.55}$$

The quantity, $\mathcal{L}$ is called the action integral of the EM field. Weyl [36] demonstrates that the action integral is a Lagrangian function, or

$$\mathcal{L} = \int dt \iiint \frac{1}{2}\left(B_x^2 + B_y^2 + B_z^2 - E_x^2 - E_y^2 - E_z^2\right) dx dy dz \tag{6.56}$$

which is of the form $\mathcal{L} = (T - v)dt$. By describing an electron in a field by Weyl's formalism one has a more general but more complicated formalism than the usual Einstein-Galilean formalism [37]. We can write a generalized Lagrangian in terms of complex quantities. For example, we form a modulus of the complex vector B as $|B|^2 = BB^* = B_{\text{Re}}^2 + B_{\text{Im}}^2$. This is the Lagrangian form for the real components of E and **B** in four-space. We can again consider $E = E_{\text{Re}} + iE_{\text{Im}}$ and $B = B_{\text{Re}} + iB_{\text{Im}}$ for the complex forms of $\underline{E}$ and $\underline{B}$. The complex Lagrangian in complex eight-space becomes

$$\mathcal{L} = \iint dt_{\text{Re}} dt_{\text{Im}} \iiint_{\text{Re}} \iiint_{\text{Im}} \frac{1}{2}\left(B_{\text{Re}}^2 - E_{\text{Re}}^2 + B_{\text{Im}}^2 - E_{\text{Im}}^2\right) \tag{6.57}$$
$$dx_{\text{Re}} dy_{\text{Re}} dz_{\text{Re}} dx_{\text{Im}} dy_{\text{Im}} dz_{\text{Im}} \,.$$

Note that this is an 8D integral, six over space. Also all quantities of the integrand are real because they are squared quantities. We can also write a generalized Poynting vector and energy relationship. We also have two equations which define a vector quantity A_μ in EM theory which corresponds to the gravitational potential $g_{\mu\nu}$. We have

$$\frac{\partial}{\partial g_{\mu\nu}}\left(\frac{1}{4} F^{\mu\nu} F_{\mu\nu}\right) = \frac{1}{2} E^{\mu\nu} \tag{6.58}$$

and

$$\frac{\partial}{\partial A_\mu}\left(\frac{1}{4} F^{\mu\nu} F_{\mu\nu}\right) = -J^\mu \tag{6.59}$$

where $E^{\mu\nu}$ is the energy tensor and J^μ is the charge and current vector. Two specific cases are for a region free from electrons, or $T^{\mu\nu} - E^{\mu\nu} = 0$, or a region free of the gravitational potential or in the weak Weyl limit of the gravitational field, $\Box F_{\mu\nu} = J_{\mu\nu} - J_{\nu\mu}$ where $\Box$ is the four-space D'Alembertian operator. The solution for this latter case is for the tensor potential $A_{\mu\nu}$,

$$F_{\mu\nu} = \frac{1}{4\pi\gamma}\left(A_{\mu\nu} - A_{\nu\mu}\right)\int\frac{de}{r} \tag{6.60}$$

if all parts of the electron are the same or uniform in charge. For the proper charge ρ_0, we have $J^{\mu} = \rho_0 A^{\mu}$.

In the limit of $A_{\mu}^{\mu} = 0$, then ρ_0, the proper density, is given as $\rho_0 = -\frac{\gamma^2}{12\pi}J_{\mu}J^{\mu}$ for $\gamma = \left(1-\beta^2\right)^{-\frac{1}{2}}$. In Weyl's 4D world then, matter cannot be constituted without electric charge and current. But since the density of matter is always positive the electric charge and current inside an electron must be a space-like vector, the square of its length being negative. To quote from Eddington:

> It would seem to follow that the electron cannot be built up of elementary electrostatic charges but resolves into something more akin to magnetic charges [38].

Perhaps we can use the structure of Maxwell's equations in complex form to demonstrate that this magnetic structure is indeed the complex part of the field.

In considering $F_{\mu\nu}$ and $A_{\mu\nu}$ as complex entities rather than four-space real forms, we may need complex forms of the current density. Also the relationship between $F_{\mu\nu}$ and $A_{\mu\nu}$ has a spatial integral over charge. If we consider $F_{\mu\nu}$ and $A_{\mu\nu}$ as complex quantities, we see possible implications for the charge e or differential charge de being a complex quantity. Perhaps the expression $e = e_{\text{Re}} + ie_{\text{Im}}$ is not appropriate, but a form for the charge integral is, such as: $\int\frac{de_{\text{Re}}de_{\text{Im}}}{\underline{r}}$ where $r = r_{\text{Re}} + ir_{\text{Im}}$ is more appropriate. Fractional charges such as for quarks, the issue of the source of charge (in an elementary particle) and its fundamental relationship to magnetic phenomena (magnetic domains) are essential considerations and may be illuminated by this or a similar formalism. Neither the source of electrics or magnetics is known, although a great deal is known about their properties.

Faraday's conclusion of the identical nature of the magnetic field of a loadstone and a moving current may need reexamination as well as the issue of Hertzian and non-Hertzian waves. Again, a possible description of such phenomena may come from a complex geometric model [39]. As discussed, one can generalize Maxwell's equations and look at real and imaginary components which comprise a symmetry in the form of the equations. We can examine in detail what the implications of the complex electric and magnetic components have in deriving a Coulomb equation and examine the possible way, given a rotational coordinate, this formalism ties in with the 5D geometries of Kaluza and Klein.

Starting $F^{\mu\nu}$, A_μ and J^μ, Maxwell's equations can be compactly written as $\frac{\partial F^{\mu\nu}}{\partial x_\nu} = J^\mu$ and again, $F_{\mu\nu} = \frac{\partial A_\mu}{\partial x_\nu} - \frac{\partial A_\nu}{\partial x_\mu}$ and $F^{\mu\nu}_\nu = J^\mu$. Now suppose that an electron moves in such a way that its own field on the average just neutralizes an applied external field $F'_{\mu\nu}$ in the region occupied by the electron. The value of $F_{\mu\nu}$ averaged for all the elements of change constituting the electron is given by

$$eF_{\mu\nu} = \frac{1}{4\pi}\left(A_{\mu\nu} - A_{\nu\mu}\right)\iint \frac{de_1 de_2}{r_{12}}$$

and (6.61)

$$eF_{\mu\nu} = \frac{1}{4\pi}\left(A_{\mu\nu} - A_{\nu\mu}\right)\frac{e^2}{a}$$

where $1/a$ is the average value of $1/r_{12}$ for every pair of points in the electron and a will then be a length comparable to the radius of the sphere throughout which the charge is spread. The mass of the electron is $m = e^2/4\pi a$. We thus have a form of Coulomb's law, as we have shown the complex form of $F^{\mu\nu}$ to be consistent with this and Maxwell's equations and that we will have a real and an imaginary Coulomb's law.

Self-consistency can be obtained in the model by assuming that all physical variables are complex. Thus, as before, we assumed that space, time, matter, energy, charge, etc. were on an equal footing as coordinates of a Cartesian space quantized variable model. It is reasonable then to complexity space and time as well as the electric and magnetic fields and to determine the relationship of the equations governing standard physical phenomena. Also to be examined in detail is any unifying properties of the model in terms of complexifying physical quantities as well as examining any new predictions that can be made.

Faraday discusses some possible implications of considering $A_{\mu\nu}$, rather than $F^{\mu\nu}$ as fundamental in such a way that $A_{\mu\nu}$ may act in a domain where $F^{\mu\nu}$ is not observed [39]. In a later section we present a complexification of $A_{\mu\nu}$ rather than $\underline{E}$ and $\underline{B}$ (in $F^{\mu\nu}$).

Continuing with the relationship of $F^{\mu\nu}$, the vector A^{μ}, and scalar potential ϕ, and the metric space, s^{μ} let us relate our complex EM field, $F^{\mu\nu}$, to complex spacetime. We have the volume element $d\tau = \sqrt{g}\,dxdydz$ for

$$ds^2 = g_{\mu\nu} dx^{\mu} dx^{\nu} \tag{6.62}$$

and for a particular vector component of $F_{\mu} = \sqrt{g_{\mu\nu}}\, f^{\mu}$.

Then we have

$$\nabla \cdot \underline{F} = \frac{1}{\sqrt{g}} \frac{\partial f^{\mu} \sqrt{g}}{\partial x^{\mu}} \tag{6.63}$$

For $F = \nabla \phi$ the function f^{μ} is related to the EM potential and gravitational potential as $f^{\mu} = g^{\mu\nu} \dfrac{\partial \phi}{\partial x^{\nu}}$. As before, $\dfrac{\partial F_{\mu\nu}}{\partial x_{\nu}} = J_{\mu}$ and $\gamma_{\mu\nu} \dfrac{\partial F_{\mu\nu}}{\partial x_{\nu}} = J_{\mu}$. As before we also had $\left(F_{41}, F_{42}, F_{43}\right) = i\underline{E}$ and $\left(F_{23}, F_{31}, F_{12}\right) = \underline{B}$ then the generalized complex form of $F^{\mu\nu}$, is

$$F^{\mu\nu} = \begin{pmatrix} 0 & B_z & -B_y & -\frac{i}{c}E_x \\ -B_z & 0 & B_x & -\frac{i}{c}E_y \\ B_y & -B_x & 0 & -\frac{i}{c}E_z \\ \frac{iE_x}{c} & \frac{iE_y}{c} & \frac{iE_z}{c} & 0 \end{pmatrix} \tag{6.64}$$

which we can denote as

$$\underline{F} = \left(B, -\frac{i}{c}\underline{E} \right) \quad \text{or} \quad \underline{F}^* = \left(-\frac{iE}{c}, B \right). \tag{6.65}$$

We can now relate the complex E and B fields of the complex spacetime coordinates.

Returning to the compact notation for the two homogeneous equations, $\nabla \times \underline{E} + \underline{1}\frac{\partial \underline{B}}{\partial t} = 0$ and $\nabla \cdot \underline{B} = 0$ as

$$\frac{\partial F_{\mu\nu}}{\partial x_{\mathrm{K}}} + \frac{\partial F_{\mathrm{K}\mu}}{\partial x_\nu} + \frac{\partial F_{\nu\mathrm{K}}}{\partial x_\mu} = 0 \tag{6.66}$$

It is very clear that introducing the imaginary components into these equations as $\partial / \partial(ix_\mu)$ and $\partial / \partial(it)$ leaves them unchanged.

Now let us look at the inhomogeneous equations $\nabla \cdot \underline{E} = 4\pi\rho$ and $\nabla \times \underline{B} = \frac{1}{c}\frac{\partial E}{\partial t} + J_e$. Consider then

$$F_{\mu\nu} = \frac{\partial A_\nu}{\partial x^\mu} - \frac{\partial A_\mu}{\partial x^\nu} \tag{6.67}$$

or

$$\underline{F} = \underline{\nabla} \times \underline{A} \quad \text{for} \sim \underline{A_\mu} = \left(\underline{A_j, \phi} \right)$$

for j runs 1 to 3 and all Greek indices run 1 to 4, as before. Then the inhomogeneous equations become in general form, $\partial F^{\mu\nu} / \partial x^\nu = s^\mu$ which sets the criterion on s for using $\partial / \partial(ix_{\mathrm{Im}})$; that is, $s' \to is$. To be consistent [40], we can use $A_\mu = \left(A_i, \frac{1}{c}\phi \right)$.

We can consider the group of affine connections for a linear transformation from one system S to another S' where S and S' are two frames of reference and

$$x'_\mu = a_{\mu\nu} x_\nu \tag{6.68}$$

where $a_{\mu\nu} a_\mu^{K} = \delta_{\nu K}$ and $\det|a_{\mu\nu}| = 1$. In general we can form a 4 x 4 coefficient matrix for the usual diagonal condition where, $a_{11} = 1$, $a_{22} = 1, a_{33} = 1$ and $a_{44} = -1$, all the other elements are zero, i.e. the signature (+++-). We can choose arrays of $a_{\mu\nu}$'s both real and imaginary for the general case so that we obtain forms for space and time components as being complex; for example,

$$x'_s = \gamma\left(x_3 + i\beta x_4\right) \tag{6.69}$$

for $x_4 = t$, $\gamma = \left(1-\beta^2\right)^{-1/2}$ *and* $\beta = v/c$. Other examples involve other combinations of complex space and time which must also be consistent with unitarity.

Let us briefly examine the effect of a gravitational field on an electron. Then we will discuss some multidimensional models in which attempts are made to relate the gravitational and EM forces. Some of these multidimensional models are real and some are complex. The structure of the metric may well be determined by the geometric constraints set up by the coupling of the gravitational and EM forces. These geometric constraints govern allowable conditions on such phenomena as types of allowable wave transmission and the manner in which remote space-times are connected. Nonlocality or remote space-time connections have implications for EM phenomena such as Young's double slit experiment and Bell's theorem.

In fact, these experiments are more general than just the properties of the photon, that is, both experiments can be and have been conducted with photons and other particles; and therefore what are exhibited are general quantum mechanical properties. Remote connection and/or transmission and nonlocality are more general than just EM phenomena but certainly have their application in electrodynamics and the nonlocal properties of the space-time metric can be tested by experiments involving classical and quantum electrodynamic properties.

6.7.2 *Complex E & B in real 4-space & the complex Lorentz condition*

Another attempt to relate the relativistic and electro-magnetic theories is the approach of Wyler in his controversial work at Princeton. The model of Kaluza and Klein use a fifth rotational dimension to develop a model to relate EM and gravitational phenomena. This geometry is one-to-one mappable to our complex Minkowski space. Wyler introduces a complex Lorentz group with similar motives to those of Kaluza and Klein [41,42]. Wyler's formalism appears to relate to our complex Maxwell formalism and to that of Kaluza and Klein. The actual fundamental formalism for the calculation of the fine-structure constant, α , is most interesting but perhaps not definitive.

$$\alpha = \frac{e^2}{\hbar c 4\pi\varepsilon_0} = \frac{e^2 c\mu_0}{2h} \tag{6.70}$$

where e is elementary charge, ε_0 vacuum permittivity and μ_0 the magnetic constant or vacuum permeability. An anthropic explanation has been given as the basis for the value of the fine-structure constant by Barrow and Tipler. They suggest that stable matter and intelligent living systems would not exist if α were much different because carbon would not be produced in stellar fusion [43].

Wyler [44] introduces a complex description of spacetime by introducing complex generators of the Lorentz group. He shows the Minkowski, M^n group is conformally isomorphic to the SO(n,2) group and then introduces a Lie algebra of M^4 which is isomorphic to SO(5,2). From his five and four spaces he generates a set of coefficients that generate the value of the fine structure constant, α. It is through introducing the complex form of the Lorentz group, $L(T^n)$ that he forms an isomorphism to SO(n,2).

Wyler calculates the EM coupling constant in terms of geometric group representations. He expands the generators of the set of linear transformations, T^n, of the group $L(T^n)$. By definition, $L(T^n)$ is isomorphic to the Poincaré group $P(M^n)$, where M^n is the Minkowski space with signature (+++-) or, more generally, (1, n-l). The conformal group $C(M^n)$ is then isomorphic to the SO(n,2) group, which is of quadratic form and signature (n,2).

Wyler then chooses the complex form

$$T^n = R^n + iV^n \tag{6.71}$$

(where R^n represents T_{Re} , and V^n represents T_{Im}) for $y \in R^n$, or y is an element of R^n and all y's are y > 0. The Poincaré group, $P(M^n)$ is the semidirect product of the Lorentz group SO(l, n-l) and the group of transformations R_n then is $g \in$ SO(n,2).

Then $C(M^4) \cong SO(4,2)$ is the invariance group of Maxwell's equations. The hyperboloids of the 4-mass shell momentum operators are $p_1^2, \ldots, p_4^2 = m^2$ from the representation of the Lie group geometry of M^4 isomorphic to SO(5,2). The intersection of the D5 (five-dimensional) hyperspace with D4 gives a structure reduced on D4 which is colinear to the reduction of a Casimir operator function, f(z) harmonic in D4.

The coefficients of the Poisson group D^n as D^4 and D^5 give the value of $\alpha \sim 1/137.036$. Actually, it is the coefficients of the Poisson nucleus $P^n(z, \xi)$ harmonic in D^n which gives the value of α in terms of z where z is, in general, a complex function and ξ is a spinor. The value is obtained from the isomorphic groups SO(5) x SO(2) and SO(4) x SO(2) which gives $(9/8\pi 4)(V(D^5)) = 1/137.037$ where $V(D^5)$ is a Euclidean value of the D5 domain [45].

The expression for the Poisson nucleus is given by Hau [45]. Note that the Wyler calculation is another example of the relationship between a fifth dimension and a complex "space" of Lorentz transformation. The Wyler theory appears to strongly support the fundamental nature of geometric models. If one can calculate the fine structure constant or any other force field coupling constants from first principles, this gives great impetus to the concept that geometric constraints are extremely significant and may potentially be able to explain the origin of scientific law. In particular, we may be able to at least describe the major force fields (nuclear, EM, weak, and gravitational) in terms of a geometric structure and, perhaps, by this formalism demonstrate the unifying aspects of major forces of nature [46].

Wyler also associates the conformal group $C(M^n) \cong SO(4,2)$ with the invariant group of Maxwell equations. The mass shell conditions on the hyperboloids of mass form the representation of the Lie algebras of M^4. Isomorphism to SO(5,2) and S(4,2) intersection lead to a model of the intersection of Maxwell's field and the elementary particle field, i.e. a possible unification of EM and weak interactions [47].

In the presence of an external gravitational field, the cosmological term is small and finite and depends on the state of vacuum state polarization. In fact, the cosmological term is given by the sum of all vacuum diagrams. In the supersymmetry then, the cosmological term vanishes and therefore the total zero-point energy density of the free fields vanishes [48].

Let us return to our complex $\underline{E}$ and $\underline{B}$ fields and suggest the relation of our formalism to the Wyler formulation. Using the invariance of line elements $s = X^2 - c^2t^2$ for $r = ct = \sqrt{X^2}$ for $X^2 = x^2 + y^2 + z^2$, to measure the distance from a test charge to an electron charge, we can write for the imaginary part of the complex Maxwell equation $\nabla \times \left(iE_{\text{Im}} \right) = \frac{1}{c}\frac{\partial\left(iB_{\text{Im}} \right)}{\partial t} + iJ_{\text{Im}}$ then for $E_{\text{Im}} = 0$.

$$\nabla \times \left(iE_{\text{Im}} \right) = 0 \quad or \quad \frac{1}{c}\frac{\partial\left(iB_{\text{Im}} \right)}{\partial t} = iJ_{\text{Im}} \tag{6.72}$$

or

$$\frac{\partial\left(i\underline{B}_{\text{Im}} \right)}{\partial r} = ic\underline{J}_{\text{Im}} \quad or \quad \frac{\partial \underline{B}_{\text{Im}}}{\partial r} = c\underline{J}_{\text{Im}} \tag{6.73}$$

for the assumed i, B_{Im} commutator relation.

Now let us examine the energy associated with the imaginary part of the magnetic field, $\underline{B}_{\text{Im}}$. We can calculate an energy invariant by squaring and integrating the above equation as [30,49]

$$\mathcal{E} = -\int_\tau J_m^2 R d\tau = -\int_\tau \left(\frac{\partial B_z}{\partial r} \right)^2 R d\tau \le 0 \tag{6.74}$$

The distance function $R(r)$ over the volume element $d\tau$ is assumed to be point-symmetrical and vanishes for positive real energy states. The volume $d\tau$ is constructed to include a small real domain where a point charge is located, avoiding possible divergences. The negative value of the energy integral leads us to hypothesize about what the source of this energy may be. Perhaps it can be related to vacuum state polarization in a Fermi sea model, as we have presented before [10]. Another possible association is with advanced potential models such as those of de

Beauregard [50,51]. A third and perhaps the most interesting association would be with the complex coordinate space [52,53].

In Weyl's non-Riemannian geometry, [36] he presents a model that does not apply to actual spacetime but to a graphic representation of that relational structure, which is the basis in which both EM and metric variables are interrelated [38]. This is the deep significance of the geometry and relates to work of Hanson and Newman [54] on the complex Minkowski space as well as Wyler's work [44] on complex group theories, such as complex Lorentz invariance, where he attempts to reconcile Maxwell's equations and relativity theory. The examination of the hyperspheres of the de Sitter space is presented by Ellis, where he attempts to unify EM and gravitational theory [55]. Eddington has suggested that the Weyl formalism, developed around 1923, is one of the major advances in the work of Einstein.

There is a significant difference between Einstein's generalization of Galilean geometry and Weyl's generalization of Riemannian geometry. The gravitational force field renders Galilean geometry useless and therefore the move to Riemannian geometry was made. In terms of Weyl's geometry, we find that the EM force, $F_{\mu\nu}$, is comparable to the surface of an electron of 4 x 10^{18} volts/cm, [38] and the size of the charge was compatible with the radius of curvature of space.

For the EM mass, $m_e = e^2 / 4\pi a,$ we have

$$m_g ds = \frac{1}{8\pi} G\sqrt{g}d\tau \tag{6.75}$$

where we denote the curvature R by G for the general case of both gravitational and EM field. The ratio of the masses m_g / m_e relates to the ratio of field strengths.

6.7.3 *Complex EM forces in a gravitational field*

We have considered the weak Weyl limit of the gravitational force in previous calculations of this chapter. We will briefly outline how the complexification of $F_{\mu\nu}$ can be formulated geometrically. We show that we obtain the same results for the relationship of mass and charge.

Let v^{μ} denote the velocity vector as $v^{\mu} = dX_{\mu} / dS$ of the electron in the field, and ρ_0 denote the proper density of charge, called e. Then the current is given by $J^{\mu} = \rho_0 v^{\mu}$. Let $F_{\mu\nu}$ refer to the applied external force of the electron. Returning to Eddington's calculation [38], we then have

$$mA^{\nu} A_{\mu\nu} = -F_{\mu\nu}\rho_0 A^{\nu}. \tag{6.76}$$

We can also write ρ_0 as e in the above equation.

In the limit of our gravitational field we can neglect the gravitational field as an external field or also the gravitational energy of the electron. To discuss the presence of an electron in a gravitational field we start from the field equations with R_{μ}^{ν} the Ricci curvature tensor and g_{μ}^{ν} the metric tensor for the case where no matter is present we have:

$$G_{\mu}^{\nu} = R_{\mu}^{\nu} - \frac{1}{2} g_{\mu}^{\nu} R = -\frac{8\pi}{c^4} G E_{\mu}^{\nu} \tag{6.77}$$

using the scalar curvature, $R = \dfrac{8\pi GE}{c^4} = 0$. Then this equation simplifies to

$$R_{\mu\nu} = -8\pi E_{\mu\nu}. \tag{6.78}$$

This equation applies to regions that contain EM fields but no matter and no electron charges in the region.

For the only surviving component in the energy considerations, we have

$$F_{41} = -F_{14} = \frac{\partial \phi}{\partial r} \tag{6.79}$$

where r is the radial separation. Then $F^{41} = g^{44} F_{41}$ and $\dfrac{\partial \phi}{\partial r} \propto \dfrac{\varepsilon}{r^2}$ and

$$E_1^1 = +E_2^2 = E_3^3 = -E_4^4 = \frac{1}{2}\frac{\partial \phi}{\partial r} = \frac{1}{2}\frac{\varepsilon^2}{r^4}.$$

Jeffrey associates m, the mass of the electron, with $4\pi\varepsilon$, giving $\alpha = \dfrac{2\pi\varepsilon^2}{m} \sim 1.5\times 10^{-13}$ cm and justifies identifying $4\pi\varepsilon$ with the electrical charge e or

$$F_{41} = \frac{\partial \phi}{\partial r} = \frac{1}{4\pi}\frac{e}{r^2} \tag{6.80}$$

We can then use $\Box F_{\mu\nu} = J_{\mu\nu} - J_{\nu\mu}$ for $A^{\mu} = \dfrac{de}{4\pi r}$ and then

$$\begin{aligned} F_{\mu\nu} &= \int \frac{de\left(A_{\mu\nu} - A_{\nu\mu}\right)}{(4\pi\gamma)r} \\ &= \frac{1}{4\pi\gamma}\left(A_{\mu\nu} - A_{\nu\mu}\right)\int \frac{de}{r} \end{aligned} \tag{6.81}$$

because all parts of the electron have the same relativity where

$$\frac{\partial^2 A^{\mu}}{\partial t^2} - \nabla^2 A^{\mu} = J^{\mu}$$

and

$$A^{\mu} = \frac{1}{4\pi}\frac{ds}{dt} v^{\mu}\int_{\tau} \frac{\rho d\tau}{r}, \tag{6.82}$$

for velocity, v^{μ}, we will drop the γ since all measurements will be assumed to be proper time measurements. Now integrating over the electron between pairs of points on the electron surface,

$$\begin{aligned} eF_{\mu\nu} &= \frac{1}{4\pi}\left(A_{\mu\nu} - A_{\nu\mu}\right)\iint \frac{de_1 de_2}{r_{12}} \\ &= \frac{1}{4\pi}\left(A_{\mu\nu} - A_{\nu\mu}\right)\frac{e^2}{a} \end{aligned} \tag{6.83}$$

where 1/a is the average value of $1/r_{12}$. We can write Eq. (6.83) as

$$-eA^{\nu}F_{\mu\nu} = \frac{1}{4\pi}A^{\nu}\left(A_{\mu\nu} - A_{\nu\mu}\right)\frac{e^2}{a} \tag{6.84}$$

and using the equation from before, relating v^{ν}, $A_{\mu\nu}$, $F_{\mu\nu}$ and A^{ν}, $mv^{\nu}A_{\mu\nu} = -F_{\mu\nu}eA^{\nu}$, so that $m = e^2/4\pi a$ as before.

How does this relate to the de Sitter spaces? In the de Sitter algebras the proper time in all inertia! frames of intervals is the same (or equivalent). This is the powerful absolute of the de Sitter space. The proper time interval $d\tau$ on its geodesic world-line in the de Sitter picture is given as

$$d\tau^2 = dt^2 - e^{2t}\left(dX^2\right) \tag{6.85}$$

for $dX^2 = dx^2 + dy^2 + dz^2$ in Euclidean coordinates and t is the cosmic time. The metric form of the de Sitter universe represents the metric form consistent with the observed (approximately flat, low density) universe that we observe. It is constant with Einstein dynamic equations and is therefore consistent with the Hubble's expansion [56].

Figure 6.2 Plotted are the geodesies of the de Sitter space which represent the field lines of the EM field. Various conditions for signal propagation are given.

Ellis [55] suggests that geometry and EM can be unified by a rigorous analysis of time. The hyperspheres of de Sitter space can be represented as a five-dimensional metric manifold which tie the geometric models of gravity and electromagnetism to the structure of matter, and time is not primary but a property of the matter (elementary particles). If $\tau = t$ is allowed in the de Sitter space, then the typical geodesies represent what appears to be EM field lines. This is the manner in which Ellis attempts to describe the EM phenomena as geometric!

The conformal invariant is given as

$$ds^2 = \frac{1}{R^2}\left(dx^2 + dy^2 + dz^2 - dR^2\right) \tag{6.86}$$

which depends only on the ratios of distances and is thus independent of scale. Let $t = -\ell nR$, then $R = e^{-t}$ and $ds^2 = e^{2t}\left(dx^2 + dy^2 + dz^2\right) - dt^2$ which is the de Sitter metric element. Ellis' geodesies of his angle metric correspond to geodesies of the de Sitter space (Figure 6.3a). In Figure 6.3b, they are time-like subluminal geodesies, and in 3c they are luminal, and in 6.3d they are space-like superluminal. The figures also contain Euclidean space planes as spheres of infinite radii.

Feinberg [57] suggests that the first step in the test of multi-dimensional geometric models is to predict some simple phenomena such as the Coulomb attraction-repulsion; note that Figure 6.3 may point a way to do this, because if we can relate this five-dimensional geometry to the complex geometry, then we can relate this complex geometry to Coulomb interactions.

The curvature of space may then be related to a rotation or angular momentum component as a Kaluza-Klein 5th dimension. We can form an isomorphism of this geometry to an 8D real-complex coordinate geometry which appears to not only unify EM theory and gravitational theory but may also resolve some other apparent paradoxes [58,59].

We have seen that the introduction of the complex *E* and *B* fields or complexifying the field, $F^{\mu\nu}$, can be handled in such a way as to not distort the electric charge on the electron. We also find consistency with the five-dimensional geometry of Kaluza and Klein, the 8D Minkowski space, and the de Sitter space where the geodesic represents the EM field lines. We can also maintain Lorentz invariance conditions for both real and complex transforms on the line element.

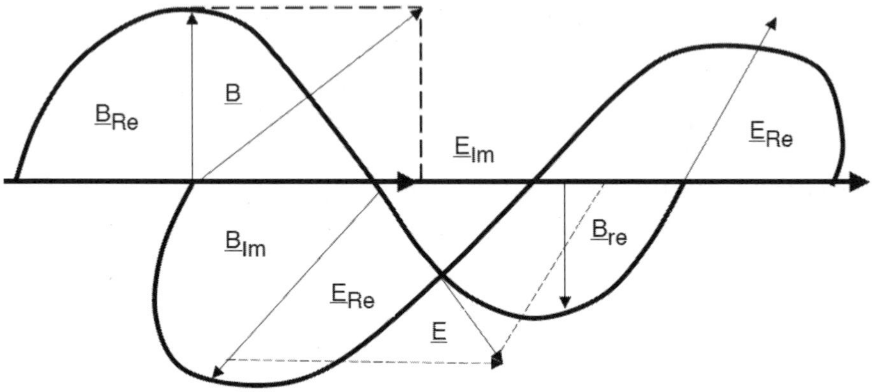

Electric and magnetic Hertzian (transverse), for B and E real and non-Hertzian (longitudinal waves for the imaginary components of B and E.

Figure 6.3 Hertzian and non-Hertzian waves.

6.8 Summation and Conclusions

This model exploits:

a) the analogy (underlined by Puthoff) between the four-vector density representation of gravity and electromagnetism in flat space-time [4].
b) the possibility of describing the causality of quantum mechanical phenomena in terms of extended solitons piloted i.e. by quantum mechanical potentials, by real guiding collective waves on a chaotic, polarizable Dirac-type aether - both moving in a flat space-time [28].
c) the representation of this real vacuum (Dirac aether) in terms of the chaotic distribution of real extended elements moving in the flat space-time.
d) the introduction of internal motions within extended sub-elements and their relation with local collective motions i.e. the $E = mc^2 = h\nu$ relation.
e) the representation of the electron (and its associated pilot-wave) in terms of extended elements with a point-like charge rotating around a center of mass [28].

These assumptions yield realistic physical characteristics to known empirical properties and predict new testable relations besides known properties of elementary particles. The present model must thus be extended, by associating new internal motions to these known properties and interpret them in terms of new strong spin-spin and spin-orbit interactions.

Our attempt is justified by the existence of EM phenomena not explained by Maxwell's equations. Barrett [28] has stated that Maxwell's theory does not explain the Aharonov-Bohm (AB) effect and Altahuler-Aharonov-Spivak (AAS) effects. It does not cover the topological phase question i.e. the Berry-Aharonov-Anandan, Pancharatnam and Chio-Wu phase-rotation effects. An inclusion of Stoke's theorem is necessary and results of Ehrenberg and Siday must be analyzed. The quantum results of Josephson, Hall, de Haas and van Alphen Sagnac-type experiments also need clarification.

The integration of gravity and electromagnetism however, is not finished, because unification is so far only accomplished in terms of bumps and holes rotating on the stochastic surface of the polarized Dirac Vacuum. Unification must also occur in terms of the richer Higher Dimensional (HD) structure of vacuum topology where one would show the geometric origin of charge and how bumps and holes transform into each other through quasi-particle like transitions piloted by advanced and retarded potentials of the fundamental unitary field itself.

References

[1] Einstein, A (1922) Geometry and Experience Sidelights on Relativity, Denver.

[2] Bass, L. & Schrödinger, E. (1955) Proc. Roy. Soc. A 232.

[3] Maxwell, J.C. (1954) A Treatise on Electricity and Magnetism, New York: Dover.

[4] Puthoff, H.E. (2002) Polarizable-Vacuum, Representation of General Relativity, in R.L. Amoroso , G. Hunter, M. Kafatos & J-P Vigier (eds.), Gravitation and Cosmology: From the Hubble Radius to the Planck Scale, Dordrecht: Kluwer.

[5] Sakharov, A.D. (1967) Vacuum quantum fluctuations in curved space and the theory of gravitation, Sov.Phys.Dokl.12:1040-1041;

[6] Sakharov, A.D. (1968) Dokl. Akad. Nauk Ser.Fiz.,177:70-71;

[7] Sakharov, A.D. (1967) Sov.Phys.Usp.34:394, (1991) Reprinted in Sov. Phys. Usp. Fiz. Nauk 161, No. 5 64-66.

[8] Halbwachs, F., Souriau, J.M. & Vigier, J-P (1922*) Journal de Physique et de Radium*, 22, p. 26.
[9] Halbwachs, F. (1960) *Théorie Relativiste Fluides*, Paris, Gauthier-Villars.
[10] Andrews, T. B. (1999) Observational tests of the static universe model and the derivation of the Hubble redshift, The Hy-Redshift Universe: Galaxy Formation and Evolution at High Redshift, Berkeley, 21-24 June, 1999. ASP Conference Proceedings, A.J. Bunker & W.J. M. van Breugel (eds.) Vol. 193, p.407.
[11] Amoroso, R.L., Kafatos, M. & Ecimovic, P. (1998) The origin of cosmological redshift in spin exchange vacuum compactification and nonzero restmass photon anisotropy, in G. Hunter, S. Jeffers & J-P Vigier (eds.) Causality & Locality in Modern Physics, pp. 23-28, Dordrecht: Kluwer.
[12] Aharanov, Y. & Bohm, D. (1959) Physical Review, 115, 485.
[13] Dirac, P.A.M. (1951) Nature, 906.
[14] Amoroso, R.L. & Vigier, J-P (2002) The origin of CMBR as intrinsic blackbody cavity-QED resonance inherent in the dynamics of the continuous-state topology of the Dirac vacuum, in R.L. Amoroso , G. Hunter, M. Kafatos & J-P Vigier (eds.) Gravitation and Cosmology: From the Hubble Radius to the Planck Scale, Dordrecht: Kluwer.
[15] Arp, H. (1987) Quasars, Redshifts and Controversies, Berkeley: Interstellar Media.
[16] Weyssenhof, J.V. & Raabi, A.(1947) *Acta Phys. Polon.*, 9, 7-18. 3.
[17] Vigier, J-P (1997) Phys. Lett. A., 235.
[18] Kholmetsky, A.L. (1995) On relativistic kinematics in the Galilean space, Galilean electrodynamics, 6:3S; 43-50.
[19] Nakano, I. (1956) Progress in Theoretical Physics, 15.
[20] Roscoe, D.F. (2006) Maxwell's equations: New light on old problems, Apeiron, Vol. 13, No. 2.
[21] Krogh, K. (2006) Gravitation without Curved Spacetime, arXiv:astro-ph/9910325v23.
[22] Graneau, P., Graneau, N., Hathaway, G. & Hull, R. (2002) Why does lightning explode and generate MHD power? Infinite Energy, 5:25; 9-11.
[23] Saumont, R. (1998) Undermining the foundations of relativity, 21st Century Science & Technology, Summer, pp 83-87. Proc. Cold Fusion and New Energy Symposium, Manchester.
[24] Evans, M.W. & Vigier, J-P (1996) The Enigmatic Photon, Vol. 1-4,

Dordrecht: Kluwer.
[25] Lehnert, B & Roy, S. (2000) Extended Electromagnetic Theory, Singapore: World Scientific.
[26] Ghosh, A. (2000) Origin of Inertia, Apeiron, Montreal.
[27] Assis, A.K. (1994) Weber's Electrodynamics, Dordrecht: Kluwer.
[28] Barrett, T.W. (1993) Electromagnetic Phenomena not Explained by Maxwell's Equations: Essays on the Formal Aspects of Electromagnetic Theory, Singapore: World Scientific.
[29] Einstein, A. (1955) Relativity, Princeton : Princeton Univ. Press.
[30] Inomata, I. (1976) Consciousness and complex EM fields, Electrotechnical Laboratory, MITI, Tokyo.
[31] Minkowski, H. (1915*) Ann, Phys. Lpz*. 47; 927; *Jber. Deutsche Mat. Vev*. 24; 372.
[32] Rauscher, E.A. (1982) Theoretical and field studies of extremely low frequency radiation and coherent linear phenomena, PSRL-726A.
[33] Frank, P. (1911) *Ann. Phys. Lpz.* 34; 599.
[34] Pauli, W. (1962) Theory of Relativity, Chicago: Permagon Press.
[35] de Broglie, L. (1955) *Mécanique Ondulatoire du Photon*, Paris: Gauthier-Villas.
[36] Wehl, H, (1918) *Math. Z.* 20, 384; *Pruss. Akad. Wiss*. 465.
[37] Wehl, H, (1920) Phys. Z. 21, 649.
[38] Eddington, A. (1922) The Mathematical theory of Relativity, London: Chelsea Publishing Co.
[39] Faraday, M. (1855) Experimental Researches in Electricity, Vols. 1-3, R. Taylor & W. Francis (eds.) London: Univ. London Press.
[40] Stratton, J.A. (1941) Electromagnetic Theory, NY: McGraw-Hill.
[41] Kaluza, T. (1921) *Sitz. Berlin Preuss. Akad. Wiss.*, 966.
[42] Klein, O. (1926) *Z. Phys*. 37, 895.
[43] Tipler, F.J. (1974) Phys. Rev. D9, 2203.
[44] Wyler, A. (1969) *Acad. Sc.* Paris Series A, 269, 743; (1971) *Acad. Sc.* Paris Series A, 271, 186.
[45] Hua, L.K. (1963) Harmonic Analysis in the Classical Domain, Providence: American Mathematical Society.
[46] Gell-Mann, M., Raymond, P. & Slonsky, R. (1979) Supergravity, P. Van Nieuwenhuizen & D.Z. Freedman (eds.) Amsterdam: N. Holland.
[47] Wyler, A. (1968) Arch. for Rational Mechanics & Analysis, 31, 35.
[48] Wess, W. & Zumino, B. (1974) Nucl. Phys. B78, 1.
[49] Wehl, H. (1957) Space-Time-Matter, New York: Dover.
[50] Costa de Beauregard, O. (1976) Acad. Sci. Paris, 104, 1251.

[51] Costa de Beauregard, O. (1977) Phys. Gelt. 60A, 93.
[52] Rauscher, E.A. (1973) in The Iceland Papers: Frontiers of Physics Conf. Amherst: Essentia Research Associates.
[53] Ramon, C & Rauscher, E.A. (1980) Found. Phys. 10, 661.
[54] Hanson, R.O. & Newman, E.T. (1975) Gen. Rel. & Grav. 6, 216.
[55] Ellis, H.E. (1974) Found. Phys. 4, 311.
[56] Rauscher, E.A. (1973) Early universe cosmological models, Bull. Amer. Phys. Soc. 18,1570.
[57] Feinberg, G. (1967) Phys. Rev. 159, 1089.
[58] Rauscher, E.A.& Targ, R. (1973) Why only 4D will not explain nonlocality, J Sci. Explor. 16, 655.
[59] Ramon, C. & Rauscher, E.A. (1982) Remote connectedness in complex geometries, PSRL-4105.

Chapter 7

Redshift/CMBR as Intrinsic Blackbody Cavity-QED Absorption/Emission Equilibrium Dynamics

The microwave 'background' makes more sense as the limiting temperature of space heated by starlight than as the remnant of a fireball - Sir Arthur Eddington [1].

Cosmologies that do not include the Big Bang have not produced any plausible alternative interpretation of the background radiation - J. Silk [2].

Alternative interpretations for the two main pillars of Big Bang cosmology are formally introduced. A redshift / CMBR complementarity is delineated as complex blackbody equilibrium conditions intrinsic to the Cavity-QED resonance dynamics of the spin exchange coupling inherent in extended spacetime hyperstructure oscillations rotating relativistically within the topology of a higher dimensional (HD) form of a covariant polarized Dirac vacuum, with correspondence to the usual asymptotically flat Einstein/Minkowski energy-dependent spacetime metric, $\hat{M}_4$. In this frame a Vigier style dissipative redshift mechanism is described as absorption and Cosmic Microwave Background Radiation (CMBR) as emission within the context of an extended de Broglie-Bohm-Vigier causal interpretation of quantum theory that includes extended EM theory and photon mass anisotropy.

7.1 Introductory Cosmological Parameters

Historically the Vigier vacuum-induced dissipative redshift mechanism has been considered the most highly developed and viable alternative to the Doppler recessional velocity model. What remains to complete the model is to introduce a more sophisticated delineation of the vacuum mechanics and a coupling to the Cosmic Microwave Background Radiation (CMBR) emission process. Astrophysicists empirically claim that the CMBR represents a near perfect blackbody spectrum. Most age of the universe measurements have been performed utilizing Hubble's Law interpreted through a Doppler expansion of the universe. Most recently age of the universe measures have been calculated utilizing data from the Wilkinson Microwave Anisotropy Probe (WMAP) satellite launched in 2001 to measure CMBR temperature. The WMAP data age of the universe is 13.37 billion years, for a cosmos composed of 4.6% ordinary baryonic matter; 23% dark matter and 72% dark energy.

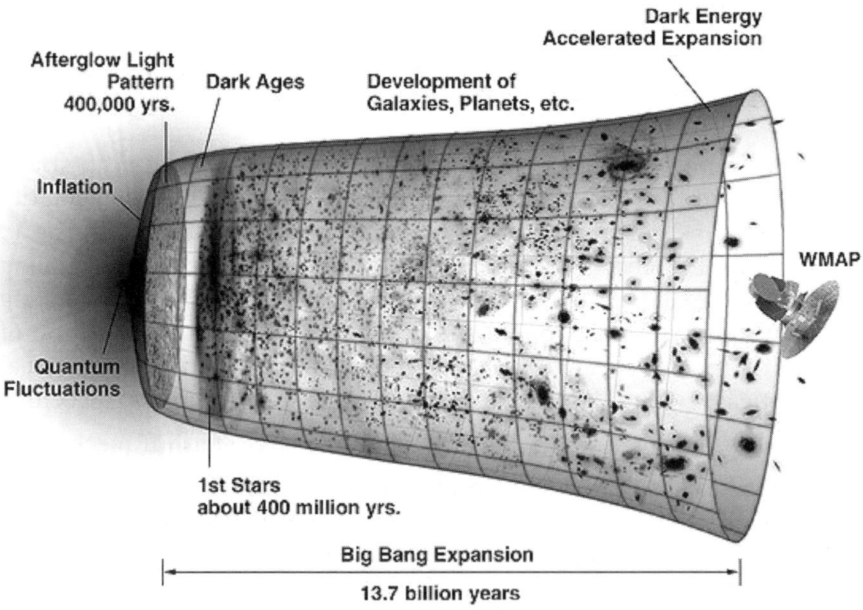

Figure 7.1 Interpretation of data from the 2001 Wilkinson Microwave Anisotropy Probe (WMAP) satellite, a follow-up to the 1992 COBE satellite from a Big Bang perspective. Figure courtesy of NASA.

If the Hubble radius, H_R instead represents an observational limit based on a 'tired-light' photon energy attenuation by vacuum coupling in a static universe one would obtain the exact same result by calculations based on the same Hubble distance relation stemming from a Hubble constant of 70.1 ± 1.3 km·s^{-1}·Mpc^{-1}. Therefore the existing 'über-bias' for one interpretation over the other is myopic and unscientific. It is true that until now a sufficiently pragmatic understanding of the nature of the photon, electromagnetic (EM) field theory, quantum theory, gravitation and vacuum structure have been elusive making the 'alternative' interpretation difficult to rigorously delineate. One cannot therefore be too 'über-critical' in exchanging one bias for another other than to complain of the puerility of human nature.

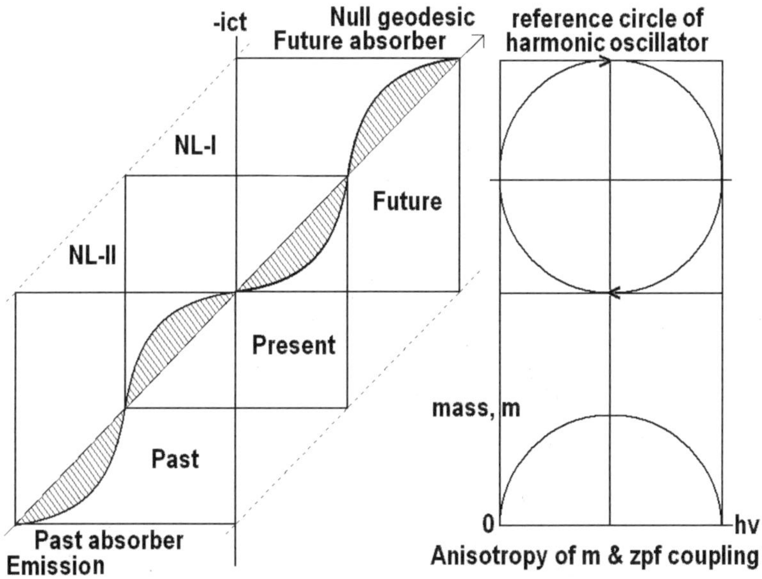

Figure 7.2 a) 2D drawing of a 3D view of a 4D hyperstructure. A Minkowski spacetime diagram of the electric vector only in terms of a present moment of 'tiled' Planck units utilizing the Wheeler-Feynman theory of radiation. The vertices represent absorption & emission. The observable present is represented by bold lines, and nonlocal components by standard line. Each event is a hyperstructure of Past, Present, and Future interactions, ultimately governed by the quantum potential. b) In the reference circle photon mass and energy fluctuate harmonically during propagation of the wave envelope (wave) and internal rotation of the ZPF during coupling (particle).

The expression 'the temperature of space' is the title of chapter 13 of Sir Arthur Eddington's famous 1926 work [1]. Eddington calculated the minimum temperature any body in space would cool to, given that it is immersed in the radiation of distant starlight. With no adjustable parameters, he obtained 3° K (later refined to 2.8° K), essentially the same as the observed, so-called 'background,' temperature [3]." Instead of being a relic of an initial hot, dense, primordial singularity, a putative model of CMBR/Redshift as blackbody emission/absorption equilibrium is predicted to occur in the context of the de Broglie-Bohm-Vigier Causal Interpretation of quantum theory where the wave function, ψ describing individual quantum particles is not a mathematical artifice as often considered in the standard Copenhagen Interpretation, but represents physically real elements piloted within a real chaotic ether by a quantum potential [4-6]. Cramer's Transactional Interpretation [7], based on the Wheeler-Feynman absorber theory of radiation [8,9] claims that any present instant [10] is a standing-wave 'transaction' of advanced-retarded future-past elements that are also physically real [7]. See Figs. 7.4 below, and also Figs. 10.2, and 10.3.

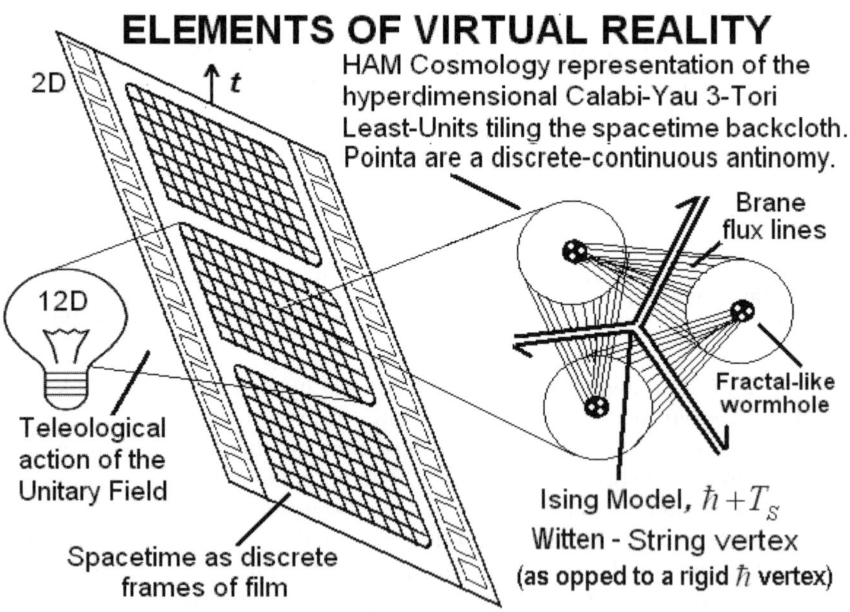

Figure 7.3 Spacetime is virtual in HAM cosmology and the least cosmological units tiling its backcloth are driven by a teleological anthropic action principle. Each 'point' is a continuous-discrete antinomy.

Although quantum theory itself is silent on the matter [11]; the theory of wave-function collapse is well known; that any measurement or interruption of its evolution results in collapse or production of a new wave-function. In HAM cosmology reality itself is a continuous wave-function collapse of HD elements. See Chap. 4 on reality and the arrow of time, the topology of this virtual standing-wave structure of an event for any instant of the eternal present [10] is extended to include hyperdimensional SUSY symmetry breaking dynamics to complete the general framework as seen especially in Chap. 3, also Chap. 5 and later in this chapter. At the time of writing no formal evidence for supersymmetry exists and no Standard Model superpartners have been found, suggesting supersymmetry is a broken symmetry with heavy 'sparticles'. However we postulate this theoretical projection is a result of Gauge Theory being only an approximation and therefore ultimately in reality neither superpartners, a Higgs Boson or graviton will be found. Supersymmetry is possible in additional dimensionality because spinor properties vary with dimensionality. In N dimensions, the size of a spinor is approximately $2^{N/2}$ or $2^{(N-1)/2}$. The maximum number of supersymmetries is 32, so that the largest dimensionality for supersymmetry is eleven. What we hope to show is that brane tension and coupling mechanisms in some form of extended Wheeler wormhole model provides the fundamental origin of mass. The best indicia for this concept of course is the Dirac spherical rotation electron model.

In the Big Bang scenario redshift and CMBR arise in a straight-forward manner – Doppler expansion and cool down from a hot initial singularity. In the HAM cosmological model the basis relies on numerous open questions and entirely new concepts such an *eternal present* [10] that is a complex self-organized standing-wave of a unique form of SUSY *future-past,* [10,12] continuous-state symmetry breaking parameters. In this context one must look for an inherent Cavity-QED (C-QED) spacetime topology within the covariant polarized Dirac vacuum [13,14] where the Planck Blackbody spectrum can be described as an equilibrium condition of cosmic redshift, as absorption and CMBR, as emission. In order for Redshift to be non-Doppler, i.e. not signifying an expanding universe, the next challenge is to rely on the implications of extended Electromagnetic Theory [15,16] especially the Proca equation, $\Box A^{\nu} - \partial^{\nu}\left(\partial\mu\partial^{\mu}\right) + m^2 A^{\nu} = j^{\nu}$ [17] that allows one to introduce a relativistic massive spin 1 particle, in this case the photon is suggested to have a small finite mass, m_γ purported to arise by internal motion dynamics of the $B^{(3)}$ longitudinal *EM* field [18-20]. Maxwell's equations are known to cut off at the vacuum; what the Proca equation is all about is to make them continuous into the vacuum [21] – the Dirac polarized vacuum we have been considering. In order to put all this together into a complete model we have extend the so-called 'tired light' mechanism developed by Vigier. In the tired-light model a massive photon couples to this Dirac covariant polarized vacuum through harmonic oscillation of its internal motion [22-24], a wave-particle duality mass anisotropy effect, where $0 \leftrightarrow m_\gamma \leftrightarrow 10^{-65}$ g respectively.

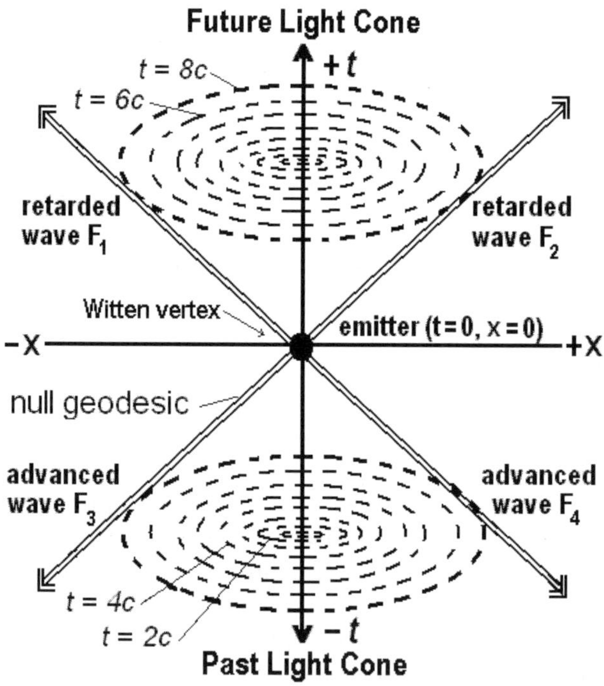

Figure 7.4 4D Minkowski light-cone of advanced and retarded waves (Eq. 7.1) emitted from a locus at $(x,t) = (0,0)$. Adapted from concepts of Cramer [7].

Retarded: $$F_1 = F_0 e^{-ikx} e^{-2\pi ift}, \quad F_2 = F_0 e^{ikx} e^{-2\pi ift} \tag{7.1a}$$

Advanced: $$F_3 = F_0 e^{-ikx} e^{2\pi ift}, \quad F_4 = F_0 e^{ikx} e^{2\pi ift} \tag{7.1b}$$

As part of the symmetry breaking process the continuous-state spin-exchange compactification dynamics of the vacuum hyperstructure is shown to gives rise naturally to a 2.735° K degree Hawking type radiation from the topology of Planck scale (albeit a whole new consideration of how the Planck regime operates) micro-black hole hypersurfaces. All prior considerations of 'tired-light mechanisms have been considered from the perspective of 4D Minkowski space [25-34]. This new process arises from a richer *open* (non-compactified) Kaluza-Klein dimensional structure of a continuous-state cosmology in an M-Theory context with duality-mirror symmetry; also supporting the complex standing-wave postulate of the model.

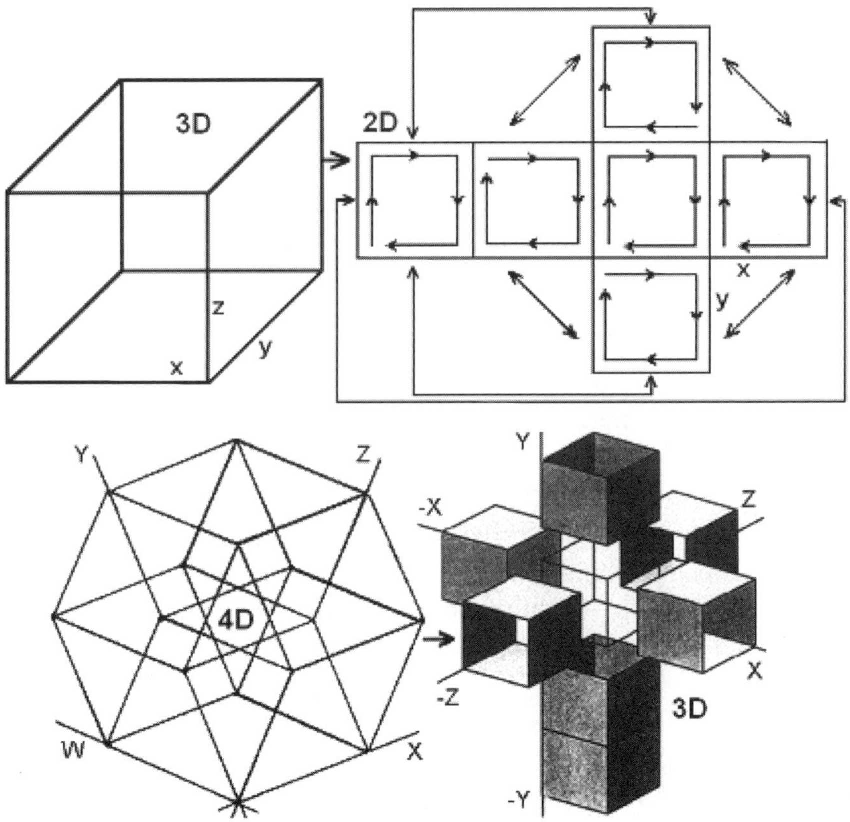

Figure 7.5 2D view of the HD geometry of space conceptualized in unfolded 3D & 4D views to aid visualization.

An additional note to keep in mind for the global nature of this cosmology: The Einstein-Hubble 3-sphere in HAM cosmology is a self-organized complex system; one of an infinite number of nested Hubble spheres, each with their own fine-tuned laws of physics. The $360^{\circ}-720^{\circ}$ Dirac spinor rotation of the electron by covariance also reflects the continuous-state transformation of HD spacetime itself as the basis for our 'virtual reality'. Each H_R is closed and finite in time, but open and infinite in the bulk of the multiverse. Complex systems are driven by an external force [35-37]; this allows the putative anthropic principle to drive the evolution (a super quantum potential) of each nested $H_{R'}$. What we are trying to emphasize is that this covariant scale-invariant structure applies to the microscopic C-QED structures we wish to model for BB equilibrium complementarity.

In Fig. 7.4a & b a 3D cube unfolds into the 2D plane, aiding the visualization of HD space. In Fig. 7.4c,d a 4D hypercube unfolds into 8 component 3D cubes as in 7.4b. If a 5D hypercube were unfolded the 8 cubes forming the 3D cross (7.4d) would be 4D hypercubes (tesseracts as in Fig. 7.4c). The translucent cube in the center of 7.4d, called the *central cube*, represents observed Euclidean reality, E_3. This central cube is surrounded by six *adjacent cubes*. The 8th cube, the *satellite cube*, is placed arbitrarily on any adjacent cube. Carried to 12D the central cube and 12D satellite causally separate as a 'mirror image of a mirror image' is separated from the initial object.

If this 'reality transformation' of HAM cosmology (Fig. 7.4) is carried to a 12D superspace, 12D can be said to describe 'eternity' because 12D is the minimum number of dimensions to be causally free of the virtual E_3/M_4 complex HD 'standing-wave' present, $\hat{M}_4 \pm C_4$ [35-38]. This is commensurate with some type of dual Calabi-Yau 3-forms which in the Wheeler-Feynman formalism can be simplistically written (as adapted from Narlikar [39]) in unexpanded form as

$$F_{symM_4}^{S^N} = \frac{1}{2}\left[R_{retC_4}^{S_N} + R_{advC_4}^{S_N} \right] \tag{7.2}$$

This 3-torus cosmological least-unit [40] singularity structure of the *advanced-retarded future-past standing-wave dynamics* is a foundational principle of the continuous-state anthropic multiverse.

The Big Bang can be obviated by invalidating Einstein-Friedmann universes upon which it is based. The field equations of General

Relativity (GR) allow for singularities, the existence of which has been used to predict black holes and Big Bang inflationary origins of a temporal universe; but both GR and quantum theory (QT) are known to be incomplete. The Big Bang although highly successful cannot claim logical consistency. This was the state of Newtonian mechanics before it was superseded by quantum mechanics and GR. It is not possible for an event to initialize inflation from an era of infinite entropy without violating the law of conservation of energy. Therefore a scientific justification for a Big Bang era is not possible [41,42]. Many other inconsistencies with the Big Bang interpretation are passionately debated in the literature [3]. The standard Big Bang model is founded on strong observational data; prompting many to accept it unconditionally. While the empirical data are correct; the interpretation relies on an incorrect metaphysical basis. The crisis facing scientific epistemology has come full circle to a similitude of Galileo's time when the logic of sound philosophical deduction failed deduce natural law. Einstein's refinement of Newtonian gravitation will be repeated for General Relativity (GR) by post-quantum anthropic cosmology, requiring inflationary models of the universe to be critically reevaluated. This chapter introduces a radical new view of compactification dynamics for a Dirac vacuum hyper-structure utilizing spin-spin coupling to build on the 'tired-light' model developed by Vigier [34].

7.2 Origin of Redshift in Nonzero Restmass Photon Anisotropy in Photon Propagation and the Vigier Tired-Light Hypothesis

The self-referential flavor of GR's equivalence principle induced conformal map between a curved Einstein-Riemannian 4-space and a locally conformally flat Lorentzian spacetime manifold shelved the propagation problem inherent in a 'Maxwellian ether' after the null results of the Michelson-Morley experiment; but Einstein said relativity did not compel us to exclude the possibility of an ether – namely spacetime itself. Since GR endows space with physical qualities; "space without ether is unthinkable" [43]. Photon anisotropy requires vacuum zero point coupling, and its propagation can no longer be considered independent of the Dirac vacuum [44]. The fluctuation of the vacuum zero point field is consistent with the Sakharov-Puthoff model of gravitation [45,46].

Einstein, Schrödinger, and de Broglie have attested to the significance of nonzero photon rest mass. Frequency dependent anisotropy results from a putative $10^{-65}g$ periodic nonzero photon restmass according to

$$E = hv - mc^2\left[1 - v^2/c^2\right]^{-1/2} \text{[35].} \quad (7.3)$$

Of critical importance to our utility of the Dirac vacuum is the consequences non-zero photon mass, $m_\gamma \neq 0$ has for quantum electrodynamics where it becomes possible to split the corresponding EM spin 1 waves into transverse, $J_3 = \pm 1$ and longitudinal $J_3 = 0$ parts [34,47-49] where the latter relates to a decoupled Yukawa action-at-a-distance scalar potential that replaces the Coulomb field [34]. This field of course vanishes when the mass of the photon is zero. This photon polarization condition has also been noticed by Sundrum in relation to a 5D string vacuum where the 3rd polarization of $m_\gamma \neq 0$ adds an additional degree of freedom allowing a form of vacuum superconductivity [49].

The Wheeler-Feynman absorber theory of radiation [8] refined by Cramer [7] and by Chu [9] is utilized for our refinement of the Vigier Dirac vacuum conductivity model because the symmetry conditions of the emitter-absorber transaction is logically consistent with both C-QED requirements of HAM cosmology and our extension of the de Broglie, Bohm, Vigier causal stochastic interpretation of quantum theory which provides a vacuum model with the inherent physical existence of these vacuum displacement currents. The dissipative mechanism is also related to general relativity. The fluctuation in photon mass although tiny is sufficient to create an oscillation in spacetime curvature which as we shall see later creates a deficit angle in the parallel transport of vacuum charge allowing the coupling and uncoupling process to operate, i.e. according to general relativity and action and reaction occurs between the $g_{\mu\nu}$ field and particles moving in the Dirac medium characterized by the energy momentum distribution $T_{\mu\nu}$ because

$$G_{\mu\nu} = R_{\mu\nu} - \frac{1}{2}g_{\mu\nu}R + \Lambda g_{\mu\nu} = 8\pi G\left(T_{\mu\nu} + g_{\mu\nu}\rho_{vac}\right) \quad (7.4)$$

meaning that photon propagation is modified by the $g_{\mu\nu}$ medium and the $g_{\mu\nu}$ medium is modified by photon propagation [50].

Dissipative redshift mechanisms have remained *ad hoc* curiosities because of little empirical support and conflict with the apparent strident success of the standard Big Bang model. Most physicists today believe the photon is massless because a massive photon would destroy the mathematics of gauge theories and would violate Einstein's theory of special relativity because mass would go to infinity since $u = c$.

$$M = M_0 \frac{1}{\sqrt{1-\mu^2/c^2}} \tag{7.5}$$

However the existence of light pressure which has been known for a long time [51] a function of irradiance I over c ($p = I/c$ for absorbed photons and $2I/c$ for reflected photons) suggests that photons carry linear momentum and energy which can readily be calculated using Einstein's mass energy relation, $h\nu = mc^2$. The de Broglie wavelength relationship for massive particles, taking the accepted value for R applied to the Vigier mass, m_γ of the photon is:

$$m_\gamma = \frac{h}{\lambda c} \tag{7.6}$$

taking $\lambda = R \approx 10^{28} cm$ for the de Broglie wavelength, λ of the photon then $m_\gamma \approx 2.2\times 10^{-65} g$ which is the value for photon restmass obtained by a number of researchers [34,52]. Where R is the radial size of the universe; and by the uncertainty relation this is the smallest possible photon mass. Further $m \to 0$ only if $R \to \infty$. The de Broglie hypothesis was verified by [53,54] for the wavelength of a material particle. A photon mass of $10^{-65} g$ is in total agreement with Vigier's tired-light hypothesis [34].

From the redshift-distance relation, $z = f(d)$ (for static or expanding universe models) following [33] photons with restmass, m_γ interact with vacuum particles of mass, m_{vac} with acceleration

$$\int \frac{d^2 y}{dt^2} dt = -\frac{2\lambda\rho\omega}{y\left[\left(\frac{1}{2}\right)^2 + y^2\right]^{1/2}}. \tag{7.7}$$

The momentum transfer per vacuum particle, m_{vac} is

$$\int m_{vac} \frac{d^2 y}{dt^2} dt = -\frac{2m_{vac} m_\gamma \omega}{y\left[\left(\frac{1}{2}\right)^2 + y^2\right]^{1/2}} \tag{7.8}$$

With t the time, y the coordinate intersecting the path, ω. Producing a 'tired-light' redshift-distance law

$$\frac{\Delta v}{v} = e^{kd} - 1, \tag{7.9}$$

where k is determined by m_γ estimated to have a value of $10^{-65} g$ [34,52,55], m_{vac} which is currently unknown and may not be completely relevant other than the putative fact that vacuum coupling occurs.

It is inherently obvious that the photon is annihilated when brought to rest; therefore it is suggested that the photon has a rest mass with a half life on the order of the Planck time of 10^{-44} s, which would still preserve gauge in the domain of the standard model of elementary particles and allow for anisotropic vacuum zero point coupling of the photon which if it also occurs in the limit of the Planck time can be a virtual interaction.

7.3 Weak-Field Gravitational Approach of a Finite Light-Pencil and Derivation of the Gravitational Field of Radiation

For the linearized Weak-Field Approximation (WFA) approach [56] assume $m_\gamma = 0$, is point-like and the usual notation $c = G = 1$. Then for Einstein's field equations:

- $R_{ik} = (1/2) g_{ik} R = 8\pi T_{ik}$ and
- $g_{ik} = \eta_{ik} + 2h_{ik}$,
- $\eta_{ik} = diag(1, -1, -1, -1)$, and
- $(h_{ik})^2 = 0$ yield the linearized field equations:

$$\Box \psi^{ik} = 8\pi T^{ik}, \quad \psi^{ik} = -\frac{1}{2}\eta^{ik} h_l^l \tag{7.10}$$

The mass of the photon proposed by Vigier [34] is derived here utilizing the Tolman, Ehrenfest, Podolsky (TEP) [50] model of spacetime curvature induced by a finite light pencil. The TEP equations are summarized below; and include Einstein's weak field approximation (WFA) applied to a mass-free radiation field. Accordingly the WPA is linear, deviating only to first order in the Galilean case suggesting that the model is local, i.e., describing spacetime curvature induced by the light pencil in its immediate vicinity. The notation used is within the context of classical GR theory.

Only the non-zero components of the energy momentum tensor, T_μ^ν are those in energy density, ρ. Since the line element integral diverges for an infinitely-long light pencil, (L_p) and energy density, ρ, the pencil length is taken to a finite value L_p with ρ also finite. Then the expression for the Galilean deviation yields an elementary function: $h_\mu^\nu := \delta^{\nu\alpha} h_{\mu\alpha}$ with for a $h := h_\alpha^\alpha$ for a L_p traveling along the positive axis of an orthogonal Lorentzian 3-sphere. The linearized WFA from [56] is:

$$\left[h_\mu^\nu - \frac{1}{2}\delta_\mu^\nu h\right](x,y,x,t) = -4\iiint \frac{\left[T_\mu^\nu\right](\overline{x},\overline{y},\overline{z},t-r)}{r} d\overline{x}, d\overline{y}, d\overline{z} \tag{7.11}$$

Which coupling the metric distribution of matter and energy taken over all elements of spatial volume $d\overline{x}d\overline{y}d\overline{z}$ for time *r*.

To determine the gravitational field of light the momentum energy tensor of an electromagnetic radiation field is formulated in natural coordinates and in Weyl's form in this manner [57]

$$T_1^1 = \frac{1}{2}\left(X^2 + Y^2 + Z^2\right) + \frac{1}{2}\left(\alpha^2 + \beta^2 + \gamma^2\right) \tag{7.12}$$

$$T_1^2 = XY + \alpha\beta \tag{7.13}$$

$$T_1^4 = \beta Z + \gamma Y \tag{7.14}$$

$$T_4^4 = \frac{1}{2}\left(X^2 + Y^2 + Z^2\right) + \frac{1}{2}\left(\alpha^2 + \beta^2\gamma^2\right) \tag{7.15}$$

For pure electromagnetic radiation all other components of *T* vanish. Here *X, Y* and *Z*; and α, β and γ are components of the electric and magnetic field strength respectively at (*XYZ*).

Using the above WFA solution for the energy momentum tensor, T_μ^ν

for incoherent electromagnetic radiation (polarized or nonpolarized) for a L_p parallel to the positive x-axis, in natural coordinates with constant linear energy density, the only density components, ρ in (7.6) reduces to

$$T_1^1 = -\rho, \quad T_4^4 = \rho, \quad T_1^4 = -\rho, \quad T_4^1 = \rho, [56] \qquad (7.16)$$

where effects at the beam boundary are neglected. Substituting $T_1^1 = -\rho$ into (7.2) gives

$$\begin{aligned} h_1^1 - \frac{1}{2}h &= 4\int\frac{[\rho]dV}{r} \\ h_2^2 - \frac{1}{2}h &= 0 \\ h_3^3 - \frac{1}{2}h &= 0 \\ h_4^4 - \frac{1}{2}h &= -4\int\frac{[\rho]dV}{r} \\ h_1^4 - h_4^1 &= 4\int\frac{[\rho]dV}{r} \end{aligned} \qquad (7.17)$$

Solving $h_1^1 - \frac{1}{2}h = 4\int\frac{[\rho]dV}{r}$ for $h_{\mu\nu}$ gives

$$h_{11} = h_{44} = -h_{14} = -h_{41} = 4\int\frac{[\rho]dV}{r} \qquad (7.18)$$

The values of $h_{\mu\nu}$ in (7.18) fixes the form of the line element due to pure electromagnetic radiation traveling along the X-axis of a system of natural coordinates.

7.4 Gravitational Action of a Light Pencil

The gravitational field in the neighborhood of a finite L_p with constant linear energy density ρ passing along the x axis between a source at $x = 0$ and an absorber at $x = l$ [7-9,56]; contributes to the radiation by

$$4\int\frac{[\rho]dV}{r} = -h_{11} = -h_{44} = h_{14} = h_{41} =$$

$$4\rho\log\frac{\left[(l-x)^2+y^2+z^2\right]^{1/2}+l-x}{\left[x^2+y^2+z^2\right]^{1/2}-x} \quad (7.19)$$

Equation (7.19) describes the gravitational contribution only in L_p neglecting any contribution from the source or absorber [56] also any internal conditions, vacuum zero point coupling, or other spin exchange which also effect propagation.

Finally for the acceleration of a test particle towards the L_p along the negative y direction determined by a geodesic originating midway between the two ends of the pencil, [56] arrive at the simple result in 4.4. This is significant because the equivalency of the gravitational and inertial mass of a L_p justifies the application of the de Broglie relationship in (7.3) to the photon verifying the Vigier hypothesis of $m_\gamma = 10^{-65} g$!

$$-\frac{d^2y}{dt^2} = \frac{2pl}{y\left[(l/2)^2+y^2\right]^{1/2}} \quad . \quad (7.20)$$

The de Broglie relationship applied above in equation (7.3) determine the Vigier mass, m_γ of $10^{-65} g$. The important characteristic achieved is that conservation of momentum is preserved because as expected the acceleration is exactly twice that calculated from Newtonian theory by taking the equivalence of gravitational and inertial mass!

7.5 Internal Motion Structure of the Photon

All these fifty years of conscious brooding have brought me no nearer to the answer to the question, 'What are light quanta?' Nowadays every Tom, Dick and Harry thinks he knows it, but he is mistaken. ... I consider it quite possible that physics cannot be based on the field concept, i.e., on continuous structures. In that case, nothing remains

> *of my entire castle in the air, gravitation theory included, [and of] the rest of modern physics.* - Albert Einstein [57]

According to Einstein rest mass results from external or internal structural motion of a particle. Vigier has also discussed this extensively in terms of the causal-stochastic interpretation of quantum theory [6]. Unlike Fermi materials that are localized in all spatial dimensions and maintain a well-developed internal kinetic structure even when at rest, photons immediately release their more open spin structure when brought to rest and immediately dissipate their energy. For photons this internal transformation oscillates. We postulate the photon rest mass fluctuates harmonically in a manner like $0 \Leftrightarrow \sim 10^{-65} g$ which signifies according to $E = mc^2$ a change in energy from inward reflection and interaction with the Dirac polarized vacuum to outward displacement through space. We believe if this were not so the speed of light would be infinite; and that this variance is key to the fine-tuning of each nested Hubble sphere, $H_{R\alpha}, H_{R\beta}, H_{R\chi} ... H_{Rn}$ (see Chap. 13). Fluctuation in mass-energy is not mysterious as it is generally known that inertial and gravitational masses are an aspect of this movement. At the DESY laboratory recent experimental results have shown that the photon has extra layers of activity [58]. This is represented in Fig. 7.1.

> In other words, the transformation of 'matter' into 'energy' is just a change from one form of movement (inwardly, reflecting, to and fro) into another form, e.g. outward displacement through space. The possibility for objects of zero rest mass exists provided that they are moving at the speed of light. For if rest mass is 'inner' movement, taking place even when an object is visibly at rest, it follows that something without 'rest mass' has no such inner movement, and *all* its movement is outward, in the sense that it is involved in displacement through space. So light does not have the possibility of being 'at rest' since it does not possess any such inner movements [59].

This does not preclude a massive photon, only points out the difference in structure between Bosons and Fermions. It has been suggested that the definition of restmass be refined [60] or perhaps some sort of a photon 'lifetime' related to frequency could be considered.

7.6 Introduction to Spin Exchange Compactification Dynamics

Photon mass anisotropy is a major requirement of the model. Our C-QED BB theory is based on the fundamental premise that the energetic interplay of mass, inertia, gravitation and spacetime topology is based on a unified symmetry of internal spin-spin coupling and spin exchange compactification with a 'super quantum potential' ultimately being the action and control principle. Spin exchange symmetry through the interplay of a unique topological package orders compactification providing a template from which superstring or twistor theory could be clarified. One purpose of compactification dynamics is to allow the 3-sphere of temporal reality to stochastically 'surf' on the superstructure of HD eternity creating our virtual reality and the perceived arrow of time allowing nonlocal interactions not possible in a Newtonian absolute space or completely described by Copenhagen quantum theory. Stated another way, the domain of quantum uncertainty stochastically separates the classical regime from the unitary regime. This allows the subspace of temporal reality to surf as it were on the face of eternity.

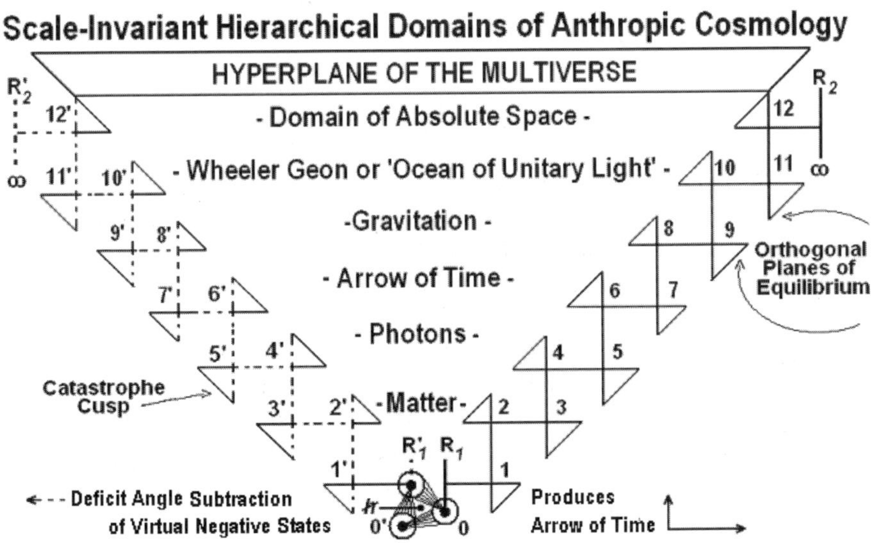

Figure 7.6 Conceptualization of the covariant scale-invariant hierarchical structure and function of HAM dimensionality from zero to 12D. For application to the arrow of time see Chap. 5.

Considering the structural-phenomenology of the array of least cosmological units tiling the spacetime backcloth to be BB cavities, an inherent inertial force in the hysteresis loop of the continuous-state compactification dynamics is the determinant of the perceived arrow of time. This pertains to the photon and quadrupole photon-graviton complex and the HD 'ocean of unitary light' it originates in, structure not observed when enfolded in HD reality.

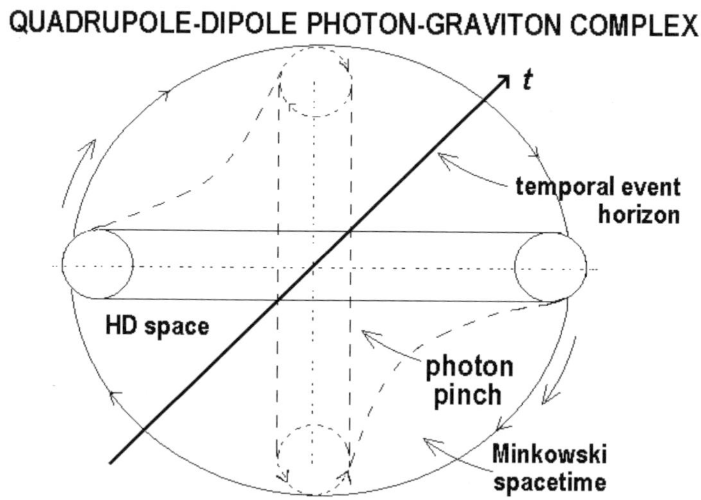

Figure 7.7 Conceptualized view of the HD quadrupole photon-graviton complex for *quadrupole* ⇔ *dipole* interactions as elements of the unitary field and the event horizon of the Minkowski spacetime leading lightcone singularity and inherent arrow of time for an Earth observer. Compare Fig. 7.11.

The localized appearance of compactification has been interpreted as a structure fixed in an early Big Bang era, but spin-exchange delocalizes compactification in a rich dynamic HD hyperstructure of continuous spacetime symmetry transformation of constant N-Dimensional collapse to the 3-sphere of Minkowski space for the reality of the observer. If we apply Mach's principle[1,2] [61,62] to the perspective of HAM cosmology we consider the inertial force and Einstein's equivalence principle to be the same and can be applied as a quantum space density of space waves

[1] Local physical laws are determined by the large-scale structure of the universe.
[2] "inertia originates in a kind of interaction between bodies" - Einstein [61].

combined from all particles in the universe to every space point:

$$\text{Mach density} \propto mc^2 = hf \propto \sum_1^N \left[\frac{\Phi_n}{r_n}\right]^2 \text{ [62].} \tag{7.21}$$

The continuous-state compactification process contains a central inherent hysteresis loop that entails an inertial drag; the oscillating boundary conditions of which determine the speed of light, c. This is its fine tuning (See Chap. 13) and the reason $c \equiv c$ and is not infinite. The continuous-state acceleration parameters of which balances the gravitational potential which along with the Stoney modulated Planck's constant, $\hbar + T_0$ (Chap. 4) and the SUSY modulated cosmological constant, Λ together balance dark energy and the minute oscillation of the curvature of spacetime, which as shown in Chap. 5 orders the arrow of time. This form of inertial drag is compatible with the Sakharov-Puthoff model of gravitational theory [45,46,63,64] which are compatible with a Dirac ether model.

Spacetime is quantized as a discontinuous Planck scale raster determined by the fundamental constants *c, G* and *h.* This comprises a basic unit of the Dirac polarized vacuum with the properties like a rotating microcosmic Klein bottle with properties like a Planck scale black hole. The Planck constant h is a product of the uncertainty principle; a complement of the Planck length, l_{pl} and Planck time, t_{pl} comprising the virtual event horizon of nonlocality. See Chaps. 9 and 11.

Compactification appears as localized scalar potentials to standard quantum measurement, but nonlocally, in the Wheeler-Feynman-Cramer-HAM model [7-9,12], are a continuous transformation of QED or SED hyperdimensional mirror symmetry Calabi-Yau 3-tori cavities in blackbody equilibrium. Delocalized compactification dynamics produce a periodic mass equivalency by oscillations of the gravitational potential (GP) providing the action principle for absorption and emission (see section 7). Theoretical feasibility of Planck scale black holes has long been demonstrated [65-67]. Thus the CMBR could be considered a form of Hawking radiation [66-68] from the hypertiling of covariant polarized Dirac sea microcavities. Planck scale black hole microcavities have been criticized as unphysically hot, but this would not be the case in continuous-state cosmology because the inherent spin exchange SUSY breaking mechanism makes any heat buildup impossible because of the constant roiling of the energy in the hysteresis looping.

The HAM is modeled as a type of dynamically transforming hyperdimensional Klein bottle, topologically representative of Kant's antinomy of an open/closed spacetime [69] with near-field far-field photon mass anisotropy leap-frog conditions; the hypergeometry of which translates in a metric of co-moving Birkhoff spheres [70] where $\dot{R} = c$ is preserved through all levels of scale [71,72]. This is part of the continuous-state future-past advanced-retarded dimensional reduction standing-wave spin-exchange compactification process. Taking the Hubble sphere as the arbitrary radius of the temporally finite observable universe, the Gravitational Potential is opposed within the 3-sphere, not by inflation but by a nonlocal equivalence to the Gravitational Potential, i.e *dark energy* of the bulk of the multiverse [12] which appears in the large scale as Einstein's cosmological constant, Λ and in the small scale as the new string tension, T_S modulated Planck constant, λ. See Chap. 4. Both the cavity-QED CMBR-emission and Redshift-absorption arise from an 'electromotive torque', an inertia in the hysteresis loop of the temporal-eternal antinomy of the continuous state process that arises as the stochastic background wake of photon-graviton propagation [12,35,36]. This is the ontological flux of the unitary field; and the source of Mach's principle. The unified field exchange mediator in this model has been dubbed the noeon [12,35,36]. The exchange is performed by topological switching [73] and therefore is an energyless ontological process. Wheeler said 'charge is topology'. This process is where this 'virtual charge' comes from. Its quote-unquote 'force' arises in the variance of the curvature of the topology, during the continuous-state, i.e. it is a force of coherence which is the great attractor of the anthropic principle guiding evolution, a super quantum potential, as described by the fundamental noetic equation, $F_{(N)} = E/R$, see Chap. 4. As introduced below we relate the Gravitational Potential equivalent acceleration of the continuous-state translation of these co-moving topologies of higher and lower fluctuating flat-curved spacetime dimensions as fundamentally equivalent to a Planck scale black body exciplex hypersurface [44].

7.7 Blackbody Exciplex Radiation - Cosmological Constraints

Employing to the tensor field equations of Einstein's general relativity, $G_{\mu\nu} + \Lambda g_{\mu\nu} = \left(8\pi G / c^4\right) T_{\mu\nu}$, especially for the Schwarzchild line element,

$$ds^2 = -\left(1 - \frac{2M}{r}\right)dt^2 + \frac{dr^2}{1-2M/r} + r^2\left(d\theta^2 + \sin^2\theta d\phi^2\right), \quad (7.22)$$

a gravitational interaction between a domain wall and a black hole might be valid when the symmetry breaking scale of the scalar field is near the Planck scale, but the assumption is the effects of gravity can be ignored near the black hole horizon and we would essentially ignore any consideration of Planck scale black bodies, The boundary conditions of a black body cavity radiates at every possible frequency and is dependent only on the temperature of the walls of the cavity. In thermodynamic equilibrium the amount of energy, $U(v)$ depends only on temperature and is independent of the material of the walls or shape of the container. The crux of the matter is that the radiation field and boundary conditions behaves like a collection of simple harmonic oscillators that can arbitrarily be chosen to have a set of boundary conditions of dimension L [40,74] which is repeated periodically through spacetime with spherical symmetry in all directions. These boundary conditions will yield the same equilibrium radiation as any other boundary conditions, and with this result no walls are actually required because the walls thermodynamically only serve in the conservation of energy [59]; allowing the putative feasibility of our C-QED exciplex model for BB CMBR/redshift equilibrium to be compatible with natural law. This seems to relate somehow to Birkhoff's theorem [70] for the gravitational potential in a spherical universe (the Einstein-Hubble 3-sphere); it seems to be this theorem that allows the 'container' and its walls to be essentially irrelevant especially in terms of the symmetry of the covariant scale invariance. Perhaps it may be better said as a nothing-everything configuration of infinite *potentia*. This is the background setting with parameters providing delicate balanced equilibrium conditions. It would appear that emissions is the simpler of the two conditions – an internal or external 'bump or hole' (Chap. 6; Fig. 7.13) coupling-uncoupling allows a boundary condition change facilitating emission. Dirac hole theory and Bohr-Summerfield conditions may have some relevance. In any case this is all governed by the boundary conditions described by noetic field equation, $F_N = E/R$ as illustrated in Fig. 7.7.

Defining the observable universe as an Einstein 3-sphere, any spherical distribution of matter of arbitrary size (according to the general theorem developed by Birkhoff [70]) maintains a uniform contribution of the GP with any particle in the volume. Metaphorically the Wheeler-Feynman-Cramer-HAM model [7,8] defines the radius of the universe, R in terms of a co-moving Hubble sphere with the topology of a hyper-Klein bottle (dual mirror symmetry Calabi-Yau 3-tori). This relation maintains itself through all levels of scale. Therefore Birkhoff's theorem [70] can apply hyperdimensionally to all matter in the multiverse. This can explain the origin of the cosmological constant [75], why space appears universally flat and why 3-sphere dark matter is not required to explain galactic rotation since in HAM cosmology [12], it is instead balanced by a multiversal *dark energy* from the 'infinite number of causally separated nested Hubble spheres.

This arbitrary cavity putatively modeling the structure of the universe, as drawn from current astrophysical data, is generally accepted to be a perfect BB radiator of 2.735° K. Einstein introduced the cosmological constant to balance the GP in a *static universe*. Which he then retracted when Hubble discovered what was erroneously thought to be a Doppler recessional redshift, which Einstein apparently thought obviated the need for a cosmological constant. Further Einstein postulated the existence of singularities derived from the field equations of general relativity; from which Friedman suggested that the universe itself originated in a temporal singularity giving rise to the Big Bang model of recent history.

It turns out there is a temporal singularity but it relates to continuous-state parameters of string tension and recession of the advanced mode of the Planck constant as it recedes into the past from the stationary locus of the eternal present [10]. It has been shown in [55] that redshift is intrinsic to photon mass anisotropy; suggesting that recession is an observational illusion of 'tired light' rather than a physically real Doppler recessional velocity indicative of a Big Bang effect.

> Let us assume that photons of rest mass, m_p interact with the vacuum particle, having mass m_0. There is, along the interaction path, w, a transfer of energy and momentum from the traveling photon to the vacuum particles which gives the vacuum particles a motion toward the trajectory (a pinch effect). The loss of photon energy and of photon momentum can be computed...The effect has a perfect geometrical symmetry, being in essence the result of an interaction between a photon along its trajectory with a strictly symmetrical potential. The redshift-distance law is obviously a 'tired-light' [33].

7.8 Blackbody Microcavity-QED Constraints

Specialized Dirac vacuum C-QED boundary conditions are taken to represent the walls of Birkhoff black body – black hole microcavities comprised of a tiled stochastic hyperstructure of Planck scale, S_N phase cells with the lower limit of dimensional size determined by the Heisenberg uncertainty principle with the cavity volume defined by $\delta x \delta y \delta z \delta t \delta p_x \delta p_y \delta p_z = |h|^3$ and the energy for each coordinate defined by $\sum_{S_N} \delta E_N \delta t \sim h$ [44]. But now we know from string theory that the string tension factor (Chap. 4) modulates the size of the cavity. During the continuous cycles of dimensional reduction the energy, E is parallel transported by an *energyless Topological Switching*[3] of higher to lower dimensionality, $D-(\delta E_x \delta t)$ without distorting the smoothness of perceived macroscopic realism because of the standing wave spin exchange process. Although in HAM reality the Planck backcloth is a 11(12)D hypertiling of topologically comoving hyperstructures, not a rigid tiling of 3D cubes with primal fixed compactification as in Big Bang theory.

7.9 CMBR Energy Damping by Vacuum Conductivity

Planck's radiation law for a harmonic oscillator is energy per unit time per unit volume. An order of magnitude calculation for the energy of a single transverse CMBR cavity wave mode for the energy density is $\omega = \frac{1}{2}\varepsilon_0 E^2 + \frac{1}{2}\mu_0 B^2 \approx \varepsilon_0 E^2$. According to Lehnert & Roy [21] energy, $E = E_0(r - c_0 t) \cdot \exp(-\frac{1}{2} R\sigma r)$ where R is radius of the universe and r is direction of propagation. This implies that the energy density has an e-folding decay length, $L_{decay} = 1/r\sigma$ where $\sigma \equiv$

---

[3] *Topological Switching* refers to the optical illusion occurring when fixating on a face of a Necker cubes where a background vertex switches to a foreground vertex; here utilized as a metaphor of how parameters of a higher dimensional topology may interplay harmonically by parallel transport into lower dimensional structures.

conductivity of the vacuum because the conductivity is extremely small. The corresponding energy decay time (damping time for E to decay from original value) would be $t_{decay} = L_{decay} / c = 1/ R\sigma c \equiv$ absorption time of the 'tired light' redshift absorption effect [21]. This applies to all waves where R is radius of universe.

Lakes found an interesting way to measure photon mass using a form of Cavendish balance [76]. See Fig. 7.8 below. His experimental design evaluated the product of photon mass squared, m_γ^2 and the energy density of the ambient cosmic magnetic vector potential, A not the usual measurement of the magnetic field. His apparatus is more sensitive than in other experiments because it measures large-scale cosmic magnetic fields associated with huge vector potentials [76].

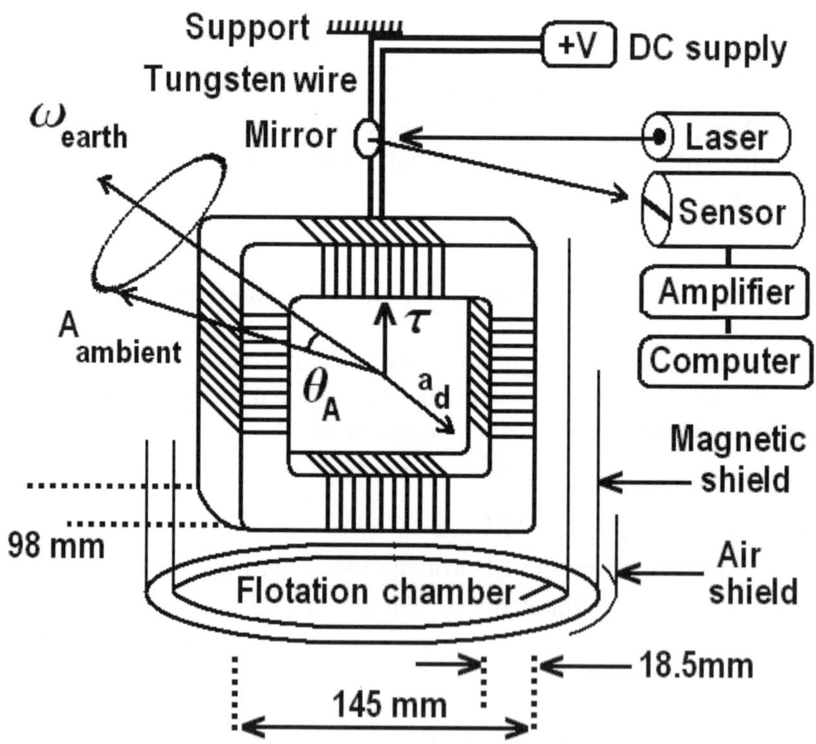

Figure 7.8 An electric current in a toroid produces a dipole field which interacts with the ambient vector potential producing a torque on the toroid which varies with the Earth's rotation. Fig. redrawn from [76].

Perhaps laser trapping techniques could be utilized to enhance the energy baseline and improve the accuracy by several orders of magnitude. Lake states: "nonzero photon mass would give rise to a wavelength dependence of the speed of light in free space, the possibility of longitudinal electromagnetic waves, a leakage of static electric signals into conductive enclosures, and a more rapid falloff ... of magnetic dipole fields with distance than the usual inverse cube dependence" [76]. We have noticed a naturally occurring case of 'leakage of static electric fields'. We have been told numerous times by automotive and marine battery distributors over the years that they cannot store them on the floor of they are discharged and damaged quickly.

We postulate an 'exciplex' C-QED black body tiling of the Dirac polarized vacuum such that redshift and CMBR are absorption-emission equilibrium conditions. The functionality of this model is facilitated by the Vigier Causal Interpretation of quantum theory and extended electromagnetic theory described by the Proca equation which includes photon mass. "The conventional form of Maxwell´s equations in the vacuum, with a vanishing electric field divergence, leads to the vanishing parameters spin, rest mass and longitudinal magnetic field of the individual photon. With a nonzero photon mass such divergence in the vacuum state, and with the requirement of Lorentz invariance, all these parameters become nonzero. For the phase and group velocities of a photon wave packet still to remain close to the experimental value of the velocity of light, and for the spin to have its experimentally determined value, the rest mass and the longitudinal magnetic field component then must become very small but nonzero. Thus the rest mass of the photon does not have to be included ad hoc and occurs from the beginning in the basic Maxwell-Proca field equations [17], but comes out from the nonzero electric field divergence. This is one of the results of my revised quantum electrodynamic theory" [15,16,21,77-79].

7.10 Possible Black Hole Considerations for Discussion

Any number of bosons may cohere in a phase cell while Fermions must have energy to occupy the same domain because of the Pauli exclusion principle and therefore must be degenerate in black holes. These Planck volumes considered as the boundary conditions of the cavity ground state, cohere stochastically to embody any required energy configuration.

The general expression for BB radiation derived by Planck takes the form:

$$M_\lambda^b = 2\pi hc^2\lambda^{-5}\left(e^{hc/\lambda kT}\right)^{-1} \tag{7.23}$$

where $M_\lambda^b(T)$ is spectral emittance, and k is the Boltzmann constant. Hawking found a similar relationship for the hypersurface of a black hole [66,67]. The topology of the Planck backcloth has been considered to be a latticework of micro black holes by some researchers; but perhaps a better postulate would be a backcloth tiling that is a form of mirror symmetry Calabi-Yau Wheeler wormhole dual 3-forms. The best indicia for such a scenario is the Dirac $360^\circ - 720^\circ$ spinor rotation of the electron; it appears such a scenario could only occur in a topology with some form of Klein-bottle hyperstructure. The thermodynamic relationship between the area of a black hole and entropy $E_{\deg raded} = \left(\sum Area/16\pi\right)^{1/2} = \left(\sum M_{tired}^2\right)^{1/2}$ and emittivity [66-68,80,81] found to occur at the hyperstructure surface of a black hole is putatively developed here as one possible example for similar emittivity for CMBR black body emission intrinsic to the C-QED features of spacetime topology.

7.11 Size Temperature Relationship of Kerr Black Holes

Bekenstein, [80] suggested a relationship between the thermodynamics of heat flow and the surface temperature of a BH, which led Hawking, 1974a to the finding that all BH's can radiate energy in BB equilibrium because the entropy of a black hole, S_{bh} is related to the surface area, A of its event horizon, where k is Boltzmann's constant, $S_{bh} = M^2 2\pi[kcG/(h/2\pi)]$ [74]. This leads to the expression for the surface temperature of a black hole:

$$T(^\circ K) = (h/2\pi)D/[32\pi hM(M - 1/2Q^2)/M + D] \tag{7.24}$$

where $D = (M^2 - Q^2 - L^2/M^2)^{1/2}$, Q = charge, and L = momentum [74]. This shows that the <u>BB temperature of a BH is the inverse of its mass</u>, which for a typical Kerr BH represents a temperature of one $^\circ K$ for a BH a little larger than the moon or for each 10^{26} g.

Accordingly the Beckenstein - Hawking relationship, while a stellar mass BH has the expected fractional degree temperature, the predicted temperature for microcavity Planck scale BH would be about $1.9\times10^{31}\,^{\circ}\mathrm{K}$. Therefore the additional physics of the Wheeler-Feynman-Cramer-HAM spin exchange dynamics must be added to account for the difference in the compressed geometry of a black hole having a fixed internal singularity structure with a lifetime of billions of years and a Planck scale black hole with an open singularity that [12,44] by rotating at the speed of light, *c* with a Planck time lifetime of 10^{-44} sec and therefore able to dissipate this heat if its theoretical prediction were otherwise true.

So while a micro-BH might be considered to have a temperature of billions of degrees Kelvin if the nature of its internal singularity and total entropy is derived through the predictions of GR and Big Bang cosmology; because according to GR a singularity occupies no volume and has infinite energy density. But GR breaks down and is known to be incomplete at the quantum level; requiring new physics to describe spacetime quantization. Further, although Einstein said 'spacetime is the ether' [43] radiation was still considered to be independent of the vacuum, which is now known not to be the case [55].

7.12 Temperature Relationship of Dirac QED Cavity 'Black Holes'

In the transition from the Newtonian Euclidean continuum to quantum theory, what still remains to be properly addressed is the ultimate nature of a discrete point. The infinite density Einstein singularity is still too classically rooted. In terms of the Wheeler-Feynman-Cramer-Chu-HAM model the energy density is delocalized in terms of the equivalent GP of compactification dynamics. Planck scale black body cavities are topologically open nonlocally. They spin exchange entropy through a continuous flux of energy; and are not scalar compactified singularities that originated in a Big Bang, but continuous-state transforms accelerating toward an open propagating ground that is never reached nonlocally as if swimming upstream with the same velocity as the flow so that the swimmers relative position is in stasis relative to a point on the shore. The inertia inherent in this dynamic results in the intrinsic $2.75^{\circ}\,\mathrm{K}$ CMBR. This is a reality of conformal scale-invariance, Chap. 4.

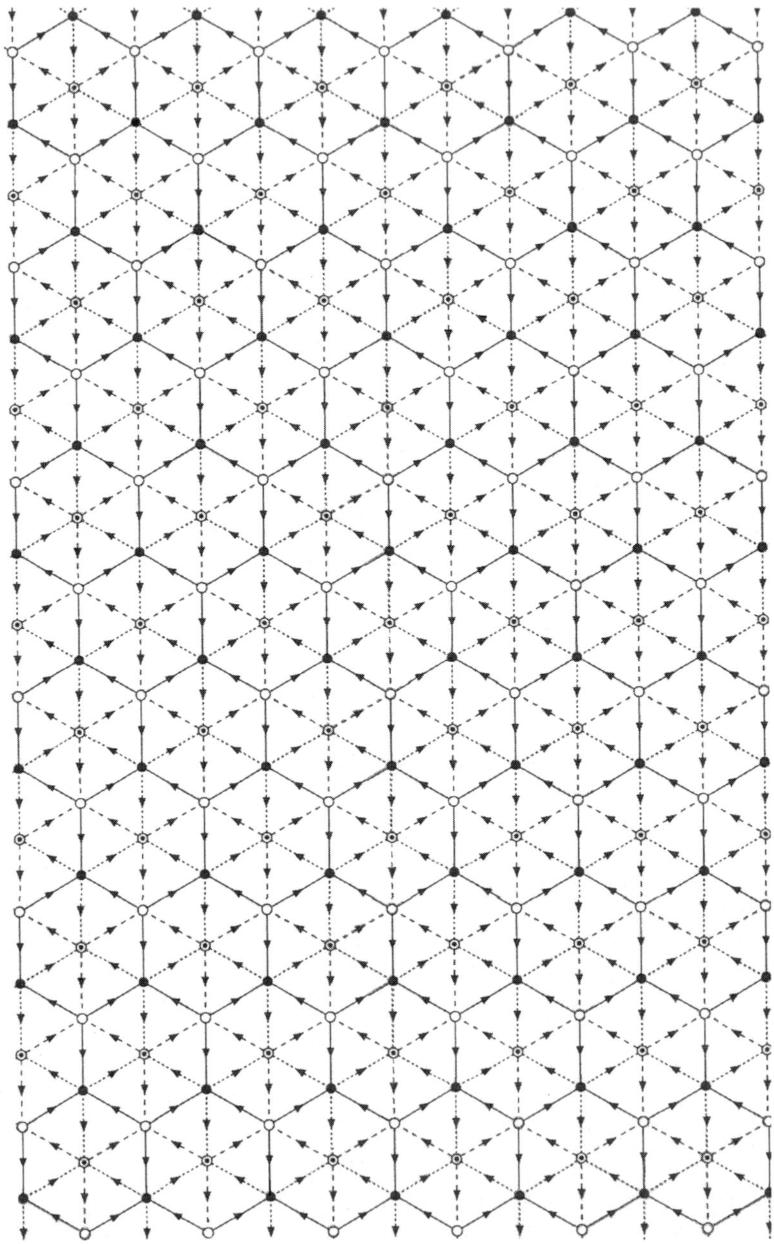

Figure 7.9 Least-unit exciplex C-QED backcloth able to accommodate any geometry and any transform by topological switching. Fig. adapted from [73].

7.13 Spin Exchange Parameters of Spacetime-Photon Coupling

Starting with the Hawking radiation modification of the Planck BB relationship as applied to BH surface dynamics, the requirement for application to a quantum BB C-QED cavity generally defined as the phase space of $|h_c|^4$ in (7.25) is the addition of spin exchange parameters, where

$$\sum_{i}^{Z^a Z_a} N_i \left|P_l / P_t\right|^4 = \left|h_c\right|^4 \Leftrightarrow C_\gamma \ . \tag{7.25}$$

N is the complex sum of Planck hyperunits comprising one BB C-QED microcavity. Spin dynamics can be readily described using the density matrix formalism. Spin states are represented as linear combinations of α and β states corresponding to the spin eigenvalues; and can be used in terms of the wave function to determine the value of spin characteristics Q.

$$Q = \left|S_{c1}\right|^2 Q_{\alpha\alpha} + S_{c1} S_{c2}^* Q_{\alpha\beta} + S_{c1}^* S_{c2} Q_{\beta\alpha} + \left|S_{c2}\right| Q_{\beta\beta}$$

$$\rho = \begin{bmatrix} \left|S_{c1}\right|^2 & S_{c1}^* S_{c2} \\ S_{c1} S_{c2}^* & \left|S_{c2}\right|^2 \end{bmatrix} \tag{7.26}$$

The density matrix ρ is made up of the spin coupling coefficients S_{c1} and S_{c2}. The diagonal elements correspond to real local spin orientations, and the non-diagonal elements correspond to complex quantities representing spin projection on planes perpendicular to axes of quantization. For the purposes of discussion any arbitrary axis may be chosen as an axis of quantization; but in the spin exchange process the geometry of the complex topology of the Argand plane transforms from real to complex in the retiling of compactification dynamics. The variance in the diagonal elements effects the longitudinal spin polarization along the axis of quantization; and the non-diagonal variances effect transverse spin polarizations. It is the phase of the elements that determine the angle of spin coupling with each dimensional axis. This relates CMBR emission/absorption to the cycle of torque moments.

The mass equivalent inertial properties comprising the linear and angular momentum components of spin exchanged in the nonlocal continuous compactification structure allow the Dirac vacuum to maintain perfect BB equilibrium inside the scale invariant Hubble Birkhoff sphere.

7.14 Spontaneous CMBR Emission by Spacetime Cavity-QED

This preliminary model for continuous spontaneous emission of STCMBR directly from C-QED dynamics of the stochastic properties of the Dirac sea, obviates CMBR origin as the relic of an initial state Big Bang cosmology as the standard model has predicted. In this model we make one speculative new assumption that is not based on the published body of empirical data for C-QED. Spontaneous emission by atomic coupling to vacuum zero-point fluctuations of the Dirac sea is already an integral part of C-QED both in the laboratory and theory; here we postulate that a similar process can occur in free space. In classical electrodynamics the vacuum has no fluctuation; by contrast quantum radiation can be viewed as partly due to emission stimulated by vacuum zero-point fluctuations.

The literature on C-QED is rich in descriptions of the nature of spontaneous emission of radiation by atoms in a cavity [82-84]. We begin development by choosing, for historical reasons, the upper limit of the number of atoms in the vacuum of space to the figure of one atom per cubic centimeter as derived by Eddington, [85]. This figure could be considered arbitrary, but for our purposes it is sufficient to note that *there are sufficient free atomic particles moving in space for spontaneous* C-QED CMBR emission.

> Charged particles are coupled to the electromagnetic radiation field at a fundamental level. Even in a vacuum, an atom is perturbed by the zero-point field, and this coupling is responsible for some basic phenomena such as the Lamb shift and spontaneous radiative decay [86].

Recent developments in C-QED have included descriptions of emission by Rydberg atoms in microwave cavities that include optical frequencies [87-95]. The Rydberg formula for atomic spectra is related to the binding energy of an electron by:

$$R = \mu_0^{\ 2} m e^4 c^3 / 8h^3 \tag{7.27}$$

where μ_0 is the magnetic permeability which is the ratio of the magnetic flux density, B of an atom to an external field strength, H. $\mu = B / H$ which is also related to the permeability of free space, μ_0, the Coulomb constant k and the magnetic constant k_m by

$$c = \sqrt{\frac{k}{k_m}} = \frac{1}{\sqrt{\mu_0 / \varepsilon_0}} = 3 \times 10^8 m / s \tag{7.28}$$

where ε_0 is the vacuum permittivity of free space; m and e are mass and charge of an electron respectively, c the speed of light and h Planck's constant. In the non-perturbative regime strength of the dipole coupling is larger than the dissipation rate and quantum mechanical effects have been shown to include multi-photon resonance, frequency shifts and atomic two state behavior at vacuum Rabi resonance, the latter of which will be of most interest in our discussion [89].

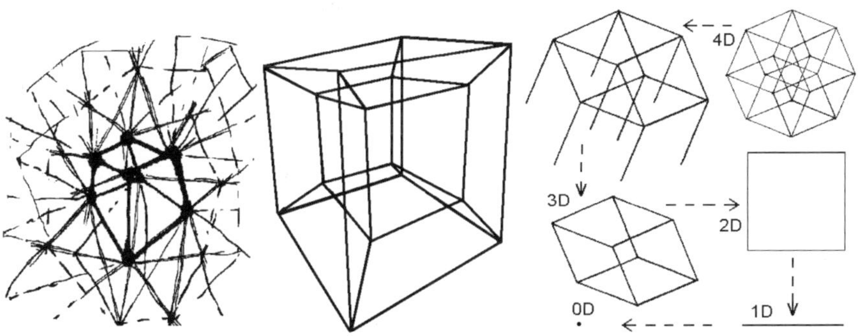

Figure 7.10 a) A CMBR photon emission from the Planck C-QED backcloth *exciplex torque modes* of the *future-past* compactification cycle. b) In HAM cosmology Euclidean space is a subspace of complex HD space (The reverse of Big Bang theory); such that each 3(4)D scale invariant 'cell' is covered by the hyper-geon of the unified field and it's associated action. c) Illustration of continuous D reduction; Not observable from a Euclidean orientation because it is imbedded in complex space $(\pm C_4)$.

Spontaneous emission requires only a single quantum so the internal state of the atom-vacuum coupled cavity system may be described by the simple quantum basis.

$$|0\rangle|-\rangle, |0\rangle|+\rangle, |1\rangle|-\rangle \tag{7.29}$$

where $|0\rangle$ and $|1\rangle$ are the Fock photon states and $|-\rangle$ and $|+\rangle$ are two states of the Rabi/Rydberg atom. Momentum operators $x(p)$ and $y(p)$ relate center of mass and atom ground state $|-\rangle$ dynamics where a master equation can describe the two-state atom interacting with the mode of the vacuum cavity momentum distribution after spontaneous emission and the emission spectra [89,96].

$$\begin{aligned}\dot{\rho} &= (1/i\hbar)\left[\hat{H},\rho\right] + K\left(2\hat{a}\rho\hat{a} - \hat{a}\hat{a}\rho - \rho\hat{a}\right) + \\ &(\gamma_I/2)\left(2\hat{\sigma} - \rho\hat{\sigma}_+ - \hat{\sigma}_+\hat{\sigma}_-\rho - \rho\hat{\sigma}_+\hat{\sigma}_-\right)\end{aligned} \tag{7.30}$$

where the a's are the boson creation and annihilation operators and the σ's the raising and lowering operators for the atom [89].

We assume that the atom acts classically as a free wave-packet where $\rho_{\text{int}}(t)$ describes the internal state of the system which can be described by

$$\rho_{\text{int}}(t) = w(t)(|0\rangle|-\rangle)(\langle-|\langle 0|) + |E_{\text{int}}(t)\rangle\langle E_{\text{int}}(t)|, \tag{7.31}$$

with

$$|E_{\text{int}}(t)\rangle = x(t)|1\rangle|0\rangle + y(t)|0\rangle|+\rangle, \tag{7.32}$$

where

$$\frac{dx}{dt} = -(\kappa + i\omega_0)x + g \quad \cos(\Omega t + \Phi)y, \tag{7.33}$$

and

$$\frac{dy}{dt} = -\left(\gamma_I/2 + +i\omega_0\right)y - g \quad \cos(\Omega t + \phi)x. \tag{7.34}$$

In addition to the atoms classical motions as a free wave-packet, the vacuum coupled system when excited, has two harmonic potentials related to the atoms motion and spontaneous emission process as in the following from Carmichael [89].

$$|\mu\rangle = \left(1/\sqrt{2}\right)\left(|0\rangle|+\rangle + i|1\rangle|-\rangle\right) \tag{7.35}$$

$$|l\rangle = \left(1/\sqrt{2}\right)\left(|0\rangle|+\rangle - i|1\rangle|-\rangle\right) \tag{7.36}$$

Vacuum Rabi atomic orbital splitting is the normal mode splitting of the coupled harmonic oscillators ; one mode describing the atomic dipole and the other the cavity field mode. This system of coupled harmonic oscillation is extremely versatile and can be applied to describe Dirac vacuum cavity QED emission of the CMBR when driven by the vacuum quantum mechanical stochastic field. Our application to the CMBR is based on the work of Agarwal [83] and Carmichael [89] on the nature of stochastic driving fields in C-QED.

Starting with the Hamiltonian for a coupled harmonic oscillator

$$H(t) = \tfrac{1}{2}\left(p_A^2 + p_C^2\right) + \tfrac{1}{2}\omega_0^2\left(q_A^2 + q_C^2\right) + 2\omega_0 g\cos\left(\Omega t + \phi\right) q_A q_C, \quad (7.37)$$

where q_A, q_C, p_A, p_C are the coordinates and momenta of the one dimensional oscillator ; with the subscripts *A* and *C* referring to atomic dipole and cavity modes respectively of the Rabi/Rydberg atom in free space. The oscillator coupling is modulated by the Doppler frequency Ω, with phase ϕ modulating the dipole coupling constant for atomic motion ; the equations of which take the form of equations (7.12) [89]. This has been a non-perturbative formalism much simpler to interpret than a QED perturbative expansion deemed sufficient for this stage of development.

7.15 Possibility of Blackbody Emission from Continuous Spacetime Compactification

It is also suggested that further development of the C-QED model of CMBR emission could be extended to include spontaneous emission from the continuous dimensional reduction process of compactification. This would follow from modeling spacetime cavity dynamics in a manner similar to that in atomic theory for Bohr orbitals. As is well known photon emission results from electromagnetic dipole oscillations in boundary transitions of atomic Bohr orbitals. Bohr's quantization of atomic energy levels is applied to the topology of Spacetime C-QED boundary conditions in accordance with equation (7.1) where spacetime QED cavities of energy, E_i undergo continuous harmonic transition to a higher state, $E_j\left(> E_{iH}\right)$ (redshift-absorption mode) or to a lower state

$E_k\left(<E_{iL}\right)$ (CMBR-emission) according to the relation $hv = E_j - E_{iL} = E_{iH} - E_k$. Thus we postulate that boundary conditions inherent in continuous standing-wave spacetime spin exchange cavity compactification dynamics of vacuum topology also satisfy the requirements for photon emission. In metaphorical terms, periodic phases or modes in the continuous spacetime transformation occur where *future-past exciplex*[4] *states* act as *torque moments* of CMBR/Redshift BB emission/absorption equilibrium.

In reviewing atomic theory Bohm, [59] states:

> Inside an atom, in a state of definite energy, the wave function is large only in a toroidal region surrounding the radius predicted by the Bohr orbit for that energy level. Of course the toroid is not sharply bounded, but ψ reaches maximum in this region and rapidly becomes negligible outside it. The next Bohr orbit would appear the same but would have a larger radius confining ψ and propagated with wave vector $k = \rho / h$ with the probability of finding a particle at a given region proportional to $|\psi|^2 = |f(x,y,z)|^2$. Since f is uniform in value over the toroid it is highly probable to find the particle where the Bohr orbit says it should be [59].

---

[4] An exciplex (a form of excimer - short for excited dimer), usually chemistry nomenclature, used to describe an excited, transient, combined state, of two different atomic species (like XeCl) that dissociate back into the constituent atoms rather than reversion to some ground state after photon emission. An excimer is a short-lived dimeric or heterodimeric molecule formed from two species, at least one of which is in an electronic excited state. Excimers are often diatomic and are formed between two atoms or molecules that would not bond if both were in the ground state. The lifetime of an excimer is very short, on the order of nanoseconds. Binding of a larger number of excited atoms form Rydberg matter clusters, the lifetime of which can exceed many seconds. An exciplex is an electronically excited complex of definite stoichiometry, 'non-bonding' in the ground state. For example, a complex formed by the interaction of an excited molecular entity with a ground state counterpart of a different structure. When it hits ground the photon emitted is a Quasiparticle soliton.

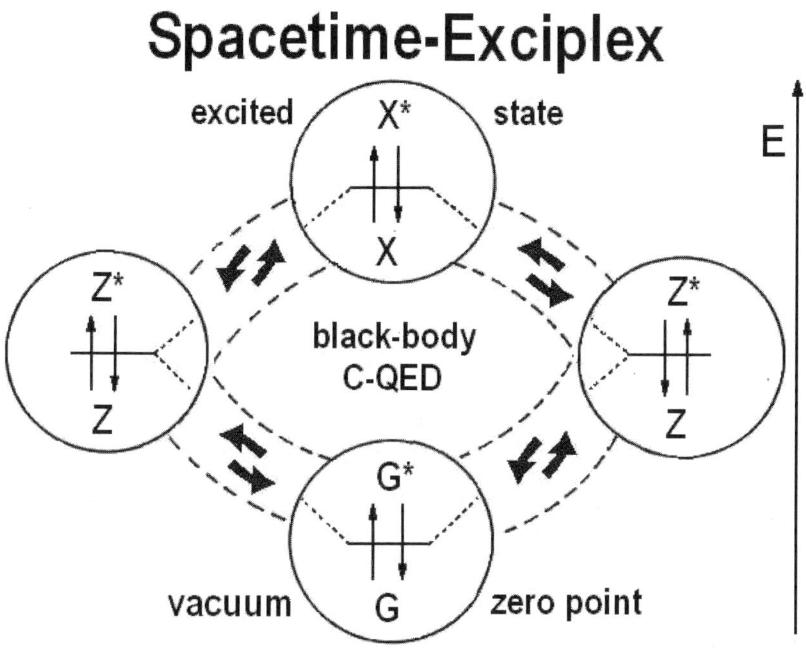

Figure 7.11 Geometric model for a spacetime C-QED black body Exciplex for red-shift-CMBR absorption-emission equilibrium dynamics.

The general equations for a putative spacetime exciplex are:

$$
\begin{gathered}
G^* + G^* \Leftrightarrow Z^*; \quad Z^* + m_\gamma \Leftrightarrow X^* \\
X^* - m_\gamma \xrightarrow{emission} Z^* \quad or \quad G^* \qquad (7.38) \\
X^* + m_\gamma \rightarrow Z^* \quad or \quad G^*
\end{gathered}
$$

where G is the ZPF ground, Z black body cavity excited states and X the spacetime C-QED exciplex coupling. The numerous configurations plus the large variety of photon frequencies absorbed allow for a full black body absorption-emission equilibrium spectrum. We believe the spacetime exciplex model also has sufficient parameters to allow for the spontaneous emission of protons by a process similar to the photoelectric effect but from spacetime C-QED spallation rather than from metallic surfaces.

A torus is generated by rotating a circle about an extended line in its plane where the circles become a continuous ring. According to the equation for a torus, $\left[\left(\sqrt{x^2+y^2}\right)-R\right]^2+z^2=r^2$, where r is the radius of the rotating circle and R is the distance between the center of the circle and the axis of rotation. The volume of the torus is $2\pi^2 Rr^2$ and the surface area is $4\pi^2 Rr$, in the above Cartesian formula the z axis is the axis of rotation.

Electron charged particle spherical domains fill the toroidal volume of the atomic orbit by their wave motion. If a photon of specific quanta is emitted while an electron is resident in an upper more excited Bohr orbit, the radius of the orbit drops back down to the next lower energy level decreasing the volume of the torus in the emission process.

We suggest that these toroidal orbital domains have properties similar to QED cavities and apply this structure to *topological switching* during dimensional reduction in the continuous state universe (HAM) model [12,35]. To summarize pertinent aspects of HAM cosmology:

- Compactification did not occur immediately after a big bang singularity, but is a continuous process of dimensional reduction by *topological switching* in view of the Wheeler-Feynman absorber model where the present is continuously recreated out of the *future-past*. Singularities in the HAM are not point like, but dynamic wormhole like objects able to translate extension, time and energy.
- The higher or compactified dimensions are not a subspace of our Minkowski 3(4)D reality, but our reality is a subspace of a higher 12D multiverse of three 3(4)D Minkowski spacetime packages.

During the spin-exchange process of dimensional reduction by topological switching two things pertinent to the discussion at hand:

- There is a transmutation of dimensional form from *extension to time to energy*; in a sense like squeezing out a sponge as the current Minkowski spacetime package recedes into the past down to the Planck scale; or like an accordion in terms of the *future-past* recreating the present.
- A tension in this process (string tension, T_0 in superstring theory) allows only specific loci or pathways to the dimensional reduction process during creation of the transient Planck scale domain. Even

though there are discrete aspects to this process it appears continuous from the macroscopic level (like the film of a movie); the dynamics of which are like a harmonic oscillator.

With the brief outline of HAM parameters in mind, the theory proposes that at specific modes in the periodicity of the Planck scale pinch effect, cavities of specific volume reminiscent of Bohr toroidal atomic orbits occur. It is proposed rather speculatively at present that these cavities, when energized by stochastically driven modes in the Dirac ether or during the *torque moment* of excess energy during the continuous compactification process, or a combination of the two as in standard C-QED theory of Rabi/Rydberg spontaneous emission, microwave photons of the CMBR type could be emitted spontaneously from the vacuum during *exciplex* torque moments. This obviously suggests that Bohr atomic orbital state reduction is not the only process of photon emission; (or spacetime modes are more fundamental) but that the process is also possible within toroidal boundary conditions in spacetime itself when in a phase mode acting like an atomic volume. A conceptualization of a Planck scale cavity during photon emission is represented in figure 7.1c with nine dimensions suppressed.

7.16 New Background Conditions of the Dirac Vacuum

If one assumes in conjunction with the de Broglie-Bohm-Vigier Causal Stochastic Interpretation (CSI) of quantum theory [4,6,22,97] that de Broglie matter-waves describe a wave-particle duality built up with real extended space structures with internal oscillations of particle-like spin, it is possible to justify Bohr's physical assumptions and predict new properties of a real Dirac covariant polarized vacuum [6,13].

Bohr's major contribution to modern physics was the model of photon emission-absorption in Hydrogen in terms of random energy jumps between stable quantum states and atomic nuclei. This discovery was one of the starting points for the Copenhagen Interpretation of quantum theory. We suggest this structural-phenomenology by general covariance applies equally as well to the symmetry conditions of the Dirac vacuum backcloth also; but as one knows the purely random description of quantum jumps suggested by Bohr is obviated by the CSI of quantum mechanics [4,6,22,98] suggesting this interaction is piloted.

We feel the CSI interpretation is required for our exciplex model to work because it is the internal motion of a massive photon that enables coupling to the Dirac vacuum.

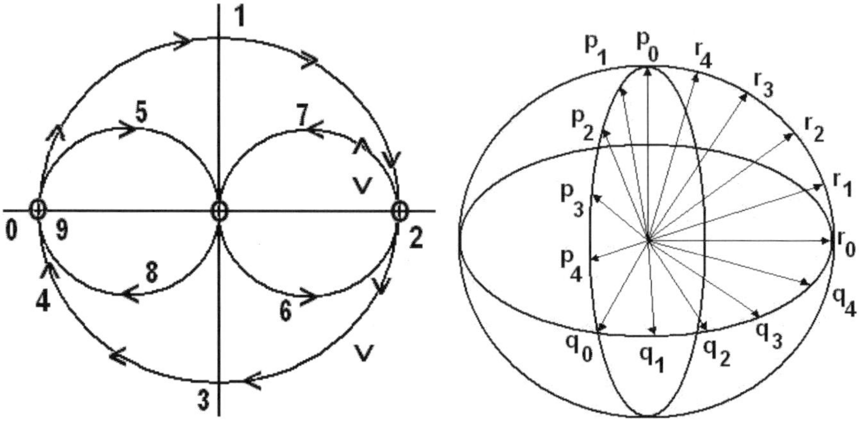

Figure 7.12 a) 2D simplistic view of 3D Dirac rotation map. b) 2D rendition of 4D view of Dirac hyperspherical rotation for raising and lowering the topological annihilation-creation vectors.

Some experimental evidence has been found to support this view [98,99] showing the possibility that the interaction of these extended structures in space involve real physical vacuum couplings by resonance with the subquantum Dirac ether. Because of photon mass the CSI model, any causal description implies that for photons carrying energy and momentum one must add to the restoring force of the harmonic oscillator an additional radiation (decelerating) resistance derived from the EM (force) field of the emitted photon by the action-equal-reaction law. Kowalski has shown that emission and absorption between atomic states take place within a time interval equal to one period of the emitted or absorbed photon wave. The corresponding transition time corresponds to the time required to travel one full orbit around the nucleus. Individual photons are extended spacetime structures containing two opposite point-like charges rotating at a velocity near *c*, at the opposite sides of a rotating diameter with a mass, $m_\gamma \simeq 10^{-65} g$ and with an internal oscillation $E = mc^2 = h\nu$. Thus a new causal description implies the addition of a new component to the Coulomb force acting randomly and may be related to quantum fluctuations. We believe this new relationship has some significance for our model of vacuum C-QED blackbody absorption/emission equilibrium.

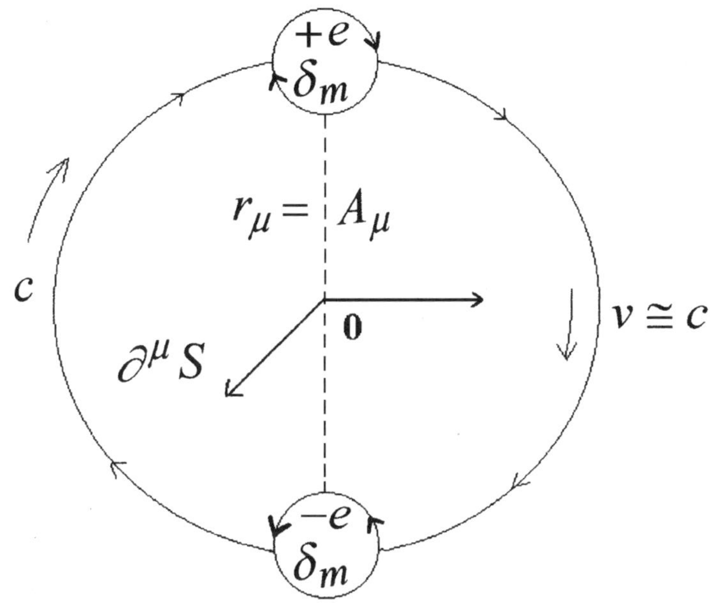

Figure 7.13 Rotating surface charges (Bumps and Holes) on the surface of the polarized Dirac ether signifying the integration EM and G. Compare Fig. 7.7.

The real ether has a covariant Dirac type stochastic surface regime with a distribution of extended photons which carry EM waves built with sets of such extended photons beating in phase; thus constituting subliminal and superluminal collective EM fields detected in the Casimir effect, so that a 'Bohr transition' with one photon absorption occurs when a nonradiating Bohr orbital electron collides and beats in phase with an ether photon. In that case a photon is emitted and Bohr electron's charge e^- spirals in one rotation in an atom towards a lower level. (But for CMBR-redshift the exciplex charge topology undergoes instead a Dirac spherical rotation of 720° which allows a 'piloting' mechanism to control the BB equilibrium C-QED domain.) Kowalski's calculations from the laser experiments have demonstrated such an orbiting charge can emit or absorb a photon within the transition time corresponding to the time interval needed to travel one full orbit [98] in terms of the CSI of quantum theory where electrons and photons are considered to contain extended structures in space and their interactions within extended time intervals.

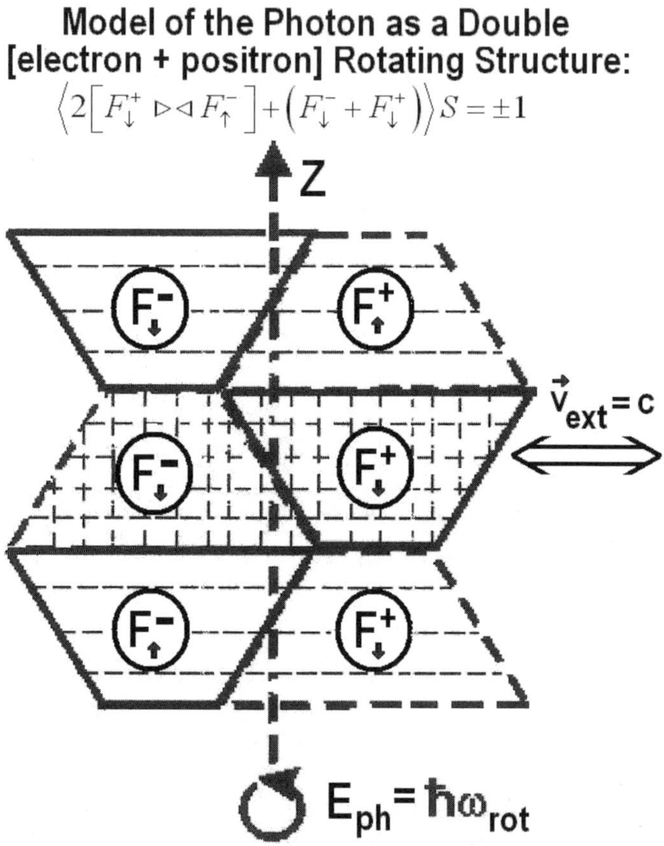

Figure 7.14 Model of the photon $\left\langle 2\left[F_{\downarrow}^{+} \rhd\!\lhd F_{\uparrow}^{-}\right]+\left(F_{\downarrow}^{-}+F_{\downarrow}^{+}\right)\right\rangle S=\pm 1$, as result of fusion of electron and positron-like triplets of subelementary fermions. The resulting symmetry shift of such structure is equal to zero, providing the absence or very close to zero photon rest mass and its propagation in the vacuum with light velocity or very close to it in the asymmetric secondary Bivacuum [100]. Figure adapted from [100].

We could think in way of an illustrative example of the high energy interaction of the photon in HAM cosmology along the lines of the

Kaivaranan bivacuum model [100] but in general we consider the photon dipole as an element of the photon-graviton complex of the unitary field. We include reference to the Kaivaranian model because we think it is a good example of the richness of vacuum structure still little understood especially as we continue to study its HD Dirac properties. Wheeler considered 'charge as topology' where lines of force in a wormhole can thread through a handle and emerge through each mouth to give the appearance of charge in an otherwise charge free spacetime [101]. We include it as a lead into Sect. 7.18 where since charge is topology, following our recalculation of the Planck constant in Chap. 4 a richer exciplex structure could be developed to show a format for Dirac vacuum exciplex proton spallation.

To summarize our conflict with the Copenhagen interpretation we reexamine Bohr's starting point for the emission and absorption of photons between jumps in stable atomic orbits in terms of the CSI to account for the recent experiments reviewed by Kowalski [98] which he interpreted to be based on extended structures in space and their interactions within extended time intervals with a real physical 'vacuum coupling' by resonance from a physically real Dirac aether which takes place during the time interval of orbit around the nucleus:

- That electrons like all other massive particles (including photons) are not point-like but extended spacetime structures in a physically real aether.
- That these structures contain internal harmonic oscillations of point-like quantum mechanical charges around the corresponding gravitational center of mass, Y_μ so that individual electrons or photons have different centers of mass and EM charge when particulate and piloted fields.
- That the Compton radius [102] of mass is significantly larger than the radius of the charge distribution.
- That the centers of charge, e, x_μ rotates around the center of mass, Y_μ with a velocity close to the velocity of light, c so that individual electrons (and photons during the centroid anisotropic mass coupling moment – see Chap. 5) are real harmonic oscillators with de Broglie type internal oscillations. See Fig. 7.13.
- That individual photons are also extended spacetime structures

containing two opposite point-like charges, $e^{\pm}$ rotating near the velocity of light, c at the opposite sides of a rotating diameter with a mass, $m_{\gamma} \simeq 10^{-65} g$ with an internal oscillation, $E = mc^2 = \hbar$.

- That the Dirac covariant polarized stochastic vacuum is a real aether distribution of these extended photons carrying EM waves built with sets of these extended photons oscillating in phase and thus constituting subluminal and superluminal collective EM fields detectable in the Casimir effect such that a Bohr transition with one photon absorption occurs when a non radiating Bohr orbital electron collides and beats in phase with an aether photon such that a photon is emitted and a Bohr electron's charge, e^- spirals in one rotation into the lower level.

In Kowalski's calculations the orbiting electron can emit or absorb a photon in the interval of one rotation [98]. We hope this discussion is sufficient for the reader to see that if these same atomic CSI conditions are applied to C-QED exciplex parameters (Fig. 7.11) black body absorption-emission redshift-CMBR equilibrium entails the same processes.

7.17 Deriving the Topological Action Principle for CMBR Emission

Well-known forms of the Schrödinger equation central to quantum theory have correspondence to Newton's second law of motion, $\sum f = ma$; which is also chosen as the formal basis for HAM CMBR emission theory. A more rigorous defense of the logic for this choice will be given elsewhere. Here only the postulate that CMBR emission is governed by a unified electro-gravitation action principle is stated. Neither Newtonian $F = Gm_1m_2 / r^2$ (although it was derived from $f = ma$) nor Einsteinian gravitation, $G = 8\pi T$ is utilized for deriving the *advanced/retarded* description of CMBR emission because the related structural-phenomenological boundary conditions of the cavities topology has no relation to classical dynamics which both of these theories do. Newton's gravitation law also contains a constant of undesired dimensionality; whereas $f = ma$ is without dimensionality. For similar reasons Einstein's gravity is also not chosen.

Since relativistic energy momentum and not mass is required, first we substitute Einstein's mass energy relation, $E = mc^2$ into Newton's second law and obtain: $F_{(N)} = E / c^2 a$. Where $F_{(N)}$ will become the unitary emission/absorption force and E arises from the complex self-organized electro-gravitational 'Geon energy' related to S_N of the HAM complex energy dependent Minkowski metric, $\hat{M}_4 \pm C_4$ as defined in the basic symmetry premises of HAM theory [12,35] where, $S_0 = \hat{M}_4$, $S_1 = -C_{4(ret)}$ and $S_2 = +C_{4(adv)}$ for the triune 12D least unit:

$$S_N = S_0 + S_1 + S_2 \tag{7.39}$$

E is scale invariant through all levels of HAM cosmology beginning at the highest level in the supralocal 12D Multiverse as a hyperdimensional Wheeler Geon [103] or 'ocean of light' of the unitary field. According to Wheeler a Geon is a ball of photons of sufficient mass that it will self cohere through gravitational action. At the micro level the Geon becomes synonymous with the E term and quantized as a unit of *Einstein's* , the fundamental physical quantity defined as a 'mole or Avogadro number of photons'. Next the equation is generalized for the HAM as derived from the work of [71].

Taking an axiomatic approach to cosmological scaling, such that all lengths in the universe are scale invariant, we begin with the heuristic relation that $c \equiv \dot{R}$ or $\dot{R} = l / t = c$ where $\dot{R}$ represents the rate of change of scale in the universe [71]. This corresponds to the Hubble relation for perceived expansion of the universe where $H_0 = \dot{R} / R$ and $a = \dot{R} \times H_0$ or substituting $\dot{R}^2 / R$. So continuing for final substitution we have $F_{(N)} = E / c^2 a = E / c^2 \times \dot{R}^2 / R$. Since $c \equiv \dot{R}$ the c^2 and $\dot{R}^2$ terms cancel and we are left with:

$$F_{(N)} = \pm E / R_t \tag{7.40}$$

Which is the unexpanded formalism for the fundamental unitary anthropic action equilibrium conditions as delineated in Chap. 3 in terms of string tension, T_0. It should be noted that R_t is a complex rotational length and could also be derived in terms of angular momentum, spacetime spinors, Penrose twistors, SUSY branes and most importantly as a complementarity of static-dynamic Casimir boundary conditions for mirror symmetry/brane duality at higher levels closer to domains described by conventional theory. But the derivation above is more fundamental to HAM CMBR. The Hubble Einstein 3-sphere, a subspace in HAM cosmology (or Calabi-Yau dual 3-tori), is *covered* by the scale invariant hyper-geon (unified) field. The spin exchange mechanism of continuous dimensional reduction-compactification dissipates the putative heat predicted by gauge theory for the Planck scale BH backcloth [65,74].

Schema of a Few Configurations of the SUSY 'Pin-Raster' Least-Unit Potentia

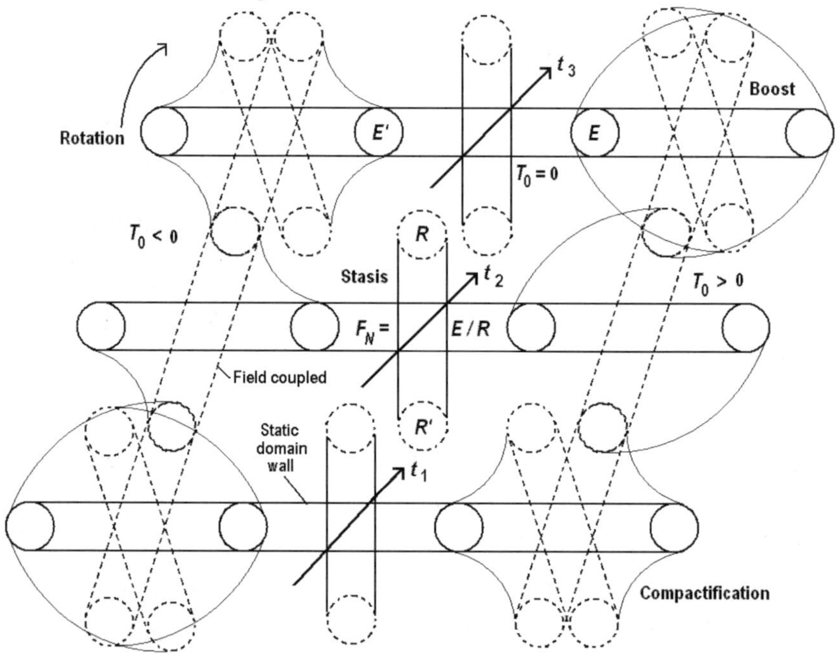

Figure 7.15 Geometric schema of the unexpanded noetic field equation. Where the central locus represents the x-axis. Loci where coupling is shown (superposed circles) would uncouple and recouple depending on whether the parallel transport mode of the cycle is at the deficit angle position or not.

The free energy for CMBR emission during the periodic *exciplex* moment arises by parallel transport during continuous dimensional reduction. Spatial dimensions, by the boundary of a boundary = 0 condition (Bianchi identities) [104] , first parallel transport to temporal dimensionality, d_t [37] and then to noetic or anthropic unitary energy, $E_{(N)}$ [105] $d_s \rightarrow d_t \rightarrow E_{(N)}$ This boost concept is key to the completion of quantum theory and unifying geometrodynamics with unitarity.

7.18 A Putative Model of Exciplex Proton Nucleosynthesis

In recent decades four types of nucleosynthesis have been considered: 1) Big Bang nucleosynthesis during the putative first three minutes of creation, 2) Stellar fission/fusion nucleosynthesis, 3) Explosive Supernova nucleosynthesis and 4) Cosmic ray spallation against the interstellar medium of gas and dust mostly by high energy protons. Spallation is also known to occur in meteor rock, the Earth's atmosphere and lava [106-117]. Here we introduce a 5th form of gentle nucleosynthesis by spacetime exciplex spallation utilizing the Vigier causal stochastic interpretation of quantum theory because of its legitimacy in dealing with the internal motion and structure of matter [4,6,22]. 3/4 mass of universe is attributed to hydrogen. If the Big Bang is incorrect as we and a few other cosmologists propose, there must be a mechanism for the 'creation' of protons [117-119].

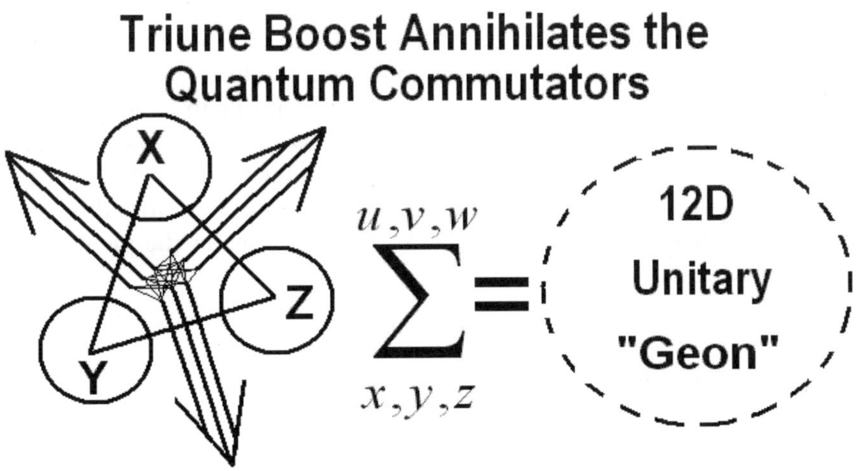

Figure 7.16 The continuous-state boost of $s \leftrightarrow t \leftrightarrow e$, signifying a new set of Noetic Transformations beyond the Lorentz-Poincaré where states that ordinarily do not commute are able to commute in the HD regime.

Chatterjee and Banerjee have developed an XD model for Hoyle and Narlikar's C-field cosmology [118]. Hoyle and Narlikar added an additional term to Einstein's field equations to introduce the C-field

$$R_{ik} - \frac{1}{2} g_{ik} R = -8\pi\left({}^{m}T_{ik} + {}^{C}T_{ik} \right) \tag{7.41}$$

where ${}^{C}T_{ik}$ is the C-field term,

$${}^{C}T_{ik} = -f\left(C_i C_k - \frac{1}{2} g_{ik} C^{\alpha} C_{\alpha} \right), \tag{7.42}$$

but Hoyle and Narlikar [119] formalized their C-Field with a negative energy density that drives expansion of the universe and is therefore not compatible with continuous-state HAM cosmology. For interest to HAM cosmology Chatterjee and Banerjee find a spontaneous compactification process in their HD derivation of C-field solutions utilizing the $\dot{R}$ scaling factor key to the continuous-state of HAM cosmology; but they also align there formalism with an expanding universe. Another point of interest of the Chatterjee and Banerjee model [118] is that introduction of the C-Field is not *ad hoc* as in the Gold and Bondi or Hoyle and Narlikar models by the compactification process that also allows for the HD conservation of matter. We do not have time to develop this model to a rigorous formalism for this volume, but we hope to or that other works will utilize the richness of the exciplex paradigm to complete the model. The other factor we have ignored in this discussion is that the oscillation of Planck's constant up to the size of the Larmour radius of the hydrogen atom provides many additional C-QED parameters for this work especially when the plethora of SUSY parameters enter the picture. And don't forget the new noetic transform…

7.19 Summary and Conclusions

An anisotropic photon rest mass calculated from both the WFA of classical GR, and the Einstein-de Broglie relationship confirms the Vigier hypothesis of $m_\gamma \neq 0$. Photon zero point coupling, as required by quantum gravity, has major cosmological implications obviating the big bang by removing the need for an initial singularity in time and still preserves gauge. The GP is equalized by compactification, enabling rigorous calculation of the cosmological constant revealing the arrow of time. Unitarity by its nature must provide pervasive application.

When the CMBR was discovered it was interpreted as definitive proof that the Big Bang was the correct model of creation. However, the same observational data may be also interpreted in the manner here. HAM Gravity, which models compactification as a rich dynamic hyperstructure provides an inherent mechanism to balance the GP in a *static universe* where the CMBR is not a remnant of adiabatic inflation but intrinsic to the equilibrium conditions of Planck scale spacetime CQED or CSED.

A preliminary formalism for CMBR-emission and *tired-light* redshift-absorption as BB equilibrium from the continuous state topological dynamics of the Dirac vacuum in a HAM has been presented. This has taken two possible forms:

1. A stochastically driven C-QED effect on Eddington free space Rabi/Rydberg atoms coupled to vacuum zero-point field fluctuations.

2. A composite *exciplex* of advanced - retarded spacetime topological cavity modes which may act as an atom-cavity « molecule » formed on the basis of gravito-quantum coherence effects by unitary action of $F_{(N)}$. Both postulated by only two new theoretical concepts, from already observed CQED effects in the laboratory:

- A Dirac type vacuum coupling between the atom and vacuum cavities of the structure of spacetime itself, and
- CMBR photon emission can also occur from the Bohr-type boundary conditions of spacetime topology without the presence of an atom with E transport by topological switching in D-reduction of $d_s \rightarrow d_t \rightarrow E_{(N)}$.

BH's have been demonstrated by Hawking to emit BB radiation in the quasiclassical limit, and the lower limit has been shown to be the Planck, mass providing a firm theoretical foundation for intrinsic vacuum emmitivity. A non inflationary origin of CMBR obviates the Big Bang requiring reinterpretation of the standard cosmological model with profound implications for the future of cosmological theory.

References

[1] Eddington, Sir Arthur (1926) Internal Constitution of the Stars, Cambridge University Press, reprinted (1988).
[2] Silk, J. (1989) The Big Bang, New York: W.H. Freeman & Co.
[3]Van Flandern, T. (2002) The top 30 problems with the Big Bang, Apeiron, Vol. 9, No. 2, pp. 72-90.
[4] Bohm, D. & Vigier, J.-P. (1954) Model of the causal interpretation of quantum theory in terms of a fluid with irregular fluctuations, Phys. Rev. 96:1; 208-217.
[5] Holland, P.R. (2000) The Quantum Theory of Motion: An Account of the de Broglie-Bohm Causal Interpretation of Quantum Mechanics Cambridge: Cambridge Univ. Press.
[6] Vigier, J-P (2000) selected papers, in S. Jeffers, B. Lehnert, N. Abramson, & L. Chebotarev (eds.) Jean-Pierre Vigier and the Stochastic Interpretation of Quantum Mechanics, Montreal: Aperion.
[7] Cramer, JG (1986) The transactional interpretation of quantum mechanics, Rev Mod Physics, 58:3, 647-687.
[8] Wheeler, J.A. & Feynman, R.P. (1945) Rev. Mod Physics, 17, 157.
[9] Chu, S-Y (1993) Phys. Rev. Lett. 71, 2847.
[10] Franck, G. (2000) Time and presence, in R.L. Amoroso et al, (eds.) Science and The Primacy of Consciousness, Orinda: Noetic Press.
[11] Penrose, R. (1996) On gravity's role in quantum state reduction, General Relativity and Gravitation, 28:5; 581-600.
[12] Amoroso, R.L. (2002) Developing the cosmology of a continuous state universe, in R.L. Amoroso, G. Hunter, M. Kafatos & J-P Vigier (eds.) Gravitation & Cosmology: From the Hubble Radius to the Planck Scale, Dordrecht: Kluwer Academic.
[13] Dirac, P.A.M. (1952) Nature (London) 169, 702.
[14] Cufaro, N., Petroni, C. & Vigier, J-P (1983) Dirac's Ether in Relativistic Quantum Mechanics, Found. Phys., 13, 253-286.
[15] Lehnert, B. (2002) New developments in electromagnetic field theory, in R.L. Amoroso, G. Hunter, M. Kafatos & J-P Vigier (eds.) Gravitation & Cosmology: From the Hubble Radius to the Planck Scale, Dordrecht: Kluwer Academic.
[16] Lehnert, B. (1998) Electromagnetic theory with space-charges in vacuo, in G. Hunter, S. Jeffers & J-P Vigier (eds.) Causality and Locality in Modern Physics, Dordrecht: Kluwer Academic.
[17] Proca, A, (1936) *Compt. Rend.*, 202, 1420.

[18] Evans, M. & Vigier, J.-P. (1994) The Enigmatic Photon Vol. 1,2, Dordrecht: Kluwer Academic.
[19] Evans, M.W. (1994) Derivation of the vacuum longitudinal field $B^{(3)}$ from the Dirac equation of the electron in the electromagnetic field, Foundations Physics Let,, 7:6; 577-583.
[20] Comay, E. (1996) Comment on the longitudinal magnetic field of circularly polarized electromagnetic waves, Chem. Phys. Lett., 261:4-5; 601-604.
[21] Roy, S. & Lehnert, B. (1998) Extended Electromagnetic Theory; Space Charge in Vacuo and the Rest Mass of the Photon, Singapore: World Scientific.
[22] Vigier, J-P (1997) Possible consequences of an extended charged particle model in electromagnetic theory, Phys. Lett. A, 235:5; 419-431.
[23] Bass, L. & Schrödinger, E. (1955) Proc. R. Soc. Lon A 232,1-6.
[24] Einstein, A. (1917) Ann. Phys., vol. 18, p. 121
[25] Zwicky, F. (1929) Proc. Nat. Ac. Sc., Washington, 15, 773.
[26] Zwicky, F. (1929). On the red shift of spectral lines through interstellar space, Proceedings of the National Academy of Sciences, 15:10; 773–779.
[27] Barnes, A. & Scargle, J.D. (1975) Improved upper limit on the photon mass, Phys. Rev. 35: 17; 1117-1120.
[28] Finlay-Freundlich, E. (1953) *Göttinger Nachrichten*, 7, 95-102.
[29] Finlay-Freundlich, E. (1954) Philosophical Magazine, 45, 303-319.
[30] Finlay-Freundlich, E. (1954) Proceedings of the Physical Society A 67, 192-193.
[31] Born, M. (1954) *Nachr. Ak. Wiss. Göttingen*, 7, 102.
[32] Pecker, J-C., Roberts, A.P., & Vigier, J-P (1972) Non-velocity redshifts and photon-photon interactions, Nature, v. 237, p. 227-229).
[33] Pecker, J-C & Vigier, J-P (1988) A Possible Tired-Light Mechanism, Apeiron, No. 2, pp 19-23.
[34] Vigier, J-P (1990) Evidence for nonzero mass photons associated with a vacuum-induced dissipative redshift mechanism, IEEE Trans Plasma Sci, 18:1;64-72.
[35] Amoroso, R.L. & Pribram, K.H. (2009) The Complementarity of Mind and Body: Realizing the Dream of Descartes, Einstein and Eccles, Cambridge: MIT Univ. Press.
[36] Amoroso, R.L. & Amoroso, P.J. (2004) The Fundamental Limit and Origin of Complexity in Biological Systems: A New Model for the

Origin of Life, in D.M. Dubois (ed.) Computing Anticipatory Systems, AIP Conf. Proceedings V. 718, pp. 144-159, Melville: Am Inst. Phys.
[37] Ramon, C. and Rauscher, E.A. (1980) Superluminal transformations in complex Minkowski spaces, Found. Phys. 10:7/8, 661-669.
[38] Ceon, R & Rauscher, E.A. (1980) Found. Phys.
[39] Narlikar, J.V. (2002) in R.L. Amoroso, G. Hunter, M. Kafatos & J-P Vigier (eds.) Gravitation & Cosmology: From the Hubble Radius to the Planck Scale, Dordrecht: Kluwer Academic.
[40] Stevens, H.H. (1989) Size of a least-unit, in M. Kafatos (ed.) Bell's Theorem, Quantum Theory and Conceptions of the Universe, Dordrecht: Kluwer Academic.
[41] Narlikar, J.V. (1996) The Lighter Side of Gravity, Cambridge: Cambridge Univ. Press.
[42] Hoyle, F. & Narlikar, J.V. (1996) Lectures on Cosmology & Action at a Distance Electrodynamics, Singapore: World Scientific.
[43] Einstein, A (1922) Sidelights on Relativity, London: Methuen & Co.
[44] Amoroso, R.L. & Vigier, J-P (2002) The origin of CMBR as intrinsic blackbody cavity-QED resonance inherent in the dynamics of the continuous state topology of the Dirac vacuum, in R.L. Amoroso , G. Hunter, M. Kafatos & J-P Vigier (eds.), Gravitation and Cosmology: From the Hubble Radius to the Planck Scale, Dordrecht: Kluwer.
[45] Puthoff, H.E. 1989, Gravity as a zero-point-fluction force, Physics Review A, 39, 2333-2342.
[46] Sakharov, A.D. (1967) Vacuum quantum fluctuations in curved space and the theory of gravitation, Sov.Phys.Dokl.12:11; 1040-1041; (1968) *Dokl. Akad. Nauk Ser.Fiz.*,177:70-71; (1967) Sov.Phys.Usp.34:394, (1991) Reprinted in Sov. Phys. *Usp. Fiz. Nauk* 161, No. 5 64-66.
[47] Deser, S. (1972) *Ann Inst Henri Poincairé*, Vol. XVI, pp. 79, Paris: Gautier-Villors.
[48] de Broglie, (1940) *La mécanique ondulatioiredu photon, in Une Nouvelle Théorie de la Lumiere, Tome* 1, pp. 121-145, Paris: Gautier-Villors.
[49] Sundrum, R. (2005) SLAC Summer Institute, Lecture notes-1, Gravity in the Quantum World and the Cosmos, www-conf.slac.stanford.edu/ssi/2005/lec_notes/default.htm
[50] Tolman, R.C., Ehrenfest, P & Podolsky, B. (1931) On the gravitational field produced by light, Phys. Rev., 37, 602-615.
[51] Nicols, E.F. & Hull, G.F. (1901) Phys. Rev., 13, 307.

[52] Narlikar, J.V., Pecker, J.C. & Vigier, J.P. (1991) Does a possible laboratory observation of a frequency anisotropy of light result from a non-zero photon mass, m_γ? Phys. Lett. A, 154, 5,6, pp. 203-209.

[53] Davisson, CJ & Germer , LH (1927) Diffraction of electrons by a crystal of nickel, Phys. Rev. 30:6, 705-740.

[54] Fowles, G.R. (1989) Introduction to Modern Optics, NY: Dover.

[55] Amoroso, R.L., Kafatos, M. & Ecimovic, P. (1998) The origin of cosmological redshift in spin exchange vacuum compactification and nonzero rest mass photon anisotrophy, in G. Hunter, S. Jeffers & J-P Vigier (eds.) Causality and Locality in Modern Physics, Dordrecht: Kluwer.

[56] Aichelburg, P.C., Ecker, G. & Sexl, R.U. (1971) Lorentz-covariant Langrangians and causality, Nuovo Cimento B, V. 2B, N.1 p. 63-76.[52]

[57] Einstein, A. (1954) Ideas and Opinions, C. Seelig, (ed.) S. Bargmann, (Trans.)

[58] Gribbin, J. (1995). Schrödinger's Kittens, Boston: Little Brown.

[59] Bohm, D (1951) Quantum Theory, Englewood Cliffs: Prentice-Hall.

[60] Hunter, G. (2005) Personal communication.

[61] Hawking, S.W. & Ellis, G.F.R. (1973) The Large Scale Structure of Space-Time, Cambridge: Cambridge University Press.

[62] Wolff, M. (2008) Schrödinger's Universe, Einstein, Waves and the Origin of Natural Laws, Parker: Outskirts Press.

[63] Einstein, A. (1923) Quote from a letter to Ernst Mach, in Misner, C.W., Thorne, K.S. & Wheeler, J.A. (1973) Gravitation, San Francisco: W. H. Freeman.

[64] Puthoff, H.E. (2002) Polarizable-vacuum, representation of general relativity, in R.L. Amoroso, G. Hunter, M. Kafatos & J.-P. Vigier (eds.), Gravitation and Cosmology: From the Hubble Radius to the Planck Scale, Dordrecht: Kluwer.

[65] Markov, M.A. (1966) *Zh. Eksp. Theor. Fiz.* v51, p. 878.

[66] Hawking, S.W. (1976) Black holes and thermodynamics, Physical Review D, V 13, No.2, 191-197.

[67] Hawking, S.W. (1974) Black hole explosions? Nature, v 248. 30-31.

[68] Berezin, V. (1997) Quantum black hole model and Hawking's radiation, Phys. Rev. D, V.55, N4, 2139-2151.

[69] Kant, I. (1878) The Critique of Pure Reason, JMD Meiklejohn, (Trans.) London: G. Bell & Sons, Pdf - http://books.google.com/books.

[70] Birkhoff, G.D. (1923,) Relativity and Modern Physics, Cambridge: Harvard Univ. Press.

[71] Kafatos, M., Roy, S. & Amoroso, R.L. (2000) Scaling in Cosmology & the Arrow of Time, in Buccheri, di Gesu & Saniga, (eds.) Studies on Time, Dordrecht: Kluwer Academic.
[72] Kafatos, M. (1996) Knowledge limits in cosmology, In M. Kafatos & Y. Condo (eds.) Examining the Big Bang and Diffuse Background Radiations, Netherlands.
[732] Kotigua, R.P. & Toffoli, T. (1998) Potential for computing in micromagnetics via topological conservation laws, Physica D, 120:1-2, pp. 139-161.
[74] Sung, J.C. (1993) Pixels of Space-Time, Woburn: Scientific Publications.
[75] Mukohyama, S. & Randall, L. (2004) A Dynamical Approach to the Cosmological Constant, arxiv.org/abs/hep-th/0306108.
[76] Lakes, R. (1998) Experimental limits on the photon mass and cosmic magnetic vector potential, Phys. Rev. Lett. 80, 1826-1829.
[77] Lehnert, B. (2009) personal communication.
[78] Lehnert B. (2007) Revised quantum electrodynamics with fundamental applications, in Proceedings of 2007 ICTP Summer College on Plasma Physics, P. K. Shukla, L. Stenflo & B. Eliasson (eds.) Singapore: World Scientific.
[79] Lehnert B. (2007) A Revised Electromagnetic Theory with Fundamental Applications. Swedish Physics Archive, D. Rabounski (ed.), The National Library of Sweden, Stockholm; and Bogolyubov Institute for Theoretical Physics, A. Zagorodny (ed.) Kiev.
[80] Beckenstein, J. D. (1973) Phys. Rev., D7, 2333–2346.
[81] Hawking, S.W.(1974), The anisotropy of the universe at large times, in IAU Symposium No. 63 on Confrontation of Cosmological Theories with Observational Data, M.S. Longair, (ed.) Dordrecht: Kluwer.
[82] Berman, P.R. (ed.) (1994). Cavity Quantum Electrodynamics, New York, Academic Press.
[83] Agarwal, G.S. (1991) Additional vacuum-field Rabi splittings in cavity QED, Phys. Rev. A, 43:5; 2595-2598.
[84] Feynman, R.P. (1961) Quantum Electrodynamics. New York: Benjamin.
[85] Eddington, A.S. (1930) The Rotation of the Galaxy, Halley Lecture, Oxford: Clarendon Press..
[86] Hinds, E.A. (1993) Perturbative Cavity Quantum Electrodynamics. Adv. Atom. Mol. and Opt. Phys., 2, 1-56.
[87] Al-Awfi, S. & Babiker, M. (1998) Atom dynamics between

conducting plates, Phys. Rev. A 58, pp. 2274-2281.
[88] Sukenik, C.I., Boshier, M.G., Cho, D., Sandoghdar, V. & Hinds, E.A. (1993) Measurement of the Casimir-Polder force, Phys. Rev. Lett., 70, pp. 560-563.
[89] Carmichael, H.J. (1993) Phys. Rev. Lett. 70:15, 2273-2276.
[90] Heinzen, D.J., Feld, M.S. (1987) Phys. Rev. Lett. 59:23; 2623-2626.
[91] Jhe, W., Anderson, A. Hinds, E.A., Meschede, D., Haroche, S, (1987) Phys. Rev. Lett. 58:7; 666-669.
[92] Raizezn, M.G., Thompson, R.J., Brecha, R.J., Kimble, H.J. & Carmichael, H.J.(1989) Phys. Rev. L. 63:3, 240-3.
[93] Rempe, G. (1993) Contemporary Phys., 34:3, 119-129.
[94] Thompson, R.J., Rempe, G. & Kimble, H.J. (1992) Phys. Rev. Let. 68:8, 1132-1135.
[95] Zhu, Y. (1990) Phys. Rev. Lett. 64, 2499.
[96] Ren, W, Cresser, J.D, & Carmichael, JH, (1992) Phys. Rev. A, 46, 7162
[97] de Broglie, L. & Vigier, J-P (1972) Phys. Rev. Lett. 28, 1001-1004.
[98] Kowalski, M. (1999) Photon Emission from Atomic Hydrogen, Physics Essays, Vol.12, 312-331.
[99] Franca, H.M., Marshall, T.W. & Santos, E. (1992) Spontaneous emission in confined space according to stochastic electrodynamics Phys. Rev. A 45:9; 6436-6422.
[100] Kaivarainen, A. (2000) Unified theory of bivacuum, particles duality, fields & time, new fundamental bivacuum mediated interaction, arXiv:physics/0003001v14 [physics.gen-ph].
[101] Wheeler, J.A. (1977) Gravitational and Electromagnetic wave flux compared and contrasted, Phys. Rev. D, 16:12, 3384-3389.
[102] Compton, A.H. (1923) A quantum theory of the scattering of x-rays by light elements, Phys. Rev. 21:5; 483-502.
[103] Wheeler, J.A. (1955) Geons, Phys. Rev., 97:2;. 511-536.
[104] Misner, C.W., Thorne, K.S. & Wheeler, J.A. (1973) Gravitation, San Francisco: W. H. Freeman.
[105] Cardone, F., Francaviglia, M.& Mignani, R. (1999) Energy as a fifth dimension, Found. Phys. L. 12:4, 347-69.
[106] Klein, O; Nishina, Y (1929) *Z. f. Phys*. 52: 853 and 869.
[107] Hoyle, F. (1946) Monthly Notices Roy. Astron. Soc. 106, 366.
[108] Hoyle, F. (1954) Astrophys. J. Suppl. 1, 121.
[109] Clayton, D.D. (1968) Principles of Stellar Evolution and Nucleosynthesis, McGraw-Hill; University of Chicago Press (1983).
[110] Rolfs, C.E. & Rodney, W. S. (1988) Cauldrons in the Cosmos,

Chicago: Univ. of Chicago Press.
[111] Marti, K. & Craig , H. (1987) Cosmic-ray-produced neon and helium in the summit lavas of Maui, Nature 325, pp. 335-337.
[112] Nadyozhin, D.K. & Panov, I.V. (1997) Nucleosynthesis induced by neutrino spallation of helium, Nuclear Physics A, 621: 1-2; 359-362.
[113] Nomoto1, K., Maeda1, K., Umeda1, H., Tominaga1, N., Ohkubo1, T., Deng1, J. & Mazzali,, P. A. (2003) Nucleosynthesis in black-hole-forming supernovae and abundance patterns of extremely metal-poor stars, arXiv:astro-ph/0306412v2.
[114] Hoyle, F., Burbidge, G. & Narlikar, J.V. (1993) A quasi-steady state cosmological model with creation of matter, Astrophysical J, 410: 437-457.
[115] Ramanand, Jha (1992) A new creation cosmology, General Relativity and Gravitation, Vol 24, Number 1, pp. 87-92.
[116] Lord, E.A. (1974) Creation-field theory from dimensional analysis, Pramana, 3:1: 35-43.
[117] Burbidge, E.M., Burbidge, G.R., Fowler, W.A. & Hoyle, F. (1957) Synthesis of the elements in stars, Rev. Mod. Phys. 29; 547.
[118] Chatterjee , S. & Banerjee, A. (2004) C-field cosmology in higher dimensions, General Relativity & Gravitation, 36:2; 303-313.
[119] Narlikar, J.V. (2002) An introduction to Cosmology, Cambridge: Cambridge University Press.

Chapter 8

Implications of Multidimensional Geometries and Measurement

Complexification of Maxwell's equations with an extension of the gauge condition to non-Abelian algebras, yields a putative metrical unification of relativity, electromagnetism and quantum theory. This unique new approach also yields a universal nonlocality with implications for Bell's Theorem and the possibility of instantaneous quantum connections because spatial separation can vanish by utilizing the complex space.

8.1 Introduction

In this chapter we develop non-Abelian gauge groups for real and complex amended Maxwell's equations in a complex 8-Dimensional Minkowski space in order to describe nonlocality in quantum theory and relativity which has implications for extending gravitational theory to the unitary regime. We demonstrate a mapping between the twistor algebra of the complex 8-space and the spinor calculus of 5D Kaluza-Klein geometry which attempts to unify Gravitational and EM theory (Chap. 6). Our quantum formalism demonstrates that solving the Schrödinger equation in a complex 8D geometry yields coherent collective state phenomena with soliton wave solutions. The model shows that standard quantum theory is a linear approximation of a higher dimensional complex space. Through this formalism we can assess that complex systems can be defined within conventional quantum theory as long as we express that theory in a hyper-geometric space. We utilize our complex dimensional geometry to formulate nonlocal correlated phenomena, including the quantum description of the 1935 EPR paradox formulated with Bell's theorem. Tests by Clauser, Aspect, and Gisin have demonstrated that particles emitted with approximate simultaneity at the speed of light, *c* remain correlated nonlocally over meter and kilometer distances. As Stapp has said, Bell's theorem and its experimental verification is one of the most profound discoveries of the 20th century. We will demonstrate the application of our formalism for complex systems and review the history of our model from 1974.

We have analyzed, calculated and extended the modification of Maxwell's equations in a complex Minkowski metric, M_4 in a C_2 space using the SU_2 gauge, SL(2,c) and other gauge groups, such as SU_n for n > 2 expanding the U_1 gauge theories of Weyl. This work yields additional predictions beyond the electroweak unification scheme. Some of these are: 1) modified gauge invariant conditions, 2) short range non-Abelian force terms and Abelian long range force terms in Maxwell's equations, 3) finite but small rest mass of the photon, and 4) a magnetic monopole like term and 5) longitudinal as well as transverse magnetic and electromagnetic field components in a complex Minkowski metric M_4 in a C_2 space.

This is an 8D complex Minkowski space, M_4+C_4 composed of 4 real and 4 imaginary dimensions consistent with Lorentz invariance and analytic continuation in the complex plane [1-6]. The unique feature of this geometry is that it admits nonlocality consistent with Bell's theorem, (EPR paradox), possibly Young's double slit experiment, the Aharonov-Bohm effect and multi mirrored interferometric experiment.

Also, expressing Maxwell's EM equations in complex 8-space, leads to some new and interesting predictions in physics, including possible detailed explanation of some of the previously mentioned nonlocality experiments [7-11]. Complexification of Maxwell's equations requires a non-Abelian gauge group which amends the usual theory, which utilizes the usual unimodular Weyl U_1 group. We have examined the modification of gauge conditions using higher symmetry groups such as SU_2, SU_n and other groups such as the SL(2,c) double cover group of the rotational group SO(3,1) related to Shipov's Ricci curvature tensor [12,13] and a possible neo-aether picture. Thus we are led to new and interesting physics involving extended metrical space constraints, the usual transverse and also longitudinal, non-Hertzian electric and magnetic field solutions to Maxwell's equations, possibly leading to new communication systems and antennae theory, nonzero solutions to $\nabla \cdot \underline{B}$, and a possible finite but small rest mass of the photon.

Comparison of our theoretical approach is made to the work on amended Maxwell's theory [14-17]. We compare our predictions such as our longitudinal field to the $B^{(3)}$ term of Vigier, and our non-Abelian gauge groups to that of Barrett and Harmuth. This author interprets this work as leading to new and interesting physics, including a possible reinterpretation of a neo-aether with nonlocal information transmission properties.

8.2 Complexified EM Fields in Local and Nonlocal Minkowski Space

We expand the usual line element metric $ds^2 = g_{\nu\mu} dx^\nu dx^\mu$ in the following manner. We consider a complex eight dimensional space, M_4 constructed so that $Z^u = x^u_{Re} + i x^u_{Im}$ and likewise for Z^ν where the indices ν and μ run 1 to 4 yielding (1, 1, 1, -1). Hence, we now have a new complex eight space metric as $ds^2 = \eta_{\nu\mu} dZ^\nu dZ^\mu$. We have developed this space and other extended complex spaces and examined their relationship with the twistor algebras and asymptotic twistor space and the spinor calculus and other implications of the theory [18-21]. The Penrose twistor SU(2,2) or U_4 is constructed from four-spacetime, $U_2 \otimes \widetilde{U}_2$ where U_2 is the real part of the space and $\widetilde{U}_2$ is the imaginary part of the space, this metric appears to be a fruitful area to explore.

The twistor Z can be a pair of spinors U^A and π_A which are said to represent the twistor. The condition for these representations are 1) the null infinity condition for a zero spin field is $Z^\mu \bar{Z}_\mu = 0$, 2) conformal invariance and 3) independence of the origin. The twistor is derived from the imaginary part of the spinor field. The underlying concept of twistor theory is that of conformally invariance fields occupy a fundamental role in physics and may yield some new physics. Since the twistor algebra falls naturally out of the complex space.

Other researchers have examined complex dimensional Minkowski spaces. In [2], Newman demonstrates that M_4 space does not generate any major 'weird physics' or anomalous physics predictions and is consistent with an expanded or amended special and general relativity. In fact the Kerr metric falls naturally out of this formalism as demonstrated by Newman [4,5].

As we know twistors and spinors are related by the general Lorentz conditions in such a manner that all signals are luminal in the usual four

N Minkowski space but this does not preclude super or trans luminal signals in spaces where N > 4. Stapp, for example, has interpreted the Bell's theorem experimental results in terms of transluminal signals to address the nonlocality issue of the Clauser, et al. and Aspect experiments [22]. Kozameh and Newman demonstrate the role of nonlocal fields in complex 8-space.

We believe that there are some very interesting properties of the M_4 space which include the nonlocality properties of the metric applicable in the non-Abelian algebras related to the quantum theory and the conformal invariance in relativity as well as new properties of Maxwell's equations. In addition, complexification of Maxwell's equations in M_4 space yields some interesting predictions, yet we find the usual conditions on the manifold hold [23-25]. Some of these new predictions come out of the complexification of four space 2 and appear to relate to the work of Vigier, Barrett, Harmuth and others [14]. Also we find that the twistor algebra of the complex eight dimensional, M_4 space is mappable 1 to 1 with the twistor algebra, C_2 space of the Kaluza-Klein five dimensional electromagnetic - gravitational metric [12,13].

Some of the predictions of the complexified form of Maxwell's equations are 1) a finite but small rest mass of the photon, 2) a possible magnetic monopole, $\nabla \cdot \underline{\beta \neq 0}$, 3) transverse as well as longitudinal $B^{(3)}$ like components of $\underline{E}$ and $\underline{B}$, 4) new extended gauge invariance conditions to include non-Abelian algebras and 5) an inherent fundamental nonlocality property on the manifold. Vigier also explores longitudinal $\underline{E}$ and $\underline{B}$ components in detail and finite rest mass of the photon [26].

Considering both the electric and magnetic fields to be complexified as $\underline{E} = \underline{E}_{\text{Re}} + i\underline{E}_{\text{Im}}$ and $\underline{B} = \underline{B}_{\text{Re}} + i\underline{B}_{im}$ for $E_{\text{Re}}, E_{\text{Im}}, B_{\text{Re}}$ and B_{Im} are real quantities. Then substitution of these two equations into the complex form of Maxwell's equations above yields, upon separation of real and imaginary parts, two sets of Maxwell-like equations. The first set is

$$\nabla \cdot \underline{E}_{\text{Re}} = 4\pi\rho_e, \ \nabla \times E_{\text{Re}} = -\frac{1}{c}\frac{\partial \underline{B}_{\text{Re}}}{\partial t}; \ \nabla \cdot \underline{B}_{\text{Re}} = 0,$$

$$\nabla \times \underline{B}_{\text{Re}} = \frac{1}{c}\frac{\partial \underline{E}_{\text{Re}}}{\partial t} = \underline{J}_e \tag{8.1}$$

the second set is

$$\nabla \cdot \left(i\underline{B}_{\text{Im}}\right) = 4\pi i \rho_m \,, \nabla \times \left(i\underline{B}_{\text{Im}}\right) = \frac{1}{c}\frac{\partial\left(i\underline{E}_{\text{Im}}\right)}{\partial t};$$

$$\nabla \cdot \left(iE_{\text{Im}}\right) = 0 \,, \nabla \times \left(i\underline{E}\right) = \frac{1}{c}\frac{\partial\left(iB_{\text{Im}}\right)}{\partial t} = i\underline{J}m \tag{8.2}$$

The real part of the electric and magnetic fields yield the usual Maxwell's equations and complex parts generate 'mirror' equations; for example, the divergence of the real component of the magnetic field is zero, but the divergence of the imaginary part of the electric field is zero, and so forth. The structure of the real and imaginary parts of the fields is parallel with the electric real components being substituted by the imaginary part of the magnetic fields and the real part of the magnetic field being substituted by the imaginary part of the electric field.

In the second set of equations, (8.2), the *i*'s, 'go out' so that the quantities in the equations are real, hence $\nabla \cdot \underline{B}_{\text{Im}} = 4\pi\rho_m$, and not zero, yielding a term that may be associated with some classes of monopole theories. See references in [16,17]. We express the charge density and current density as complex quantities based on the separation of Maxwell's equations above. Then, in generalized form $\rho = \rho_e = i\rho_m$ and $J=J_e+iJ_m$ where it may be possible to associate the imaginary complex charge with the magnetic monopole and conversely the electric current has an associated imaginary magnetic current.

The alternate of defining and using, which Evans does $\underline{E} = \underline{E}_{Re} + i\underline{B}_{Im}$ and $\underline{B} = \underline{B}_{Re} + i\underline{E}_{Im}$ would not yield a description of the magnetic monopole in terms of complex quantities but would yield, for example $\nabla \cdot \left(i\underline{B}_{\text{Im}}\right) = 0$ in the second set of equations. Using the invariance of the line element $s^2 = x^2 - c^2t^2$ for $r = ct = \sqrt{x}^{\,2}$ and for $s^2 = x^2 + y^2 + z^2$ for the distance from an electron charge, we can write the relation, $\frac{1}{c}\frac{\partial\left(iB_{im}\right)}{\partial t} = iJm$ or $\frac{1}{c}\frac{\partial B_{im}}{\partial t} = J_m$; $\nabla \times \left(iE_{\text{Im}}\right) = 0$ for $\underline{E}_{\text{Im}} = 0$ or

$$\frac{1}{c}\frac{\partial\left(iB_{\text{Im}}\right)}{\partial t} = iJm \tag{8.3}$$

8.3 Complex Minkowski Space: Implications for Physics

In a series of papers, Barrett, Harmuth and Rauscher have examined the modification of gauge conditions in modified or amended Maxwell theory. The Rauscher approach, as briefly explained in the preceding section is to write complexified Maxwell's equation in consistent form to complex Minkowski space [17].

The Barrett amended Maxwell theory utilizes non-Abelian algebras and leads to some very interesting predictions which have interested me for some years. He utilizes the non-commutative SU_2 gauge symmetry rather than the U_1 symmetry. Although the Glashow electroweak theory utilizes U_1 and SU_2, but in a different manner, but his theory does not lead to the interesting and unique predictions of the Barrett theory.

Barrett, in his amended Maxwell theory, predicts that the velocity of the propagation of signals is not the velocity of light. He presents the magnetic monopole concept resulting from the amended Maxwell picture. His motive goes beyond standard Maxwell formalism and generate new physics utilizing a non-Abelian gauge theory.

The SU_2 group gives us symmetry breaking to the U_1 group which can act to create a mass splitting symmetry that yield a photon of finite (but necessarily small) rest mass which may be created as self energy produced by the existence of the vacuum. This finite rest mass photon can constitute a propagation signal carrier less than the velocity of light.

We can construct the generators of the SU_2 algebra in terms of the fields $\underline{E}$, $\underline{B}$, and $\underline{A}$. The usual potentials, A_μ is the important four-vector quality $A_\mu = (\underline{A}, \phi)$ where the index runs 1 to 4. One of the major purposes of introducing the vector and scalar potentials and also to subscribe to their physicality is the desire by physicists to avoid action at a distance. In fact in gauge theories A_μ is all there is! Yet, it appears that, in fact, these potentials yield a basis for a fundamental nonlocality!

Let us address the specific case of the SU_2 group and consider the elements of a non-Abelian algebra such as the fields with SU_2 (or even SU_n) symmetry then we have the commutation relations where $XY-YX \neq 0$ or $[X,Y] \neq 0$. Which is reminiscent of the Heisenberg uncertainty principle non-Abelian gauge. Barrett does explain that SU_2 fields can be transformed into U_1 fields by symmetry breaking. For the SU_2 gauge amended Maxwell theory additional terms appear in terms of operations such as $A \cdot E, A \cdot B$ and $A \times B$ and their non-Abelian converses. For example $\nabla \cdot B$ no longer equals zero but is given as $\nabla \cdot B = -jg(A \cdot B - B \cdot A) \neq 0$ where $[A,B] \neq 0$ for the dot product of A and B and hence we have a magnetic monopole term and j is the current and g is a constant. Also Barrett gives references to the Dirac, Schwinger and 't Hooft monopole work. Further commentary on the SU_2 gauge conjecture of Mamuth that under symmetry breaking, electric charge is considered but magnetic charges are not. Barrett further states that the symmetry breaking conditions chosen are to be determined by the physics of the problem. These non-Abelian algebras have consistence to quantum theory.

In our analysis, using the SU_2 group there is the automatic introduction of short range forces in addition to the long range force of the U_1 group. U_1 is one dimensional and Abelian and SU_2 is three dimensional and is non-Abelian. U_1 is also a subgroup of SU_2. The U_1 group is associated with the long range $1/r^2$ force and SU_2, such as for its application to the weak force yields short range associated fields. Also SU_2 is a subgroup of the useful SL(2,c) group of non compact operations on the manifold. SL(2,c) is a semi simple four dimensional Lie group and is a spinor group relevant to the relativistic formalism and is isomorphic to the connected Lorentz group associated with the Lorentz transformations. It is a conjugate group to the SU_2 group and contains an inverse. The double cover group of SU_2 is SL(2,c) where SL(2,c) is a complexification of SU_2. Also SL(2,c) is the double cover group of SU_3 related to the set of rotations in three dimensional space. Topologically, SU_2 is associated with isomorphic to the three dimensional spherical, O_3^+ (or three dimensional rotations) and U_1 is associated with the O_2 group of rotations in two dimensions. The ratio of Abelian to non-Abelian components, moving from U_1 to SU_2, gauge is 1 to 2 so that the short range components are twice as many as the long range components.

Instead of using the SU_2 gauge condition we use SL (2,c) we have a non-Abelian gauge and hence quantum theory and since this group is a spinor and is the double cover group of the Lorentz group (for spin ½) we have the conditions for a relativistic formalism. The Barrett formalism is non-relativistic. SL (2,c) is the double cover group of SU_2 but utilizing a similar approach using twistor algebras yields relativistic physics.

It appears that complex geometry can yield a new complementary unification of quantum theory, relativity and allow a domain of action for nonlocality phenomena, such as displayed in the results of the Bell's theorem tests of the EPR paradox, and in which the principles of the quantum theory hold universally. The properties of the nonlocal connections in complex four-space may be mediated by non- or low dispersive loss solutions. We solved Schrödinger equation in complex Minkowski space [27,28].

In progress is research involving other extended gauge theory models, with particular interest in the nonlocality properties on the S pact-time manifold, quantum properties such as expressed in the EPR paradox and coherent states in matter.

Utilizing Coxeter graphs or Dynkin diagrams, Sirag lays out a comprehensive program in terms of the A_n, D_n and E_6, E_7 and E_8 Lie algebras constructing a hyper-dimensional geometry for as a classification scheme for elementary particles. Inherently, this theory utilizes complexified spaces involving twistors and Kaluza-Klein geometries. This space incorporates string theory and GUT models [29].

8.4 Complex Vector and Advanced Potentials and Bell's Inequality

The issue of whether Bell's theorem and other remote connectedness phenomena, such as Young's double slit experiment, demands superluminal or space-like signals or prior luminal signals is an area of hot debate. Also, the issue of advanced vs. retarded potentials is of interest in this regard.

Using the complex model of A^{μ} we will examine the issue of the nonlocality of Bell's theorem as quantum mechanical 'transactions' providing a microscopic communication path between detectors across space-like intervals, which violate the EPR locality postulate. This picture appears to be consistent with the remote connectedness properties of complex Minkowski space. Also there are implications for macroscopic communications channels; another area of hot debate. Detailed discussions of Bell's theorem are given in [30].

We will formulate fields in terms of A or $\underset{\sim}{A} = (A^{j}, \phi)$ where A^{j} is $\underline{A}$ rather than the tensor $F_{\mu\nu}$ or $\underline{E}$ or $\underline{B}$. We can proceed from the continuity equation $\nabla \cdot J + \partial\rho / \partial t = 0$ and the expression $F_{\mu\nu} = \partial A_{\nu} / \partial X_{\mu} - \partial A_{\mu} / \partial X_{\nu}$. For the usual restored potentials then, we have the Lorentz condition

$$\nabla \cdot A + \mu\varepsilon \frac{\partial \phi}{\partial t} = 0 \text{ and also } \nabla^2 A - \mu\varepsilon \frac{\partial^2 A}{\partial t^2} = -\mu J \tag{8.4}$$

We can also derive $$\nabla^2 \phi - \mu\varepsilon \frac{\partial^2 \phi}{\partial t^2} = -\frac{1}{\varepsilon}\rho \tag{8.5}$$

These equations possess a restored potential solution. The radiation field in quantum electrodynamics is usually quantized in terms of (A, ϕ). [We can also convert back to the $\underline{E}$ and $\underline{B}$ fields using $E = -\nabla\phi - \partial A / \partial t$ and $B = \nabla \times A$.] Quantization of the field consists of regarding the coordinates (x, k) or (q, p) as quantum mechanical coordinates of a set of equivalent harmonic oscillators. In the second quantized method treating k_r, q_r and A_r as quantum numbers then we have quantized allowable energy levels such as $W = \sum_r (n_r + \eta_2)\hbar\omega_r$.

Solutions are given in the form

$$\Psi \propto \sum_{n_r} n_r e \exp - \frac{iW(n_r)}{\hbar} \tag{8.6}$$

and we have a Hamiltonian equation of motion $\dot{p}_{ab} + (ck)^2 q_{ab} = 0$ or $\dot{q}_{ab} = p_{ab}$ and

$$\mathcal{H} = \tfrac{1}{2} \sum [p_{ab}^2 + (ck)^2 q^2 q_{ab}^2]. \tag{8.7}$$

The electromagnetic field energy of the volume integral $(E^2 + B^2)/8\pi$ is just equal to the Hamiltonian.

We can examine such things as absorption and polarization in terms of the complexification of $\underline{E}$ and $\underline{B}$ or $\underline{A}$ and ϕ. We define the usual $D = \varepsilon E$ (or displacement field) and $B = \mu H$ for a homogeneous isotopic media. If we introduce p_0 and m_0 as independent of $\underline{E}$ and $\underline{H}$ where the induced polarizations of the media are absorbed into the parameters ε and μ, we have

$$D = \varepsilon E + p_0 \text{ and } H = \tfrac{1}{\mu} B - m_0 \tag{8.8}$$

Then we define a complex field as

$$\underline{\underline{Q}} \equiv \underline{B} + i\sqrt{\varepsilon\mu}\,\underline{E} \tag{8.9}$$

so that we have Maxwell's equations now written as

$$\nabla \times Q + i\sqrt{\varepsilon\mu}\,{\partial Q}/{\partial t} = \mu J \text{ and } \nabla \cdot Q = i\sqrt{{\mu}/{\varepsilon}}\;\rho \tag{8.10}$$

Using vector identities [23-25] and resolving into real and imaginary parts, we have

$$\nabla^2 H - \varepsilon\mu\frac{\partial^2 H}{\partial t^2} = -\nabla \times J \text{ and } \nabla^2 E - \varepsilon\mu\frac{\partial E}{\partial t^2} = \mu\frac{\partial J}{\partial t} + \frac{1}{\varepsilon}\nabla\rho \tag{8.11}$$

We define Q in terms of the complex vector potential that $A_{\mathrm{Re}} \to L_{complex}$ and $\phi_{\mathrm{Re}} \to \phi_{complex}$. Then

$$Q = \nabla \times L - i\sqrt{\varepsilon\mu}\frac{\partial L}{\partial t} - i\sqrt{\varepsilon\mu}\nabla\phi \tag{8.12}$$

subject to the condition similar to before, $\nabla \cdot \underline{L} + \varepsilon\mu\,{\partial\phi}/{\partial t} = 0$. Then we have

$$\nabla^2 L - \varepsilon\mu\,{\partial^2 L}/{\partial t^2} = -\mu J \text{ and } \nabla^2\phi - \varepsilon\mu\,{\partial^2\phi}/{\partial t^2} = -{1}/{\varepsilon}\;\rho \tag{8.13}$$

Separation into real and imaginary parts of these potentials, L and ϕ can be written as

$$L = A_{\mathrm{Re}} - i\sqrt{{\mu}/{\varepsilon}}\;A_{\mathrm{Im}} \text{ and } \phi = \phi_{\mathrm{Re}} - i\sqrt{{\mu}/{\varepsilon}}\;\phi_{\mathrm{Im}} \tag{8.14}$$

Upon substitution into the equation for Q and separation into real and imaginary parts we have

$$\underline{B}_{\mathrm{Re}} = \nabla \times \underline{A}_{\mathrm{Re}} - \frac{\mu\partial \underline{A}_{\mathrm{Im}}}{\partial t} - \mu\nabla\phi_{\mathrm{Im}}\,;$$

$$\underline{E}_{\mathrm{Re}} = -\nabla\phi_{\mathrm{Re}} - \frac{\partial \underline{A}_{\mathrm{Re}}}{\partial t} - \frac{1}{c}\nabla \times \underline{A}_{\mathrm{Im}} \tag{8.15}$$

The usual equations are allowed when A_{Im} and ϕ_{Im} are taken as zero.

If free currents and charges are everywhere zero in the region under consideration, then we have

$$\nabla \times Q + i\sqrt{\varepsilon\mu}\,{\partial Q}/{\partial t} = 0\,; \quad \nabla Q = 0 \tag{8.16}$$

and we can express the field in terms of a single complex Hertzian vector $\underline{\Gamma}$ as the solution of

$$\nabla^2 \underline{\Gamma} - \varepsilon\mu \; {\partial^2 \underline{\Gamma}} \big/ {\partial t^2} = 0 \tag{8.17}$$

We can define Γ by

$$\Gamma \equiv \pi_{\mathrm{Re}} - i\sqrt{\mu / \varepsilon} \;\; \pi_{\mathrm{Im}} \tag{8.18}$$

where $\phi_{\mathrm{Re}} = -\nabla \cdot \underline{\pi}$ and we can write such expressions as

$$A_{\mathrm{Im}} = \mu\varepsilon \, {\partial \pi_{\mathrm{Im}}} \big/ {\partial t} \;\text{and}\quad \phi_{\mathrm{Im}} = \nabla \cdot \underline{\pi}_{\mathrm{Im}} \tag{8.19}$$

This formalism works for a dielectric media; but if the media is conducting the field equations are no longer symmetric, then the method fails. Symmetry can be maintained by introducing a complex induced capacity $\varepsilon' = \varepsilon_{\mathrm{Re}} \pm i \;\; \sigma_{\mathrm{Im}} / \omega$. The vector B is in a solenoid charge-free region; this method works. Calculation of states of polarization by the complex method demonstrates its usefulness and validity. Also, absorption can be considered in terms of complex fields. We will apply this method to solutions that can be described as restored and advanced and may explain Bell's theorem of nonlocality. Linear and circular polarization can be expressed in terms of complex vectors $A = A_{\mathrm{Re}} + iA_{\mathrm{Im}}$. The light quanta undergoing this polarization is given as $\hbar\omega\hat{n} = \hbar\sigma = \hbar k$. Complex unit vectors are introduced so that real and imaginary components are considered orthogonal. We have a form such as $A = (A \cdot \hat{\ell}_{\mathrm{Im}})\hat{\ell}_{\mathrm{Re}} + (A \cdot \hat{j}_{\mathrm{Im}})\hat{j}_{\mathrm{Re}}$. The linearly polarized wave at angle θ is

$$A = \frac{A}{\sqrt{2}}(\ell_{\mathrm{Re}} \;\; e^{-i\theta} - ij_{\mathrm{Re}} \;\; e^{i\theta}) . \tag{8.20}$$

Now let us consider using this polarization formalism to describe the polarization-detection process in the calcium source photon experiment of J. Clauser et al. [31]. Let us first look at solutions to the field equations for time-like and space-like events. The nonlocality of Bell's theorem appears to be related to the remote connectedness of the complex geometry and the stability of the soliton over space and time.

We will consider periodically varying fields which move along the x-axis. For source-free space, we can write

$$c^2 \nabla^2 \underline{F} = \phi^2 \underline{F} \big/ \phi t^2 \tag{8.21}$$

where $\underline{F}$ represents either $\underline{E}$ or $\underline{B}$. The two independent solutions for this equation are [32] $\underline{E}_{\pm}(x,t) = E_0 \sin(2\pi kx \pm vt)$ and

$$\underline{B}_{\pm}(x,t) = B_0 \sin 2\pi(kx \pm vt) \tag{8.22}$$

and k is the wave number and v the frequency of the wave. The ± sign refers to the two independent solutions to the above second order equation in space and time. The wave corresponding to E_+ and B_+ will exist only when t < 0 (past lightcone) and the wave corresponding to $\underline{E}$ and $\underline{B}$ will exist for $t > 0$ (future lightcone). Then the $\underline{E}$ wave arrives at a point x in a time t after emission, while E_+ wave arrive at x in time, t before emission (like a tachyon).

Using Maxwell's equations for one spatial dimension, x, and the Poynting vector which indicates the direction of energy and momentum flow of the electromagnetic wave, we find that E_+ and B_+ correspond to a wave emitted in the $+x$ direction but with energy flowing in the $-x$ direction. For example, $E_+(x,\ t)$ is a negative-energy and negative-frequency solution. The wave signal will arrive at $t = x/c$ before it is emitted, and is termed an advanced wave. The solution $\underline{E}(x,t)$ is the normal positive-energy solution and arrives at x in time, $t = x/c$, after the instant of emission and is called the retarded potential, which is the usual potential.

The negative energy solutions can be interpreted in the quantum picture in quantum electrodynamics as virtual quantum states such as vacuum states in the Fermi-sea model. 'These virtual states are not fully realizable as a single real state but can definitely effect real physical processes to a significant testable extent'. The causality conditions in S-matrix theory, as expressed by analytic continuation in the complex plane, relate real and virtual states [28,29]. Virtual states can operate as a polarizable media leading to modification of real physical states. In fact, coherent collective excitations of a real media can be explained through the operations in an underlying virtual media.

Four solutions emerge: Two <u>retarded</u> (F_1 and F_2) connecting processes in the forward light cone and two <u>advanced</u>, (F_3 and F_4) connecting processes in the backward slight cone [33]. These four solutions are

$$F_1 = F_0 e^{-i(-kx-\omega t)}, \quad F_2 = F_0 e^{i(kx-\omega t)};$$
$$F_3 = F_0 e^{i(-kx+\omega t)}, \quad F_4 = e^{i(kx+\omega t)} \tag{8.23}$$

where F_1 is for a wave moving in the (-x, + t) direction, F_2 is for a (+x, +t) moving wave, F_3 is for a (-x, -t) moving wave, and F_4 is a (+x, -t) moving wave. F_1 and F_4 are complex conjugates of each other and F_2 and F_3, are complex conjugates of each other, so that $F_1^+ = F_4$, and $F_2^+ = F_3$. Then the usual solutions to Maxwell's equations are retarded plane wave solutions.

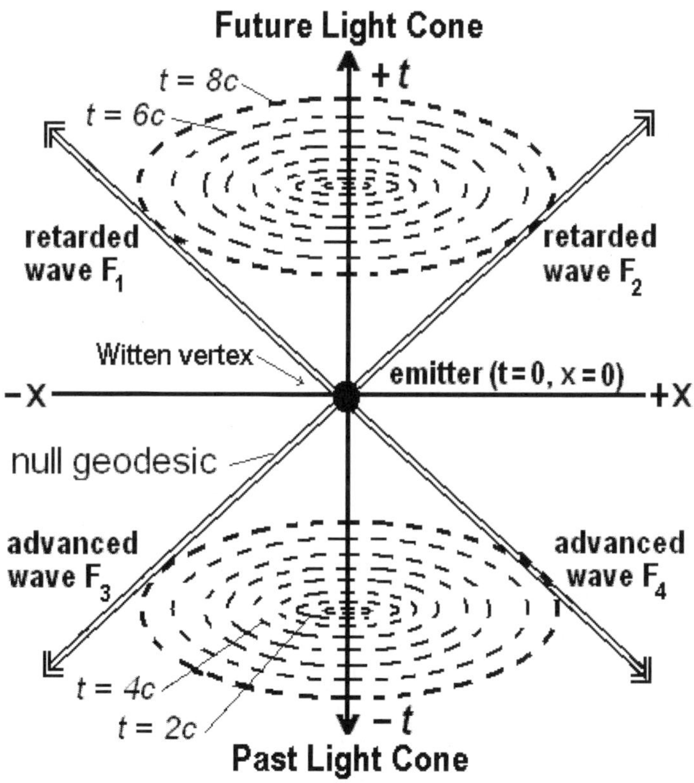

Figure 8.1 Adaptation of a complex Minkowski light-cone showing advanced-retarded future-past Cramer wavefront transactions with a central Witten Ising lattice string vertex able to undergo symmetry transformations.

The proper formulation of nonlocal correlations, which appear to come out of complex geometries may provide a conceptual framework for a number of quantum mechanical paradoxes and appear to be explained by Bell's nonlocality, Young's double slit experiment, the Schrödinger cat paradox, superconductivity, superfluidity, and plasma 'instabilities' including Wheeler's 'delayed choice experiment'. Interpretation of these phenomena is made in terms of their implications about the lack of locality and the decomposition of the wave function which arises from the action of advanced waves which 'verify' the quantum-mechanical transactions or communications.

Cramer [33] has demonstrated that the communication path between detectors in the Bell inequality experiments can be represented by space-like intervals and produce the quantum mechanical result. By the addition of two time-like four vectors having time components of opposite signs which demonstrate the locality violations of Bell's theorem and is consistent with the Clauser, Fry and Aspect experimental results. This model essentially is an 'action-at-a-distance' formalism.

One can think of the emitter (in Bell's or Young's quantum condition) as sending out a pilot or probe 'wave' in various allowed directions to seek a 'transaction' or collapse of the wave function. A receiver or absorber detects or senses one of these probe waves, 'sets its state' and sends a 'verifying wave' back to the emitter confirming the transaction and arranging for the transfer of actual energy and momentum. This process comprises the nonlocal collapse of the wave function. The question now becomes: does such a principle have macroscopic effects? Bell's nonlocality theorem can be effective over a matter of distance.

An attempt to examine such a possible macroscopic effect over large distances has been made by Partridge [34]. Using 9.7 GHz microwave transmitted by a conical horn antenna so that waves were beamed in various directions. Partridge found that there was little evidence for decreased emission intensities in any direction for an accuracy of a few parts per 10^9. Interpretation of such a process is made in terms of advanced potentials. Previously mentioned complex dimensional geom.-etries give rise to solutions of equations that form subluminal and superluminal signal propagations or solitons.

The possibility of a remote transmitter-absorber communicator now appears to be a possibility. The key to this end is an experiment by Pflelgov and Mandel [35]. Interference effects have been demonstrated, according to the authors, in the superposition of two light beams from two independent lasers. Intensity is kept so low that, to high probability, one photon is absorbed before the next one is emitted. The analogy to Young's double slit experiment is enormous.

In Wheeler's recent paper, he presents a detailed discussion of the physics of delayed choice proton interference and the double slit experiment (from the Solvay conference, Bohr-Einstein dialogue). Wheeler discusses the so-called Bohm 'hidden variables' as a possible determinant that nonlocality collapses the wave function [36]. Further theoretical and experimental investigation is indicated; but there appears to be a vast potential for remote nonlocal communication and perhaps even energy transfer (Chaps. 5 and 12). In the next section we detail the forms of transformations of the vector and scalar potentials at rest and in moving frames, continuing our formulation in terms of $(\underline{A,\phi})$. The issues of sub and superluminal transformations of $\underline{A}$ and ϕ are given in a complex Minkowski space. Both damped and oscillatory solutions are found and conditions for advanced and restored potentials are given.

8.5 Superluminal Vector and Scalar Potential Transformation Laws

For simplicity we will consider superluminal boost $v_x = \infty$ along the positive x direction. The space and time vectors in the real 4D Minkowski space transform as follows [37]

$$x' = +t,\ y' = -iy,\ z' = iz,\ t' = x \tag{8.24}$$

for real and imaginary parts separately, where x, y, z, t are real quantities in the laboratory (S) frame, and x',y',z',t' are the real quantities in the moving (S') frame. Now in the 6D (M^6) complex Minkowski space, the above transformation laws for a superluminal boost $(v_x = +\infty)$ in the positive x direction become [38]

$$\begin{aligned}
&x'_{\text{Re}} + ix'_{\text{Im}} = t_{x,\text{Re}} + it_{x,\text{Im}}, \quad y'_{\text{Re}} + iy'_{\text{Im}} = y_{\text{Im}} - iy_{\text{Re}}, \\
&z'_{\text{Re}} + iz'_{\text{Im}} = z_{\text{Im}} - iz_{\text{Re}}; \quad t'_{x,\text{Re}} + it'_{x,\text{Im}} = x_{\text{Re}} + ix_{\text{Im}}, \\
&t'_{y,\text{Re}} + it'_{y,\text{Im}} = t_{y,\text{Im}} - it_{y,\text{Re}}, \quad t'_{z,\text{Re}} + it'_{z,\text{Im}} = t_{z,\text{Im}} - it_{z,\text{Re}}
\end{aligned} \tag{8.25}$$

The transformation laws given by (8.25) preserve the magnitude of the line element but not the sign as in:

$$-x'^{\mu}x'^{\nu} = x^{\mu}x^{\nu} \tag{8.26}$$

where index μ and ν run over 1,2,3,4 representing 1 as time vector and 2,3,4 as spatial vectors. Therefore we have the signature (+++-). Similar to the transformation laws for space and time vectors as given by (8.25) we can write the transformation laws for the vector and scalar potential. For a superluminal boost in positive x direction, the transformation laws for $(\underline{A}, \phi)$ are:

$$A'_x = \gamma\left(A_x - \frac{v_x^2}{c^2}\phi\right), \quad A'_y = A_y, \quad A'_z = A_z, \quad \phi' = \gamma(\phi - v_x A_x) \tag{8.27}$$

where ϕ is the scalar potential and γ is the usual Lorentz term

$$\gamma' \equiv 1 \Big/ \left(\frac{v_x^2}{c^2} - 1\right)^{\frac{1}{2}} \tag{8.28}$$

We consider A'_x, etc., transforming as a gauge. In Eq. (8.27), the vector potential A is considered to be a four-vector real quantity, A_μ or $\underset{\sim}{A} = (A_x, A_y, A_z, \frac{i\phi}{c})$, which preserves the length of the line element but not the sign, i.e. we have

$$A_\mu A_\mu = -A'_\mu A'_\mu \tag{8.29}$$

Eq. (8.27) then simplifies to the following relationships for the velocities approaching infinity, $v_x = \infty$.

We can write the transformation laws for scalar and vector potentials under the superluminal boost in the positive x direction for $v_x = +\infty$. From the rest frame, S, to the moving frame, S', for unaccelerated vector and scalar potentials, we have

$$A_x = -\phi', \quad A_y = A'_y, \quad A_z = A'_z, \quad \phi = -A'_x \tag{8.30}$$

From the moving frame, S', to the rest frame, S, for the unaccelerated vector and scalar potentials we obtain

$$A'_x = -\phi, \quad A'_y = A_y, \quad A'_z = A_z, \quad \phi' = -A_x \tag{8.31}$$

Equation (8.31) is valid for real or complex vector and scalar potentials. Real and imaginary parts are easily separable in a complex quantity and they will transform according to Eq. (8.31) under the influence of a superluminal boost in the positive x direction. Now if these are the retarded (or accelerated or advanced) vector and scalar potentials then the transformation laws under the superluminal boosts will be different from the ones given by Eq. (8.31). These will be given by the combination of Eq. (8.31) and the transformation laws of the complex space and time vectors as given by Eq. (8.25).

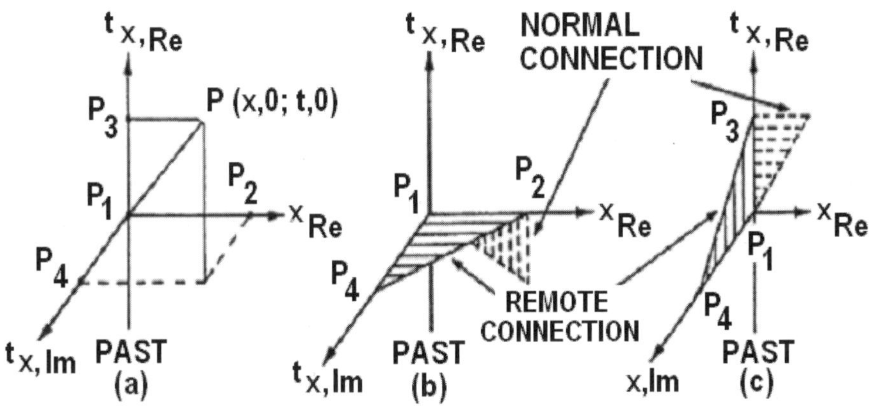

Figure 8.2 We represent the location of four points in the complex manifold. In Fig. 1a, point P_1 is the origin, and P is a generalized point which is spatially and temporally separated from P_1. In Fig. 8.2b, the points P_1 and P_2 are separated in space but synchronous in time. This could be a representation of real-time nonlocal spatial separation. In Fig. 8.2c, points P_1 and P_3 are separated temporally and spatially contiguous. This represents an anticipatory temporal connection.

These conditions are illustrated in Fig. 8.2. In Fig. 8.2a we represent a generalized point $P(x_{Re}, t_{Re}, t_{Im})$, displaced from the origin which is denoted as P_1. This point can be projected on each dimension x_{Re}, t_{Re} and t_{Im} as points P_2, P_3, and P_4 respectively. In Fig. 8.2b, we denote the case where a real-time *spatial* separation exists between points, P_1 and P_2 on the x_{Re} axis, so that $\Delta x_{Re} \neq 0$, and there is no anticipation, so that $t_{Re} = 0$, and access to imaginary time t_{Im}, nonlocality can occur between the P_1

to P_4 interval, so that $\Delta t_{\mathrm{Im}} \neq 0$. Then, our metric gives us $\Delta s^2 = 0$, where nonlocality is the contiguity between P_1 and P_2 by its access to the path to P_4. By using this complex path, the physical spatial separation between P_1 and P_2 becomes equal to zero, allowing direct nonlocal connectedness of distant spatial locations, observed as a fundamental nonlocality of remote connectedness on the spacetime manifold.

Figure 8.2c represents the case where anticipation occurs between P_1 and an apparent future anticipatory accessed event, P_3 on the t_{Re} axis. In this case, no physical spatial separation between observer and event is represented in the figure. Often such separation on the x_{Re} exists. In the case where $x_{\mathrm{Re}} = 0$, then access to anticipatory information, along t_{Re} can be achieved by access to the imaginary temporal component, t_{Im}. Hence, remote, nonlocal events in four space or the usual Minkowski space, appear contiguous in the complex eight space and nonlocal temporal events in the four space appear as anticipatory in the complex eight space metric. Both nonlocality and anticipatory systems occur in experimental tests of Bell's Theorem and perhaps in all quantum measurement processes.

The propagation constant is considered to be isotropic in vacuum and defined as $d_x = \omega / v_\phi$, where v_ϕ, is the phase velocity and ω is the radian frequency of the propagating signal. Usually in most cases the phase velocity of propagation in vacuum is a constant $v_\phi = c$, where c is the velocity of light in vacuum. For the purpose of this paper, we will consider a tachyon traveling faster than light emitting an electromagnetic signal at frequency ω which propagates at the velocity of light. This assumption will simplify the subject matter of this paper. Later on, in a separate paper, we will examine the faster than light electromagnetic signals emitted by a traveling tachyon which might lead into a Doppler effect at velocities faster than light.

Considering only the advanced potential solution from Eq. (8.24), Eq. (8.24) can now be rewritten as two separate terms, so that in the S frame,

$$\begin{aligned} A_x = (A_{0x,\mathrm{Re}} + iA_{0x,\mathrm{Im}})\{e \exp i[\omega t_{x,\mathrm{Re}} - kx_{\mathrm{Re}}] \times \\ e \exp -[\omega t_{x,\mathrm{Im}} - kx_{\mathrm{Im}}]\} \end{aligned} \tag{8.32}$$

where the first exponent represents the usual type of oscillatory terms and the second exponent represents a decaying component which is not present in the usual 4D spacetime model. Note also that we have used the isotropy of the vector k in Eq. (8.32) as examined in the previous section.

Now let us examine the complex exponential of Eq. (8.32) using the transformations of Eq. (8.24) as follows so that we have for the exponents

$$e\exp i[\omega x'_{\mathrm{Re}} - kt'_{x,\mathrm{Re}}]; \quad e\exp-[\omega x'_{\mathrm{Im}} - kt'_{x,\mathrm{Im}}] \tag{8.33}$$

We regroup terms in ω and k so that we have

$$e\exp i[\omega(x'_{\mathrm{Re}} + ix'_{\mathrm{Im}}) - k(t'_{x,\mathrm{Re}} - it'_{x,\mathrm{Im}})] \tag{8.34}$$

Now using equations for $\underline{x}' = x'_{\mathrm{Re}} + ix'_{\mathrm{Im}}$ we have

$$e\exp i[\omega x' - k(t'_{z,\mathrm{Re}} - it'_{x,\mathrm{Im}})] \tag{8.35}$$

Note that the second part of the exponent for the k term does not reduce to t' since there is a minus sign before $it'_{x,\mathrm{Im}}$. Thus for the boost $v_x \to \infty$ or $v > c$, we obtain for $e\exp i[\omega t + kx]$ from Eq. (8.24) under this transformation going to

$$e\exp i[\omega x']; \quad e\exp- k[t'_{x,\mathrm{Re}} - it'_{x,\mathrm{Im}}] \tag{8.36}$$

Let us look at the example of the transformation from A'_x (in the moving frame S') to its form in the restframe, S. We find a mixing vector and scalar potential. In the SLT from the restframe S to the moving S' frames we have a change of length of the time component vector in Eq. (8.36). The vector potential term A_{0x} transforms as

$$A'_x = \gamma\left(A_x - \frac{v_x^2}{c^2}\phi\right) \tag{8.37}$$

which is the same as Eq. (8.28), so that for the superluminal boost $v_x \to \infty$, implies that

$$\gamma \equiv \frac{1}{\sqrt{\frac{v_x^2}{c^2} - 1}} = \frac{1}{\frac{v_x}{c}\sqrt{1 - \frac{c^2}{v_x^2}}} \cong \frac{c}{v_x} \tag{8.38}$$

where the $\sqrt{1 - c^2 / v_x^2}$ term approaches unity as $v_x \to \infty$. Then we rewrite the transformed vector potential as

$$A_x^{'} = \frac{1}{\sqrt{\frac{v_x^2}{c^2} - 1}}, \qquad A_x - \frac{\frac{v_x}{c}}{\sqrt{\frac{v_x^2}{c^2} - 1}}\phi \tag{8.39}$$

Then for $v_x \to \infty$ and from Eqs. (8.38) and (8.39),

$$A_x^{'} = \frac{cA_x}{v_x} - \frac{v_x}{c^2}\frac{c}{v_x}\phi = 0 - \frac{1}{c\phi} \equiv -\phi \tag{8.40}$$

for units in which $c = 1$. Therefore $A_x^{'} = -\phi$ for a superluminal boost, $v_x \to \infty$.

For the transformation of the scalar potential, in analogy to Eq. (8.28), we have

$$\phi' = \gamma(\phi - v_x A_x) \tag{8.41}$$

and for $v_x \to \infty$, we have $\gamma \cong c / v_x$ so that in the limit of the SLT,

$$\phi' \lim_{v\to\infty} = \frac{c}{v_x}\phi - cA_x = -cA_x \tag{8.42}$$

and for the units of $c = 1$, then $\phi' - A_x$. Compare this equation to Eq. (8.40). Also for $A_y^{'} = A_y$ and $A_z^{'} = A_z$ we can now write

$$\begin{aligned} A_x &= [A_{0x,\mathrm{Re}} + iA_{0x,\mathrm{Im}}]e\exp i[\omega t + kx] = \\ &[-\phi_{\mathrm{Re}}^{'} - i\phi_{\mathrm{Im}}^{'}]e\exp i\omega x' \quad e\exp\pm k_x[t_{x,\mathrm{Re}}^{'} - it_{x,\mathrm{Im}}^{'}] \end{aligned} \tag{8.43}$$

where $x' = x_{\mathrm{Re}}^{'} + ix_{\mathrm{Im}}^{'}$.Using the result of Eqs. (8.40) and (8.42) for the non-exponent part and the exponential term given in Eq. (8.35). Equation (8.43) gives us the vector and scalar form in the moving *S'* frame.

If we consider only the accelerated potential, then we consider only the plus sign in Eq. (8.43). By using the definition of complex quantities, Eq. (8.43) can be rewritten in a compact, simplified form:

$$A_x = -\phi_{0x}^{'}\exp(i\omega x')\cdot\exp(ik_x t_x^{'}). \tag{8.44}$$

Then by use of Eq. (8.44) we can describe the *x* component of the complex vector potential in moving frame *S'* after a superluminal boost in the positive *x* direction. The same vector potential in the rest frame is defined.

The transformation of the A_y and A_z components of the complex vector potential under a superluminal boost in the positive x direction can similarly be written

$$\text{As}\quad \begin{aligned} A_y &= A'_{0y}\exp[-\omega(t'_{y,\mathrm{Re}} + it'_{y,\mathrm{Im}})]\cdot\exp[-ky(z'_{\mathrm{Re}} + iy'_{\mathrm{Im}})] \\ &= A'_{0z}\exp[-\omega(t'_{z,\mathrm{Re}} + it'_{z,\mathrm{Im}})]\cdot\exp[-ky(z'_{\mathrm{Re}} + iz'_{\mathrm{Im}})] \end{aligned} \qquad (8.45)$$

We will now consider the scalar potential as defined by a complex quantity, so that

$$\phi' = \phi'_{\mathrm{Re}} + i\phi'_{\mathrm{Im}} \qquad (8.46)$$

which we use for the non-exponential term of Eq. (8.45) which then becomes

$$A_x = -\phi' e \exp i\omega x' \quad e\exp k[t_{x,\mathrm{Re}} - it_{x',\mathrm{Im}}] \qquad (8.47)$$

Let us now compare the vector potential forms of A_x in Eq. (8.42) in the S or laboratory frame, and A_x of Eq. (8.47) in the S' frame or moving frame (see Table 8.1).

TABLE 8.1 Comparison of The Exponential Part of the Vector Potential A_x In The *S* and *S'* Frames Of Reference

	OSCILLATORY	DAMPED
S Frame	$A_{0x} \propto e\exp i[\omega t_{x,\mathrm{Re}} - kx_{\mathrm{Re}}]$	$e\exp-[\omega t_{x,\mathrm{Im}} - kx_{\mathrm{Im}}]$
S' Frame	$\phi' \propto \exp i[\omega x']$	$e\exp k[t'_{x,\mathrm{Re}} - it'_{x,\mathrm{Im}}]$

In the oscillatory solution of the S' frame for ϕ', we find no dependence on the wave number factor k and hence we have apparent media independence, recalling $x' = x_{\mathrm{Re}} + ix_{\mathrm{Im}}$, whereas in the S frame for A_{ox}, we have dependence on ω and k.

For the damped solution, we have ω and k dependence in the S frame for A_{ox}, which is a pure real exponential and hence not oscillatory. In the S' frame then, ϕ' sometimes has a damped solution dependent on k which has a real and imaginary component. The exponential factor can be written as

$$t'_{x,\mathrm{Re}} - it'_{x,\mathrm{Im}} = x_{\mathrm{Re}} - ix_{\mathrm{Im}} \tag{8.48}$$

Time dilation and vector length are modified in the complex twelve dimensional space [38]. We find that a superluminal, unidimensional (x-dimensional) boost in complex Minkowski space not only modifies space and time (as well as mass) by the γ factor, it also modifies $\underset{\sim}{A} = (\underline{A}, \phi)$ and we find a mixing of $\underline{A}$ and ϕ for $\underline{A} = A_j$ where j runs 1 to 3 (or spacelike quantities) and ϕ transforms as a temporal quantity for subluminal transformations.

8.6 Insights into Dirac and Penrose Spinor Calculus

The spinor calculus of the Kaluza-Klein geometry [11,12] mappable one to one with the twistor space of the complex eight space. The Dirac equation is based on the fundamental properties of spinors. The complexification of four-space by the Rauscher [39] and Newman [2-5] method yields a manner to relate Maxwell's equations to the relativistic spacetime metric, as shown above. In this section we detail the Dirac spinor formalism with the twistor topology.

The Penrose and other twistor approaches have been in an attempt to quantize gravity in order to unify the physics of the micro-cosmos and macro-cosmos. Such an approach has been taken by Penrose et al. and is based on the concept of a more general theory that has limits in the quantum theory and the relativistic theory [40]. In addition, there have been approaches to the underlying structure of space-time in the quantum and structural regime [40-43]. A structured and/or quantized space-time may allow a formalism that unequally relates the electromagnetic fields with the gravitational metric. Feynman and Penrose graphs were developed in an attempt to overcome the divergences of such an approach. In order to translate the equations of motion and Lagrangians from spinors to twistors, one can use the eigenfunctions of the Casimir operators of the Lie algebra of $U(2,2)$.

The simplest case of a zero rest mass field is the simplest and can represent the photon for $n/2$ spin where $n \neq 0$, and we can write

$$\nabla_{AA'}\, \varphi^{A...N} = 0 \tag{8.49}$$

for $A,\ldots,N$ written in terms of n indices, and for $n=1$, we have the Dirac equation for massless particles. For a spin zero field, we have the Klein-Gordon equation

$$\nabla^{AA'}\, \nabla_{AA'}\, \varphi = 0 \tag{8.50}$$

and for $n=2$, we have the source-free Maxwell equation $\square F^{\mu\nu} = 0$ for spin 1 or U_1 for the electromagnetic fields, and for $n=4$, we have the spin Einstein free field equations, $R_{\mu\nu} = 0$. The indices μ and ν run 0 to 3. For a system with charge, then $\square F^{\mu\nu} = J_{\mu\nu} - J_{\mu\nu}$, or this can be written as $\dfrac{F_{\mu\nu}}{\partial x_\nu} = J_\mu$ and then we can write

$$\gamma_{\mu\nu} \frac{\partial F_{\mu\nu}}{\partial x_\nu} = J_\mu \tag{8.51}$$

We present an approach to relate the twistor topology to the spinor space and specifically to the Dirac spinors. Both Fermi-Dirac and Bose-Einstein statistics are considered. The twistor theory and Dirac models can be related to electrodynamics, and gravitation. The Penrose spin approach is designed to facilitate the calculation of angular momentum states for SL(2,2). The spinor formalism, in the Dirac equation, utilize spinors within the quantum theory. The twistor formalisms are related to the structure of space-time and the relation of the spinors and twistors is also of interest because it may yield a relationship between quantum mechanics and relativity. The twistor theory has been related to conformal field theory and the string theory [44]. Also, twistor theory has been related to quaternions and complex quaterionic manifolds [45]. The projective twistor space, PT, corresponds to two copies of the associated complex projective space of CP^3 or $CP^3 \times CP^3$. It is through the conformal geometry of surfaces in S^4, utilizing the fact that CP^3 is an S^2 bundle over S^4, that can be related to quaternions [44].

The complex 8-space and the Penrose twistor topology are fundamentally related since the twistor is derived from the imaginary part of the spinor field. The Kerr Theorem results naturally from this approach in which twisting is shear free in the limit of asymptotic flat space. The twistor is described as a two-plane in complex Minkowski space, M^4. Twistors define the conformal invariance of the tensor field, which can be identified with spin or spinless particles. For particles with a specific intrinsic spin, s, we have $Z^\alpha \overline{Z}_\alpha = 2s$, and for zero spin, such as the photon, $Z^\alpha \overline{Z}_\alpha = 0$ where $\overline{Z}_\alpha$ is the Hermitian conjugate of Z^α, and Z^α and Z_α can be regarded as canonical variables such as $\underline{x}$, $\underline{p}$ in the quantum theory phase space analysis. Note that these fields are independent of the origin [59]. The twist free conditions, $Z^\alpha \overline{Z}_\alpha$, hold precisely when Z^α is a null twistor. The upper case Latin indices are used for spinors, and the Greek indices for twistors. The spinor field of a twistor is conformally invariant and independent of the choice of origin [45]. For the spinor, the indexes A and A' take on values 1, 2 [44]. We briefly follow along the lines of Hanson and Newman in the formalism relating the complex Minkowski space to the twistor algebra. Spinors and twistors are related by the general Lorentz conditions in such a manner as to retain the fact that all signals are luminal in the real four-space, which does not preclude superluminal signals in an $N > 4$ dimensional space. The twistor Z^α can be expressed in terms of a pair of spinors, ω^A and π_A, which are said to represent the twistor. We write

$$Z^\alpha = \left(\omega^A, \pi_{A'}\right) \tag{8.52}$$

where $\omega^A = i\, r^{AA'} \pi_{A'}$

Every twistor Z^α is associated with a point in complex Minkowski space, which yields an associated spinor, ω^A, $\pi_{A'}$. The spinor is associated with a tensor which can be Hermitian, but is not necessarily Hermitian. The spinor can be written equivalently as a bivector forming antisymmetry. In terms of spinors ω^A and $\pi_{A'}$, they are said to represent the twistor Z^α as $Z^\alpha = \left(\omega^A, \pi_{A'}\right)$. In terms of components of the twistor space in Hermitian form, φ for $\varphi_{AA'} = \varphi_{A'A}$, we have,

$$\varphi\left(Z^\alpha Z^\beta\right) = \overline{Z^0 Z^2} + \overline{Z^1 Z^3} + \overline{Z^2 Z^0} + \overline{Z^3 Z^1} \tag{8.53}$$

where the α index runs 0 to 3. The components of Z^α are Z^0, Z^1, Z^2, Z^3 and are identifiable with a pair of spinors, ω^A and $\pi_{A'}$, so that

$$\omega' = Z^1, \quad \pi_{0'} = Z^2, \quad \pi_{1'} = Z^3 \tag{8.54}$$

so that we have

$$Z\bar{Z}_{\mu} = \mu^{\circ}\bar{\pi}_{\circ} + \mu'\bar{\pi}_{1} + \pi_{\circ}'\mu^{\circ'} + \pi_{1'}\bar{\mu}^{1'} \tag{8.55}$$

Note that the spinor ω^{A} is the more general case of μ^{A}. This approach ensures that the transformations on the spin space preserve the linear transformations on twistor space, which preserves the Hermitian form, φ.

The underlying concept of twistor theory is that of conformal invariance or the invariance of certain fields under different scalings of the metric under the general relativistic space-time metric, $g_{\mu\nu}$. Related to the Kerr theorem, for asymptotic shear-free null flat space, the analytic functions in the complex space of twistors may be considered a twisting of shear-free geodesics. In certain specific cases, shear inclusive geodesics can be accommodated. Twistors are formally connected to the topology of certain surfaces in complex Minkowski space M^{4}. This space, the complex space C^{4}, is the cover space of R^{4}, the four dimensional Riemannian space. On the Riemann surface, one can interpret spinors as roots of the conformal tangent plane of a Riemann surface into R^{3}. This approach is significant because it ensures the diffeomorphism of the manifold. Complexification is formulated as $Z^{\mu} = X^{\mu}_{\text{Re}} + X^{\mu}_{\text{Im}}$, which constitutes the complexification of the Minkowski space, M^{4}. The usual form Minkowski space is a submanifold of complex Minkowski space. Twistors are space-time structures in Minkowski space, which is based upon the representation of twistors in terms of a pair of spinors. Twistors provide a unique formulation of complexification. The interpretation of twistors in terms of asymptotic continuation accommodate curved space-time.

The spinor representation of a twistor makes it possible to interpret a twistor as a two-plane in complex Minkowski space, M^{4}. Then we can related ω^{A} and $\pi_{B'}$ so that $\xi^{AA'}$ is a solution as

$$\omega^{A} = i\xi^{AB'}\pi_{B'} \tag{8.56}$$

for the position vector $\xi^{AB'}$ in the complex Minkowski space. We can also consider the relationship of $Z^{AA'}$ and $\pi^{A'}$ to a complex position vector as

$$Z^{AA'} = \xi^{AA'} + \omega^{A}\pi^{A'} \tag{8.57}$$

where ω^A is a variable spinor. Just as in the conformal group on Minkowski space, spin space forms a two-valued representation of the Lorentz group. Note that SU_2 is the four value covering group of $C(1, 2)$, the conformal group of Minkowski space. The element of a four dimensional space can be carried over to the complex eight space. The Dirac spinor space for spin, n is a covering group of SO_n where this cohomology theory will allow us to admit spin structure and can be related to the SU_2 Lie group. Now let us consider the spin conditions associated with the Dirac equation and formulate the Dirac 'string trick' that describes the electron spin path. The requirement for a 720° twist or rotation results from the electron spin and chirality where the spin is aligned or anti-aligned along the particle's direction of motion.

For a spin, $s = \frac{1}{2}$ particle, the spin vector $u(p)$ is written as $\begin{pmatrix} 1 \\ 0 \end{pmatrix}$ and $\begin{pmatrix} 0 \\ 1 \end{pmatrix}$ for spin up and spin down and p is momentum. For a particle with mass we have for $c \neq 1$,

$$\left(-i\hbar c \alpha_\mu \frac{\partial}{\partial x_\mu} + \beta m c^2 \right) \psi = 0 \tag{8.58}$$

for the time independent equation, and we can divide Eq. (8.58) by $i\hbar c$ and have,

$$\left(\gamma_\mu \frac{\partial}{\partial x_\mu} + \frac{mc}{\hbar} \right) \psi = 0 \tag{8.59}$$

where $k_\circ = mc/\hbar$ and $\gamma_\mu = ic\hbar\alpha\mu$ where indices μ run 0 to 3. The dependent Dirac equation is given as,

$$\left(-i\hbar c \alpha_\mu \frac{\partial}{\partial x_\mu} + \beta m c^2 \right) \psi + \frac{i}{\hbar} \frac{\partial \psi}{\partial t} = 0 \tag{8.60}$$

The solution to the Dirac equation is in terms of spin $u(p)$ as

$$\psi = u(p) e \frac{i}{\hbar} (p \cdot \underline{x} - Et) \tag{8.61}$$

the Dirac spin matrices $\gamma_{\mu} = ic\hbar\alpha_{\mu}$. The spinor calculus is related to the twistor algebra, which relates a 2-space to an associated complex 8-space.

The Dirac equation and spinors are fundamentally connected. For example, we have the Dirac spin matrices, $\gamma_{\mu} = \begin{pmatrix} 0 & \sigma_{\mu} \\ \sigma_{\mu} & 0 \end{pmatrix} = -i\,\beta\alpha_{\kappa}$ where terms such as $\gamma_{\mu}(1-\gamma_5)$ come into the electroweak vector - axial vector formalism. The three Dirac spinors, which are also related to the Pauli spin matrices, are given as

$$\sigma_x = \begin{vmatrix} 0 & 1 \\ 1 & 0 \end{vmatrix}, \quad \sigma_y = \begin{vmatrix} 0 & -i \\ i & 0 \end{vmatrix} \text{and } \sigma_z = \begin{vmatrix} 1 & 0 \\ 0 & -1 \end{vmatrix} \tag{8.62}$$

and $\gamma_5 \equiv i\gamma_0\gamma_1\gamma_2\gamma_3 = i\gamma^0\gamma^1\gamma^2\gamma^3$ for $\gamma_0 = \beta$ is given as,

$$\gamma_0 = \begin{pmatrix} 1 & 0 & 0 & 0 \\ 0 & 1 & 0 & 0 \\ 0 & 0 & -1 & 0 \\ 0 & 0 & 0 & -1 \end{pmatrix} \tag{8.63}$$

for trace $tr\beta = 0$, that is, Eq. (8.63) can be written as,

$$\gamma_0 = \beta = \begin{pmatrix} I_2 & 0 \\ 0 & -I_2 \end{pmatrix} \tag{8.64}$$

where we have the 2 x 2 spin matrix as $I_2 = \begin{vmatrix} 1 & 0 \\ 0 & 1 \end{vmatrix}$. Note that the Dirac spinors are the standard generators of the Lie algebra of SU_2.

The commutation relations of the Dirac spin matrices is given as

$$\{\gamma^{\mu}, \gamma^{\nu}\}_{+} = \gamma^{\mu}\gamma^{\nu} + \gamma^{\mu}\gamma^{\nu} = ig^{\mu\nu}\,\underset{\sim}{I} \tag{8.65}$$

and $\det|\gamma_{\mu\nu} = \det\|g_{\mu\nu}|$ where $g_{\mu\nu}$ is the metric tensor. The Dirac spin matrices come into use in the electroweak vector-axial vector model as $\gamma_{\mu}(1-\gamma_5)$ for γ_5 as,

$$\gamma_5 = i\,\gamma_0\gamma_1\gamma_2\gamma_3 = i\,\gamma^0\gamma^1\gamma^2\gamma^3 \tag{8.66}$$

where indices run 0 to 3. We can also write,

$$\gamma_{\mu\nu}\left(x^5, x^\mu\right) = \sum_{n=-\infty}^{\infty} \gamma_{\mu\nu}^{(n)}\left(x^\nu\right) e^{inx^5} \tag{8.67}$$

which expresses some of the properties of a five dimensional space having $\gamma_0, \gamma_{1,} \gamma_2, \gamma_3$ and γ_5. Note that γ_5 is associated with a five dimensional metric tensor. This five-dimensional space passes exactly one geodesic curve which returns to the same point with a continuous direction which is similar to the formation of the Dirac string trick which requires a 720° path of an electron to return to its exact original quantum state. The electromagnetic potential; and the metric of the Kaluza-Klein geometry are related where we express $\gamma_{\mu 5}$ in terms of a potential φ_μ so that we have

$$\gamma_{\mu 5} = \sqrt{2\kappa}\phi_\mu \tag{8.68}$$

where $\kappa \equiv {8\pi}/{F}$ and where $F = {c^4}/{G}$ or the Rauscher quantized cosmological force. Then we have a five-space vector as,

$$\gamma_{\nu 5} = \begin{pmatrix} 0 \\ 0 \\ 0 \\ 0 \\ 1 \end{pmatrix} \tag{8.69}$$

Through this approach, we can relate covariance and gauge invariance.

For the Poisson's equation we have,

$$\nabla \varphi_\mu = \frac{1}{2}\kappa c^4 \mu_\circ \tag{8.70}$$

where again $\kappa \equiv {8\pi}/{F}$ as above. The electromagnetic field, $F_{\mu\nu}$, can be expressed as,

$$F_{\mu\nu} = \frac{\partial \varphi_\mu}{\partial x^\nu} - \frac{\partial \varphi_\nu}{\partial x^\mu} \tag{8.71}$$

which yields an interesting relation of the gravitational metric to the electromagnetic field. Also the Lagrangian is given as $L = \frac{1}{2} F^{\mu\nu} F_{\mu\nu}$ so that $\mathcal{L} = L\sqrt{-g}$ for the space-time metric g. Note that we have

$L = \int \sqrt{g}\, d\tau$, where $d\tau$ represents a four-space. Now let us return to our discussion of the twistor algebra and relate it to the spinor calculus. The Penrose twistor space also yields a five dimensional formalism similar to that formulated by the Kaluza-Klein theory.

The quanta are associated with a quantum field of particles that carry both momentum and energy. The total energy Hamiltonian can be defined in terms of a number of simple phonon states which can be expressed in terms of a_n^+ creation and a_n destruction operator states. Since all creation operators commute, these states are completely symmetric and satisfy Bose-Einstein statistics. Such phonon states, having a definite number of phonons, are called Fock states, which is the vector sum of the momentum of each of the photons in the state. The ground state $|0\rangle$ can be considered the photon vacuum state or Fock state where the photon is taken as a phonon state. The creation and destruction operators commute as $\{a_n, a_n^+\} = \delta_{nn'}$ for the delta function $\delta_{nn'}$. Both projective and non-projective twistors are considered as images in a complex Riemannian manifold in its strong conformal field condition. In analogy to the Hartree-Fock spaces, or Fock space, using the appropriate spin statistics, Bose-Einstein or Fermi-Dirac; duality, analytic continuation, unitary and other symmetry principles. Particles can be considered as states as the Fock space elements or the 'end' of each disconnected portion of the boundary of the manifold [47].

We can consider an n-function as a 'twistor wave' function for a state of n-particles. In the first order consideration, Penrose considers a set of n-massless particles as a first order approximation. We form a series on a complex manifold as elements of the space C_n as

$$f_0,\ f_1(z^\alpha),\ f_2(z^\alpha, y^\alpha),\ f_3(z^\alpha, y^\alpha, x^\alpha), \ldots \qquad (8.72)$$

which are, respectively, the 0th function, 1st function, 2nd function, and 3rd function, etc. of the twistor space, which are also elements of C_n. We can also consider $f_0,\ f_1,\ f_2,\ f_3, \ldots$ as the functions of several nested twistors in which f_0 is the central term of the wave of the twistor space. We can say that these nested tori can act as a recursive sequence. Penrose relates the twistor to particle physics by suggesting that, to a first approximation, f_1 corresponds to the amplitude of a massless, spin 1 particle, f_2 to a lepton spin ½ particle, and f_3 to Hadron particle states, and f_4 to higher energy and exotic Hadron particle states. Mass results from the breaking of conformal invariance for f_n for $n = 2$ or greater, similar to the *S*-matrix approach [48]. The harmonic functions, f_n , form a harmonic sequence, where f_n for $n = 2$ form the Fermion states, and f_n for $n = 3$ form the Hadron twistor states. Essentially, in the twistor space, we have a center state f_0 around which f_1, f_2, ... occur. Each of these sequences waves forms a torus-like topology, hence, f_1 and f_2 form a double nested tori set consistent with both spin 1 and spin ½ particle states where all n states are elements of the twistor, z , as $n \in z$. In the specific case of a massless particle with spin for f_1 , the two-surface in complex Minkowski space corresponding to the twistor represents the center of mass of the system so that the surface does not intersect the real Minkowski space. This reflects the system's intrinsic spin. We see an analogy to the 3-tori Calabi-Yau string theory. The higher order f_n may describe higher order string modes or oscillations of $Z^{\alpha}\overline{Z}_{\alpha} = 0$ or f_0 . This occurs for the case using f_1, f_2, and f_3 and, hence, all known particle states.

The topology of the first three Penrose projective twistor states are PT , PT^+ , and PT^- . The PT^+ , and PT^- are the domain of the projective twistor space, PT , where we denote these two states in which $(-1,1)$ are elements of t where ε is small. We denote two line elements which are denoted in terms of twistors as a surface on the sphere S^3 as $PT^{\pm}$ which corresponds to $\underset{-t}{Z^{\alpha}}\,\underset{-t}{\overline{Z_{\alpha}}} = 0$ and $\underset{t}{Z^{\alpha}}\,\underset{t}{\overline{Z_{\alpha}}} = 0$ for $t = 1 - \varepsilon$ for PT^+ , and PT^- gives $t = 1 - \varepsilon = \varepsilon - 1$. These two branches correspond to a transformation matrix,

$$\begin{pmatrix} 1 & 0 & t & 0 \\ 0 & 1 & 0 & t \\ t & 0 & 1 & 0 \\ 0 & t & 0 & 1 \end{pmatrix} \tag{8.73}$$

This gives us a translation formulation for vectors into the states of spinors in terms of t, in terms of the spinors

$$\begin{pmatrix} \omega^{0}_{t\,1} \\ \omega^{1}_{t}{}' \\ \pi_{t\,0}{}' \\ \pi_{t\,1} \end{pmatrix} = \begin{pmatrix} 1 & 0 & t & 0 \\ 0 & 1 & 0 & t \\ t & 0 & 1 & 0 \\ 0 & t & 0 & 1 \end{pmatrix} \begin{pmatrix} \omega^{0} \\ \omega^{1}{}' \\ \pi_{0}{}' \\ \pi_{1} \end{pmatrix} \tag{8.74}$$

which is Z_t^{α} and $t \sim \pm 1$ since ε is small. Then in terms of twistors,

$$\hat{\omega}^{A} = \omega^{A} + \varepsilon\xi^{AB} \frac{\partial f}{\partial \omega^{B}} \tag{8.75}$$

for $\hat{\pi}_{A'} = \pi_{A'}$ where ω and π are orthogonal spinors. The term $\varepsilon\xi^{AB} \dfrac{\partial f}{\partial \omega^{B}}$ is small compared to ω^{A} and π_{A} since ε is small. The unit spinors or vectors are $\hat{\omega}^{A}$ and $\hat{\pi}_{A'}$ for both $A, B = 1,2$.

A 5D surface of projective twistors in a spin free state, which can have genus $g = 0$ for a spherical, no 'hole' surface to $g \neq 0$ for $S^{3} \times R$. Penrose has formulated the relations between the conformal geometry of Minkowski space, complex analysis, and hence, analytic continuation, and the solutions to certain conformally invariant differential equations such as Maxwell's equations. Gauge theory in this context also allows the formalism of the Yang-Mills equations, which have become a major tool in four-dimensional differential topology. The Yang-Mills theory is a non-Abelian gauge group theory, which is the basis of modern quantum particle field theory. Invariance under the local gauge group SU_2 can be extended to larger groups SU_n for $n > 2$. A theory which is invariant under the local gauge group SU_2 is referred to as a Yang-Mills theory. For example, chromodynamics is a Yang-Mills theory with the gauge group SU_3. The exploration of conformally invariant conditions on Minkowski space is formulated for contour integral formulation process solutions to the Dirac equation. The contour integral methods allow integrability and are used to deal with the 'holes' or singularities in real and complex manifolds [49,50].

Work is in progress to complete the complexification of the Dirac equation [51] in the complex-8 space.

8.7 Conclusions

It appears that utilizing a complexification of Maxwell's equations with the extension of the gauge condition to non-Abelian algebras, yields a possible metrical unification of relativity, electromagnetism and quantum theory. This unique new approach yields a universal nonlocality. No radical spurious predictions result from the theory, but some new predictions are made which can be experimentally examined. Also, this unique approach in terms of the twistor algebras may lead to a broader understanding of macro and micro nonlocality and possible transverse electromagnetic fields observed as nonlocality in collective plasma state and other media. In the next chapter we demonstrate application of the model to complex 12-space and develop correspondence to M-Theory and F-Theory.

References

[1] Penrose, R. & Newman, E.J. (1978) Proc. Roy. Soc. A363, 445.
[2] Newman, E.T. (1973) J. Math. Phys.14, 774.
[3] Hansen, R.O. & Newman, E.T. (1975) Gen. Rel. and Grav. 6, 216.
[4] Newman, E.T. (1976) Gen. Rel. and Grav. 7, 107.
[5] Newman, E.T. (1983) 3rd MG meeting on Gen. Rel., Ha Nang (ed.) Amsterdam: North-Holland, pp. 51-55.
[6] Rauscher, E.A. (1996) The physics of nonlocality, in: The Iceland Papers, Frontiers of Physics, p 93, (ed.) B.J. Josephson, Essentia Res. Associates, Amherst, Wisconsin, 54406; and (1979) P.A.C.E., Ottowa.
[7] Stapp, H.P. (1993) Phys. Rev. A47, 847.
[8] Bell, J.S. (1964) Physics 1, 195.
[9] Clauser, J.F. & Horne, W.A. (1971) Phys. Rev. 10D, 526.
[10] Aspect, A. et. al. (1982) Phys. Rev. 49, 1804.
[11] Einstein, A., Podolsky, B. & Rosen, N. (1935) Phys. Rev. 47, 777.
[12] Kaluza, Th. (1921) Sitz. Berlin Press, *A. Kad. Wiss*, 968.
[13] Klein, O. (1926) *Z. Phys*. 37, 805; (1927) *Z. Phys*. 41, 407.
[14] Vigier, J-P (1991) Foundations of Physics, 21, 125.

[15] Evans, M.W. & Vigier, J-P (1996) The Enigmatic Photon Vols. 1 & 2, Dordrecht: Kluwer Academic.
[16] Rauscher, E.A. (1993) Electromagnetic Phenomena in Complex Geometries and Nonlinear Phenomena, Non-Hertzian Waves and Magnetic Monopoles, Millbrae: Tesla Books.
[17] Rauscher, E.A. (2008) Electromagnetic Phenomena in Complex Geometries and Nonlinear Phenomena, Oakland: Noetic Press.
[18] Baston, R. & Eastwood, M.B. (1985) The Penrose transformation for complex homogenous spaces, Twistor Newsletter, 20, 34.
[19] Eastwood, M.G. (1985) Complex quaternionic Kahler manifolds, Twistor Newsletter 20, 63.
[20] Penrose, R. (1986) Pretzel twistor spaces, Twistor Newsl 21, 7-11.
[21] Hughston L.P. & Ward , R.S.(1979) Adv in Twistor Theory, Pitman.
[22] Aspect, A., Granger, P. & Rogier, G. (1982) Phys Rev Let, 49, 91; Aspect, A. (1976) Phys. Rev. D14, 1944.
[23] Rauscher, E.A. & Targ, R. (2006) Investigation of a complex space-time metric, pp. 121-146, Frontiers of Time, New York: AIP Conference Proceedings #863.
[24] Rauscher, E.A. (1976) Bull. Am. Phys. Soc. 21, 1305.
[25] Rauscher, E.A. (1968) J. Plasma Phys. 2, 517.
[26] Amoroso, R.L., Hunter, G., Kafatos, M. & Vigier, J-P (eds.) Gravitation & Cosmology: From The Hubble Radius to the Planck Scale, Dordrecht: Kluwer.
[27] Rauscher, E.A. (1987) Soliton solutions to the Schrödinger equation in complex Minkowski space, pp. 89-105, Proceeding of the First International Conference, Berkeley.
[28] Rauscher, E.A. & Amoroso, R.L.(2005) The Schrödinger equation in complex Minkowski, nonlocal and anticipatory systems, Unified Theories, Budapest, Hungary, in R. L. Amoroso, I. Dienes & C. Varges (eds) Oakland: Noetic Press.
[29] Sirag, S.P. (1996) A mathematical strategy for a theory of particles, pp. 579-588, The 1st Tucson Conference, S.R. Hameroff, A.W. Kasniak & A.C. Scott (eds.) Cambridge: MIT Univ. Press.
[30] Rauscher, E.A. (1999) D and R Spaces, Lattice Groups and Lie Algebras and their Structure, LBL Report.
[31] Clauser, J.F. & Shimoney, A. (1978) Pep. Prog. Phys. 41, 1881.
[32] Jackson, J.D. (1975) Classical Electrodynamics, NY: Wiley & Sons.
[33] Cramer, J.G. (1980) Phys. Rev, D22, 362.
[34] Partridge, R.B. (1973) Nature, 244, 263.

[35] Pflelgov, R.L. & Mandel, L. (1967) Phys. Lett. 24A, 766.
[36] Wheeler, J.A. (1982) Int. J. Theor. Phys., 21, 557.
[37] Ramon, C. & Rauscher, E.A. (1980) Found. Physics, 10, 661.
[38] Amoroso, R.L. (2002) Developing the cosmology of a continuous state universe, in R.L. Amoroso et al (eds.) Gravitation & Cosmology: From The Hubble Radius to the Planck Scale, Dordrecht: Kluwer.
[39] Ramon, C. & Rauscher, E.A. (1980) Superluminal transformations in complex Minkowski spaces, Foundations of Physics 10, 661.
[40] Rauscher, E.A. (1971) A unifying theory of fundamental processes, Lawrence Berkeley National Laboratory Report (UCRL-20808; and (1968) Bull. Am. Phys. Soc. 13, 1643.
[41] Rauscher, E.A. (1972) Closed cosmological solutions to Einstein's field equations, Lett. Nuovo Cimento 3, 661.
[42] Rauscher, E.A. (1981) Conceptual Changes in Reality Models and New Discoveries in Physics: Soliton Solutions to the Schrödinger Wave Equations, Tecnic Research Lab of California Report PSRL-1076, in the Proceedings of the 1st Intl Energy Conference, Ontario; Univ of Toronto.
[43] Rauscher, E.A. (1983) Coherent Nondispersive Solutions to the Schrödinger and Dirac Equations, Tecnic Res. Labs, CA Report, PSRL-6941.
[44] Isham, J., Penrose, R. & Sciomci, D.W (eds.) (1975) Quantum Gravity, Oxford: Clarendon Press.
[45] Nash, C. & Sen, S. (2000) Topology and Geometry for Physicists, New York: Academic Press.
[46] Rauscher, E.A., Rasmussen, J.D. & Harada, K.(1967) Coupled channel alpha decay rate theory applied to Po^{212m}, Nucl. Phys. A94, 33.
[47] Rauscher, E.A. (1968) Electron interactions and quantum plasma physics, J. Plasma Phys. 2, 517.
[48] Chew, G. (1964) The Analytic S-matrix, New York: Benjamin.
[49] Dang, Y., Kauffman, L.H. & Sandin, D. (2002) Hypercomplex Iterations: Distance Estimation And Higher Dimensional Fractals, Singapore: World Scientific.
[50] Rauscher, E.A. (1968) A unified theory of fundamental processes, Univ. California, Los Angeles, Bull. Am. Phys. Soc., 13, 1643.
[51] Amoroso, R.L. & Rauscher, E.A. (2009) The complexified Dirac equation and the completion of quantum theory (In Progress).

Chapter 9

Probability $\equiv 1$: An Empirical Protocol for Surmounting Quantum Uncertainty

This amazing technique is still in its infancy - J.G. Cramer [1].

It is easy to show using standard theory that if a system starts in an eigenstate of some observable, and measurements are made of that observable N times a second, then, even if the state is not a stationary one, the probability that the system will be in the same state after, say, one second, tends to one as N tends to infinity; that is, that continual observations will prevent motion...- Alan Turing [2].

Interaction-Free Measurement (IFM) is a quantum mechanical procedure for detecting the state of an object without a phenomenological interaction occurring with the measuring device. A radical extension of the various experimental protocols spawned recently by the Elitzur-Vaidman IFM thought experiment is proposed to ontologically surmount (without interaction or collapse) the quantum uncertainty principle with probability$\equiv 1$ through utility of the additional degrees of freedom inherent in the supersymmetric regime of string/brane theory. Just as the UV catastrophe provided a clue for the immanent transition from Classical to Quantum Mechanics, the duality in the Quantum Zeno Effect hints at another new horizon. IFM provides an intermediate indicia of this developing scenario. The quantum Zeno paradox experimentally implemented in IFM protocols hints at the duality between the regular phenomenological quantum theory and a completed unitary or ontological model beyond the formalism of the standard Copenhagen interpretation. Utilizing extended theoretical elements associated with a new formulation for the topological transformation of a 'cosmological least unit', a putative empirical protocol for producing IFM with probability$\equiv 1$ is introduced in a manner representing a direct causal violation or absolute surmounting of the quantum Uncertainty Principle.

9.1 Introduction – Philosophy of Phenomenology Versus Ontology

In the 1970's the concept of quantum non-demolition (QND) [3] arose as a process for performing extremely sensitive measurements without disturbing an extremely weak signal which lead to the Weber approach for gravitational interferometry. But there was a trade-off between the accuracy of a QND measurement and its inevitable back-action on the conjugate observable to that being measured. Recently myriad terms were introduced for programs exploring manipulation of the quantum uncertainty principle for non-collapse of the wave function: Negative Result Measurement (NRM) [4], Quantum Non-Demolition (QND) [3,5,6], Interaction Free Measurement (IFM) [7-15], Quantum Zeno Effect (QZE) [16-19], Bang-Bang Decoupling (BBD) [20], Quantum Error Correction (QEC) [21], Quantum Interrogation Measurement (QIM) [22,23], Counter Factual Computing (CFC) [24,25], Absorption-Free Measurement (AFM) [26,27], Quantum Seeing in the Dark (QSD) [28], Quantum Erasure Experiment (QEE) [29,30], Interaction Free Imaging (IFI) [31] and the Bomb Testing Experiment (BTE) [7].

An interaction is any action, generally a force, mediated by an exchange particle like the photon in electromagnetic interactions. This physical concept of a fundamental interaction regards phenomenological properties of matter (Fermions) mediated by the exchange of an energy / momentum field (Bosons) as described by the Galilean, Lorentz or Poincaré groups of transformations. "There has been some controversy and misunderstanding of the IFM system concerning what is meant by 'interaction' in the context of 'interaction-free' measurements. In particular, we stress that there must be a coupling (interaction) term in any Hamiltonian description." [32]. Here we wish to introduce a new ontological type of homeomorphic transformation with no exchange particle mediated by an interactionless or energyless topological switching [33].

It is impossible by definition to violate the uncertainty principle, $\Delta x \Delta p_x \geq \hbar/2$ or $\Delta E \Delta t_x \geq \hbar/2$ within the framework of Copenhagen phenomenology arising from operation of a 'Heisenberg Microscope'. This is a fundamental empirical fact demonstrated by the Stern-Gerlach experiment where space quantization is produced along the z axis by continuous application of a non-uniform magnetic field to atomic spin structure [34], or by Young's double-slit experiment [35] for example. Recent work stemming from the Elitzur-Vaidman bomb-test thought experiment [7] has begun to change this immutable law. The Elitzur-

Vaidman bomb-test experiment was first demonstrated experimentally in 1994 [36] using a Mach-Zender interferometer (Fig. 9.1); and soon led to two main improved procedures:

1) Multiple recycled Measurements and
2) Multiple Interferometers.

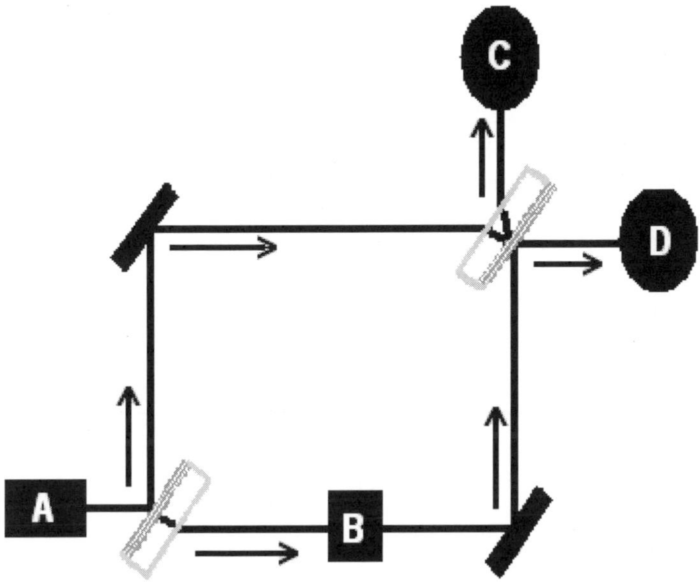

Figure 9.1 General form of the Mach-Zehnder interferometer used to determine the phase shift caused by a sample placed in the path of one of two collimated laser beams. A is the beam source, B the sample and C & D the detectors. Note the two types of mirrors.

The Mach-Zehnder interferometer [37] works by using pairs of correlated photons produced by spontaneous parametric-down conversion from a molecular crystal such as $LiIO_3$. Initially in the first experiments for a 50-50 beam splitter for a one time measurement cycle the IFM probability was 25% according to the formula in Eq. (9.1) [36]; but for repeated measurements and/or various forms of multiple interferometers the IFM probability can be arbitrarily increased toward unity as shown in Fig. 9.2.

$$\eta = \frac{P(Det2)}{P(Det2) + P(Bomb)} \tag{9.1}$$

The probability for the IFM model was suggested to occur in powers of $\pi / 2N$ by $P_{IFM} = [1 - 1/2(\pi / 2N)^2 + ...]^{2N}$ where N is the number of beam splitters in the Max-Zender interferometer. In his seminal paper (A thought experiment) Elitzur suggested a maximum IFM of 50%. Kwiat's team developed a method to improve the model to 80% with $P_{QSD} = 1 - (\pi^2 / 4N) + O(1/N^2)$ where in this case N is the number of photon cycles through the apparatus [36]. In regards to the Elitzur and Vaidmann consideration that their model could be explained by the 'Many-Worlds' interpretation Cramer has proposed, "they suggest that the information indicating the presence of the opaque object can be considered to come from an interaction that occurs in a separate Everett-Wheeler universe and to be transferred to our universe through the absence of interference" [38].

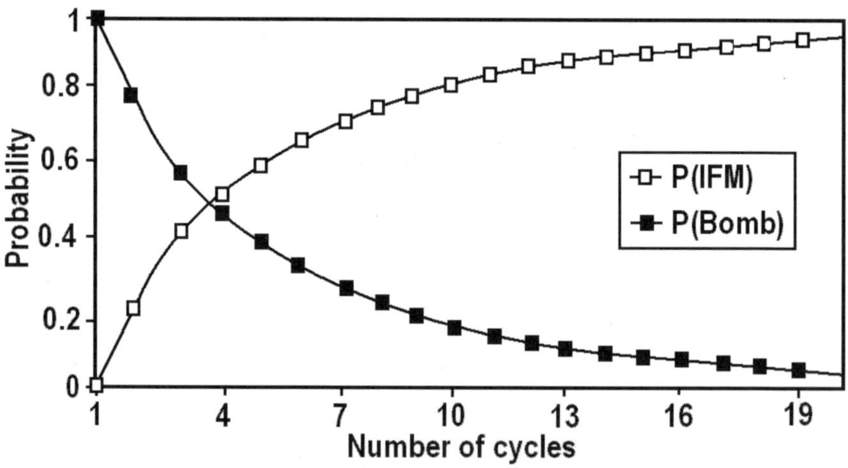

Figure 9.2 IFM probability can be arbitrarily increased toward unity by repeated measurements. Figure adapted from [36].

In this chapter a putative protocol is delineated not for another sophisticated improvement of the varied stepwise degrees of violating the uncertainty relation by the several IFM protocols; but for completely surmounting the uncertainty relation directly, in a straightforward manner, for any and every single action with Probability ≡ 1. In an unexpected way our model has similarities to IFM/QSD but uses extended theory to fully complete the task of uncertainty violation. One could say the new noetic protocol sort of turns the IFM methodology upside down and inside out. The HD regime of the noetic protocol is like the complete "hall of mirrors" where the whole battery of interferometers and multiple cycling routine is inherent in the HD regime, such that only one 'measurement' is required for probability $\equiv 1$. The methodology of this new empirical protocol is fully ontological (rather than the usual phenomenological) and action in the new HD SUSY regime in causal violation of Copenhagen phenomenology not in an Everett 'many-worlds' sense [39] but in a manner that extends to completion the de Broglie-Bohm-Vigier causal interpretation of quantum theory [40]. The ontological basis is realized utilizing the additional degrees of freedom of a 12D F-Theory iteration of M-Theory [41] along with the key supposition of conformal scale-invariance pertaining to the state of quantum of information.

While considerations of the vacuum are of paramount concern for string theory, much of its putative essential parameters are ignored in the avid exploration of other details. The $P \equiv 1$ model relies heavily on the existence of a Dirac polarized vacuum [42-44]. Of primary concern at this point of our development is its inclusion of extended electromagnetic theory [45-47] which is a key element in manipulating the structural-phenomenology of spacetime.

An experimental design, relying on the utility of a new fundamental teleological action principle inherent in the topological geometry of a covariant polarized Dirac vacuum putatively driving the evolution of self-organization in spacetime as an autopoietic complex system, is developed to elucidate the methodology for surmounting uncertainty. The experimental apparatus, a multi-level interferometer, is designed to focus this noetic unitary field.

As we shall see the protocol relies on the symmetry conditions of new self-organized cosmological parameters amenable to a resonant hierarchy of coherently controlled topological interactions able to undergo what Toffoli calls 'topological switching' [48] as the energyless basis for the Micromagnetics of information exchange. Finally to complete the concatenation we utilize theoretical concepts associated with the putative covariant polarized Dirac vacuum [42-44] forming a string theoretic spacetime background [49-51] also making correspondence between our ontological view of quantum theory and an extension of Cramer's Transactional Interpretation [52].

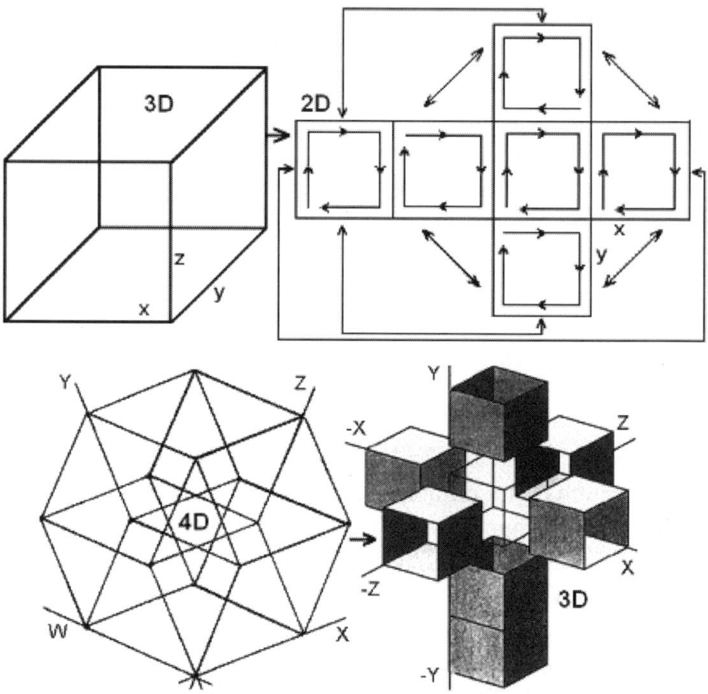

Figure 9.3 The suggestion is that the central translucent cube in the lower right represents a 'particle in a box' quantum state that through conformal scale-invariance remains physically real when the metaphor is carried to 12D where it becomes like the 'mirror image of a mirror image' and in that sense is causally free of the E_3 quantum state and thereby open to ontological information transfer in violation of Copenhagen uncertainty.

9.2 The Proper Cosmological Perspective is Key

When physicists embraced the 3D Newtonian world view about a hundred years ago the universe was considered to be mechanical and predictable like a clockwork. Since the advent of QT reality is believed to be quantum and statistical or uncertain. Following this line of reasoning when a Theory of Everything (TOE) is realistically discovered based on a unitary field, should some form of monism be embraced? We postulate that cosmology is not uniquely based on any of these three conditions, but a continuous-state dynamic transformation of the three regimes comprised of a Wheeler-Feynman-Cramer complementarity [52-55] as outlined in Chap. 3. Physics has long resisted the role of the observer in physical theory; but in an anthropic cosmology the observer is an inherent key element or better said, the basis of observation [55]. This conundrum of the observer can be avoided here as its effects only become critical for process needing to control a much deeper region of spacetime (see Chap. 12).

Einstein stated that 'all of physics is based on measurements of *duration* and *extension*'. Until now this has occurred within the parameters of a 4D Minkowski-Riemann spacetime metric under Gauge conditions utilizing various forms of the $E_3 / \hat{M}_4$ Galilean-Lorentz-Poincaré transformations describing classical, quantum and relativistic conditions. These criteria are no longer sufficient and indeed our protocol for surmounting the uncertainty principle requires description of a new cosmological regime described by a new set of 12D transformations [56,57] we hope to call the Noetic Transformation because of its relevance to anthropic considerations (see Chap. 5). In this regard in spite of Bell's theorem, following Einstein's conundrum, we restate his complaint that quantum theory is incomplete and therefore wholly inadequate for some processes.

Cramer's transactional model of QT [52] has been ignored by most physicists for a variety of reasons we will not take the time to address here. This just means that when we bandy it about here as a key foundation of HAM cosmology it is foreign and not well understood. A *Cramer transaction* entails Wheeler-Feynman [53] future-past, standing-wave symmetry conditions which when extended to the HD SUSY regime readily lend themselves to extension to mirror symmetry conditions inherent in the 12D F-Theory iteration of M-Theory [41]. Further we suggest that the new 12D noetic transform adds additional piloting super-quantum potential [58] parameters, suggesting two forms, levels or regimes for quantum mechanics – that of the observed 4D phenomenological interaction associated with the uncertainty principle; and the new HD ontological 'piloting' or anthropic guidance regime. As discussed in Chap. 5 reality itself is a transaction (see Figs. 5.2, 5.3).

Because the external world we observe is a limited subspace [55,59] of a larger contiguous reality some elements are removed from perception by subtractive interferometry.

In the standard Copenhagen Interpretation of QT an event emerges only as a result of measurement and objective reality is considered to be a probabilistic illusion. Cramer considers 'all off diagonal elements of the line element physically real' during the process of the offer-wave-confirmation-wave process preceding a transaction (event) [52]. We may call the final event a resultant of the conditions of Heisenberg Potentia. Here we still wish to consider reality illusory to the Minkowski observer.

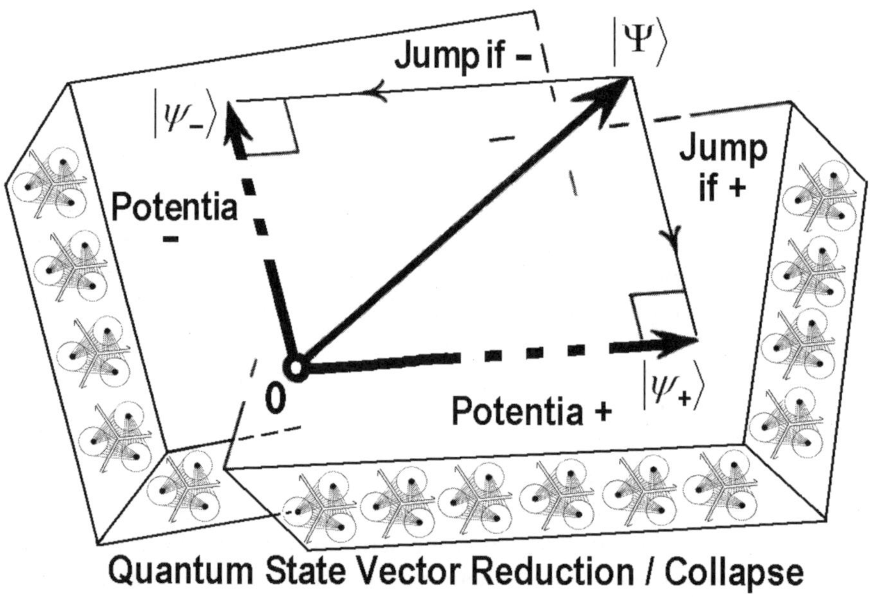

Figure 9.4 A way to look at a transaction as a collapse, $|\Psi\rangle$ to the 2D Euclidean plane from, in this case, an HD potentia of two possible orthogonal states, $|\psi_+\rangle, |\psi_-\rangle$.

Issues of the nature of the fundamental cosmological background continue to be debated with disparate views jockeying for philosophical supremacy; a scenario remaining tenable because experimental avenues for testing physics beyond the standard model have remained elusive. Here a putative empirical protocol is devised for manipulating the so-called covariant Dirac polarized vacuum (DPV) providing a methodology for both surmounting uncertainty and low energy protocols for testing string theory. The DPV has a sixty-year history in the physics literature [42-44] which has for the most part been ignored by the main stream physics community for a number of philosophical conflicts. The problem of surmounting uncertainty is solved by the utility of additional degrees of freedom introduced by utilizing a multiverse cosmology and the associated extended theoretical elements.

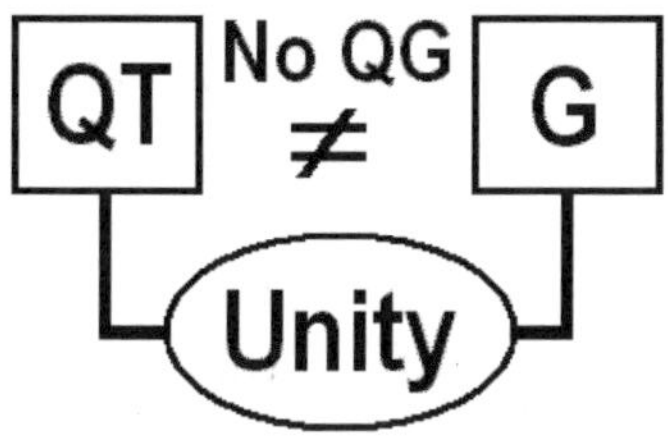

Figure 9.5 The domain for the unification of quantum theory and gravitation occurs in the unitary regime; not with each other because quantization has a cutoff similar to the limit discovered for Newtonian mechanics.

9.3 Micromagnetics of Spacetime Conformation

An extensive body of literature exists for phenomena related to the zero-point field; but relative to noetic theory this work is considered metaphorically descriptive only of the 'fog over the ocean' rather than the structural-phenomenology of the ocean itself. Instead the deep structure of a real covariant Dirac polarized vacuum is utilized [42-44,60]. The Casimir, Zeeman, Aharanov-Bohm and Sagnac effects are considered evidence for a Dirac vacuum. New assumptions are made concerning the Dirac polarized vacuum relating to the topology of spacetime and the structure of matter cast in a 12D form of Relativistic Quantum Field Theory (RQFT) in the context of the new HAM cosmological paradigm [61-63]. In this anthropic cosmology the observed Euclidean-Minkowski spacetime present, $E_3 - \hat{M}_4$ is a virtual standing wave of highly ordered Wheeler-Feynman-Cramer retarded-advanced future-past parameters respectively [52,53]. See Figs. 9.4 and 9.19, 9.20 for a graphic illustration of this paradigm. An essential ingredient of HAM cosmology is that a new action principle synonymous with the unified field arises naturally and is postulated to drive self-organization and evolution through all levels of scale [64-66].

In this context an experimental design [57,67] is introduced to isolate and utilize the new anthropic action to test empirically its putative ability to effect conformational structure of the topology of spacetime to surmount the usual phenomenologically based uncertainty in an ontological matter with probability $\equiv 1$.

Noetic Theory postulates that spacetime topology is 'continuously transformed' by the self-organizing properties of the long-range coherence [68,69] of the anthropic, *élan vital* or unitary noetic field [64,65,70-81]. In addition to manipulating conformational change, from the experimental results we attempt to calculate the energy Hamiltonian required to manipulate the Casimir topological conformation in terms of the noetic field equation, $F_N = E / R$ (unexpanded form, see Chap. 4). This resonant coupling produced by the teleological action of the anthropic noetic field driving its hierarchical self-organization has local, nonlocal and supralocal (complex HD) parameters [64]. The Schrödinger equation, extended by the addition of the de Broglie-Bohm quantum potential-pilot wave mechanism has been used to describe an electron moving on a neural manifold; but this is not a sufficient extension to describe anthropic noetic aspects of the continuous-state symmetry breaking of spacetime topology which requires further extension to include action of the noetic unitary field in additional dimensions.

The Noetic Field [64,65,70-83] produces periodic symmetry variations with long-range coherence [67-69] that can lead to a critical Noetic Effect [64,72] of the Ising model lattice gas rotation of the Riemann sphere spacetime backcloth. This can be described by a form of double-cusp catastrophe dynamics (Fig. 9.9). Operationally the plane of equilibrium experiences sustained hyperincursion by the noetic field. The coupled modes of this process rely on a special form of the harmonic oscillator called the incursive oscillator [82-85]. There is a force of coherence [86]. For example for an Earth observer's temporal perception, railroad tracks recede into a point at the horizon. For an atemporal eternal HD observer, the tracks remain parallel. This is the origin of the coherence force which forms a kind of logic gate driving equilibrium of the Casimir boundaries to parallel or degenerate modes thus giving rise to the possibility of effecting conformational states.

This is a boundary condition problem; here probably of the Born-von Karman type where the boundary conditions restrict the wave function to periodicity on a Bravais lattice of hexagonal symmetry, stated simply as $\psi(r+N_i a_i)=\psi_r$, where i runs over the dimensions of the Bravais lattice, a_i are the lattice vectors and N_i are integers [87,88]. In this model presence of the periodic spherical rotation effects of the cyclical coherence-decoherence modes allow the action of the noetic field [64]. This Noetic Processing is governed by the fundamental equation of unitarity, $F_N = E/R$ (Fig. 9.7). Cyclotron resonance, logarithmic spiral (Fig. 11.18), Kaluza-Klein hierarchy or genus-1 helicoid 'parking garage (Fig. 11.7) may maintain piloting by the noetic field or induce an electromotive 'radiation pressure' or topological switching coherence force, the Noetic Effect (Fig. 9.7), on the topology of spacetime leading to conformational change in the static-dynamic [89-91] leapfrogging' of the Casimir boundary conditions of topological brane states.

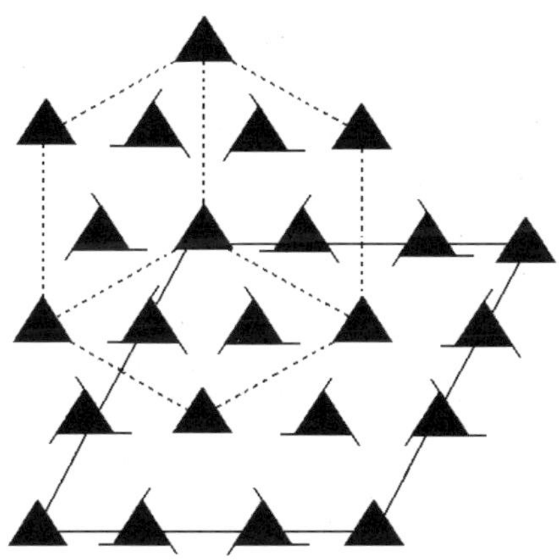

Figure 9.6 HD emergence from a LD lattice gas. If the central vertex of the cube represents a Euclidean point, the 12 satellite points represent HD control parameters

We can't be sure yet which of the hierarchical formalisms might be the physical one until some empirical work is done. Intellectually we lean toward the concept of the action of a cyclotron resonance hierarchy acting on the genus-1 helicoid parking garage structure modulated by some form of Bessel function because this seems to meld well with catastrophe theory and the future-past symmetry breaking parameters we postulate in to be inherent in the structural-phenomenology of HAM continuous-state spacetime topology. The structural-phenomenology of atoms and molecules is full of domain walls amenable to description by combinations of Gauss' and Stokes' theorems ordered in terms of Bessel Functions where boundary conditions create resonant cavities built up by alternating static and dynamic Casimir conditions [89-91]. As frequency increases central peaks occur with opposite or zero polarity at the domain edges. These properties are relevant to Ising Model [92] spin flips of the domains of the Riemann-Block Spheres effecting homeostatic planes of equilibrium (Fig. 9.7a). The noetic effect can maintain equilibrium or produce catastrophes causing conformational change in the Casimir spacetime structures [93].

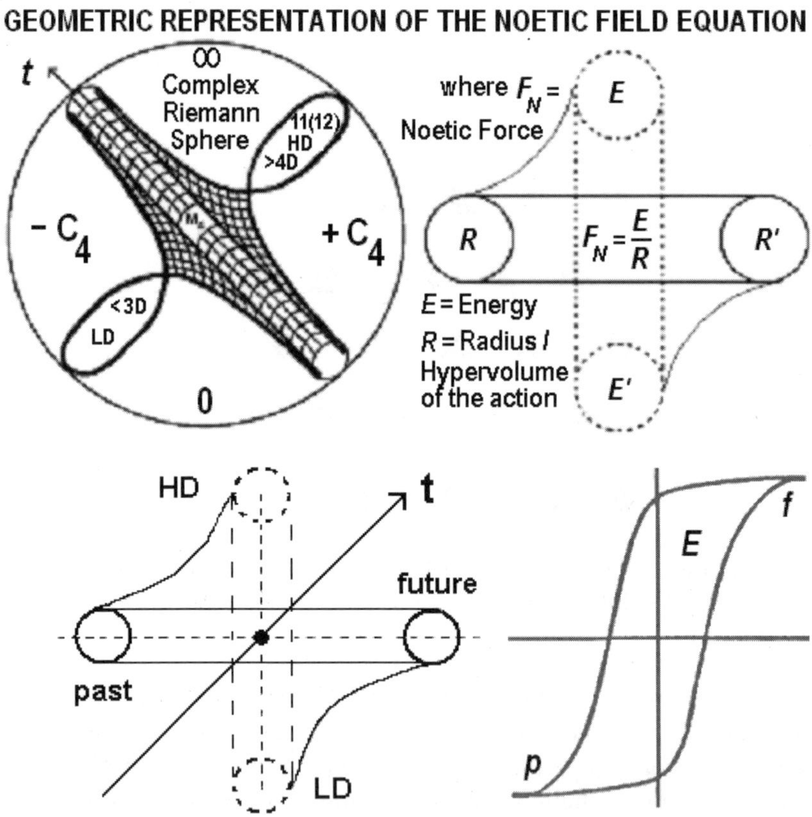

Figure 9.7 Topological and geometric idealizations of the noetic field equation describing an action of the noetic field, called the 'noetic effect', on a biological or spacetime manifold.

9.4 Catastrophe Theory and the Noetic Formalism

Regarding *dynamical systems* that generally operate in a framework of stability and equilibrium – Technically these systems have a restrictive class called gradient systems which contain singularities or points of *extrema*. Some causal action can institute a bifurcation of an extrema that can initiate a qualitative change in the physical state of the system.

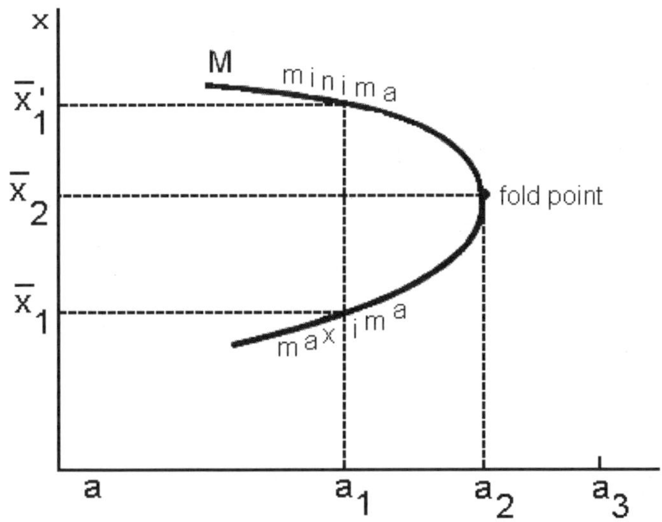

Figure 9.8 Basis of catastrophe theory.

Catastrophe theory[1] describes the breakdown of stability of any equilibrium system causing the system to jump to another state as the control parameters change. The changes in the singularities associated with the bifurcation of extrema are called elementary catastrophes [94-96] and can be described by real mathematical functions

$$f : R^N \rightarrow R. \tag{9.2}$$

The equation describing an elementary catastrophe utilizes variables representing *Control* and *State* parameters of the system and is a smooth

---

[1] The groundwork for Catastrophe Theory began with Poincaré's work in 1880 on the qualitative properties of solutions to differential equations; and became formalized in the 1950's by R. Thom's work on mapping singularities in structural stability, which he called catastrophes.

real function of r and n where R represents the resultant singularity or catastrophe

$$f : R^r \times R^n \to R \, . \tag{9.3}$$

The r variables are the control parameters of the state variables n. The function f is therefore an r-parameter family of functions of n variables.

If we let
$$f\left(a_i, ...a_r; \;\; x_j, ..x_n \right) \tag{9.4}$$

be a smooth real-valued function of $r + n$ real variables we get equation (9.3). The number of elementary catastrophes depends only on r and is finite for $r \geq 5$ totalling eleven (table 9.1) and infinite for $r \geq 6$.

r (Control Factors)	Number of Catastrophes	Name		Dimensions
r = 1	1	A_2	Fold Catastrophe	2D
r = 2	1	$A_{\pm 3}$	Cusp Catastrophe	3D
r = 3	3	A_4	Swallowtail	4D
r = 4	2	$A_{\pm 5}$	Butterfly	5D
r = 5	4	A_6	Wigwam	6D
r = 3	-	D_{-4}	Eliptic Umbilic	5D
r = 3	-	D_{+4}	Hyperbolic Umbilic	5D
r = 4	-	D_5	Parabolic Umbilic	6D
r = 5	-	D_{-6}	2nd Elliptic Umbilic	7D
r = 5	-	D_{+6}	2nd Hyperbolic Umbil	7D
r = 5	-	$E_{\pm 6}$	Symbolic Umbilic	7D
r = 6	∞	X_9	Double Cusp	9-11D

Table 9.1 The general forms of catastrophes showing how the dimensions increase as the number of control factors increase. The names bear some resemblance to the geometric pattern of the catastrophe. The double cusp catastrophe is utilized in development of Noetic Theory because it models most closely noetic superspace transitions and is compatible with the fundamental equation of consciousness.

This model can be utilized to call for a new field of vacuum engineering based on the structural-phenomenology of the noetic field and whether resultant action of the noetic effect is positive or negative.

Spacetime exhibits complex self-organization. The noetic field is the factor driving this self-organization [64]; therefore we postulate hyperincursion and anticipatory properties are inherent in the fundamental hierarchical basis of this self-organization which can be formally described in terms of Double-Cusp Catastrophe Theory.

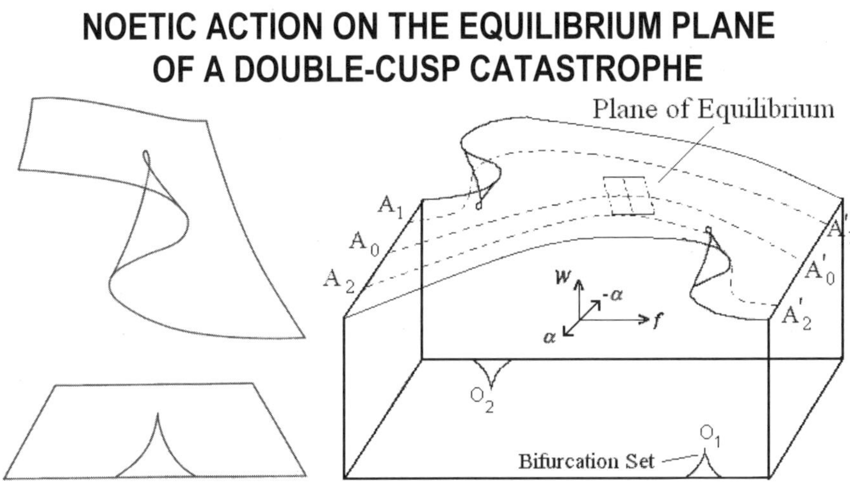

Figure 9.9 In a, the DCC is illustrated showing cusps at each end of the plane of equilibrium. The DCC is said to occur in ≥ 9 dimensions and thought to be the catastrophe form most compatible with NFT symmetry. The plane of equilibrium is a topological manifold tiled of noetic *least units*. The equilibrium manifold undergoes a 'conscious' quantum computation best described by interactive computation [67,85]. Fig. 8b graphically illustrates the fundamental scale invariant noetic equation $F_{(N)} = E / R$ of conscious action. Any internal or external stress or change in E is a nonlinear dynamic process producing stability or instability in the boundary conditions of R; an instability in E $\rightarrow$ stress $\rightarrow$ displacement $\rightarrow$ catastrophe $\rightarrow$ jump…whereas stable flux is homeostatic. 8b like noetic HAM cosmology is also a form of hysteresis loop of the Hamiltonian generalized in 8c as future-past parameters of noetic spacetime. The area E represents the energy of the noetic force, F_N.

The structural-phenomenology of Double-Cusp Catastrophe (DCC) Theory in $\geq 9D$ appears homeomorphic to the Riemannian manifold of both 10(11)D M-Theory and the topological geometry of the continuous-state spin-exchange dimensional reduction compactification process inherent in the action of the corresponding scale invariant least-unit of noetic superspace as cast in HAM cosmology [61-63]. In this general framework the double-cusp equilibrium surface is analyzed in terms of a hierarchy of Ising-like lattice gas jumps in state [92] providing a framework for considering the noetic least-unit tiling [97] of the Planck backcloth as a complex HD catastrophe manifold mediated by the unitary noetic field which because of the polarized properties of the Dirac vacuum lends itself to empirical mediation [57].

Unit Circle and Associated Flag Manifold of Temporal Evolution for Noetic Catastrophe Cycle

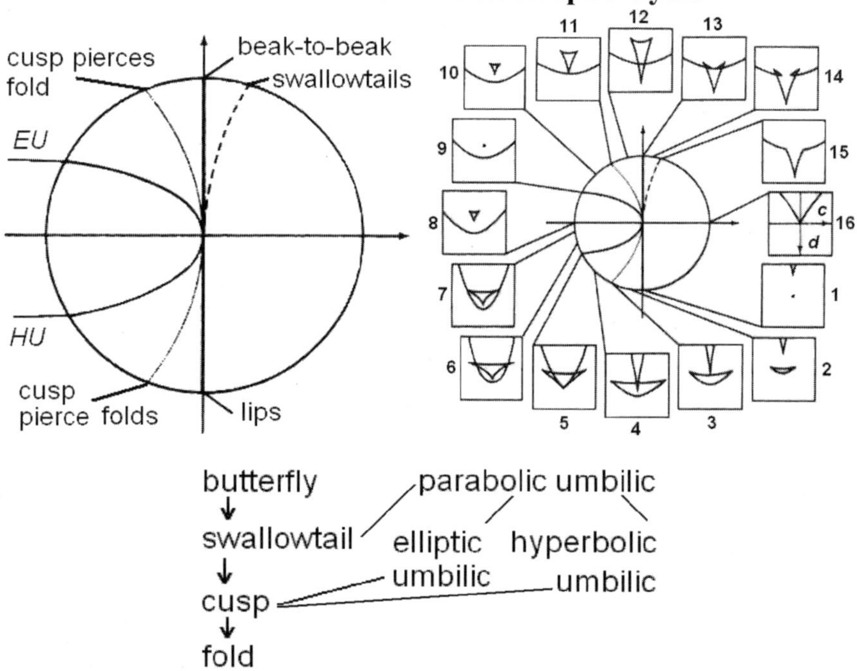

Figure 9.10 a) represents a plane of the unit circle with corresponding cross sections in b: for example shows a cusp. A single point in 1 grows to the 'lips' in 2. In 3 to 4 the original cusp 16 penetrates the mouth becoming a hyperbolic umbilic point at 5, turning into an elliptic umbilic at 6, shrinking to a point in 9. Growing again in 10 to pierce the fold line in 11 and through it in 12. A 'beak-to-beak singularity in 13 breaks in 14, collapsing to a swallowtail 15. The seven fundamental catastrophes contain 'subcatastrophes according to the diagram in 3c. Figures adapted from [94-96].

TABLE 9.2 GEOMETRIC CONTENT OF 12D SPACETIME

N-Space	Points	Lines	Squares	Cubes	Tesseracts	5T	6T	7T	8T	9T	10T	11T	12T
0	1												
1	2	1											
2	4	4	1										
3	8	12	6	1									
4	16	32	24	8	1								
5	32	80	80	40	10	1							
6	64	192	240	160	60	12	1						
7	128	448	672	560	280	84	14	1					
8	256	1,024	1,792	1,792	1,120	448	112	16	1				
9	512	2,304	4,608	5,376	4,032	2,016	672	144	18	1			
10	1,024	5,120	11,520	15,360	13,440	8,064	3,360	960	180	20	1		
11	2,048	11,264	28,160	42,240	42,240	29,568	14,784	5,280	1,320	220	22	1	
12	4,096	24,576	67,584	112,640	126,720	101,376	59,136	25,344	7,920	1,760	264	24	1

The putative significance of Table 9.2 for the application of double-cusp catastrophe theory to the noetic HAM formalism is that the structure of possible boundary conditions and the number of control points is revealed. For example, in this simplistic view, a 3D point in real spacetime might have 16 control photon-gravitons (noeons) covering it. Carrying the analogy all the way up to the 12D holoscape of the Multiverse, the same 3D point might be controlled or guided by a total of 8,176 photons. The number arrived at by summing the points of D4 to D12. No point in the universe is isolated; so this metaphor does not include the possible power factor by associated points in both the HD and LD HAM backcloth. Within the inherent continuous-state dimensional reduction compactification process, the LD domain (dimensions less than 3) might be coupled to orders of magnitude more photon-gravitons. This detail of Noetic Theory has not been completely worked out yet.

One can say that the noetic least-unit [55,97] tiling the fabric of the Planck backcloth is a complex HD catastrophe manifold with Dirac spherical rotation symmetry mediated by the unitary action of the noetic field. Any internal or external stress or change in energy, E is a nonlinear dynamical process producing stability or instability in the boundary conditions of R; a causal instability in $E \rightarrow$ stress $\rightarrow$ displacement $\rightarrow$ catastrophe $\rightarrow$ Ising jump...whereas stable flux is homeostatic. The hysteresis loop of the noetic field (Fig. 9.3b) is conformally scale invariant; the same processes occur in HAM cosmology and domains of living systems. The area represents the energy of the string tension, T_0 or *élan vital*. This energy, E_N is measured in *Einsteins*, the fundamental physical quantity defined as a 'mole'-Avogadro's number (6.02×10^{23}) of bosons, defined here as noeons, the exchange unit of the unitary field [64].

Equation (9.5) describes the equilibrium surface of the DCC [94-96] as modeled in (Fig. 9.3); where $B \pm Q$ is the state variable and μ_d and υ_d are the control parameters.

$$(B+Q)^3 + (B+Q)\mu_d + \upsilon_d = 0 \qquad (9.5)$$

The position of the two cusps is found at $\mu_d = 0$ and $\upsilon_d = 0$. At any moment temporal permutations of the noetic catastrophe cycle evolve in time from future to past and higher to lower dimensions in the same manner as the spacetime present of the least-unit of HAM cosmology for the spatial domains: $R^{12} \supseteq ...R^4 \supseteq R^3 \supseteq R^2 \supseteq R^1 \supseteq R^0$; followed by an Ising rotation where the cycle repeats.

9.5 Protocol for Experimentally Testing Noetic Cosmology

Extrapolating Einstein's energy dependent or deformed spacetime metric, $\hat{M}_4$ [98-100] to a supersymmetric 12D standing-wave future-past advanced-retarded topology of the anthropic multiverse we have designed a spacetime resonance experiment for a covariant Dirac polarized vacuum which has properties like an 'ocean of light'. If this is true spacetime acts like a 'surface wave' on the upper regime of the complex self-organized Dirac Sea and is therefore amenable to descriptive methods of nonlinear dispersive wave phenomena generally of the basic form

$$L(\mu) = \varepsilon N(\mu) \qquad (9.6)$$

where L and N are Linear and Nonlinear operators respectively in the linear limit where $\varepsilon = 0$ with elementary dispersive wave solutions $\mu_i = A_i \cos\theta_i$, $\theta_i = k_i x - \omega(k_i)t$ for one dimension plus time where nonlinearity creates resonant interactions between the μ_i solutions and the Amplitude A_i depends on t, creating potentially substantial effects where initial absent modes can become cumulative interactions producing shock wave effects.

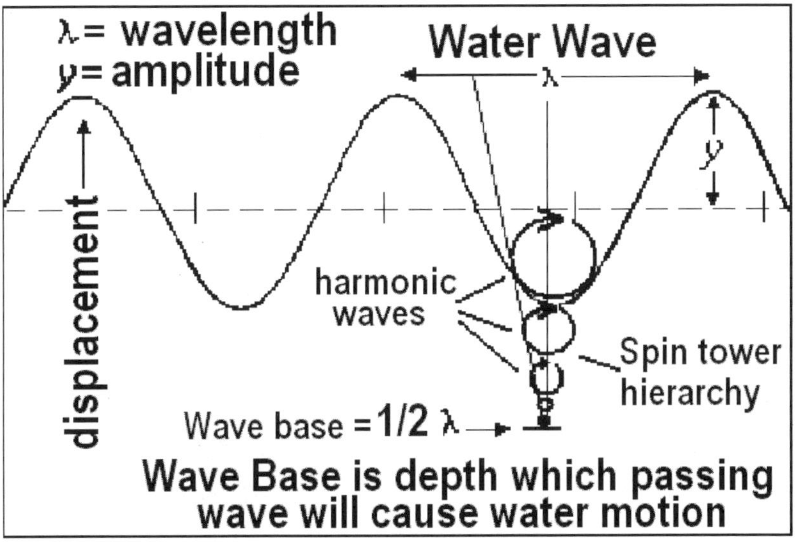

Figure 9.11 The spacetime topological hierarchy may have properties like water waves where the wave moves but the water remains stationary.

Motion of a one dimensional classical harmonic oscillator is given by $q = A\sin(\omega t + \varphi)$ and $p = m\omega A\cos(\omega t + \varphi)$ where A is the amplitude and φ is the phase constant for fixed energy $E = m\omega^2 A^2 / 2$. For state $|n\rangle$, with $n = 0,1,2...\infty$ and Hamiltonian $E_n = (n + 1/2)\hbar\omega$ the quantum harmonic oscillator becomes

$$\langle n|q^2|n\rangle = \hbar / 2m\omega \langle n|(a^\dagger a + aa^\dagger)|n\rangle = E_n / m\omega^2 \qquad (9.7)$$

and

$$\langle n|p^2|n\rangle = 1/2(m\hbar\omega)\langle n|a^\dagger a + aa^\dagger = mE_n \qquad (9.8)$$

where a and $a^\dagger$ are the annihilation and creation operators,

$$q = \sqrt{\hbar / 2m\omega}(a^\dagger + a) \text{ and } p = i\sqrt{m\hbar\omega / 2}(a^\dagger a).$$

For the 3D harmonic oscillator each equation is the same with energies

$$E_x = (n_x + 1/2)\hbar\omega_x, E_y = (n_y + 1/2)\hbar\omega_y$$

and

$$E_z = (n_z + 1/2)\hbar\omega_z \text{ [101,102].} \tag{9.9}$$

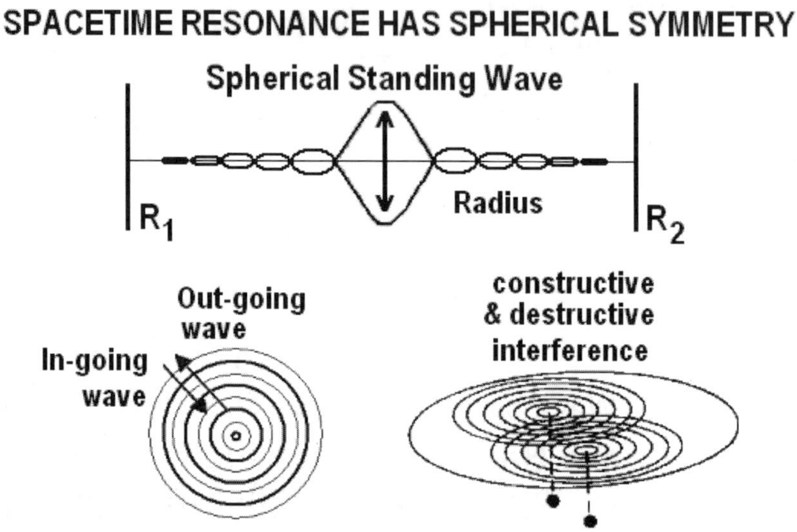

Figure 9.12 The Dirac polarized vacuum has hyperspherical symmetry. a) Metaphor for standing-wave present showing future-past elements, R_1, R_2, eleven of twelve dimensions suppressed for simplicity. b) Top view of a) a 2D spherical standing-wave. c) Manipulating the relative phase of oscillations creates nodes of destructive and constructive interference.

In Dubois' notation the classical 1D harmonic oscillator for Newton's second law in coordinates *t* and *x(t)* for a mass *m* in a potential $U(x) = 1/2(kx^2)$ takes the differential form

$$\frac{d^2x}{dt^2} + \omega^2 x = 0 \quad \text{where} \quad \omega = \sqrt{k/m} \tag{9.10}$$

which can be separated into the coupled equations [6-9]

$$\frac{dx(t)}{dt} - v(t) = 0 \quad \text{and} \quad \frac{dv(t)}{dt} + \omega^2 x = 0\,. \tag{9.11}$$

From incursive discretization, Dubois creates two solutions

$x(t+\Delta t)$, $v(t+\Delta t)$ providing a structural bifurcation of the system which together produce Hyperincursion. The effect of increasing the time interval discretizes the trajectory as in Fig. 9.13 below. This represents a background independent discretization of spacetime [82-85].

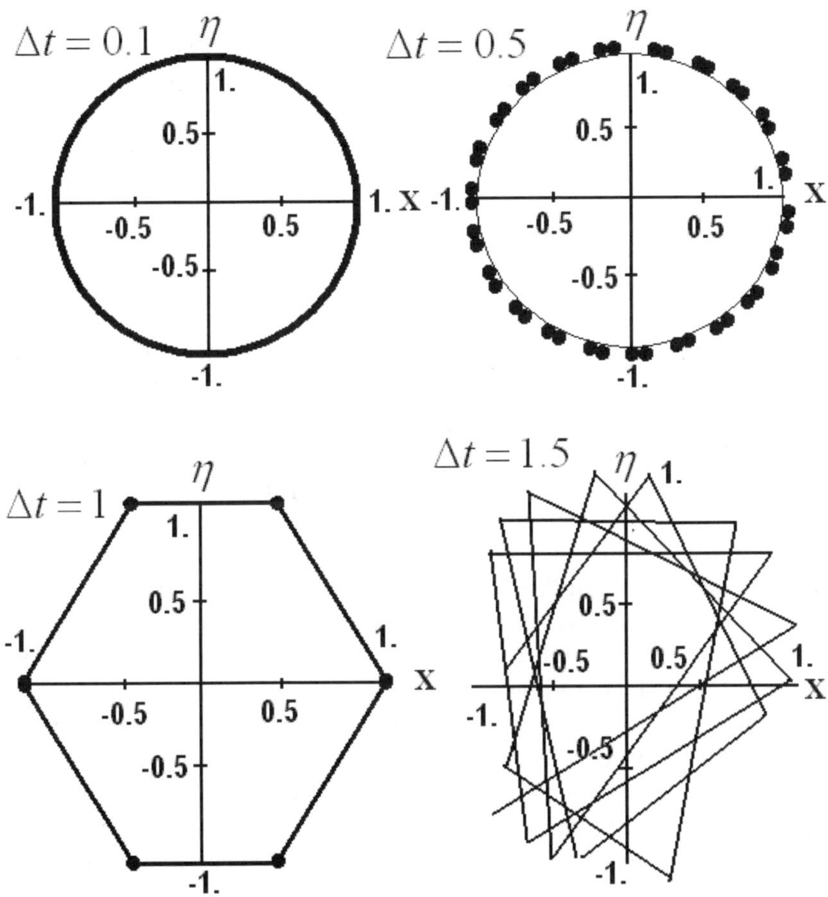

Figure 9.13 Numerical simulation of the phase space trajectory of the Dubois *superposed incursive oscillator* based on coordinates and velocities $x_n = 1/2[x_n(1)+x_n(2)]$ $v_n = 1/2[v_n(1)+v_n(2)]$ is shown in the figure for values of $\Delta\tau = \omega t$ equal to 0.1, 0.5, 1.0 and 1.5. Initial conditions are $\chi_0 = 1, \eta_0 = 0$ and $\tau_0 = 0$ with total simulation time $\tau = \omega t = 8\pi$. Figure adapted from [82-85].

9.6 Introduction to a P ≡ 1 Experimental Design

In a homogeneous magnetic field, the forces exerted on opposite ends of the dipole cancel each other out and the trajectory of the particle is unaffected. If the particles are classical "spinning" particles then the distribution of their spin angular momentum vectors is taken to be truly random and each particle would be deflected up or down by a different amount producing an even distribution on the screen of a detector. Instead, quantum mechanically, the particles passing through the device are deflected either up or down by a specific amount. This means that spin angular momentum is quantized (also called space quantization), i.e. it can only take on discrete values. There is not a continuous distribution of possible angular momenta. This is the usual fundamental basis of the standard quantum theory and where we must introduce a new experimental protocol to surmount it. This is the crux of our new methodology: If application of a homogeneous magnetic field produces quantum uncertainty upon measurement, then "do something else".

In NMR spectroscopy often it is easier to make a first order calculation for a resonant state and then vary the frequency until resonance is achieved. Among the variety of possible approaches that might work best for a specific quantum system, if we choose NMR for the Noetic Interferometer it is relatively straightforward to determine the spin-spin resonant couplings between the modulated electrons and the nucleons; but achieving a critical resonant coupling with the wave properties of matter and the spacetime backcloth is another matter. Firstly, for HAM cosmology $\hbar$ is not a rigid barrier as in Standard Model Big Bang-Copenhagen cosmology; $\hbar$ is a virtual limit of past-advanced elements of the continuous-state standing-wave present as it cyclically recedes into the past where the least unit [97] cavities tiling the spacetime backcloth can have radii $\leq$ the Larmour radius of the hydrogen atom [64,65,103,104]. This new Planck length oscillates through a limit cycle from the Larmour radius of the hydrogen atom to standard $\hbar$. This is like a wave-particle duality – Larmour radius at the future-retarded moment and $\hbar$ at the past-advanced moment. The dynamics are different for future-retarded elements which have been theorized to have the possibility of infinite radius for $D > 4$ [59]. This scenario is a postulate of string theory. Considering the domain walls of the least-unit structure, the $\hbar$-Larmour regime is considered internal-nonlocal and the Larmour-infinity regime considered external-supralocal

[61-63].

For simplicity we introduce our review of NMR concepts for the hydrogen atom, a single proton with magnetic moment μ, angular momentum J related by the vector $\mu = \gamma J$ where γ is the gyromagnetic ratio and $J = \hbar I$ where I is the nuclear spin. The magnetic energy $U = -\mu \cdot B$ of the nucleus in an external magnetic field in the z direction is $U = -\mu_z B_0 = -\gamma \hbar I_z B_0$ where values of I_z, m_I are quantized according to $m_I = I, I-1, I-2, I-3, \ldots -1$ [105,106].

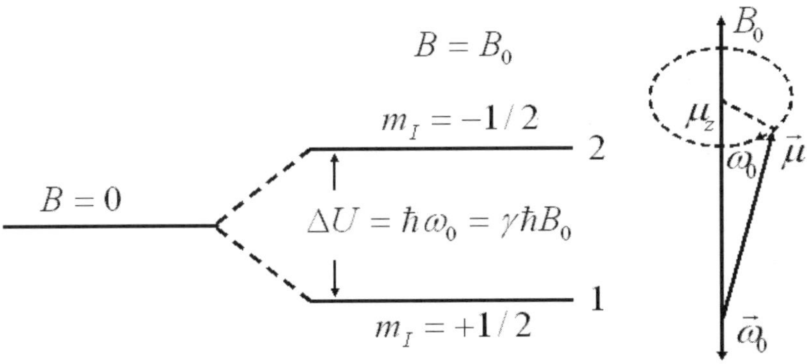

Figure 9.14 a) The two magnetic energy states for the spin, I = ½ single proton of a hydrogen atom in a magnetic field. b) Time variation of the magnetic moment of the proton in magnetic field B_0 with precession frequency $\omega_0 = \gamma B_0$, the fundamental resonant frequency from a).

For most nuclear species the z-component of the magnetization, M grows exponentially until reaching equilibrium according to $M_z(t) = M_0(1 - e\exp - t/T_1)$ where T_1 is the spin-lattice relaxation time. Of interest for the noetic interferometer is the fact that (Fig. 9.14a) as μ precesses cyclically from $m_I = -1/2$ to $m_I = +1/2$ the nucleons experience a torque, τ changing J by $\tau = dJ/dt$ or $\mu \times B = dJ/dt$. Under thermal equilibrium the x-y components are zero; but M_z can be rotated into the x-y plane creating transverse M_x and M_y components $dM/dt = \gamma M \times B$ for the entire system by applying a rotating circularly polarized oscillating magnetic field $2B_1 \cos\omega t\hat{i}$ of frequency ω in addition to the constant magnetic field $B_0\hat{k}$. Now the total time dependent field decomposes into the two counterpropagating fields

$$B_1(\cos\omega t\hat{i} + \sin\omega t\hat{j}) + B_1(\cos\omega t\hat{i} - \sin\omega t\hat{j}). \qquad (9.12)$$

This more complicated form for use with multiple applied fields is necessary, as described below, for use with the Sagnac Effect, quadrupole, and dipole dynamics [107,108] required to operate the noetic interferometer.

Nuclear Quadrupole Resonance (NQR) is a form of NMR in which quantized energy level transitions are induced by an oscillating RF magnetic field in the electric quadrupole moment of nuclear spin systems rather than the magnetic dipole moment. The nuclear quadrupole moment, Q is based on the nuclear charge distributions $\rho(r)$ departure from spherical symmetry defined as the average value of $1/2(3z^2 - r^2)\rho(r)$ over the nuclear volume. Q has the dimension of area where the nuclear angular momentum, for which $m_I = I$ where I is the nuclear spin quantum number and m_I is the quantum number for the z component of the spin $m_I = -1, +1, ..., I-1, I$. Nuclei with $I = 0$ have no magnetic moment and are therefore magnetically inert. Similarly in order for $Q = 0$ the nucleus must be spherical with spin $I \geq 0$. For spin I = 1/2 nuclei have dipole moments, μ but no Q. Q is positive for prolate nuclei and negative for oblate nuclei [109,110].

For an isolated nucleus in a constant magnetic field H_0 with nuclear spin number $I > 0$ the nucleus possess a magnetic moment. From Quantum Theory (QT) the length of the nuclear angular momentum vector is $[I(+\imath)]^{1/2}\hbar$ where measurable components are given by $m\hbar$ with m the magnetic quantum number taking any $(2I + \imath)$ value from the series $I, I-\imath, I-2, ..., -(I-\imath), -I$. For the I = 3/2 case there are four values along the direction of the applied magnetic field H_0.

Of the three types of spin-spin coupling, this experiment relies on the hyperfine interaction for electron-nucleus coupling, specifically the interaction of the nuclear electric quadrupole moment induced by an applied oscillating RF electric field to act on the nuclear magnetic dipole moment μ. When the electron and nuclear spins are strongly aligned along their *z*-components the Hamiltonian is $-m \cdot B$, and if *B* is in the *z* direction

$$H = -\gamma_N I \cdot B = -\gamma_N B I_x \tag{9.13}$$

with $m = \gamma_N I$, γ_N the magnetogyric ratio $\gamma_N = e\hbar / 2m_p$ and m_p the mass of the proton [111].

Radio frequency excitation of the nuclear magnetic moment, μ to resonance occurs for a nucleus collectively which rotates μ to some angle with respect to the applied field B_0. This produces a torque $\mu_i \times B_0$ causing the angular momentum, μ itself to precess around B_0 at the Larmour frequency $\omega_L = \gamma_N B_0$ [111-113]. This coherent precessing of μ can also induce a 'voltage' in surrounding media, an energy component of the Hamiltonian to be utilized (Fig. 9.14) to create interference in the structure of spacetime [57].

Metaphorically this is like dropping stones in a pool of water: One stone creates concentric ripples; two stones create domains of constructive and destructive interference. Such an event is not considered possible in the standard models of particle physics, quantum theory and cosmology. However Noetic science uses extended versions of these theories wherein a new teleological action principle is utilized to develop what might be called a 'transistor of the vacuum'. Just as standard transistors and copper wires provide the basis for almost all modern electronic devices; This L.O.V.E.R. using the information content of spacetime geodesics (null lines) will become the basis of many forms of Noetic Technologies. After a bit of thought we thought a little fun was warranted and came up with a name for the core of this noetic class of vacuum technologies: Laser Oscillated Vacuum Energy Resonator (L.O.V.E.R.). Wouldn't it be a kick if for the next 1,000 years noetic or anthropic technologies are 'full of love'?

Simplistically in this context, utilizing an array of modulated tunable lasers, atomic electrons are RF pulsed with a resonant frequency that couples them to the magnetic moment of the nucleons such that a cumulative interaction is created to dramatically enhance the Haisch-Rueda inertial back-reaction [114-117]. The laser beams are counter-propagating producing a Sagnac effect Interferometry to maximize the violation of Special Relativity. This is the 1st stage of a multi-tier experimental platform designed (according to the tenets of Noetic Field Theory) to 'open a hole' in the fabric of spacetime in order to isolate and utilize the force $\hat{F}_U$ of the Unitary Field.

The interferometer utilized as the basis for our vacuum engineering research platform has been dubbed the Laser Oscillated Vacuum Energy Resonator. It is a multi-tiered device. The top tier is comprised of counter-propagating Sagnac effect ring lasers that can be built into an IC array of 1,000+ ring lasers. If each microlaser in the array is designed to be counterpropagating, an interference phenomena called the Sagnac Effect occurs that violates special relativity in the small scale [118]. This array of RF modulated Sagnac-Effect ring lasers provides the top tier of the multi-tier L.O.V.E.R. Inside the ring of each laser is a cavity where quantum effects called Cavity Quantum Electrodynamics (C-QED) may occur. A specific molecule is placed inside each cavity. If the ring laser array is modulated with resonant frequency modes chosen to achieve spin-spin coupling with the molecules electrons and neutrons, by a process of Coherent Control [119] of Cumulative Interaction an inertial back-reaction is produced whereby the electrons also resonate with the spacetime backcloth in order to 'open an oscillating hole' in it. This requires a form of RQFT compatible with the 12D version of M-theory called F-Theory [41] relying on the symmetry conditions of HAM cosmology within which it is cast [61-63]. See Chap. 3.

The first step in the interference hierarchy (Fig. 9.15) is to establish an inertial back-reaction between the modulated electrons and their coupled resonance modes with the nucleons. The complete nature of inertia remains a mystery [120]. But if one follows the Sakarov [121] and Puthoff [122] conjecture, the force of gravity and inertia, the initial resistance to motion, are actions of the vacuum zero-point field. Therefore the parameter m in Newton's second law $f = ma$ is a function of the zero-point field [114-116,123-125]. Newton's third law states that 'every force has an equal and opposite reaction'. Haisch & Rueda [114-117] claim vacuum resistance arises from this reaction force, $f = -f$. We have also derived an electromagnetic interpretation of gravity and electromagnetism [126] that suggests this inertial back-reaction is like an electromotive force[2] of the de Broglie matter-wave field in the spin

[2] Electromotive force, E: The internal resistance r generated when a load is put upon an electric current I between a potential difference V, i.e. $r = (E - V)/I$.

exchange annihilation creation process inherent in a hysteresis of relativistic spacetime fabric (Fig. 9.7). In fact we go further to suggest that the energy responsible for Newton's third law is a result of the continuous-state flux of the ubiquitous noetic field [64]. For the L.O.V.E.R. we assume the Haisch-Rueda postulate is correct

$$f = \frac{d\rho}{dt} - \lim_{\Delta t \to 0} \frac{\Delta \rho}{\Delta t} \equiv \frac{d\rho_*}{dt_*} - \lim_{\Delta t_* \to 0} \frac{\Delta \rho_*}{\Delta t_*} = f_* \tag{9.14}$$

where $\Delta\rho$ is the impulse given by the accelerating agent and thus $\Delta\rho_*^{zp} = -\Delta\rho_*$ [114-117].

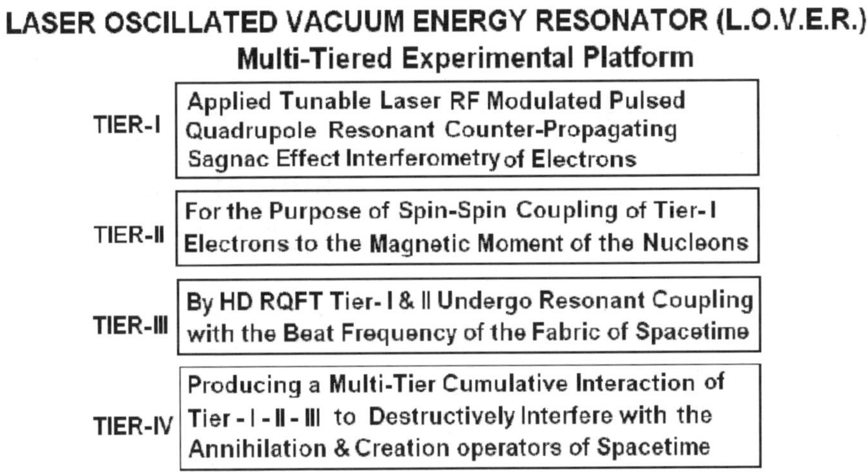

Figure 9.15 Design elements for the HD Cavity-QED trap of the Noetic Interferometer postulated to constructively-destructively interfere with the topology of the 12D spacetime manifold to manipulate the unitary field. Substantial putative effects are possible if cumulative interactions of the interference nodes of the cyclotron resonance hierarchy produce shock waves.

The cyclotron resonance hierarchy must also utilize the proper beat frequency of the continuous-state dimensional reduction spin-exchange compactification process inherent in the symmetry of noetic spacetime naturally 'tuned' to make the speed of light $c \equiv c$. With this apparatus in place noetic theory suggests that destructive-constructive C-QED interference of the spacetime fabric occurs such that the noeon eternity wave, ℵ of the unitary field, U_F is harmonically (like a holophote) released into the cavity of the detector array. Parameters of the Dubois incursive oscillator are also required for aligning the interferometer hierarchy with the beat frequency of spacetime.

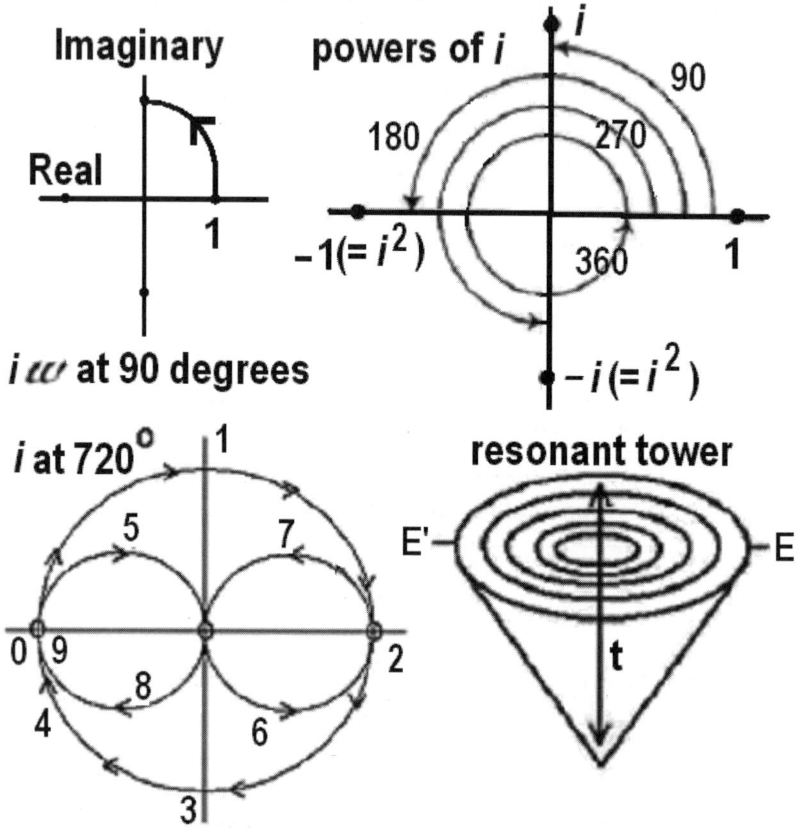

Figure 9.16 Powers of *i* in the complex plane. For 90° to 360° the concept can be readily illustrated in 2D; but for 720° and above 4D is required which cannot easily be depicted in 3D so the representation in 9c) is used, which might also be represented by a Klein bottle which was not used because the torus in 9c) more easily shows the rotation topology, which for spin 1/2 is the Dirac rotation of the electron. 9d) is a simplistic representation of a powers of *i* resonance hierarchy.

If the water wave conception for the 'Dirac sea' is correct, the continuous state compactification process contains a tower of spin states from spin 0 to spin 4. Spin 4 represents the unified field and makes cyclic correspondence with spin 0 where Ising lattice spin flips create dimensional jumps. Spin 0, 1/2, 1, and 2 remain in standard form. Spin three is suggested to relate to the orthogonal properties of atomic energy levels and space quantization. Therefore the spin tower hierarchy precesses through 0, 720°, 360°, 180°, 90° & 0 (∞) as powers of i as illustrated in Fig. 9.16.

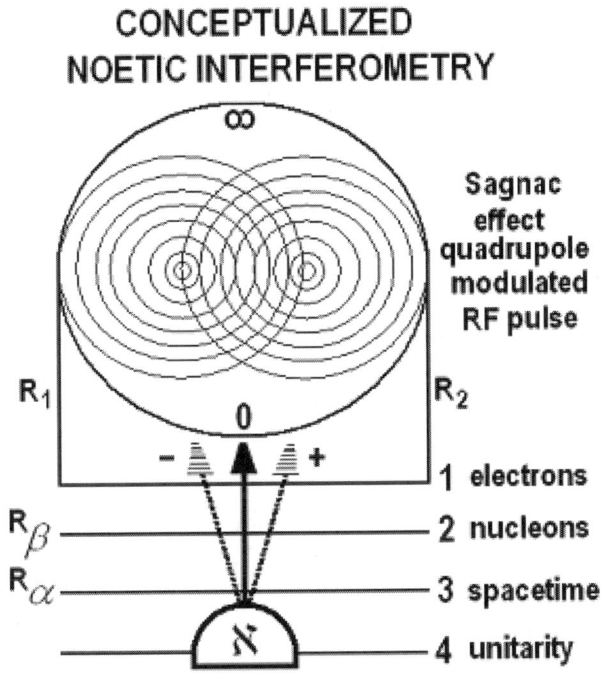

Figure 9.17 Conceptualized Ising model Riemann sphere cavity-QED multi-level Sagnac effect interferometer designed to 'penetrate' space-time to emit the 'eternity wave, א'. Experimental access to vacuum structure or for surmounting the uncertainty principle can be done by two similar methods. One is to utilize an atomic resonance hierarchy and the other a spacetime resonance hierarchy. The spheroid is a 2D representation of a HD Ising model Riemann sphere able to spin-flip from zero to infinity.

As illustrated in Fig. 9.10 the coherent control of the multi-level tier of cumulative interactions relies on full utilization of the continuous-state cycling inherent in parameters of HAM cosmology [61-63]. What putatively will allow noetic interferometry to operate is the harmonic coupling to periodic modes of Dirac spherical rotation in the symmetry of the HD geometry. The universe is no more classical than quantum as currently believed; reality rather is a continuous state cycling of nodes of classical to quantum to unitary, $C \rightarrow Q \rightarrow U$. Space does not permit detailed delineation of the parameters of HAM cosmology here; more detailed discussion can be found in [61-63]. The salient point is that cosmology, the topology of spacetime itself, has the same type of spinorial rotation and wave-particle duality Dirac postulated for the electron. Recall that the electron requires a 4D topology and 720° for one rotation instead of the usual 360° to complete a rotation in 3D. The hierarchy of noetic cosmology is cast in 12D such that the pertinent form of relativistic quantum field theory has significantly more degrees of freedom whereby the modes of resonant coupling may act on the structural-phenomenology of Dirac 'sea' itself rather than just the superficial zero-point field surface approaches to vacuum engineering common until now.

Hierarchical Harmonic Oscillator Parameters	
classical	$X = A\cos(\omega t)$
quantum	$\frac{\hbar^2}{2m}\frac{d^2\psi}{dx^2} + \left(E - \frac{kx^2}{2}\right)\psi = 0$
annihilation creation	$x(t) = x_0\left[a\exp(-i\omega t) + a^\dagger\exp(i\omega t)\right]$
future-past retarded-advanced	$F_1 = F_0 e^{-ikx} e^{-2\pi i f t}, \quad F_2 = F_0 e^{ikx} e^{-2\pi i f t}$ $F_3 = F_0 e^{-ikx} e^{2\pi i f t}, \quad F_4 = F_0 e^{ikx} e^{2\pi i f t}$
incursive	$\frac{dx(t+\Delta t)}{dt} - v(t) = 0, \frac{dv(t+\Delta t)}{dt} + \omega^2 = 0$

Figure 9.18 Basic mathematical components of the applied harmonic oscillator: classical, quantum, relativistic, transactional and incursive are required in order to achieve coherent control of the cumulative resonance coupling hierarchy in order to produce harmonic nodes of destructive and constructive interference in the spacetime backcloth.

The parameters of the noetic oscillator (Figs. 9.17,9.18) may best be implemented by RQFT using a form of de Broglie fusion. According to de Broglie a spin 1 photon can be considered a fusion of a pair of spin 1/2 corpuscles linked by an electrostatic force. Initially de Broglie thought this might be an electron-positron pair and later a neutrino and antineutrino. "A more complete theory of quanta of light must introduce polarization in such a way that to each atom of light should be linked an internal state of right and left polarization represented by an axial vector with the same direction as the propagation velocity" [127]. These prospects suggest a deeper relationship in the structure of spacetime of the Cramer type [52] (Fig. 9.19).

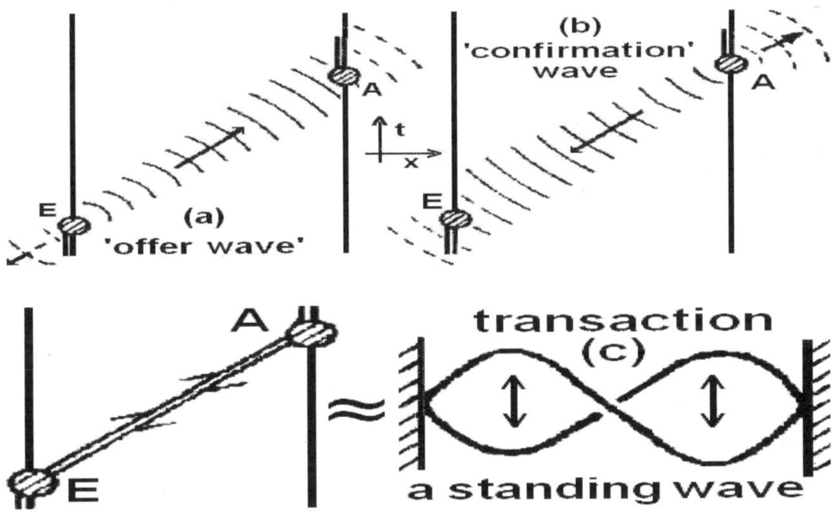

Figure 9.19 Transactional model. a) Offer-wave, b) confirmation-wave combined into the resultant transaction c) which takes the form of an HD future-past advanced-retarded standing or stationary wave. Figs. adapted from Cramer [52].

The epistemological implications of a 12D RQFT must be delineated. The empirical domain of the standard model relates to the 4D phenomenology of elementary particles. It is the intricate notion of what constitutes a particle that concerns us here – the objects emerging from the quantized fields defined on Minkowski spacetime. This domain for evaluating physical events is insufficient for our purposes. The problem is not only the additional degrees of freedom and the associated extra-dimensionality, or the fact that 'particles' can be annihilated and created but that in HAM cosmology they are continuously annihilated and recreated within the holograph as part of the annihilation and recreation of the fabric of spacetime itself. This property is inherent in the 12D Multiverse because temporality is a subspace of eternity [64,65]. This is compatible with the concept of a particle as a quantized field. What we are suggesting parallels the wave-particle duality in the propagation of an electromagnetic wave. We postulate this as a property of all matter and spacetime albeit as continuous-state standing waves.

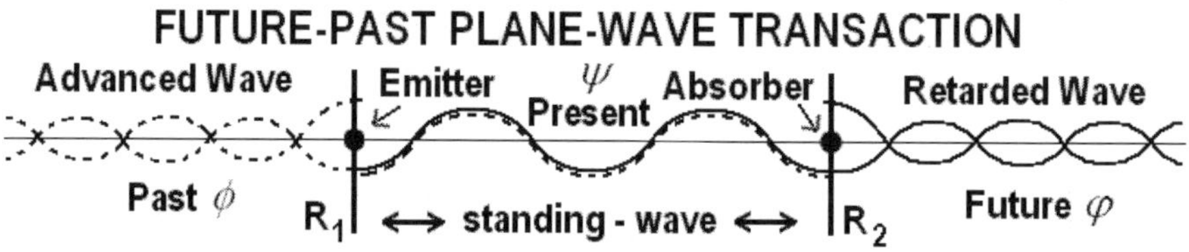

Figure 9.20 Structure of a transaction (present state or event) where the present is a standing-wave of future-past elements. The separation of these parameters in terms of de Broglie's fusion model is suggested to allow manipulation of the harmonic tier of the L.O.V.E.R. with respect to T-Duality or mirror symmetry.

For a basic description, following de Broglie's fusion concept, assume two sets of coordinates x_1, y_1, z_1 and x_2, y_2, z_2 which become

$$X = \frac{x_1 + x_2}{2}, \quad Y = \frac{y_1 + y_2}{2}, \quad Z = \frac{z_1 + z_2}{2}. \tag{9.15}$$

Then for identical particles of mass *m* without distinguishing coordinates, the Schrödinger equation (for the center of mass) is

$$-i\hbar \frac{\partial \psi}{\partial t} = \frac{1}{2M} \Delta \psi, \quad M = 2m \tag{9.16}$$

In terms of Fig. 9.20, Eq. 9.16 corresponds to the present and Eq. (9.17a) corresponds to the advanced wave and (9.17b) to the retarded wave [98].

$$-i\hbar \frac{\partial \phi}{\partial t} = \frac{1}{2M} \Delta \phi, \qquad -i\hbar \frac{\partial \varphi}{\partial t} = \frac{1}{2M} \Delta \varphi. \tag{9.17}$$

Extending Rauscher's concept for a complex eight space differential line element $dS^2 = \eta_{\mu\upsilon} dZ^{\mu} dZ^{*v}$, where the indices run 1 to 4, $\eta_{\mu v}$ is the complex eight-space metric, Z^{μ} the complex eight-space variable and where $Z^{\mu} = X^{\mu}_{\text{Re}} + iX^{\mu}_{\text{Im}}$ and Z^{*v} is the complex conjugate [128,129], to 12D continuous-state HAM spacetime; we write just the dimensions for simplicity and space constraints

$$x_{\mathrm{Re}}, y_{\mathrm{Re}}, z_{\mathrm{Re}}, t_{\mathrm{Re}}, \pm x_{\mathrm{Im}}, \pm y_{\mathrm{Im}}, \pm z_{\mathrm{Im}}, \pm t_{\mathrm{Im}} \tag{9.18}$$

where ± signifies Wheeler-Feynman/Cramer type future-past/retarded-advanced dimensions. This dimensionality provides an elementary framework for applying the hierarchical harmonic oscillator parameters suggested in Figs. 9.15 and 9.18.

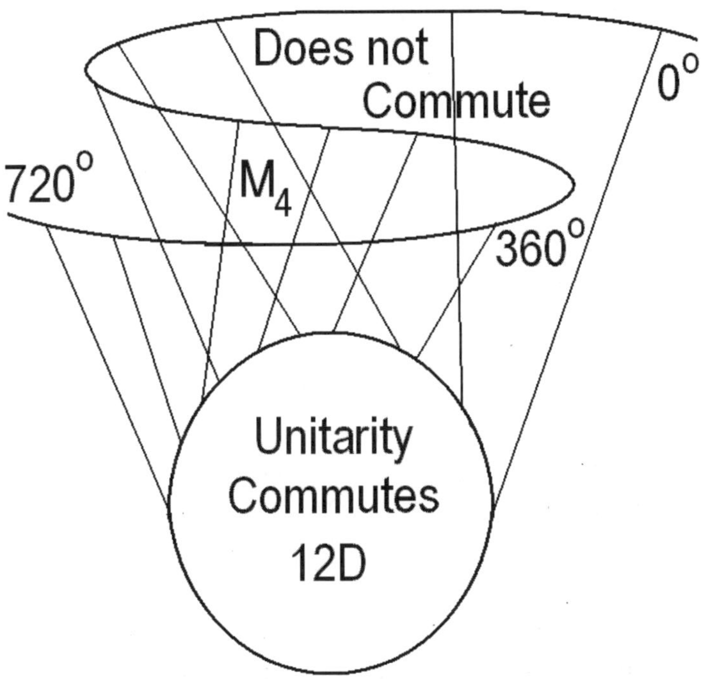

Figure 9.21 4D Minkowski space is like an HD quantum 'knot' tangled in a manner that the component phases do not commute. Conceptually this is like the observed retrograde motion of the planets. This is the same as A 3D view of a 4D Dirac rotation or 'pinch' of the 360 – 720° spinor rotation of the electron.

9.7 Conclusions

If the Noetic Interferometer is able to isolate and manipulate the eternity wave, ℵ it will become a primary research platform for developing a whole new class of vacuum based technologies; whereas one could say virtually all electronic devices up to now are based on transistors and copper wires. The L.O.V.E.R. could be called a transistor of the vacuum, where rather than copper wires, the geodesics or null lines of spacetime are utilized to transfer information topologically with no exchange particle mediating the 'interaction' which perhaps should be called a correlation or entanglement in this scenario to distinguish phenomenology versus ontology.

This brief introduction is only a primitive overview of introducing the anticipated new field of $\aleph$-wave (eternity–wave) vacuum engineering that as Cramer stated in the first sentence of this chapter will revolutionize many fields of science [130].

> When the great innovation appears, it will most certainly be in a muddled, incomplete form. To the discoverer himself it will be only half-understood; to everyone else it will be a mystery. For any speculation which does not at first glance look crazy, there is no hope [131].

Finally we stress that vacuum energy is not 'produced' by the noetic interferometer. The interferometer manipulates the boundary conditions 'insulating' or 'hiding' the unitary geodesics of spacetime by constructive and destructive interference allowing the holophotic release of unitary noeons by completing a cascading water-wheel like circuit already existing behind the usual spacetime domain walls of reality. Probably L.O.V.E.R. vacuum energy is emitted into the L.O.V.E.R. as a form of superradiance [132] of the hysteresis loop of least-unit parallel transport.

We have found already that a fair number of our colleagues want to dismiss this model right off because of its utilization of XD. This is the sort of myopic view that has consistently plagued the history of science whenever 'big-leap' innovation occurs. We hope readers here will not fall into this quagmire! The model is empirically testable which hopefully makes up for some of the lack of precision in our axiomatic approach or thin rigor in portions of our attempts at formalism. In addition to the protocols presented here we have described already an experiment to utilize the noetic $\aleph$-wave for the putative manipulation of prion protein conformation [57].

As we go to press we found recent work by Seth Lloyd called 'quantum afterlife' or 'quantum illumination' where residual effects of entanglement are purported to survive after decoherence [133-135]. We believe Lloyd's model could be used to provide a validation of the covariant scale-invariant properties inherent in our $P \equiv 1$ model because for us entanglement is hierarchical and multilevel such that entanglement still exists even when decoherence occurs at a local or other level.

References

[1] Cramer, J.G. (2000) The Alternate View: "Interaction-Free" Quantum Measurement and Imaging, Vol. CXX No. 6, pp. 78-81, Analog Science Fiction & Fact, Dell Magazines.
[2] Teuscher, C. (ed.) (2004) Alan Turing: Life and Legacy of a Great Thinker, quoted by A. Hodges, p. 54, New York: Springer.
[3] Thorne, K.S., Drever, R.W.P., Caves, C.M., Zimmermann, M. & Sandberg, V.D. (1978) Quantum nondemolition measurements of harmonic oscillators, Phys. Rev. Lett. 40;667 – 671.
[4] Gurvitz, S.A. (2003) Negative result measurements in mesoscopic systems Phys. Lett. A, 311:4-5;292-296.
[5] Braginsky, V.B., Vorontsov, Y.I. & Thorne, K.S. (1980) Quantum nondemolition measurements, Science, 209:4456, 547-557.
[6] Ralph, T.C., Bartlett, S.D., O'Brien, J.L., Pryde, G.J & Wiseman, H.M. (2004) Quantum non-demolition measurements on qubits, arXiv:quant-ph/0412149v1
[7] Elitzur, A. C. & Vaidman, L. (1993) Quantum mechanical interaction-free measurements. Found. Phys. 23; 987-997.
[8] Kwiat, P., Weinfurter, H., Herzog, T., Zeilinger, A. & Kasevich, M. (1995) Interaction-free quantum measurements. Phys. Rev. Lett. 74, 4763-4766.
[9] du Marchie Van Voorthuysen, E.H. (1996). Realization of an interaction-free measurement of the presence of an object in a light beam, Am. J. Phys. 64:12; 1504-1507; or arXiv:quant-ph/9803060 v2 26.
[10] Simon, S.H. & Platzman, P.M. (1999) Fundamental limit on "interaction free" measurements, arXiv:quant-ph/9905050v1.
[11] Vaidman, L. (1996) Interaction-free measurements, arXiv:quant-ph/9610033v1.
[12] Paraoanu, G.S. (2006) Interaction-free measurements with superconducting qubits, Phys. Rev. Lett. 97, 180406.
[13] Vaidman, L. (2001) The meaning of the interaction-free measurements arXiv:quant-ph/0103081v1.
[14] Vaidman, L. (2001) The paradoxes of the interaction-free

measurements, arXiv:quant-ph/0102049v1.
[15] Vaidman, L. (2000) Are interaction-free measurements interaction free?, arXiv:quant-ph/0006077v1.
[16] Helmer, F., Mariantoni, M., Solano, E. & Marquardt, F. (2008) Quantum, Zeno effect in the quantum non-demolition detection of itinerant photons, arXiv:0712.1908v2.
[17] Facchi, P. Lidar, D.A. & Pascazio, S. (2004) Unification of dynamical decoupling and the quantum Zeno effect, Phys. Rev. A 69, 032314; or arxiv:quant-ph/0303132.
[18] Sudarshan, E.C.G. & Misra, B. (1977) The Zeno's paradox in quantum theory, J. Math. Phys. 18:4; 756–763.
[19] Facchi, P. & Pascazio, S. (2002) Quantum Zeno subspaces, arXiv:quant-ph/0201115v2.
[20] Wocjan. P. (2004) Efficient decoupling schemes with bounded controls based on Eulerian orthogonal arrays, arXiv:quant-ph/0410107.
[21] Knill, E., Laflamme, R., Ashikhmin, A., Barnum, H., Viola, L. & Zurek, W.H. (1999) Introduction to quantum error correction, arXiv:quant-ph/0207170v1.
[22] Kwiat, P.G. White, A.G., Mitchell, J.R., Nairz, O., Weihs, G. Weinfurter, H. & Zeilinger, A. (1999) High-efficiency quantum interrogation measurements via the quantum Zeno effect, Phys. Rev. Lett. 83, 4725-4728; or arXiv:quant-ph/9909083v1.
[23] Carroll, S.M. (2006) Quantum Interrogation, reprinted in The Open Laboratory: The Best Writing on Science Blogs, B. Zivkovic (ed.) (Lulu: Morrisville, NC), p. 123.
[24] Jozsa, R. (1996) Counterfactual quantum computation, University of Plymouth preprint, unpublished.
[25] Mitchison, G. & Jozsa, R. (2000) Counterfactual computation, Proc. Roy. Soc. (Lond) A; or arXiv:quant-ph/9907007v2.
[26] Mitchison, G. & Massar, S. (2000) Absorption-free discrimination between semi-transparent objects, arXiv:quant-ph/0003140v2.
[27] Nakanishi, T., Yamane, K. & Kitano, M. (2002) Absorption-free optical control of spin systems: The quantum Zeno effect in optical pumping, arXiv:quant-ph/0103034v2.
[28] Kwiat, P.G., Weinfurter, H. & Zeilinger, A. (1996) Quantum Seeing in the Dark, P, Scientific American, Nov.
[29] Elitzur, A.C. & Dolev, S. (2000) Nonlocal effects of partial measurements & quantum erasure, arXiv:quant-ph/0012091v12000.
[30] Hiley, B.J. & Callaghan, R.E. (2006) What is erased in the quantum

erasure? Found. Phys., 36:12.
[31] White, A.G., Mitchell, J.R. Nairz, O. & Kwiat, P.G. (1998) Interaction-free imaging, Phys. Rev. A 58; 605.
[32] White, A.G., Mitchell, J.R., Nairz, O. & Kwiat, P.G. (1998) "Interaction-Free" Imaging, arXiv:quant-ph/9803060v2.
[33] Kotigua, R.P. & Toffoli, T. (1998) Potential for computing in micromagnetics via topological conservation laws, Physica D, 120:1-2, pp. 139-161.
[34] Gerlach, W & Stern, O. (1922) *Das magnetische moment des silberatoms, Zeitschrift für Physik* 9, 353-355.
[35] Young, T. (1804) Experiments and calculations relative to physical optics, Philosophical Trans. of the Royal Society of London 94, 1-16.
[36] Kwiat, P.G., Weinfurter, H., Herzog, T., Zeilinger, A. & Kasevich, M. (1995) Experimental realization of 'interaction-free' measurements, in D.M. Greenberger & A. Zelinger (eds.) Fundamental Problems in Quantum Theory, A Conference held in Honor of Professor John A. Wheeler, Annals of the New York Academy of Science, Vol. 755, p. 383, New York: New York Academy of Science.
[37] Jaekel, M.T. & S. Reynaud, S. (1990) Quantum limits in interferometric measurements, Europhys. Lett. 13, 301.
[38] Cramer, J.G. (2006) A transactional analysis of interaction-free measurements, Found. Phys. Lett. 19: 1; 63-73; or arXiv:quant-ph/0508102v23, 2008.
[39] Everett, H. (1957) Relative state formulation of quantum mechanics, Rev. Mod. Phys., Vol 29, pp 454-462.
[40] Bohm, D. & Vigier, J.-P. (1954) Model of the causal interpretation of quantum theory in terms of a fluid with irregular fluctuations, Phys. Rev. 96:1; 208-217.
[41] Kaku, M. (1999) Introduction to Superstrings and M-Theory, New York: Springer.
[42] Dirac, P.A.M. (1952) Is there an ether? Nature, 169: 172.
[43] Pettroni, N.C. & Vigier, J.-P. (1983) Dirac's aether in relativistic quantum mechanics, Found. Phys., 13:2, 253-285.
[44] Vigier, J-P (1980) De Broglie Waves on Dirac Aether: A Testable Experimental Assumption, *Lettere. al Nuovo Cimento*, 29; 467-475.
[45] Lehnert, B. (2002) New developments in electromagnetic field theory, in R.L. Amoroso, G. Hunter, M. Kafatos & J.-P. Vigier (eds.) Gravitation & Cosmology: From the Hubble Radius to the Planck Scale, Dordrecht: Kluwer Academic.

[46] Lehnert, B. (1998) Electromagnetic theory with space-charges in vacuo, in G. Hunter, S. Jeffers & J-P Vigier (eds.) Causality and Locality in Modern Physics, Dordrecht: Kluwer Academic.
[47] Proca, A, (1936) *Compt. Rend.*, 202, 1420.
[48] Kotigua, R.P. & Toffoli, T. (1998) Potential for computing in micromagnetics via topological conservation laws, Physica D, 120:1-2, pp. 139-161.
[49] Witten, E. (1993) Quantum background independence In string theory, arXiv:hep-th/9306122v1
[50] Sen, A. & Zwiebach, B. (1994) Local background independence of classical closed string field theory, Nucl. Phys. B414, 649.
[51] Sen, A. & Zwiebach, B. (1994) Quantum background independence of closed string field theory, Nucl. Phys. B423, 580.
[52] Cramer, J. (1986) The Transactional Interpretation of Quantum Mechanics, Rev. Mod. Phys 58, 647-687.
[53] Wheeler, J.A., & Feynman, R. (1945) Rev. Mod. Phys., 17, 157;
[54] Chu, S-Y (1993) Phys. Rev. Lrtt., 71, 2847.
[55] Amoroso, R.L. & Pribram, K.H. (2009) The Complementarity of Mind and Body: Realizing the Dream of Descartes, Einstein and Eccles, Cambridge: MIT University Press.
[56] Amoroso, R.L. (2000) The parameters of temporal correspondence in a continuous state conscious universe, in R. Buccheri & M. Saniga (eds.) Studies in the Structure of Time: From Physics to Psycho(patho)logy, Dordrecht Kluwer Academic.
[57] Chu, M-Y.J. & Amoroso, R.L. (2008) Empirical mediation of the primary mechanism initiating protein conformation in prion propagation, in D. Dubois (ed.) Partial Proceedings of CASYS07, IJCAS, Vol. 22, Univ. Liege Belgium.
[58] Holland, P.R. (1995) The Quantum Theory of Motion: An Account of the de Broglie-Bohm Causal Interpretation of Quantum Mechanics, Cambridge: Cambridge Univ. Press.
[59] Randall, L. (2005) Warped Passages, Unraveling the Mysteries of the Universe's Hidden Dimensions, New York: Harper-Collins.
[60] Vigier, J-P (2000) selected papers, in S. Jeffers, B. Lehnert, N. Abramson, & L. Chebotarev (eds.) Jean-Pierre Vigier and the Stochastic Interpretation of Quantum Mechanics, Montreal: Aperion.
[61] Amoroso, R.L. (2002) Developing the cosmology of a continuous state universe, in R.L. Amoroso, G. Hunter, M. Kafatos & J-P Vigier (eds.) Gravitation & Cosmology: From the Hubble Radius to the Planck

Scale, Dordrecht: Kluwer Academic.
[62] Amoroso, R.L. (2005) Paradigm for a continuous-state holographic conscious multiverse, in R.L. Amoroso & B. Lehnert (eds.) Extending the Standard Model: Searching for Unity in Physics, Oakl.: Noetic Press.
[63] Amoroso, R.L. (2009) Defining a context for the cosmology of awareness, in R.L. Amoroso & K.H. Pribram (eds.) The Complementarity of Mind and Body: Realizing the Dream of Descartes, Einstein and Eccles, Cambridge: MIT University Press.
[64] Amoroso, R.L. (2004) The fundamental limit and origin of complexity in biological systems: A new model for the origin of life, in D.M. Dubois (ed.) Computing Anticipatory Systems, AIP Conf. Proceedings Vol. 718, pp. 144-159, Melville: American Inst. of Physics.
[65] Amoroso, R.L. (2009) The physical origin of the principle of self-organization driving living systems, in R.L. Amoroso & K.H. Pribram (eds.) The Complementarity of Mind and Body: Realizing the Dream of Descartes, Einstein and Eccles, Cambridge: MIT Univ. Press.
[66] Kafatos, M. Roy, S. & Amoroso, R. (2000) Scaling in Cosmology & the Arrow of Time, in Buccheri, di Gesu & Saniga, (eds.) Studies on Time, Dordrecht: Kluwer Academic.
[67] Amoroso, R.L. (1996) The production of Fröhlich and Bose-Einstein coherent states in in vitro paracrystaline oligomers using phase control laser interferometry, Bioelectrochemistry & Bioenergetics, 41:1, pp.39-42.
[68] Fröhlich, H. (1968) Long-range coherence and energy storage in biological systems, Int. J. Quantum Chem. 2:641-649.
[69] Preparata, G. (1995) QED Coherence in Matter, Singapore: World Scientific.
[70] Driesch, H. (1914) C.K. Ogden (trans.) The History and Theory of Vitalism, London: Macmillan.
[71] Amoroso, R.L. (2009) Consciousness: The philosophical foundations of noetic field theory, in R.L. Amoroso & K.H. Pribram (eds.) The Complementarity of Mind and Body: Realizing the Dream of Descartes, Einstein and Eccles, Cambridge: MIT Univ. Press.
[72] Amoroso, R.L. & Martin, B. (1995) Modeling the Heisenberg matrix: Quantum coherence & thought at the holoscape matrix and deeper complementarity, in J. King & K. H. Pribram (eds.) Scale in Conscious Experience, Mahwah: Lawrence Earlbaum.
[73] Amoroso, R.L. (1997) Consciousness a radical definition: The hard problem made easy, Noetic J. 1:1, pp. 19-27.

[74] Amoroso, R. L. (1997) A brief introduction to noetic field theory: The quantization of mind, in L. Rakic, G. Kostopoulos, D. Rakovic, & D. Koruga (eds.) Brain and Consciousness, Belgrade: ECPD.
[55] Amoroso, R.L, (2000) Consciousness, a radical definition: Substance dualism solves the hard problem, in Amoroso, R.L., Antunes, R, Coelho, C., Farias, M., Leite, A., & Soares, P. (eds.) Science & the Primacy of Consciousness, Orinda: Noetic Press.
[76] Amoroso, R.L (1999) An introduction to noetic field theory: The quantization of mind, Noetic J. 2:1, pp. 28-37.
[77] Amoroso, R.L. & Martin, B.E. (2002) Consciousness: 'A thousand points of light', The emergence of self-organization from the noumenon of the conscious universe, Noetic J., 3:4, 289-311.
[78] Amoroso R.L. (2003) Awareness: physical cosmology of the fundamental least unit, Noetic J. 4:1, 1-15
[79] Amoroso, R.L. (2003) The Fundamental Limit and Origin of Biological Systems, Noetic J. 4:1; 24-32.
[80] Amoroso, R.L. (2003) The physical basis of qualia: Overcoming the 1st person 3rd person barrier, Noetic J. 4:3, pp. 212-230.
[81] Amoroso, R.L. (2002) The physical basis of consciousness: A fundamental formalism, Part 1 Noesis, XXVI, Bucharest: Rom. Acad.
[82] Dubois, D.M. (2001) Theory of incursive synchronization and application to the anticipation of delayed linear and nonlinear systems, in D.M. Dubois (ed.) Computing Anticipatory Systems: CASYS 2001, 5th Intl Conf., Am Inst of Physics: AIP Conf. Proceedings 627, pp. 182-195.
[83] Antippa, A.F. & Dubois, D.M. (2008) The synchronous hyperincursive discrete harmonic oscillator, in D. Dubois (ed.) proceedings of CASYS07, preprint.
[84] Dubois, D.M. (2008) The quantum potential and pulsating wave packet in the harmonic oscillator, in D. Dubois (ed.) proceedings of CASYS07, preprint.
[85] Antippa, A.F. & Dubois, D.M. (2004) Anticipation, orbital stability and energy conservation in discrete harmonic oscillators, in D.M. Dubois (ed.) Computing Anticipatory Systems, AIP Conf. Proceedings Vol. 718, pp.3-44, Melville: American Inst. of Physics.
[86] Icke, V. (1995) The Force of Symmetry, Cambridge Univ. Press.
[87] Mackay, A.L. & Pawley, G.S. (1963) Bravais Lattices in Four-dimensional Space, Acta. Cryst. 16: 11–19.
[88] Calzolari, A., Marzari, N., Souza, I. & Nardelli, M.B. (2003) arXiv:cond-mat/0311018.

[89] Schwinger, J. (1992) Casimir energy for dielectrics, Proc. Nat. Acad. Sci. USA 89, 4091-3.
[90] Schwinger, J. (1993) Casimir light: The source, Proc. Nat. Acad. Sci. USA 90, 2105-6.
[91] Schwinger, J. (1994) Casimir energy for dielectrics: spherical geometry, Proc. Nat. Acad. Math. Psych. 41:64-67, San Ffrancisco: W.H. Freeman.
[92] Witten, E. (1996) Reflections on the fate of spacetime, Phys. Today, (April) pp. 24-30.
[93] Amoroso, R.L. (2005) Application of double-cusp catastrophe theory to the physical evolution of qualia: Implications for paradigm shift in medicine & psychology, in G.E. Lasker & D.M. Dubois (eds) Anticipative & Predictive Models in Systems Science, Vol. 1, pp. 19-26, Windsor: The International Institute for Advanced Studies in Systems Research & Cybernetics.
[94] Gilmore, R. (1981) Catastrophe Theory for Scientists & Engineers, New York: Dover.
[95] Poston T. & Stewart, I (1978) Catastrophe Theory & Its Applications, New York: Dover; Gilmore, R. (1981) Catastrophe Theory for Scientists & Engineers, Dover.
[96] Qin, S. et al. (2001) Int. J of Solids & Structures, 38, pp. 8093-8109.
[97] Stevens, H.H. (1989) Size of a least unit, in M. Kafatos (ed.) Bell's Theorem, Quantum Theory and Conceptions of the Universe, Dordrecht: Kluwer Academic.
[98] Cardone, F. & Mignani, R. (2004) Energy and Geometry - An Introduction to Deformed Special Relativity, Singapore: World Sci.
[99] Cardone, F., Marrani, A. & Mignani, R. (2003) The electron mass from deformed special relativity, Electromagnetic Phenomena Vol.3, N.1; 9, special number dedicated to Dirac's centenary; (2008) hep-th/0505134.
[100] Cardone, F., Marrani, A. & Mignani, R. (2008) A new pseudo-Kaluza-Klein scheme for geometrical description of interactions, arXiv:hep-th/0505149v1.
[101] Bohm, D. (1963) Quantum Theory, p. 353, Englewood Cliffs: Prentice-Hall.
[102] Messiah, A. (1999) Quantum Mechanics, p. 438, 444, NY: Dover.
[103] Hatfield, B. (1992) Quantum Field Theory of Point Particles and Strings, Reading: Addison-Wesley.
[104] Grotz, K. & Klapdor, H.V. (1990) The Weak Interaction in

Nuclear, Particle and Astrophysics, New York: Adam Hilger.
[105] Slichter, C.P. (1990) Principles of Magnetic Resonance, 3rd edition, Springer Series in Solid-State Sciences 1, New York: Springer.
[106] Schumacher, R.T. (1970) Introduction to Magnetic Resonance, Menlo Park: Benjamin-Cummings.
[107] Farrar, T.C. & Becker, E.D. (1971) Pulsed and Fourier Transform NMR, New York: Academic Press.
[108] Abragam, A. (1961) Principles of Nuclear Magnetism, Oxford: Clarendon Press.
[109] Dehmelt, H.G. (1954) Nuclear quadrupole resonance, Am. J. Phys., 22:110.
[110] Semin, G.K, Babushkina, T.A. & Yakobson, G.G. (1975) Nuclear Quadrupole Resonance in Chemistry, New York: Wiley.
[111] Atkinson, P.W. (1994) Molecular Quantum Mechanics, 2nd edition, Oxford: Oxford University Press.
[112] Hausser, O. (1974) Coulomb reorientation, in J. Cerny (ed.) Nuclear Spectroscopy and Reactions, Part C, NY: Academic Press.
[113] Humieres, D., Beasley, M.R., Huberman, B.A. & Libchaber, A. (1982) Chaotic states and routes to chaos in the forced pendulum, Phys. Rev. A, 26:6, 3483-34.
[114] Rueda, A. & Haisch, B. (1998) Contributions to inertial mass by reaction of the vacuum to accelerated motion, Found. Phys, 28:7, 1057-1108.
[115] Rueda, A. & Haisch, B. (1998) Physics Lett. A, 240, 115.
[116] Rueda, A., Haisch, B.& Puthoff, H.E. (1994) Phys. Rev. A, 49, 678.
[117] Rueda, A. & Haisch, B. (2002) The inertia reaction force and its vacuum origin, in R.L. Amoroso, G. Hunter, M. Kafatos & J-P Vigier (eds.), Gravitation & Cosmology: From the Hubble Radius to the Planck Scale, pp. 447-458, Dordrecht: Kluwer Academic.
[118] Vigier, J-P (1997) New non-zero photon mass interpretation of the Sagnac effect as direct experimental justification of the Langevin paradox, Phys. Lett. A, 234:2, 75-85.
[119] Garcia-Ripoli, J.J., Zoller, P. & Cirac, J.I. (2005) Coherent control of trapped ions using off-resonant lasers, Phys. Rev. A 71, 062309; 1-13.
[120] Vigier-J-P (1995) Derivation of inertial forces from the Einstein-de Broglie-Bohm causal stochastic interpretation of quantum mechanics, Found. Phys. 25:10, 1461-1494.
[121] Sakharov, A.D. (1968) Sov. Phys. Dokl. 12, 1040.

[122] Puthoff, H.E. (2002) Polarizable vacuum approach to General Relativity, in R.L. Amoroso, G. Hunter, M. Kafatos & J-P Vigier (eds.), Gravitation & Cosmology: From the Hubble Radius to the Planck Scale, pp. 431-446, Dordrecht: Kluwer Academic.
[123] Burns, J.E. (1998) Entropy and vacuum radiation, Found. Phys. 28 (7), 1191-1207.
[124] Burns, J.E. (2002), Vacuum radiation, entropy and the arrow of time, in R.L. Amoroso, G. Hunter, S. Jeffers & M. Kafatos, (eds.) Gravitation & Cosmology: From the Hubble Radius to the Planck Scale, Dordrecht: Kluwer Academic.
[125] Zeh, H.D. (1989) The Physical Basis of the Direction of Time, New Y. :Springer-Verlag.
[126] Vigier, J-P & Amoroso, R.L. (2002) Can one unify gravity and electromagnetic fields? in R.L. Amoroso, G. Hunter, S. Jeffers & M. Kafatos, (eds.), Gravitation & Cosmology: From the Hubble Radius to the Planck Scale, Dordrecht: Kluwer Academic.
[127] Borne, T., Lochak, G. & Stumpf, H. (2001) Nonperturbative Quantum Field Theory and the Structure of Matter, Dordrecht: Kluwer.
[128] Rauscher, E.A. (1983) Electromagnetic Phenomena in Complex Geometries and Nonlinear Phenomena, Non–Hertzian Waves and Magnetic Monopoles, Millbrae: Tesla Books; (2009) 2nd edition, Oakland: The Noetic Press, and references therein.
[129] Rauscher, E. (2002) Non-Abelian gauge groups for real & complex Maxwell's equations, in R.L. Amoroso, G. Hunter, S. Jeffers & M. Kafatos, (eds.), Gravitation & Cosmology: From the Hubble Radius to the Planck Scale, Dordrecht: Kluwer.
[130] Amoroso, R.L. (2009) How to Build a Conscious Quantum Computer: The Immanent Era of Conscious Technologies, book proposal submitted, Cambridge: MIT Univ. Press.
[131] Dyson, F.J. (1958) Innovation in Physics, Scientific American, 199, No. 3, September.
[132] Eberly, J.H. (1972) Superradiance revisited, AJP, 40; 1374-1383.
[133] Llyod, S. et al. (2008) Quantum illumination, Phys. Rev. Lett. Dec. 19, 101: 25, 253601; arXiv-quant-ph: 0803.2022v2.
[134] Llyod, S. (2008) Enhanced sensitivity of photo detection via quantum illumination, Science, Sept. 12., pp. 1463-1465.
[135] Cho, A. (2008) Quantum flashlight pierces the darkness, Science, Sept. 12., pp. 1433-1443.

Chapter 10

On the Possibility of Relativistic Shock-Wave Effects in Observations of Quasar Luminosity

A growing number of conflicts within the Standard Model call into question the fundamental interpretation of the Doppler component of the putative Hubble Expansion Law and the nature of events in spacetime associated with conventional coordinates of the line element as attached to the physical basis of the observer. Also of paramount importance is that Einstein's geometrodynamics is not a complete theory of gravity. We postulate that nonlinear effects associated with the propagation of light in an intense gravitational field produces shock waves creating 'light-booms' along boundary conditions at cosmological distances approaching the limit of observation that if correct would explain Quasi-Stellar Object (QSO) luminosity. These gravitational shock waves are considered observationally manifest in the spectrum of QSOs and Supernova as a continuous front of 'light booms' produced by superluminal boosts associated with continuous coordinate transformations relative to a distant observer. This model aligns with the view suggesting that QSOs are most likely a form of Seifert spiral galaxy with Active Galactic Nuclei (AGN) in the vicinity of the putative observational limit of the Hubble radius, H_R.

10.1 The Quasar Redshift-Distance Interpretation Controversy

As optical and radio telescopes continue to improve a vast amount of data continues to be accumulated on the large-scale structure of the universe. The popular view has been to interpret the data to support a hot Big Bang cosmological model; but as attested to in this chapter QSO's provide strong observational evidence that the Big Bang assumption is incorrect. From the early 1960's when the redshift of QSOs, and galaxies were compared with radio sources it became apparent that the redshift plot of QSOs contrasted with apparent brightness did not follow the usual Hubble correlation [1]. These redshift observations beginning with QSO 3C 273 in 1963 to more than 100 QSOs in 1963 still continued to show the same redshift apparent magnitude disparity when the number of sources was increased beyond 7,000 QSOs in the mid 1990's [2]. Most astrophysicists were not willing to accept that these redshift observations were not a measure of distance. Large redshift QSOs are not faint and typically have bolometric luminosities of ~ 100 times that of normal galaxies [1]. Woltjer [3] and Rees [4] found a way to interpret the QSO redshift as being wholly cosmological phenomena by considering the radiating surfaces as having relativistic motion [1].

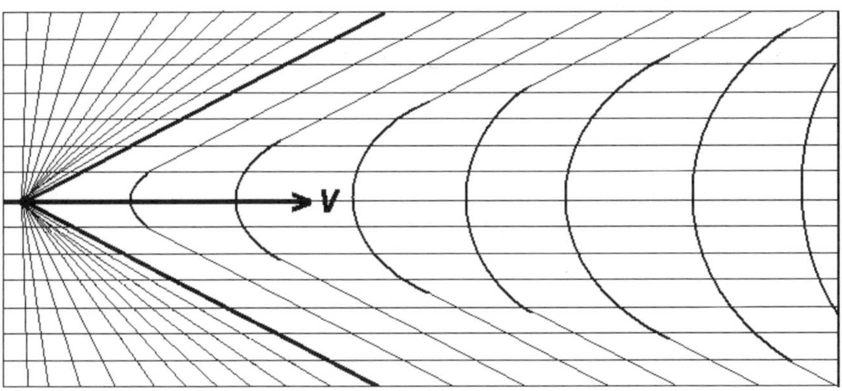

Figure 10.1 Wave front of a Doppler redshift for $Z \cong .85\ c$.

Around the same time Arp comprised a catalog of unusual galaxies [5]. He noticed a physical association between radio sources, QSOs and some of the peculiar galaxies. But the observed redshift of the central galaxies was small and the redshift of the associated QSO very large suggesting that the QSO redshift could not be of cosmological origin. Arp had clearly shown with a high level of statistical accuracy that 'there was a clear association between radio QSOs with large redshifts and galaxies with very small redshifts' [1]. The linear separation between galaxy and QSO was generally the same demonstrating a clear association between the galaxy and the QSO [6-8].

> His work was greeted with astonishment and disbelief...and heavily criticized, often very unfairly. In response he began an extensive observational program...The community has remained skeptical of these results...one argument made against the reality of these associations by a leading observer was that if these results were correct, we had no explanation of the nature of the redshift! In other words, if no known theory is able to explain the observations, it is the observations that must be in error! [1]

Arp's colleagues at Mt. Wilson and Palomar were so troubled by his results that they petitioned the observatories directors to take away all of Arp's observing time. Arp protested when the recommendation was implemented and after his appeals to the trustees were turned down he retired and relocated to the Max Planck Institute in Munich [1,9].

10.2 QSOs an Issue of the Fundamental Basis of Geometrodynamics

Newton's formulation of the gravitational force law requires each particle to respond instantaneously to every other massive particle regardless of the distance between them which he proved; but the proof is only valid in Euclidean space. Today this would be described by the Poisson equation,

$$\left(\frac{\partial^2}{\partial x^2}+\frac{\partial^2}{\partial y^2}+\frac{\partial^2}{\partial z^2}\right)\varphi(x,y,z)=f(x,y,z) \tag{10.1}$$

according to which, when the mass distribution of a system changes, its gravitational field instantaneously adjusts. Therefore the theory requires the speed of gravity to be infinite. Einstein's Geometrodynamics

$$G_{\mu\nu}+\Lambda g_{\mu\nu}=\frac{8\pi G}{c^4}T_{\mu\nu} \tag{10.2}$$

is a classical extension of Newtonian gravitation and therefore an incomplete theory. Physical theory incorporates an upper limit on the propagation speed of an interaction, maintaining that instantaneous action-at-a-distance is impossible. However quantum entanglement between separated particles enables instantaneous correlations which led to the puzzle as to whether causality or locality must be abandoned.

The recent measurement controversy of the speed in which Gravity propagates has only addressed this semi-classical component [10-28]. The search for a Quantum Gravity (QG) is misplaced by the incorrect assumption that gravitation is quantized [29]; this is not so. The interaction between gravity and Quantum Theory (QT) occurs at the level of unitarity, not within an intermediate G-QT regime. This is because of the nature of the graviton, a quadrupole photon-graviton complex not the usual spin 2 Bose quantum-graviton with any associated properties. See Chap. 5. It is instead a condition of brane topology (according to the theory presented in this volume). We hope this chapter creates some insight into solving the conceptual; basis of this puzzle. Still in either case gravity has properties beyond the local, $v_g \simeq c$ velocity propagation. An additional instantaneous correlation as in the EPR experiment, but in unitarity where this action occurs is needed to describe gravitation. I suppose EPR in that sense provides good indicia of the incompleteness of QT. This duality of the laws of gravity is indicated in a variety of astrophysical effects such as the Titius-Bode series or QSO luminosity as addressed here. Before going into that, the aim of this chapter, a discussion of the Titus-Bode relation suggests an associated relationship to the missing components of gravitation.

Astrophysicist Silk stated, "…highly redshifted sources, most notably the radio galaxies and the quasars, reveal strong evolutionary effects. Equal volumes of space contain progressively more quasars and powerful radio galaxies at greater distances. Only by disputing the interpretation of quasar redshifts as a cosmological distance indicator can this conclusion be avoided" [30].

Taking an axiomatic approach we begin with a number of postulates:

- That the Hubble redshift, H_0 is non-Doppler (no cosmological expansion or inflation).
- Redshift is caused instead by photon mass, $m_\gamma \neq 0$ anisotropy, for $m_\gamma = h\nu / c^2$ with internal motion coupling periodically to the Dirac covariant polarized vacuum [31].
- Quasars (QSOs) are most likely a form of Seifert spiral galaxy with Active Galactic Nuclei (AGN) near the limit of observation at the Hubble Radius, H_R [32].
- The spectra of QSOs, the most luminous objects in the universe, can be explained in terms of Gravitational Shock Waves (GSW).
- Spacetime is asymptotically flat [33].

- The Cosmological Principle (CP) holds within reasonable limits.
- Expansion/Inflation of the universe is an observational illusion of misinterpreting of the Hubble redshift as a Doppler effect [34].
- This illusion arises from the continuous-state dimensional reduction properties of the present instant as a virtual subspace of an HD atemporal domain [35].

10.3 Recent Refinements of the Titius-Bode Series as an Indicator of a Possible New Gravitational Dynamic

The Titius-Bode law for planetary orbitals is in an exponential function of planetary sequence out from the sun. The law relates the semi-major axis, a of each planet in units so that the Earth's semi-major axis = 10, with $a = n + 4$ where n = 0, 3, 6, 12, 24, 48... with each value of $n > 3$ twice the previous value. The resulting values can be divided by 10 converting them to astronomical units (AU). The hypothesis was discredited as a predictor of orbits after the 1846 discovery of Neptune and the discovery of Pluto in 1930. When originally published it generally satisfied by all the known planets Mercury through Saturn. Two solar planets have a number of large moons that could have been created by a process similar to that which created the planets themselves. The four large satellites of Jupiter plus the largest inner satellite Amalthea adhere to a regular, but non-Bode, spacing with the four innermost moons in orbital periods that are each twice that of the next inner satellite. The large moons of Uranus have a regular, but non-Bode, spacing.

Results from simulations of planetary formation support the idea that a randomly chosen stable planetary system will likely satisfy a Titius–Bode law. Dubrulle and Graner [36] have shown that power-law distance rules can be a consequence of collapsing-cloud models of planetary systems possessing two symmetries: rotational invariance (the cloud and its contents are axially symmetric) and scale invariance (the cloud and its contents look the same on all length scales), the latter being a feature of many phenomena considered to play a role in planetary formation, such as turbulence. To test if a similar rule applies to extrasolar planetary systems so far only 55 Cancri, a binary star approximately 41 light-years away in the constellation Cancer, has sufficient planets to make predictions. An undiscovered planet / asteroid belt is predicted at ~2 AU.

Recent new calculations have shown that the Titius-Bode Law can be accurately demonstrated by the Euler-LaGrange equation for the free energy variations of the plasma initially forming the sun and solar system [37-39]. Using a 1st order Bessel function scaled to the geometry of the solar system, Wells has shown that the Titius-Bode numbers correspond to extrema of the roots and make exact predictions for the outer planets where the Titius-Bode series originally failed [40]. These new insights stem from the seminal work of Chandrasekhar [41] on the equilibrium properties of the boundary conditions of a volume of plasma.

TITIUS-BODE LAW - SOL				
Planet	**k**	**Mass(M_E)**	**Bode Distance**	**Actual Distance- AU**
Mercury	0	0.05527	0.4	0.39
Venus	1	0.81500	0.7	0.72
Earth	2	1.0000	1.0	1.00
Mars	4	0.10745	1.6	1.52
(Ceres)	8	0.00016	2.8	2.77
Jupiter	16	317.83	5.2	5.20
Saturn	32	95.159	10.0	9.54
Uranus	64	14.500	19.6	19.2
Neptune	128	17.204	38.8	30.06

Table 10.1 Titius-Bode Law for planets orbiting Sol [42]

TITIUS-BODE LAW - 55 CANCRI				
Planet	**K**	**Mass (M_J)**	**Bode Distance**	**Actual Distance-AU**
55 CANCRI -e	0	>0.034	0.039	0.038
55 CANCRI -b	1	>0.824	0.104	0:115
55 CANCRI -c	2	>0.169	0.283	0:240
55 CANCRI -f	4	>0.144	0.768	0:781
55 CANCRI -5	8	-	2.08	(not discovered)
55 CANCRI -d	16	>3.835	5.643	5:77
55 CANCRI -7	32	-	15.3	(not discovered)

Table 10.2 Titius-Bode Law for exoplanets orbiting 55 Cancri [43,44].

This "hints at other phenomena associated with the morphology of the system" [37]; we postulate this might reveal a feedback mechanism between the two modes of operation for gravity that could be responsible for destruction of a planet that should have formed in the asteroid belt. It is sometimes suggested that the gravitational force from Jupiter disrupted the planets formation. Our idea is that this feedback mechanism might arise from a harmonic oscillation between the effects of classical gravitation operating at the speed of light, c and the operation of the as yet undiscovered effect of quantum/unitary gravitation operating instantaneously. This effect if true provides indicia for our model of gravitational shock waves which also have oscillatory parameters.

10.4 Critique of Hubble's Law as Applied to Doppler Expansion

Redshift refers generally to motion of a source relative to an observer; with blueshift for motion toward the observer, $Z < 0$ and redshift for velocity away from the observer, $Z > 0$ for an object not in the line of sight the relativistic form of the Doppler effect is

$$1+Z=\frac{1+v\cos(\theta)/c}{\sqrt{1-v^2/c^2}}\,. \tag{10.3}$$

When the motion of the source is in the line of sight, $\theta = 0$ the equation reduces to the general formula

$$1+Z=\sqrt{\frac{1+v/c}{1-v/c}} \tag{10.4}$$

where one can tabulate Z:

V	Z
.5c	.73
~.6c	1
.75c	1.64
.8c	2.00
.85c	2.51
.95c	5.24
.96c	~6
.99c	13.11

Table 10.3 Tabulation of Z compared to velocity approaching c.

The largest Z currently known is for the most distant QSO CFHQS J2329-0301 with Z $\simeq$ 6.43 [45]. A QSO with Z > 10 has been observed but is still unconfirmed. Hubble's redshift law is considered quite variable; and interpretation depends on a number of factors like the specific cosmological model utilized or if Λ is 0, + or -. The best indirect evidence supporting our thesis is that QSO's are the most luminous objects in the known universe and that an object, especially one as massive as a QSO is supposed to be, receding at ~c would indicate ~ infinite mass.

10.5 The Observer and the Cosmological Principle

In summarizing the *Cosmological Principle* (that the universe is homogeneous and isotropic on average in the large-scale) [46] events are idealized instants in spacetime defined by arbitrary time and position coordinates *t, x, y, z*, written collectively as x^i where *i* runs from 0 to 3. The standard line element is

$$ds^2 = \sum_{ij} g_{ij} dx^i dx^j = g_{ij} dx^i dx^j, \tag{10.5}$$

where the metric tensor

$$g_{ij}(x) = g_{ji}(x) \tag{10.6}$$

is symmetric [46]. In local Minkowski form all first derivatives of g_{ij} vanish at the event and equation (10.5) takes the form

$$ds^2 = dt^2 - dx^2 - dy^2 - dz^2. \tag{10.7}$$

The Cosmological Principle generally suggests that the clocks of all observers are synchronized throughout all space because of the inherent homogeneity and isotropy. Because of this synchronization of clocks for the same world time *t*, for co-moving observers the line element in (10.7) becomes

$$ds^2 = dt^2 + g_{\alpha\beta} dx^\alpha dx^\beta = dt^2 - dl^2, \tag{10.8}$$

where dl^2 represents spatial separation of events at the same world time, *t*. This spatial component of the event dl^2 can be represented as an Einstein 3-sphere (compatible with the dual 6D Calabi-Yau 3-toris)

$$dl^2 = dx^2 + dy^2 + dz^2 + dw^2 \tag{10.9}$$

which is represented by the set of points (x, y, z, w) at a fixed distance R from the origin:

$$R^2 = x^2 + y^2 + z^2 + w^2 \tag{10.10}$$

where

$$w^2 = R^2 - r^2 \text{ and } r^2 = x^2 + y^2 + z^2 \tag{10.11}$$

so finally we may write the line element of the Einstein 3-sphere from Eq. (10.9) as

$$dl^2 = dx^2 + dy^2 + dz^2 + \frac{r^2 dr^2}{R^2 - r^2}. \text{ [46]} \tag{10.12}$$

By imbedding an Einstein 3-sphere in a flat HD space, specifically as a subspace of a new complex 12D superspace, [34,47,48] new theoretical interpretations of standard cosmological principles are feasible. This is the line element we feel is most compatible with the oscillatory spacetime boundary parameters required by our model of gravitational shock waves in QSO luminosity.

10.6 Some Fundamental Insights on Shock Waves

In general a shock wave is defined as an abrupt, discontinuous, nonlinear change in the characteristics of a medium that travels at a velocity higher than an ordinary wave often through a vortex fanning out from the source of the shock. Shock energy dissipates in a short distance and the accompanying expansion wave merges with the shock wave, partially canceling it. So our putative gravitational light-boom results from the degradation and merging of the shock wave and the expansion wave produced by the oscillating spacetime boundary conditions. To get a shock wave something has to be traveling faster than the local wave speed. In this regard some segments of the light around the vortex fan are traveling at the normal speed of light, so that the waves leaving the QSO pile up on each other and a shock wave, the pressure increases, and spreads out sideways. Because of this 'constructive interference' effect, shocks are intense like an explosion.

Shock/vortex interactions and superluminal vortex breakdown occur when a superluminal vortex stream encounters a shock wave, the discontinuous pressure rise of the shock wave can be sufficient to burst the vortex with an oscillation of light booms depending upon the structure of the vortex [49-55]. The structure of shock/vortex interactions has been investigated in a series of Soviet studies using various flow visualization methods [56-58]. These studies show that shock/vortex interactions result in highly unsteady flow patterns in which the shock wave bulges forward in the upstream direction showing a decrease to minimum value on the vortex axis. The cause of this action - the Ranque-Hilsch effect is currently unknown. But Crocco's theorem [59] (Eq. 10.13) suggests a steady flow gradient's total enthalpy relates to entropy gradients and vorticity, both of which are present in a vortex core. For an over expanded nozzle flow a strong interaction is distinguished from a weak interaction by the formation of a secondary recompression shock downstream of the bubble shock suggesting that the strong interaction corresponds to supersonic vortex breakdown. Finally Delery et al. show that the strength of a shock required to burst a supersonic stream-wise vortex is inversely related to the vortex strength [51].

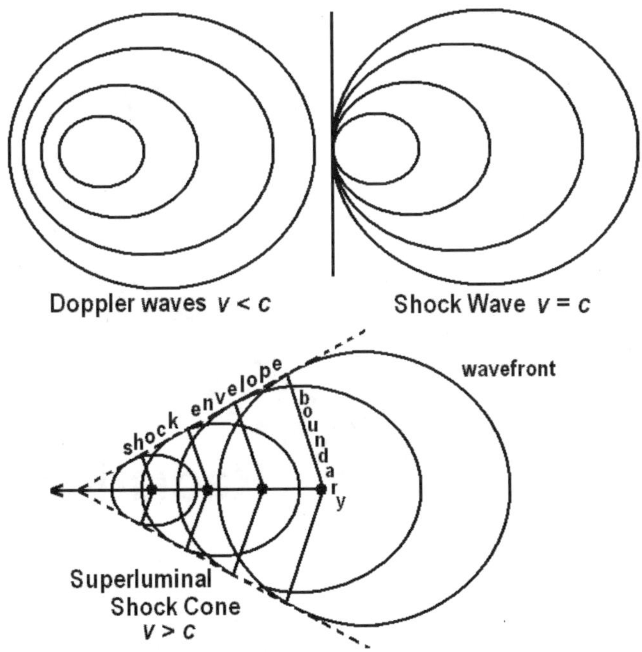

Figure 10.2 Simplistically considering a shock as originating from a point source in two dimensions the disturbance forms circular wavefronts centered at successive positions of the QSO's harmonic gravitational image source as illustrated at the bottom of the Figure. The wavefronts overlap and form the shock envelope. In 2D the shock envelope is a wedge, and in 3D it forms a cone.

In aircraft nomenclature the semi vertical angle of the cone is the Mach angle μM = arc sin (1/M), where M is the ratio of source speed to sound speed and is called the Mach number. All sound is contained in the shock envelope where for the first approximation the envelope is the location of the sonic boom [60]. Analogous phenomena exist in disciplines besides fluid mechanics. In nuclear physics particles accelerated beyond the speed of light in a refractive medium create a visible phenomenon known as Čerenkov radiation emitted when a charged particle such as an electron passes through an insulator at a speed greater than the speed of light in that medium. The characteristic 'blue glow' of nuclear reactors is Čerenkov radiation [61].

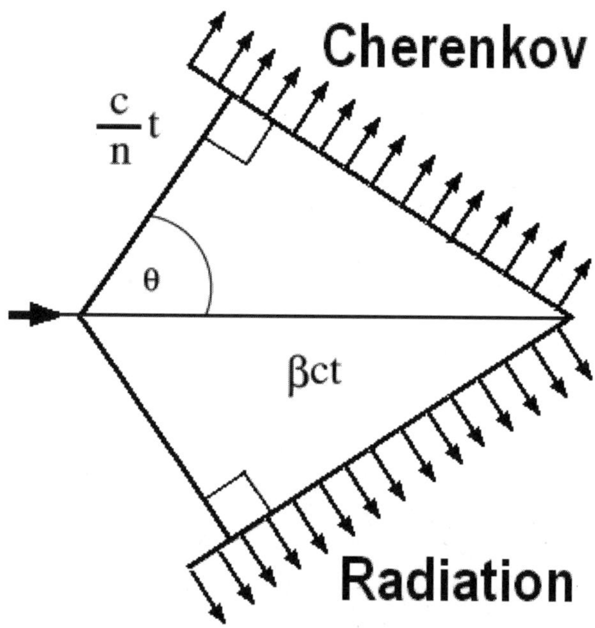

Figure 10.3 Geometry of Cherenkov radiation. In the figure, *v* is the velocity of the particle (arrow), β is *v/c* and *n* is the refractive index of the medium. The arrows are the direction of Cherenkov radiation so that $\cos\theta = 1/n\beta$.

In line with our postulate that all shock phenomena have similar characteristics Lyman & Morgenstern [52] have garnered three geometric insights into aircraft shock suppression that could also shed light on spacetime characteristics of gravitational shock waves: 1) A relation between lift force and airfoil area, 2) A volumetric shock cancellation phenomena that could give indicia to our postulate of constructive and destructive interference in the gravitational wave light cone, and 3) A directionality control by non-planar shaping that reduces centerline off-track signals.

10.7 New Cosmological Gravitational Shock Parameters

The nature of the universe has remained an open question. Kant attempted to solve the debate between Newton and Leibniz concerning whether the universe was open or closed by suggesting the antinomy [62] that the universe is both open and closed, i.e. closed and finite in the semi-classical limit within the observed temporal boundaries of the observed Hubble radius, H_R; and open and infinite into an HD atemporal holographic multiverse domain beyond, H_R. Our model is cast in such a Multiverse with a potential for an infinite number of nested Hubble Spheres in causal separation each with their own fine-tuned laws of physics. This is pertinent here in passing because the cosmology in balancing the cosmological constant, Λ gives a backcloth that predicts asymptotically flat spacetime and an interpretation for dark energy as arising from the rest of the multiverse.

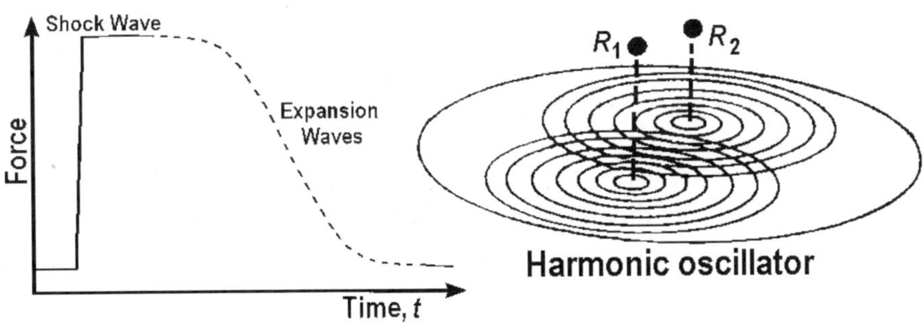

Figure 10.4 a) General shape of a simple shock wave. b) If shocks oscillate harmonically they may undergo constructive and destructive interference.

The Einstein gravitational potential oscillates the tidal gravitational field associated with the curvature of spacetime and predicts gravitational waves that propagate with a velocity of $v \simeq c$. We postulate a new cosmological principle related to the action of gravitational wave shock fronts [63]. This action arises from a duality in the nature of gravity, whereas classical general relativity propagates according to $v \simeq c$ the eventual discovery of the completed form of quantum/unitary gravitation will show an additional quantum component with similarities to the EPR condition with instantaneous nonlocal synchronicity. Our postulate here is that this action at cosmological distances and for massive objects such as AGN QSOs creates a spacetime oscillatory shock fronts in the line of sight gravitational light cones leading to an apparent 'light boom' in QSO luminosity.

10.8 Hypersonic Shock Waves

We proceed for preliminary delineation under the assumption that the equations of state for hypersonic shock waves apply generally to any compressible media with shocks such as sonic booms or gravitational shock waves on the Dirac superfluid of spacetime especially those of secondary shock waves [60] that we postulate could be extended to support our theory that QSO luminosity can be explained by gravitational shock waves arising from an oscillatory interference of boundary conditions in propagation between the dual modes of gravity, i.e. classical and quantum-unitary.

From Crocco's equation for smooth flow in an ideal gas [49,59]

$$\Delta h_0 = -\frac{\partial \mu}{\partial t} + T\Delta S + \vec{\mu} \times \vec{\omega} \tag{10.13}$$

then following Kaouri [60] we develop a circulation theorem for a flow with shocks to eventually apply to the oscillation of boundary conditions for gravitational duality because the circulation theorem can be applied to parallel and perpendicular vorticity. For the closed curve,

$$C = \sum_{i=1}^{n} C_i \tag{10.14}$$

in Fig. 10.4 where C_i is the domain from P_i to $P_i + 1$ and the circulation around C is

$$\Gamma = \sum_{i=1}^{n} \int_{C_i} u \cdot dx = \sum_{i=\Gamma}^{n} \Gamma_i \,. \tag{10.15}$$

To construct the circulation theorem one needs to evaluate

$$\frac{d\Gamma}{dt}=\sum_{i=1}^{n}\frac{d}{dt}\left(\int_{C_i}u\cdot dx\right)=\sum_{i=1}^{n}\frac{d\Gamma_i}{dt}. \tag{10.16}$$

For each C_i the expression $\frac{d\Gamma_i}{dt}=\int_{C_i}u\cdot dx$ needs to be evaluated. Applying Crocco's equation (10.13) and summing all the C_i contributions we arrive at

$$\begin{aligned}\frac{d\Gamma}{dt}&=\sum_i\int_{P_i^+}^{P_{i+1}^-}TdS+\sum_i[H]_i+\\&\sum_i\left(u\left(P_{i+1}^-\right)\cdot\dot{P}_{i+1}-u\left(P_i^+\right)\cdot\dot{P}_i\right)\end{aligned} \tag{10.17}$$

where the 2nd term on the right is the sum of $[H]_i$, the total jump at the *i*th shock [60].

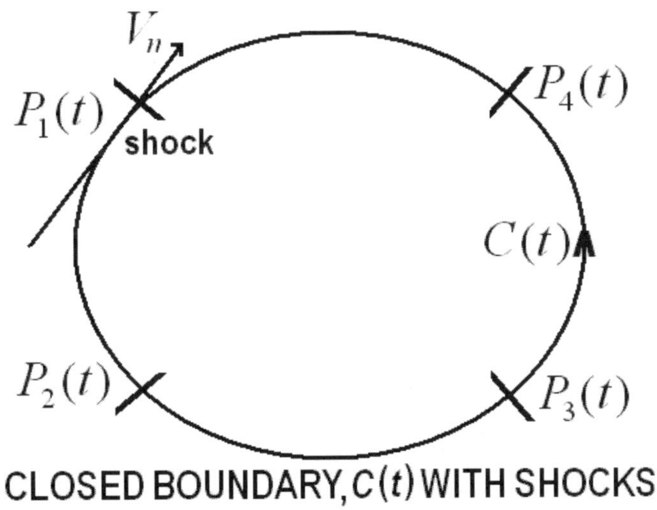

Figure 10.5 A closed curve boundary, *C*(*t*) impinged arbitrarily by 4 gravitational shocks at positions $P_i(t)$ here with $1\le i\le 4$ creating an entropic jump or constructive interference summation of gravity shock waves. Figure redrawn from [60].

Recent new work by Kaouri [60] on the dynamics of secondary sonic boom shock waves appears to provide insight into our idea of the dual nature of gravitational wave propagation.

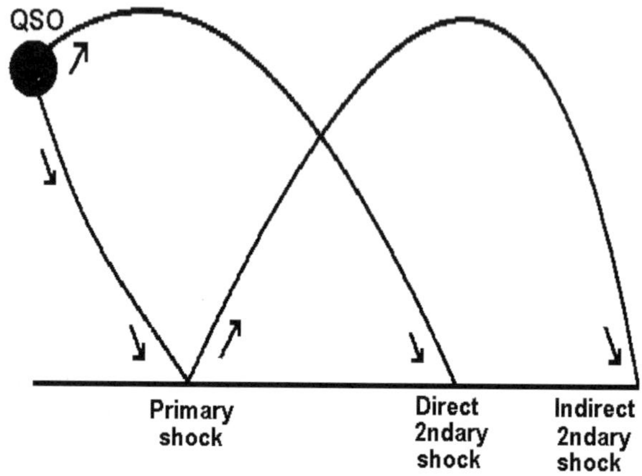

Figure 10.6 A conceptualization of a direct and indirect secondary boom. Figure redrawn from [60].

If Figs. 10.4 and 10.5 were combined one might end up with a conceptual view like that diagrammed in Fig. 10.6 but with a QSO at the center. If the physical case for a QSO contained a 2nd set of cusps in the bottom quadrant, the harmonic (Fig. 10.8) constructive and destructive interference of gravitational pressure waves could be a factor in producing 'light booms' in a manner dynamically similar to those producing Cerenkov radiation [61].

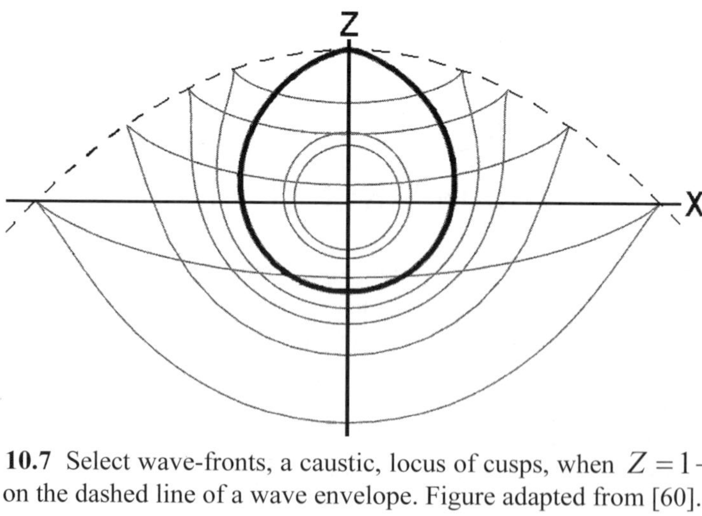

Figure 10.7 Select wave-fronts, a caustic, locus of cusps, when $Z = 1 - x^2 / 4$ plotted on the dashed line of a wave envelope. Figure adapted from [60].

The nonlinear nature of compressed fluid flow is the primary element of shock formation. If we consider a sinusoidinal gravitational influence of sufficient intensity where the curvature fluctuation across the wave is propagating adiabatically a disparity occurs in the velocity, *c* of propagation of light. The 'compressed' portion of the 'wave' will steepen to form a 'vertical' pressure front or shock as in Fig. 10.1b. The shock wave propagates because of a 'shift' in momentum transfer among flow regions of variable velocity. Shocks, being waves, only form in hyperbolic flow. The characteristic lines of flow are linear and merge into an envelope creating the shocks. Also as generally known parametric conditions can create triple shocks.

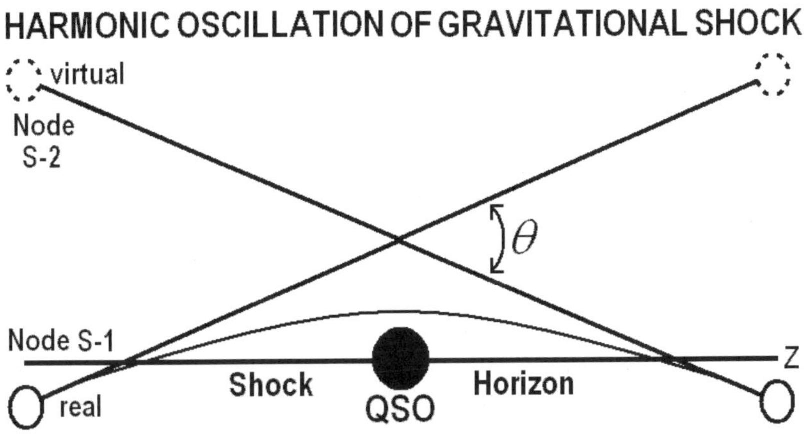

Figure 10.8 Schema of the primary postulate for a gravitational 'light boom'.

If the putative dual nature of gravitational propagation is physically real, then at cosmological distances for narrow axis large masses such as AND QSO's the coupling and uncoupling of the two principles could lead to a harmonic oscillation of the boundary conditions of the gravitational horizon such that a constructive/destructive interference occurs where at the summation nodes 'light booms' occur.

The nonlinear *x* component of gravitational shock nodes summating at collective shock fronts along $R_{E\text{-}Q}$ Adv + $R_{E\text{-}Q}$ Ret as seen by an Earth observer is shown in Fig. 10.9. The actual 'light boom' is the harmonic summation of shock nodes from the x, y and z axes and nonlinear assets arising from the interaction of the dual coupling of gravitational wave propagation.

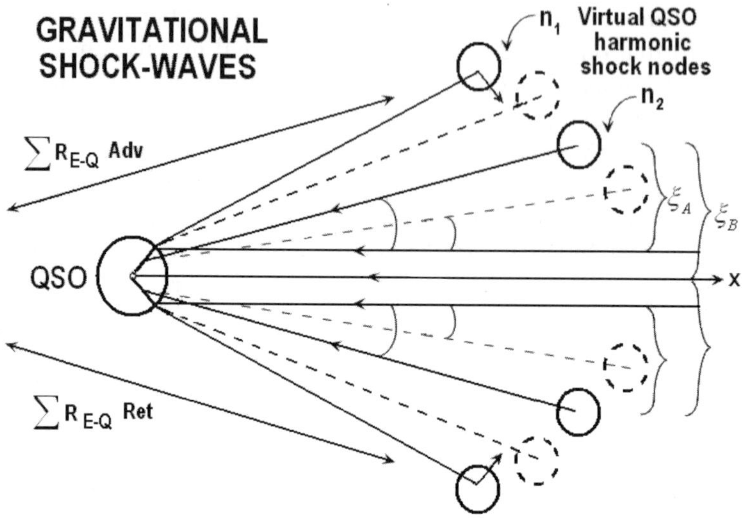

Figure 10.9 Elaboration of one quadrant of Fig. 10.8 showing a more detailed association of the oscillatory nature of the boundary conditions.

10.9 Gravitational Shock Waves

According to Misner, Thorne & Wheeler junction conditions may act as generators of gravitational shocks. They suggest that the dynamics of spacetime geometry for a 3-surface, Σ which includes the intrinsic Riemann scalar curvature invariant, R for example, also includes an extrinsic curvature tensor, K_{ij}. When imbedded in an enveloping 4-geometry hypersurface it can be applied to the change (shrinkage and deformation) in the vector, n parallel transported as junction conditions applicable to the gravitational field (spacetime curvature) and the stress-energy generating it. A discontinuity in K_{ij} across a null surface without stress-energy producing it is a geometric manifestation of a gravitational shock-wave generated by a different embedding in spacetime 'above' Σ than 'below' Σ [64].

Dray and 't Hooft [65] developed the fundamental conditions for introducing a gravitational shock wave in a particular class of vacuum solutions to Einstein's field equations by way of a coordinate shift. They outlined a model for generalizing gravitational shock waves for a massless particle moving in flat Minkowski space [66] formulated as two Schwarzschild black holes of equal masses glued together at the horizon. For a spherical shell of unequal masses moving along $u = u_0 \neq 0$ their solution [67] represents two Schwarzschild black holes glued together at $u = u_0$. By infinitely boosting the Dray-'t Hooft solutions various forms of gravitational shock waves have been found [68-73]. Sfetsos [74] extends these results to the case with matter fields and a non-vanishing cosmological constant. Using the d-dimensional spacetime metric

$$ds^2 = 2A(u,v)\,dudv + g(u,v)\,h_{ij}(x)\,dx^i dx^j \qquad (10.18)$$

with $(i, j = 1,2,...,d-2)$ he considers a string based dilatonic black hole gravitational solution [75,76] from the perspective of a conformal background field theory of coset $SL(2,\mathbb{R})/\mathbb{R}\otimes\mathbb{R}^2$ to achieve a differential shift factor

$$\left(\frac{d^2}{d\rho^2}+\frac{1}{\rho}\frac{d}{d\rho}-\varepsilon\right)f(\rho) = -16\varepsilon\rho\frac{1}{p}\delta(\rho) \qquad (10.19)$$

where $\rho^2 = x^2 + y^2$ and for the black hole singularity case with $\varepsilon = 1$ Eq. (10.19) is a modified Bessel equation. When $\varepsilon = -1$, Eq. (10.19) is interpreted as an expanding universe [74].

Spitkovsky [77] has developed a simulation for a relativistic Fermi emission shock process that could provide an alternative to or component process for our gravitational shock work. His simulations on relativistic collisionless shocks propagating in initially unmagnetized electron-positron pair plasmas showed natural production of accelerated particles as part of a shock evolution. He studied the mechanism that populates the suprathermal tail for particles gaining the most energy. The simulation showed the main acceleration occurs near the shock where for each reflection these particles gain energy, $\Delta E \sim E$ as is expected in relativistic shocks [78-80].

10.10 Conclusions

Newton's theory of gravitation required instantaneous action at a distance or the conservation of angular momentum would be violated. According to Einstein's theory of general relativity, an instantaneous influence would violate causality and the special theory of relativity and so must be mediated by a field. This is the dual nature of gravity that we have put as the basis for our model.

Shock phenomena remain a relatively little explored area of science both within and transdisciplinary. We have tried to show that it is possible with further study to relate shock phenomena to gravitational waves especially for narrow axis massive cosmological objects such as AGN QSOs that readily lend themselves to 'light-boom' effects that could therefore be used to explain QSO luminosity as further evidence of the insurmountable shortcomings of Big Bang cosmology.

Our model would appear to work best by contrasting both modes of the dual nature of gravity because a nonlinear jump in flow occurs with a discontinuity. From the 2nd Law of Thermodynamics entropy can only increase when a particle crosses a shock. The duality of the propagation of the gravitational influence is evident in the idea of Birkhoff's theorem [81] in that a spherically symmetric gravitational field is produced by a massive object such as a QSO at the origin; if there were another concentration of mass-energy somewhere else, this would disturb the spherical symmetry. This effect could occur if interference occurs between the usual modes of the gravitational influence by shock parameters.

More work needs to be done developing this model. We have only outlined what we perceive as an appealing avenue. At the close of writing we found an interesting 2009 article by Crawford suggesting new supernova data consistent with a static universe [82]. Also several more high redshift QSO's have been discovered that seem to support our shock theory for QSO luminosity [83-86].

References

[1] Hoyle, F., Burbidge, G. & Narlikar, JV. (2000) A Different Approach to Cosmology, From a Static Universe Through the Big Bang Towards Reality, Cambridge: Cambridge University Press.

[2] Hewitt, A. & Burbidge, G. (1993) A revised and updated catalog of quasi-stellar objects, Astrophys. J. Suppl. Ser., 87: 451-947.
[3] Woltjer, I. (1966) Inverse Compton radiation in quasi-stellar objects, Astrophys. J., 146: 597.
[4] Rees, M.J. Studies in radio source structure I, A relativistically expanding model for variable quasi-stellar radio sources, MNRAS 135, 345 (1967).
[5] Arp, H. (1966) Atlas of Peculiar Galaxies, Long Beach: CA Institute of Technology.
[6] Burbidge, G., O'Dell, S.L. & Strittmatter, , P.A. (1972) Astrophys. J., 175, 601.
[7] Bolton, J. & Wall, J. (1970) Aust. J. Physics, 23, 789.
[8] Bahcall, J.N., McKee, C.F. & Bahcall, N. (1971) Astrophys. Lett., 10, 147.
[9] Arp, H. (1987) Quasars Redshifts and Controversies, Berkeley: Interstellar Media.
[10] Carlip, S. (2000) Aberration and the speed of gravity, Phys. Lett. A 267: 81–87. arXiv: GR-qc/9909087.
[11] Kopeikin, S.M. (2001) Testing relativistic effect of propagation of gravity by very-long baseline interferometry, Astrophys. J. 556: L1–L6; or arXiv:gr-qc/0105060.
[12] Asada, H. (2002) The light-cone effect on the Shapiro time delay, Astrophys. J. 574: L69; or arXiv:astro-ph/0206266.
[13] Van Flandern, T.; & Vigier, J.P. (2002) Experimental repeal of the speed limit for gravitational, electrodynamic, and quantum field interactions, Found. Phys. 32: 1031–1068.
[14] Will, C.M. (2003) Propagation speed of gravity and the relativistic time delay, Astrophys. J. 590: 683–690 or arXiv:astro-ph/0301145.
[15] Fomalont, E. B., & Kopeikin, S.M. (2003) The measurement of the light deflection from Jupiter: Experimental results, Astrophys. J. 598: 704–711; or arXiv:astro-ph/0302294.
[16] Kopeikin, S.M. (2003) The measurement of the light deflection from Jupiter: Theoretical interpretation, arXiv: astro-ph/0302462.
[17] Kopeikin, S.M. (2003) The post-Newtonian treatment of the VLBI experiment on September 8, 2002. Phys. Lett. A 312: 147–157; or arXiv:gr-qc/0212121.
[18] Faber, J.A. (2003) The speed of gravity has not been measured from time delays, arXiv: astro-ph/0303346.
[19] Samuel, S. (2003) On the speed of gravity and the v/c corrections to

the Shapiro time delay, Phys. Rev. Lett. 90: 231101; or arXiv:astro-ph/0304006.
[20] Asada, H. (2003) Comments on 'measuring the gravity speed by VLBI', arXiv: astro-ph/0308343.
[21] Samuel, S. (2004) On the Speed of Gravity and the Jupiter/Quasar Measurement, Int. J. Mod. Phys. D 13: 1753–1770; or arXiv:astro-ph/0412401.
[22] Shapiro, E.I. (1964) Fourth test of General Relativity, Phys Rev Lett. 13; 789-791.
[23] Pascual-Sánchez, J-F (2004) Speed of gravity and gravitomagnetism, Int. J. Mod. Phys. D 13: 2345-2350; or arXiv:gr-qc/0405123.
[24] Carlip, S. (2004) Model-Dependence of Shapiro Time Delay and the Speed of Gravity/Speed of Light Controversy, Class. Quant. Grav. 21: 3803-3812 or arXiv:gr-qc/0403060.
[25] Kopeikin, S.M. (2004) The speed of gravity in general relativity and theoretical interpretation of the jovian deflection experiment, Classical and Quantum Gravity, 21: 3251–3286; or arXiv:gr-qc/0310059.
[26] Kopeikin, S. & Fomalont, E. (2006) On the speed of gravity and relativistic v/c corrections to the Shapiro time delay, Phys. Lett. A 355: 163–166, arXiv:gr-qc/0310065.
[27] Kopeikin, S. (2006). Gravitomagnetism and the speed of gravity, Int. J. Mod. Phys. D 15: 305–320; or arXiv:gr-qc/0507001.
[28] Kopeikin, S. & Fomalont, E. (2007) Gravimagnetism, causality, and aberration of gravity in the gravitational light-ray deflection experiments, Gen. Rel. Grav. 39: 1583–1624; or arXiv:gr-qc/0510077.
[29] Feynman, R.P., Morinigo, F.B. & Wagner, W.G. (2003) Feynman Lectures on Gravitation, Boulder: Westview Press.
[30] Silk, J.(1989) The Big Bang, New York: W.H. Freeman & Co.
[31] Amoroso, R.L., Kafatos, M. & Ecimovic, P. (1998) The origin of cosmic microwave background radiation in the intrinsic fluctuation of vacuum compactification cavity quantum electrodynamics, In Causality and Locality in Modern Physics, G. Hunter, & S. Jeffers, (eds.) Dordrecht: Kluwer.
[32] Kembhavi, A.K. & Narlikar, J.V. (1999) Quasars and Active Galactic Nuclei, Cambridge: Cambridge Univ. Press.
[33] Hawking, S. W. & Ellis, G. F. R. (1973) The Large Scale Structure of Space-Time, Cambridge: Cambridge Univ. Press.
[34] Amoroso, R.L. (2002) Developing the cosmology of a continuous

state universe, in R.L. Amoroso, G. Hunter, M. Kafatos & J-P. Vigier (eds.) Gravitation & Cosmology: From the Hubble Radius to the Planck Scale, Dordrecht: Kluwer Academic.
[35] Amoroso, R.L. (2004) The Fundamental Limit and Origin of Complexity in Biological Systems: A New Model for the Origin of Life, in D.M. Dubois (ed.) Computing Anticipatory Systems, AIP Conf. Proceedings Vol. 718, pp. 144-159, Melville: American Inst. of Physics.
[36] Graner, F. & Dubrulle, B. (1994) Titius-Bode laws in the solar system, Part I: Scale invariance explains everything, Astron. Astrophys. 282, pp. 262-268.
[37] Coles, P. & Lucchin, F. (2002) Cosmology, The Origin and Evolution of Cosmic Structure, Chichester: J. Wiley & Sons.
[38] Wells, D.R. (1986) Titius Bode and the helicity connection, IEEE Trans. Plasma Sci., 14; 865-873.
[39] Janaki, M.S. & Dasgupta, B. (1990) Eigenmodes for electromagnetic waves propagating in a toroidal cavity, IEEE Trans. Plasma Sci., 18:1; 78-85.
[40] Wells, D.R. (1990) Was the Titius-Bode series dictated by the minimum energy states of the generic solar system?, IEEE Trans. Plasma Sci. 18:1; 73-76.
[41] Chandrasekhar, S. (1958) On the equilibrium configuration of an incompressible fluid, Proc. Nat. Acad. Sci. US, 42; 273-279.
[42] Nieto M. (1972) The Titius-Bode Law of Planetary Distances: its History and Theory, Oxford: Oxford Univ. Press.
[43] Poveda, A. & Lara, P. (2008) The exo-planetary system of 55 Cancri and the Titius-Bode law, *Revista Mexicana de Astronomay Astrofsica*, 44, 243-246 or astro-ph/0803.2240v1.
[44] Kotliarov, I. (2008) The Titius-Bode law revisited but not revived, arXiv-space-ph: 0806.3532v1.
[45] Willott, C.J. et al. (2007) Four quasars above redshift 6 discovered by the Canada-France high-Z quasar survey, Astron. J. 134; 2435–2450.
[46] Peebles, P.J.E. (1993) Principles of Cosmology, Princeton: Princeton Univ. Press.
[47] Amoroso, R.L. (2006) The Holographic Conscious Multiverse, in R.L. Amoroso, B. Lehnert & J-P Vigier (eds.) Beyond the Standard Model: The Search for Unity in Physics, Orinda: Noetic Press.
[48] Amoroso, R.L & Pribram, KH (2010) The Complementarity of Mind and Body: Realizing the Dream of Descartes, Einstein and Eccles, Cambridge: MIT Univ. Press.

[49] Cattafesta, L.N. & Settles, G.S, (1992) Experiments on shock/vortex interaction, 30th Aerospace Sciences Meeting, January 6-9, Reno, NV, Washington: American Institute of Aeronautics and Astronautics www.mne.psu.edu/psgdl/AIAA92-0315.pdf.
[50] Cutler, A.D. & Levey, B.S. (1991) Vortex breakdown in a supersonic jet, AIAA-91-1815, Honolulu, Hi, June, 1991, Washington: American Institute of Aeronautics and Astronautics.
[51] Delery, J., Horowitz, E., Leuchter, O., & Solignac, J-L., Fundamental studies on vortex flows, *La Recherche Aerospatiale*, No. 1984-2.
[52] Lyman, V. & Morgenstern, J.M. (2004) Calculated and measured pressure fields for an aircraft designed for sonic-boom alleviation, 22nd Applied Aeronautics Conference, 16-19 August 2004, Providence, RI, Washington: American Institute of Aeronautics and Astronautics, www.sonicbooms.org/ReferenceTechPapers/AIAA2004_4846.pdf.
[53] Hall, M.G. (1972) Vortex breakdown, Ann. Rev. Fluid Mech., V. 4, pp. 195-218.
[54] Leibovich, S. (1978) The Structure of vortex breakdown, Ann. Rev. Fluid Mech., V. 10, pp. 221-46.
[55] Leibovich, S. (1984) Vortex stability and breakdown: survey and extension, AIAA J. V. 22, No. 9.
[56] Zatoloka, V.V., Ivanyushkin, A.K., & Nikolayev, A.V., (1978) Interference of vortices with shocks in air scoops: dissipation of vortices, Fluid Mechanics, Soviet Research, Vol. 7, No. 4.
[57] Glotov, G.F., (1989) Interference of vortex braid with free-stream shock waves & nonisobaric jets, *Uchenyye Zapiski* TsAGI, XX:5; 21-32.
[58] Gostintsev, Yu.A., Zelentsov, V.V., Ilyukhin V.S., & Pokhil, P.F., (1969) Structure of underexpanded supersonic swirling gas jet, *Izv. AN* SSSR, *Mekhanika Zhidkosti i Gaza*, 4:5; 158-162.
[59] Prim, R.C. (1948) Extension of Crocco's theorems to flows having non-uniform stagnation enthalpy, Phys. Rev. 73, 186.
[60] Kaouri, K. (2004) Secondary Sonic Boom, Phd. thesis, Oxford University, http://eprints.maths.ox.ac.uk/159/.
[61] Lehnert, R. & Potting, R. (2004) Vacuum Cherenkov radiation, arXiv:hep-ph/0406128v1.
[62] Kant, I. (1878) The Critique of Pure Reason, JMD Meiklejohn, (Trans.) London: G. Bell & Sons, Pdf - http://books.google.com/books.
[63] Amoroso, R.L. (2005) On the possibility of relativistic shock-wave effects in observations of QUASAR luminosity-preliminary introduction,

Noetic J., 6:1-4; 53-56.
[64] Misner, C.W., Thorne, K.S. & Wheeler, J.A. (1973) Gravitation, San Francisco: W. H. Freeman
[65] Dray, T. & 't Hooft, G. (1985) The gravitational shock wave of a massless particle, Nucl. Phys. B253, 173.
[66] Aichelburg. P.C., & Sexl, R.U. (1971) General Relativity and Gravitation 2, 303.
[67] Dray, T. & 't Hooft, G. (1985) Commun. Math. Phys. 99, 613.
[68] Lousto, C.O. & Sanchez, N. (1991) 915; Nuclear Physics B355, 231.
[69] Lousto, C.O. & Sanchez, N. (1992) Nuc. Phys. B383, 377.
[70] Hayashi, K. & Samura, T. (1994) Phys. Rev. D50, 3666.
[71] Hotta, M. & Tanaka, M. (1993) Class. Quant. Grav. 10, 307.
[72] Ferrari, V. & Pendenza, P. (1990) Gen. Rel. Grav. 22, 1105.
[73] H. Balasin, & H. Nachbagauer, H., The Ultrarelativistic Kerr-Geometry and its Energy-Momentum Tensor, arXix:gr-qc/9405053.
[74] Sfetsos, K. (1994) On gravitational shock waves in curved spacetimes, arXiv:hep-th/9408169v3.
[75] Gibbons, G. & Maeda, K. (1988) Nucl. Phys. B298, 741.
[76] Kallosh, R., Linde, A., Ortın, T., Peet, A. & Van Proeyen, A. (1992) Phys. Rev. D46, 5278.
[77] Spitkovsky, A. (2008) Particle acceleration in relativistic collisionless shocks: Fermi process at last? arXiv:astro-ph/0802.3216v2.
[78] Papapetrou, A. & Treder, H-J. (1962) Shock waves in general relativity, in L. Infeld (ed.) Recent Developments in General Relativity, Oxford: Pergamon Press.
[79] Zeldovich, Y.B. & Rayzer, Y.P. (1967) Physics of Shock Waves And High Temperature Hydrodynamic Phenomena, New York: Academic Press.
[80] Kaloper, N. (2005) Brane-induced gravity shocks, arXiv:hep-th/0501028v4.
[81] Birkhoff, G. (1923) Relativity & Modern Physics, Cambridge: Harvard Univ. Press.
[82] Crawford, D.F. (2009) Observations of type 1A supernovae are consistent with a static universe, arXiv:astro-ph-co/0901.4172v1.
[83] Willott, C.J., Delorme, P., Reylé, C., Albert, L., Bergeron, J., Crampton1, D., Delfosse, X., Forveille, T., Hutchings, J.B., Mclure, R.J., Omont, A. & Schade, D. (2009) Six more quasars at redshift 6

discovered by the Canada-France high-Z quasar survey, arXiv: 0901.0565v1.
[84] Lopez-Corredoira, M. (2009) Apparent discordant redshift QSO-galaxy associations, arXiv:0901.4534v1, astro-ph.CO.
[85] Komberg1, B.V. & Pashchenko, I.N. (2009) Giant radio galaxies-old long-living quasars? arXiv-astro-ph.co/0901.3721v1.
[86] Croton , D.J. (2009) A simple model to link the properties of quasars to the properties of dark matter halos out to high redshift, Monthly Notices Radio Astron. Soc.; or arXiv- astro-ph.co/0901.4104v1.

Chapter 11

The Bulk Implementation of Universal Scalable Quantum Computing

Quantum Computing (QC) has remained elusive beyond a few qubits. Feynman's recommended use of a "synchronization backbone" for achieving bulk implementation has generally been abandoned as intractable; a conundrum we believe arises from limitations imposed by the standard models of Quantum Theory (QT) and Cosmology. It is proposed that Feynman's model can be utilized to implement Universal Quantum Computing (UQC) with valid extensions of QT and cosmology. Requisite additional degrees of freedom are introduced by defining a relativistic basis for the qubit (r-qubit) in a higher dimensional (HD) conformal invariant context and defining a new anticipatory based cosmology (cosmology itself cast as a hierarchical form of complex self-organized system) making correspondence to 12D symmetries in the F-Theory incarnation of M-Theory. The causal structure of these conditions reveal an inherent new "action principle" driving self-organization and providing a basis for applying Feynman's synchronization backbone principle. Operationally a new set of transformations (beyond the standard Galilean / Lorentz-Poincaré) ontologically surmount the quantum condition, $\Delta x \Delta \rho \cong \hbar$ (producing decoherence during both initialization and measurement) by an acausal energyless topological interaction. Utilizing the structural-phenomenology of the HD regime requires new commutation rules and corresponding I/O techniques based on a coherent control process of cumulative interaction to manipulate applicable harmonic modes of HD spacetime manifolds such as those described by the genus-1 helicoid "parking garage" structure as a spin-exchange continuous-state spacetime resonance hierarchy. Finally it is suggested that this UQC model takes the same form as the mind-body interaction – that of conscious quantum computing.

11.1 Introduction – Basics of Quantum Computing

Whereas a classical Turing machine using a register of binary bits, 1 or 0 can only be in one state at a time; a quantum computer (QC) with a sequence on n qubits can be in a superposition of all the 2^n qubit states simultaneously. The simplest implementation would be a system of particles with two spin states as in Fig. 11.1.

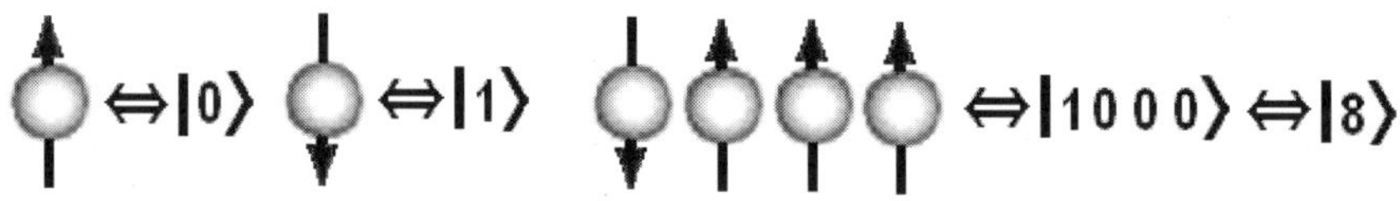

Figure 11.1 a) A qubit can be made from a particle with two spin states $|0\rangle$ and. $|1\rangle$. b) A four-bit register with $2^4 = 16$ four-bit strings 0000, 0001,…,1111 with state $|1000\rangle$ representing the number 8 shown. While a classical bit is only in the single vector state, a qubit wave function is in superposition of all 16 possibilities simultaneously.

It is easy to see the enormous power of quantum computing. A 1,000 qubit quantum register is described by 2^{1000} complex numbers which is many orders of magnitude more atoms than inside the Hubble radius (see Chap. 12). The main engineering problems for building a bulk quantum computer is decoherence during initialization and data readout, and quantum gates that operate faster than the decoherence time of the system which is often between nanoseconds and seconds. The ontological approach developed in this chapter makes decoherence on readout irrelevant because the methodology bypasses the quantum uncertainty principle (see also Chap. 8).

The predictions of Moore's law have held true over the whole history of the transistor. Just for fun we use the reciprocal parameter to predict the appearance of bulk quantum computing (Fig. 11.2) with the utility of the principles delineated here by ~2012; or will a phase effect between the classical to quantum regimes occur by Schnell's Law of refraction?

Eventually almost any quantum system could be used for various forms of bulk quantum computing. Some of the candidates currently being considered are: solid state and molecular solution nuclear magnetic resonance, superconducting, trapped ion, quantum dot, topological quasiparticle, optical lattice, cavity-QED, quantum optics, quantum spin, Bose-Einstein condensate, Adiabatic, transistor based, molecular magnets, quantum Hall effect, Fullerene electron spin resonance, and diamond spin.

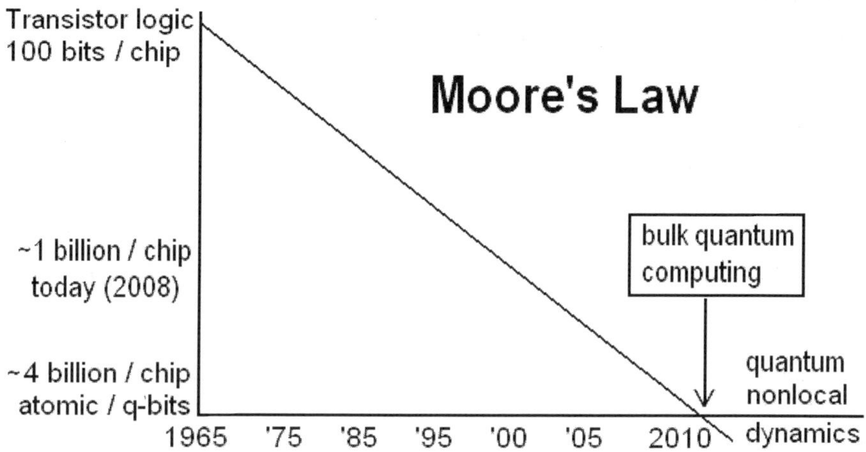

Figure 11.2 Moore's law states that since the invention of the integrated circuit in 1958, the number of transistors on a chip doubles about every two years, as well as the processing speed, accompanied by a corresponding reciprocal shrinkage in the size of a transistor [1]. Right on schedule the Intel Itanium had 1 billion transistors in 2008. The graph predicts quantum computing should appear by the year 2011 unless a phase effect delay occurs from Schnell's Law.

11.2 Overview of New Fundamental Parameters

The basis of our approach for the ontological realization of bulk or Universal Quantum Computing (UQC) is introduced conceptually utilizing an axiomatic approach to facilitate delineating the philosophy for the formalism. The theoretical model requires a new cosmology based on an extension of Quantum Theory (QT) or vice versa depending on whether one's view is top-down or bottom-up. Both the new cosmology and extension of QT are anticipatory because they take the form of complex self-organized hierarchical systems [2-8]. The extended QT is derived from a combined relativistic extension of Cramer's Transactional Interpretation [9], based on the Wheeler-Feynman Absorber Theory of Radiation [10], and an HD extension of the de

Broglie, Bohm, Vigier Causal Interpretation of QT where $m_\gamma \neq 0$ [11,12]. The cosmology is that of a 12D Holographic Anthropic Multiverse (HAM) making correspondence to the F-Theory [13] iteration of String Theory [2-5,7,8,14]. Why does UQC require an anticipatory approach cast in new QT and HAM cosmology? Salient reasons discussed are: 1. Causal conditions, 2. New Symmetry relations, 3. Utilization of Feynman's suggestion, and 4. Coherent control methods of operation. Interestingly the self-organized parameters of the cosmology entail an inherent synchronization backbone, amounting to getting half the UQC for free!

11.3 The Causal Separation of Phenomenology from Ontology

Because of the recent jump from Newtonian Mechanics to Quantum Mechanics most physicists believe we live in a quantum universe. The logical progression of this line of reasoning would suggest we live in a Unitary Universe once a unified field is empirically delineated. This is an erroneous conclusion. A better assumption based on anticipatory properties inherent in 12D HAM cosmology suggests that the universe is a continuous-state interplay of all three modes [2-5]. Because we only observe the Euclidean component of this world view, we assume reality is a complex virtual standing wave with the present a continuously created subspace of HD future-past parameters [2-10]. Our task is to demonstrate an ontological methodology for surmounting the inherent uncertainty conditions of Copenhagen regime phenomenology with a new set of transformations that utilize an "energyless topological switching" to exchange information [15].

Reality as locally observed, (Fig. 11.3a) is Euclidean, E_3 or 3(4)D Minkowskian, $\hat{M}_4$ depending on whether time is introduced in the Newtonian or Einstein sense. In Fig. 11.3b the 3D cube is shown unfolded into a cross in the 2D plane with arbitrary loss of the *z* direction. Figure 11.3c shows a 4D tesseract that includes the 4th dimension designated as *w*. In a manner analogous to Fig. 11.3b the 4D tesseract is unfolded into a 3D cross as shown in Fig. 11.3d. Loss of the *w* direction makes it easier for the human mind to visualize a hypercube. For simplicity we use the 3(4)D cross to illustrate how '12' is the minimum number of dimensions required to describe *eternity* (defined as causal separation from E_3) and conceptually reveal how to overcome the limitations of the quantum principle inherent in Copenhagen uncertainty; i.e. since the phenomenology of Copenhagen action produces uncertainty by definition - choose instead an ontological process that does not discretize the *z* field commutator as in Eq. (11.2).

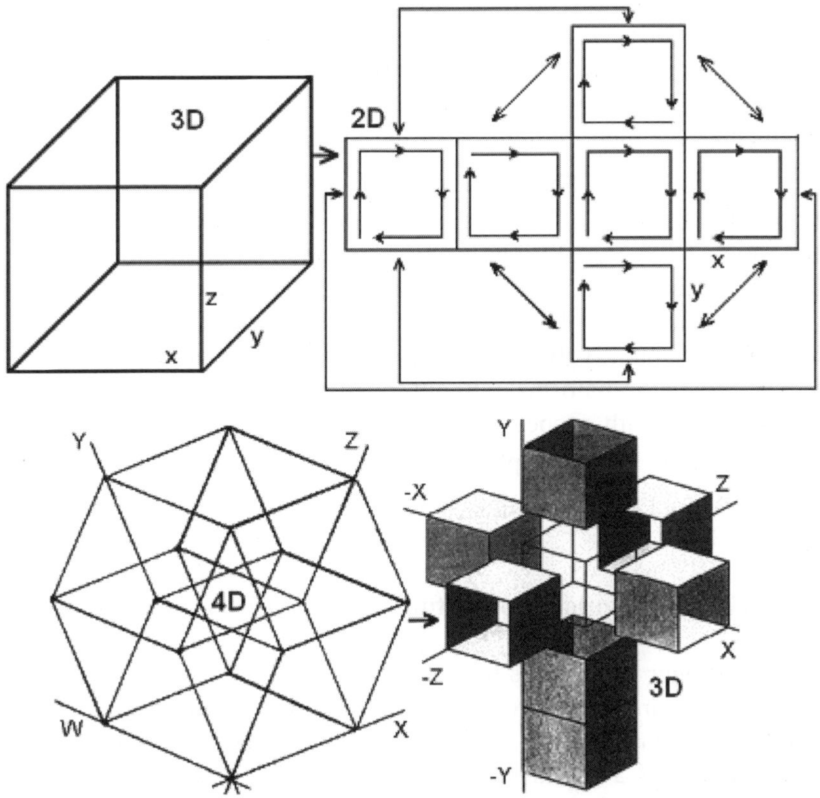

Figure 11.3 Geometry of space in 3 & 4D. (a,b) A 3D cube unfolds into the 2D plane. This metaphor aids the visualization of HD space. (c,d) A 4D hypercube unfolds into 8 component 3D cubes as in (1b). If a 5D hypercube were unfolded the 8 cubes forming the 3D cross (d) would be 4D hypercubes (tesseracts as in (c). The translucent cube, called the central cube, represents observed reality, E_3. This central cube is surrounded by six adjacent cubes. The 8th cube, the satellite cube, is placed arbitrarily on any adjacent cube. Carried to 12D the central cube and 12D satellite causally separate as a 'mirror image of a mirror image'..

To further illustrate Fig. 11.3, assume the translucent *central* cube (Fig. 11.3d) represents local E_3 reality as a subspace surrounded by six *adjacent* HD cubes that are components of the 4D tesseract in Fig. 11.3c. The 8th cube, the *satellite*, is arbitrarily attached to the y direction adjacent cube. Let the central cube hold a standard quantum state. A primary assumption of continuous-state noetic cosmology is that all eight 4D cubes (4096 in 12D) contain the information of the central cube's state by superposition (an inherent property of the conformal invariance [16,17] of HD Relativistic Quantum Field Theory (RQFT)). This is a fundamental symmetry condition of the Superspace of HAM cosmology [2-5]. The satellite cube is periodically *causally free* of the E_3 / M_4 central cube because of the relativistic transformation. This continuous-state topological transformation of the standing-wave modes is the inherent synchronization backbone in the backcloth of spacetime itself; as if half of the QC is obtained for free. In this context QC operations are ontological (if putatively performed in a specific manner described by the new noetic transform) without phenomenological collapse of the wave function with respect to quantum information contained in the central E_3 cube. This metaphor performed rigorously in a 12D context is able to surmount the uncertainty relation!

11.4 Review of Angular Momentum and Pauli-Dirac Spin Matrices

The Schrödinger equation is invariant under Galilean but not the Lorentz transformation and therefore incompatible with the principle of relativity and all phenomena relating to the interaction of light and matter leading to the concept of 2nd quantization [18]. Our 12D extension of QT goes beyond the usual Klein-Gordon and Dirac models of RQFT. This is an issue of the observers cosmology with an inherent complementarity between 1st and 2nd quantization much like wave-particle duality. This is a continuous-state property [2-5] readily described by methods similar to that attributed to Dirac spherical rotation of the electron [2].

Separation of the Schrödinger equation into spherical coordinates reveals the Hamiltonian

$$H = \frac{1}{2m}\left(p_r^2 + \frac{L^2}{r^2} \right) + V(r), \tag{11.1}$$

where p_r is the radial momentum $(m\dot{r})$ and L the angular momentum vector. As well known, the three components of angular momentum, derived from each other by cyclic permutation, are $L_z = xp_y - yp_x$, $L_x = yp_z - zp_y$, $L_y = zp_x - xp_z$, $L = r \times \rho$ where the total angular momentum $L^2 = L_x^2 + L_y^2 + L_z^2$ has commutation rules $L \times L = i\hbar L$ [18-22]. SO(3) rotation generators l_1, l_2 and l_3 satisfy $l_1 l_2 - l_2 l_1 = l_3$, $l_2 l_3 - l_3 l_2 = l_1$, $l_3 l_1 - l_1 l_3 = l_2$; are related quantum mechanically to angular momentum components L_1, L_2, L_3 with $L_x = i\hbar l_1, L_y = i\hbar l_2, L_z = i\hbar l_3$ about Cartesian axes giving the usual commutation rules $L_x L_y - L_y L_x = i\hbar L_z$, $L_y L_z - L_z L_y = i\hbar L_x$, $L_z L_x - L_x L_z = i\hbar L_y$.

Angular momentum refers to intrinsic spin about a massive particles center of mass and its magnetic moment by SO(3) Lie algebra which is non-Abelian so the elements do not all commute. The Pauli matrices satisfy these commutation rules when acting on two component spinor wave functions $\{\psi_0(x), \psi_1(x)\} \equiv \Psi_A$; but by the uncertainty relation, $\Delta x \Delta \rho \cong \hbar$ only one set of these operators may commute at a time. Non-relativistic Fermi spin $1/2\hbar$ particles with spin angular momentum operator $S = 1/2\hbar\sigma$ can be expressed as the three anticommuting Pauli *2* x *2* spin matrices Eq. (11.2) satisfying $\sigma_x \sigma_y = -\sigma_y \sigma_x = i\sigma_z$ as derived empirically from the Stern-Gerlach experiments [18,20].

$$L_x = \frac{\hbar}{2}\sigma_x = \begin{bmatrix} 0 & 1 \\ 1 & 0 \end{bmatrix},$$
$$L_y = \frac{\hbar}{2}\sigma_y = \begin{bmatrix} 0 & -i \\ i & 0 \end{bmatrix}, \qquad (11.2)$$
$$L_z = \frac{\hbar}{2}\sigma_z = \begin{bmatrix} 1 & 0 \\ 0 & -1 \end{bmatrix}.$$

Here we demonstrate a complex HD geometric-micromagnetic [15] method where all commutator relations can periodically simultaneously commute like the Casimir-like 'total spin' operator, $\mathbf{J}^2 = L_x^2 + L_y^2 + L_z^2$ commutes with all three components of L in 3D [23-25]. This is possible in HD because E^3 (Fig. 11.1) in HAM cosmology has the same properties as the Dirac spherical rotation of the electron [2,5]. The topology of these boundary conditions is described by HD expansion of the noetic field equation, $F_n = E / R$ [26]; F_N is the cyclic noetic force, E the continuous-state Lagrangian and R the complex coherence length [27].

Relativistic spin $1/2\hbar$ particles are described by Dirac's formalism for the wave equation which has been expressed by several notations $E\psi + c(\alpha \cdot p)\psi + mc^2 \beta\psi = 0$; or $i\hbar \frac{\partial \psi}{\partial t} - i\hbar c\alpha \cdot grad\psi + mc^2 \beta\psi = 0$ [28] which when expressed by Dirac's σ matrices can be expanded into the following 4 x 4 matrices:

$$\beta = \begin{bmatrix} 1 & 0 & 0 & 0 \\ 0 & 1 & 0 & 0 \\ 0 & 0 & -1 & 0 \\ 0 & 0 & 0 & -1 \end{bmatrix} \quad \alpha_x = \begin{bmatrix} 0 & \sigma_x \\ \sigma_x & 0 \end{bmatrix} = \begin{bmatrix} 0 & 0 & 0 & 1 \\ 0 & 0 & 1 & 0 \\ 0 & 1 & 0 & 0 \\ 1 & 0 & 0 & 0 \end{bmatrix}$$

$$\alpha_y = \begin{bmatrix} 0 & 0 & 0 & -i \\ 0 & 0 & i & 0 \\ 0 & -i & 0 & 0 \\ i & 0 & 0 & 0 \end{bmatrix} \quad \alpha_z = \begin{bmatrix} 0 & 0 & 1 & 0 \\ 0 & 0 & 0 & -1 \\ 1 & 0 & 0 & 0 \\ 0 & -1 & 0 & 0 \end{bmatrix} \tag{11.3}$$

which are Hermitian and readily seen to contain the *2* x *2* Pauli matrices (2) as in the center of matrix α_x for example [29]. An interesting point developed below is that in cases where $m = 0$ (or for high E where any massive particle behaves like $m = 0$) only three anticommuting matrices instead of four are required. This means the Pauli matrices will suffice and the spinor needs only 2 components which relates to the Wehl or chiral representation [30].

In another popular notation the Dirac equation is represented as

$$E + mc^2 + \frac{ke^2}{r} = \phi c + imc^2 = \vec{\phi} c + imc^2 \tag{11.4}$$

where $E \to -\frac{\hbar}{i}\frac{\partial}{\partial t}; \vec{\phi} \to \frac{\hbar}{i}\vec{\nabla}$ which has the general solution $\frac{\hbar c}{i}\left(\gamma_1 \frac{\partial}{\partial x} + \gamma_2 \frac{\partial}{\partial y} + \gamma_3 \frac{\partial}{\partial z} + \gamma_0 \frac{\partial}{\partial ict} + \right)\psi - mc^2\psi = 0$ which is easily shorthanded to $\left(\gamma^{\mu}\partial_{\mu} + i\frac{mc^2}{tc}\right)$. There are two non-relativistic limits; the first $\left(E + mc^2 + ke^2 / r\right)^2 = \left(\phi c\right)^2 + \left(mc^2\right)^2$, is the well known Klein-Gordon equation, and the second, $E + mc^2 + ke^2 / r$ $= \sqrt{(\phi c)^2 + (mc^2)^2} \cong mc^2\left[1 + \frac{1}{2}\left(\frac{\phi c}{mc^2}\right)^2 + \ldots\right] \cong mc^2 + \frac{\phi^2}{2m} + \ldots$ thus becoming $E = \frac{\phi^2}{2m} - \frac{ke^2}{r}$, the Schrödinger equation. In this notation the matrices in (11.3) $\alpha_x, \alpha_y, \alpha_z, \beta_{ict}$ correspond to $\gamma_1, \gamma_2, \gamma_3, \gamma_0$, respectively and ψ to the matrix $\psi = \begin{pmatrix} \psi_{e\uparrow} \\ \psi_{e\downarrow} \\ \psi_{\phi\uparrow} \\ \psi_{\phi\downarrow} \end{pmatrix}$.

11.5 Noumenal Reality Versus Phenomenology of Quantum Theory

Feynman has shown that reality can be considered incompatible with QT [31]. If we let *A,B,C* represent the three observables *a,b,c*, their values *P*(*a,b*), *P*(*b,c*), *P*(*c,a*), their transition probabilities $a \to b, b \to c, c \to a$ and $\phi(a,b), \phi(b,c), \phi(c,a)$ the corresponding quantum mechanical amplitudes; the transition probabilities *P*(*x,y*) (*x,y* = *a,b,c*) are measurable empirically by the classical rules of probability leading to

$$P(a,c) = \sum\nolimits_b P(a,b) \cdot P(b,c) \tag{11.5}$$

with the summation taken over all values of observable *B* [31,32].

Measuring *P*(*x,y*) in a case where the relative frequency of *x* is an ensemble prepared so that *y* is realized with certainty, the identity equation (6) can be shown to be wrong because the difference (called the interference term) between the right and left sides of (6) is found to be some orders of magnitude larger than the experimental error. If one calculates the interference terms according to the rules of QT from the empirically correct formula

$$\phi(a,c) = \sum\nolimits_b \phi(a,b)\phi(b,c) \tag{11.6}$$

and utilizing the connection between probability and amplitude

$$\left|\phi(x,y)\right|^2 = P(x,y); \quad x,y = a,b,c \text{ [31]}. \tag{11.7}$$

This contradiction between the classical probability identity (11.5) and the results of the Copenhagen interpretation (6) have been elucidated by Feynman:

> …Looking at probabilities from a frequency point of view (11.5) simply results from the statement that in each experiment giving *a* and *c, B* had some value. The only way (11.6) could be wrong is the statement, "*B* had some value", must sometimes be meaningless. Noting that (11.7) replaces (11.6) only under the circumstance that we make no attempt to measure *B*, we are led to say that the statement, "*B* has some value", may be meaningless whenever we make no attempt to measure *B*" [31].

Feynman's statement delineates Schrödinger's cat paradox. He states regarding the interference term that if we say "*B* had some value" when we make no attempt to measure it is true we have a contradiction with experiment because there is a contradiction between objective reality and the validity of QT in the orthodox Copenhagen interpretation. For our purposes here we resolve this paradox by abandoning the notion of a local absolute objective reality by stating that the observer's 3(4)D reality is virtual and that the 11(12)D HAM anticipatory reality is physically more complete. This is a key foundational element of our UQC model because we postulate firstly that the very existence of the observer discretizes reality and secondly that the application of the arbitrary *z*-field discretizes *L* such that it does not universally commute.

This is of course experimentally demonstrated as the standard interpretation of QT and is the basis for its formalism. This scenario avoided in the 12D anticipatory model of UQC is not possible by "law" if Copenhagen is applied. We demonstrate a model with zero commutator for all values of *L* where state evolution can be manipulated ontologically (from a position of causal separation) rather than the standard phenomenology of wave function collapse producing the uncertainty relation, $\delta x \delta \rho_x \geq 1/2\hbar$.

11.6 Justification for the Incursive Noetic Model

If M-theory/F-Theory subsumes the standard model of particle physics and cosmology, strings will represent the primary physical element; and $\hbar$ will no longer be considered a fundamental constant. First let's consider the continuous-state compactification of noetic superspace. The 12th D (hyperplane of Fig. 11.12) is an absolute space signifying the geometric limit of our reality from which a 9 to 11D manifold drops out (site of unitary field) as the 1st continuous-state compactification of the harmonic superspace delineated as

$$x_N = x_N + 2\pi R_N \tag{11.8}$$

where $N \rightarrow$ from 1 to 8D and R the periodic radius of space N goes from $\hbar$ to ∞. This condition exists because unit strings are not related to $\hbar$ = 1, but to string tension, denoted simplistically $T_s = 1/\pi$ [33,34]. Fields on this periodic space therefore satisfy

$$\phi(x) = \phi(x + 2\pi r) \tag{11.9}$$

which means the field ϕ can be power expanded periodically with eigenfunctions

$$\phi(x) = \sum_k \phi_k e^{ipx} \tag{11.10}$$

where $p = k \ / \ R$ and k is an arbitrary integer so that the momentum conjugate x is quantized in integers, a feature of all compactifications. Note for our purposes that compactification of a dimension quantizes the momentum corresponding to the compactified coordinate [13].

This has immediate repercussions for the anticipatory UQC model. For Copenhagen, only the z component of the angular momentum vector of a particle on a Riemann sphere is considered well defined. The Dirac equation, usually formulated in 4D, must be recast in the 11(12)D superspace [35] to include additional causal action in the symmetry of advanced–retarded potentials and heterotic splitting of the 8D resonant tower (Figs. (5,6,7), where the wave function and all off-diagonal elements are physically real and therefore accessible as in Cramer's transactional interpretation [9]. A transaction (Figs. 11.4 and 11.5) is represented as a form of standing wave both of which support the energy dependent nature [2] of the periodic 12D continuous-state superspace. The separation of these parameters in terms of de Broglie's fusion model [36,37] is used for ontological manipulation of the harmonic tier (Figs. 11.7, 11.8 and 11.9).

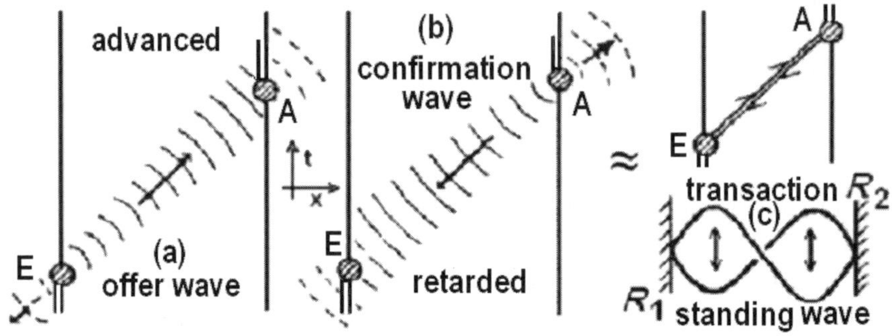

Figure 11.4 Illustration of an event or 'real' transaction in Cramer's future-past interpretation of QT, where E is emitter and A absorber. Adapted from [9].

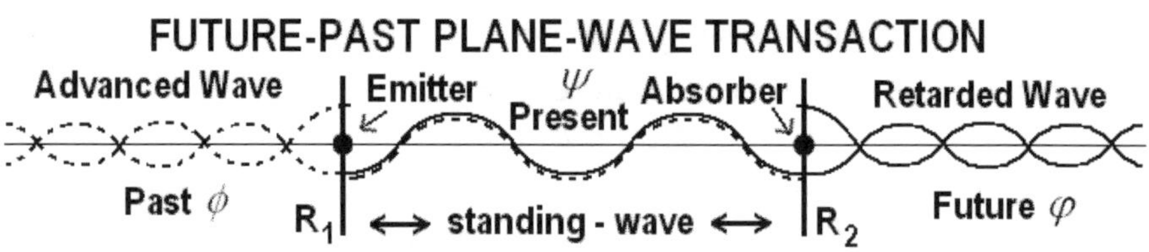

Figure 11.5 Symbolic structure of a transaction (present state or event) where the present moment is a standing-wave of physically real future-past retarded-advanced elements. See Fig. 11.4. Figure redrawn from [9].

It is suggested that continuous compactification of noetic superspace in this framework produces a singularity (a cyclic wave-particle) which is the observed E_3 reality itself; and the eight-form factorizes into two four-forms $N_8 \rightarrow X_4 \wedge X_4$, i.e. the advanced – retarded components of an HD extension of a Cramer transaction [9]. Because M_4 is Einstein's energy-dependent spacetime metric, $\widehat{M}_4$ where strings are susceptible to *EM* charges, (p, q); the tension of these heterotic strings becomes

$$T_{p,q} = pe^{-\phi} + qe^{\phi} \tag{11.11}$$

which can be used to demonstrate that string tension, T vanishes at the singularity E_3 [13,38].

11.7 Essential Properties of Complex Noetic 12 Space

The UQC model relies on a new 12D Absolute Space (AS) (ultimate arena of reality) from which properties of a Wheeler geon [39] or 'ocean of light' (the unitary noetic field) emerge. The noetic AS is an atemporal, highly ordered and symmetric harmonic superspace from which all other space relative to an Earth observer is a composite subspace. The geon domain (9 to 11D) is the first compactification regime; and because of coherence of the unitary field, railroad tracks would not recede but remain parallel.

A set of null lines (complex arrow of time), a loci of eternal points, remains hidden from local observed reality as an eternal present. This is part of the complex, $\pm C_4$ Wheeler-Feynman-Cramer duality of the future-past standing-wave comprising the continuous state present: "a relativistic spin-exchange dimensional reduction compactification process" which represents a new set of transformations beyond Galilean and Lorentz/Poincaré to describe the inherent dynamics of this unitary domain and create the arrow of time [40]. This condition results in our E^3-M^4 domain being a subspace of eternity (Fig. 11.1); and the essential process for producing the 'synchronization backbone' inherent in the backcloth of HAM cosmology [3-5].

As in special relativity where c remains constant and independent of the velocity of the source; the 12D AS remains static and absolute whether matter is stationary or in relativistic motion. In this context there is a duality in terms of conservation laws, annihilation/creation, advanced/retarded potentials or between space and energy including an asymmetry between the future-past. The new set of transformations makes correspondence with M-Theory and is conceptually considered a higher dimensional extension of Dirac Spherical Rotation [5,41-43].

Thus issues of the historical controversy between relational and AS are pushed to the new 12D domain. Within the Classical limit the former 3D Euclidean AS remains relative to the eternal present [44] of the subjective observer. Einstein demonstrated that the application of special relativity to a 3(4)D Minkowski/Riemann manifold makes space relational. The new relational space extends Einstein's view from four to eleven dimensions. In the 12D noetic superspace, S_N the 11D unitary noetic field (and the local 3(4)D $B^{(3)}$ component of the *EM* field [45]) translates longitudinally, but the space (as in water waves) remains fixed because the wave bumps against the close-packed spheres or least units [8,46] (like the water molecules) allowing only transverse displacement while the wave is locally present. This wave cyclically undergoes $m_\gamma = 0$ and $m_\gamma \neq 0$ plus $B^{(3)}$ for certain polarizations. See Chap. 6.

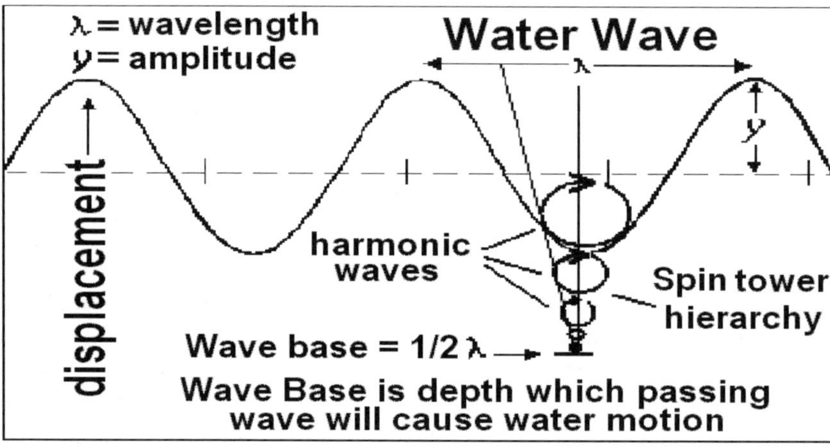

Figure 11.6 Symbolic representation of a 12D Ocean (Dirac sea) of Light (unitary field) modeled after the Wheeler Geon. The "ocean" provides a practical metaphor for 12D space in that polarization of the Dirac sea is believed to have properties similar to water waves.

Current thinking on the topology of space takes three general forms: 1) The most commonly accepted 3(4)D Minkowski/ Riemann spacetime manifold; and two putative HD superspace additions, 2) Calabi-Yau space preferred by M-Theory and 3) Dodecahedral space. Nature of the true vacuum remains an open question. The 3D absolute space of Newton became the 3(4)D relational spacetime of Einstein. The 12th D of Noetic cosmology represents a new form of absolute space, a periodic superspace where the eternal twelfth dimension has a Wheeler Geon [39] or ocean of 'light' (the unified field) as its $9 \rightleftarrows 11$D subspace. The relational 3(4)D Minkowski/Riemann spacetime manifold is a continuous state standing wave subspace of the 12D noetic superspace; it acts as a topological cover of an eternal present [44] which is not observed and continuously decays into spacetime.

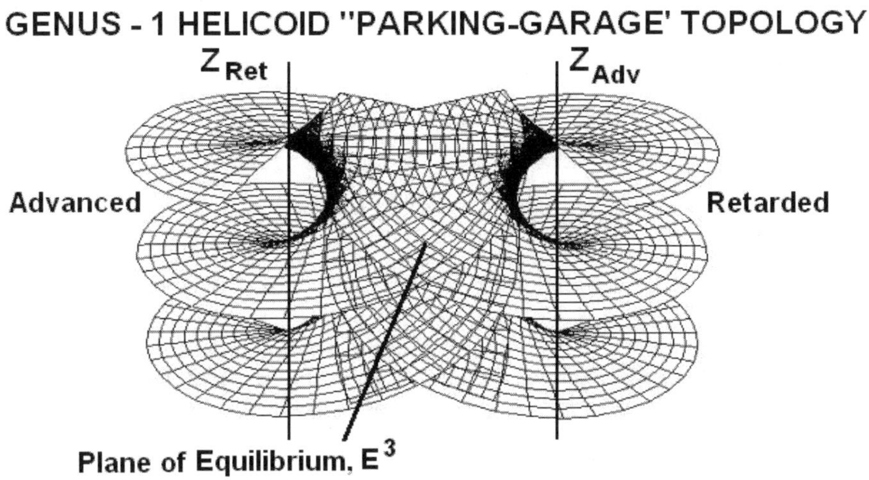

Figure 11.7 The Helicoid, a minimal embedded surface [47], is swept out by a line rotating about and moving down the z axis. Here a double Genus-1 Helicoid is joined into a "parking garage" ramp structure representing the future-past hierarchical topology of noetic space. An ordinary 2D plane can be twisted into a helicoid. Also see [16].

"Space quantization" or the quantization of orientation of atomic systems observed empirically primarily by Stern-Gerlach [18] and secondarily in other phenomena like the Zeeman Effect in an inhomogeneous magnetic field led to the basis for representing spin ½ fermions as a uniform Dirac spherical rotation through a 720° cycle [5,41-43] and the commutation relation for angular momentum in quantum theory. We explore extending these properties to 12D, 12D as required for UQC ontological operation.

If the noetic space water wave conception is correct, the continuous-state compactification process contains a tower of spin state Lie groups from spin 0 to spin 4. Spin 4 represents the unified field and makes cyclic correspondence with spin 0 where spacetime lattice Riemann sphere Ising lattice spin flips [15,38] create dimensional jumps through the helicoids topology (Fig. 11.7). Spin 0, 1/2, 1, & 2 remain in standard form. Spin three is suggested to relate to the orthogonal properties of atomic energy levels and space quantization. Therefore the spin tower hierarchy precesses through 0, 720°, 360°, 180°, 90° & 0 (∞) as powers of *I*, as conceptually illustrated in Figs. 11.8 and 11.25.

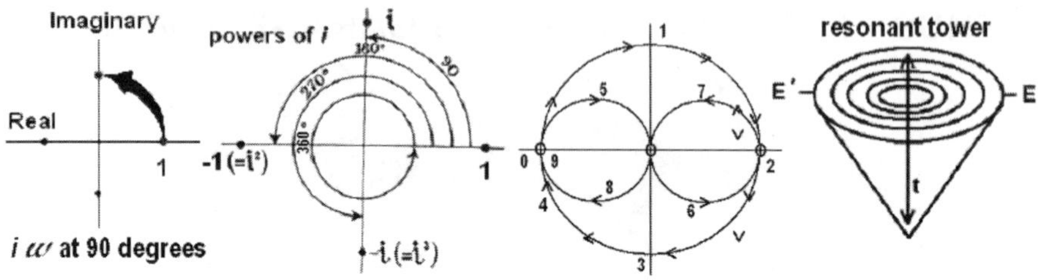

Figure 11.8 (a) Complex dimension at 90° from the real axis. (b) Powers of *i* from 90° to 360°. (c) Power of *i* at 720°. (d) Resonant hierarchy comprised of powers of *i* in conjunction with the topology of the Genus-1 helicoid "parking-Garage" of the string vacuum with either Ising model, logarithmic spiral or cyclotron resonance hierarchy parameters for applying ladder operators of the resonant modes required to ontologically operate the UQC model.

An instant *t*, for position $r \equiv (x, y, z)$ or for the light cone (Fig. 11.9) $r = xdt$, defines a point or event $d = \sqrt{x^2 + y^2 + z^2}$ in ordinary spacetime coordinates, a pseudo-Euclidean metric tensor [48] representing the sixteen points of a 4-sphere (Fig. 11.3c)

$$G_{\mu\nu} = \begin{pmatrix} 1 & 0 & 0 & 0 \\ 0 & -1 & 0 & 0 \\ 0 & 0 & -1 & 0 \\ 0 & 0 & 0 & -1 \end{pmatrix}. \tag{11.12}$$

In summarizing the observers relationship to the *Cosmological Principle* (that the universe is homogeneous and isotropic on average in the large-scale) [49,50] events are idealized instants in spacetime defined by arbitrary time and position coordinates *t, x, y, z*, written collectively as x^μ where μ runs from 0 to 3. The standard line element is

$$ds^2 = \sum_{\mu\nu} G_{\mu\nu} dx^\mu dx^\nu = G_{\mu\nu} dx^\mu dx^\nu, \tag{11.13}$$

where the metric tensor

$$G_{\mu\nu}(x) = G_{\mu\nu}(x) \qquad (11.14)$$

is symmetric [48]. In local Minkowski form all the first derivatives of g_{ij} vanish at the event and equation (11.13) takes the form

$$ds^2 = cdt^2 - dx^2 - dy^2 - dz^2. \qquad (11.15)$$

The Cosmological Principle generally suggests that the clocks of all observers are synchronized throughout all space because of the inherent homogeneity and isotropy. Because of this synchronization of clocks for the same world time *t*, for co-moving observers the line element in (11.15) becomes

$$ds^2 = dt^2 + G_{\alpha\beta}dx^\alpha dx^\beta = dt^2 - dl^2, \qquad (11.16)$$

where dl^2 represents special separation of events at the same world time *t*. This spatial component of the event dl^2 can be represented as an Einstein three-sphere

$$dl^2 = dx^2 + dy^2 + dz^2 + dw^2 \qquad (11.17)$$

which is represented by the set of points (*x, y, z, w*) at a fixed distance *R* from the origin:

$$R^2 = x^2 + y^2 + z^2 + w^2 \qquad (11.18)$$

where

$$w^2 = R^2 - r^2 \;\text{ and }\; r^2 = x^2 + y^2 + z^2 \qquad (11.19)$$

so finally we may write the line element of the Einstein three-sphere from equation (11.17) as

$$dl^2 = dx^2 + dy^2 + dz^2 + \frac{r^2dr^2}{R^2 - r^2} \text{ [51].} \qquad (11.20)$$

By imbedding Einstein's model of the three-sphere in a flat HD space, specifically as a subspace of a new complex 12D superspace, [2-5,7] new theoretical interpretations of standard cosmological principles are feasible.

Although the Newton and Coulomb potentials have similar forms the two theories have developed separately. For our purposes, following the Sakharov-Puthoff conjecture [52], that gravity is a product of fluctuation of the zero point field; we unify them with the Amoroso-Vigier methods [53,54] where both fields are represented by 4-vector field densities A_μ. Both phenomena are considered different types of motion within the same real physical field in flat spacetime as two different vacuum types of collective perturbations carried by a single vacuum field (unified). See Fig. 11.12b.

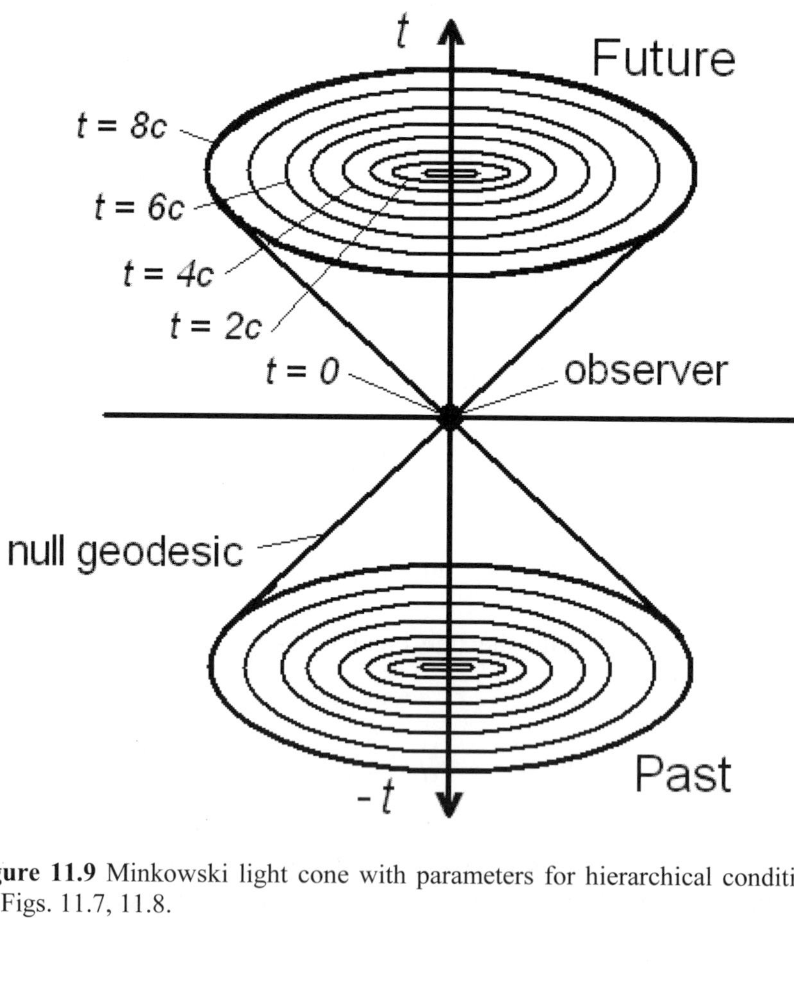

Figure 11.9 Minkowski light cone with parameters for hierarchical conditions of Figs. 11.7, 11.8.

Maxwell's equations traditionally describe only transverse elements that 'cut-off' at the vacuum. Here for HAM cosmology extended electromagnetic theory is utilized where the Einstein-de Broglie relation, $E = \hbar\nu = mc^2$ allows additional degrees of freedom such as longitudinal components $B^{(3)}$ and polarized vacuum conditions where $m_\gamma \neq 0$ suggests that the photon is piloted. These conditions suggest the need for both the standard *EM* field and extended $\mu\nu$ field coordinates; an understanding of which will be seen to be required for the ontological UQC operations.

In our original integration of *G* and *EM* we chose to fix the $\mu\nu$ field coordinates [53,54]. Here we go a step further. Dirac himself suggested by the rule of coordinate law that the pilot wave and the photon decouples [55]. The two sets of coordinates *EM* or $\mu\nu$ would normally be considered independent of each other. We integrate them in the topology of the Dirac polarized sea and alternate the fixing and decoupling of $\mu\nu$ and *EM* coordinates as an inherent 'leapfrogging' (Fig. 11.10) of the nonlocal-supralocal continuous-state standing-wave present [2-5,7]. Like wave-particle duality of matter, HAM cosmology $EM - \mu\nu$ duality extends to spacetime itself in that the unified field harmonically discretizes into spatial boundary conditions of an Ising model Euclidean point (Fig. 11.12a).Two types of computer animation in terms of 'figure' and 'ground' illustrate this. First, the animated figure crosses (arrow of time) the stationary background from left to right, disappears off the screen and reappears cyclically with an inherent frame rate. Each L-R cycle can be considered as one discrete spacetime least-unit quantum to the external observer. However as well known, our so-called quantum is actually comprised of a number of discrete frames that appear continuous to the external observer because of the refresh rate. This could be considered as the properties of quantum phase space and that material Fermi surfaces appear smooth because of the relativistic velocity of the surface electrons.

In the second case, the animated figure remains permanently fixed in the center of the screen and the background moves continuously from left to right (Arrow of time again) across the screen. For the sake of the metaphor one can say that this latter case is introspective relative to the observer and the first case is objective (quantum) or external to an observer.

Neither of these two views offer a complete description of reality; as noted above, a third case of simultaneity is required. The apparent separateness of the two views; i.e. 'we live in a quantum universe' is the root of the problem because as proposed here we live in a continuous-state universe that is classical, quantum and unitary depending on perspective. The challenge here is to show that by adopting this view a model of Universal Quantum Computing with an inherent spacetime synchronization backbone can be delineated.

Noetic Space "leapfrogs" from holographic unitarity to discretized reality. This simplifies the boundary conditions and variables needed for UQC operations. The 12D Multiverse surface is considered a new form of Absolute Space (AS) and our observed Euclidean E_3 is a pseudo-AS or subspace of this regime. Because of the leapfrogging which I suppose is a fancy form of Witten's Ising flip [38] of the covariant string vertex (Fig. 11.12a). The E_3 pseudo-AS is a periodic discretization or 'frozen moment' of one 4D set of the 12D parameters (when time is included). This gives the least unit of the superspace the geometry of a torus; or in our Wheeler-Feynman future-past model considered as two 4D advanced-retarded tori. This suggests the boundary conditions A:B; A':B' (Fig. 11.8b) are HD boundary conditions of a harmonic oscillator allowing coherent control of the UQC to be operated with 4D parameters. As well known the usual form of Maxwell's equations in vacuum with $m_\gamma \neq 0$ and $B^{(3)} = 0$ has infinite families of boundary free exact solutions with the Lorentz gauge vector potential $A_\mu = 0$; but in the noetic case with $m_\gamma \neq 0$ where Maxwell's equations do not cut off at the vacuum, there is only one family and one set of boundary conditions, a model justified empirically by the existence of the Casimir and Zeeman effects [18,23-25]. *EM* theory implies the effects of the *EM* vector four-potential A_μ on the phases, *S* of quantum mechanical waves

$$\Delta S = \frac{q}{h}\int \phi dt - \frac{q}{hc}\int \vec{A}\cdot d\vec{S}. \tag{11.21}$$

For the continuous-state integration the mass term, m_γ is introduced into Maxwell's equations. One may also describe gravity with a four-vector density A_μ^g so that the Newton and Coulomb potentials take the same form but with different coupling constants suggesting both are different aspects of the same fundamental (unified) field with $A_\mu A_\mu \to 0$ where A_μ denotes the total four-potential in a covariant polarized Dirac vacuum.

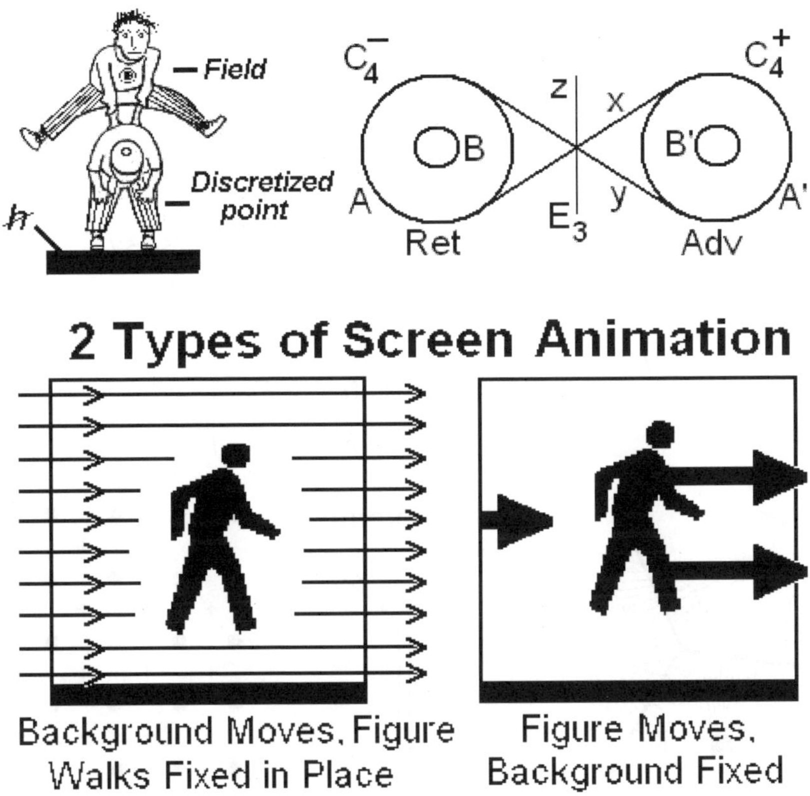

Figure 11.10 (a) Leapfrog metaphor of virtual reality. (b) This metaphor adds Ising properties to the future-past transaction. The central Euclidean point, E_3 is created and annihilated as a standing wave harmonic oscillator within the boundaries (denoted by A:B; A':B') of two complex 4D tori. (c) The leapfrog duality of the $EM - \mu\nu$ metric also includes two types of spin exchange coupling-decoupling background–foreground interactions.

From the *EM* vector potential $A^\mu(x)$ where $F_{\nu\mu} = A_{\nu,\mu} - A_{\mu,\nu}$, the components of *E* and *B* form second rank dual antisymmetric spacetime field strength tensors $F^{\mu\nu}$(Adv), $^{*}F^{\mu\nu}$(Ret) defined as $F^{\mu\nu} = \partial^{\mu} A^{\nu} - \partial^{\nu} A^{\mu}$ and $^{*}F^{\mu\nu} = \frac{1}{2}\varepsilon \exp \mu\nu\rho\sigma\, F_{\rho\sigma}$ respectively as matrices

$$F^{\mu\nu} \equiv \begin{pmatrix} 0 & -E^x & -E^y & -E^z \\ E^x & 0 & -B^z & B^y \\ E^y & B^z & 0 & -B^x \\ E^z & -B^y & B^x & 0 \end{pmatrix},$$

[56,57]. (11.22)

$$^{*}F^{\mu\nu} \equiv \begin{pmatrix} 0 & -B^x & -B^y & -B^z \\ B^x & 0 & E^z & -E^y \\ B^y & -E^z & 0 & E^x \\ B^z & E^y & -E^x & 0 \end{pmatrix}$$

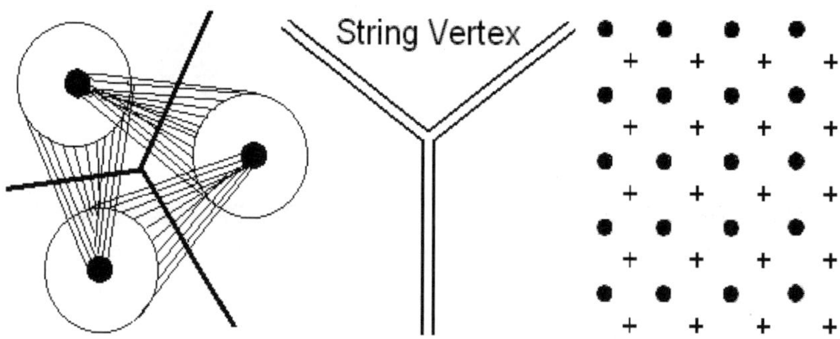

Figure 11.11 (a) Triune nature of close-packed noetic least cosmological unit with a classical discrete $\hbar$ vertex. (b) Witten Ising model string vertex. (c) Ising model spin-flip background as descriptive elements of symmetry breaking.

If properties of the Dirac vacuum are expanded to conform with noetic cosmology, Fig. 11.12b graphically represents the integration of Eqs. (11.12) and (11.13) on the top of the Dirac sea where the central point is a space-like radial 4-vector $A_{\mu} = r_{\mu} \exp(iS/\hbar)$ with frequency $\nu = m_{\gamma}c^2/\hbar$. The oppositely rotating dipoles $\pm e$ correspond to gravity and *EM* with each individual sub-element four-momentum $\partial_\mu S$. For detailed discussion see [53,54]. Figure 11.12b represents one close-packed noetic hypersphere least-unit just below this regime which is the vertex at 0 where further unification to the unitary field occurs.

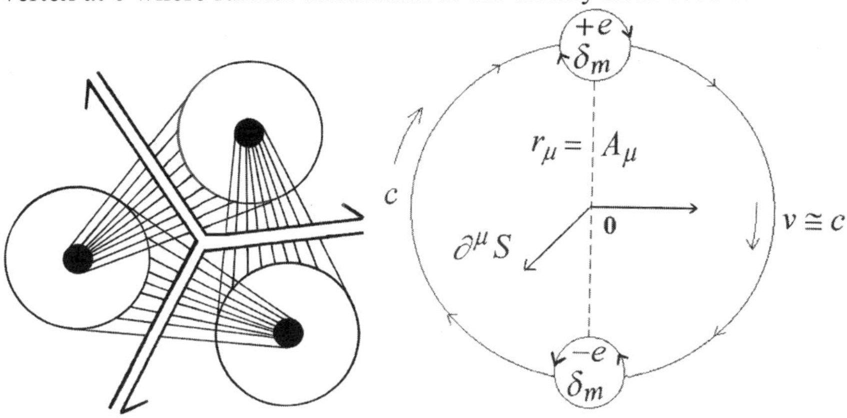

Figure 11.12 Two additional models of the close-packed least-unit points tiling the covariant Dirac backcloth (as in Fig. 11.11) in the HAM cosmology of 12D noetic superspace. (a) Fundamental continuous-state vertex as in string theory able to undergo Ising model spin-flips similar to the 0 to ∞ rotations of the HD Riemann sphere. (b) Conceptualization of two oppositely charged vacuum subelements rotating at $v \cong c$ around a central point behaving like a dipole (+*e*) *EM* 'bump' and (-*e*) *G* 'hole' on the topological surface of the covariant polarized Dirac vacuum.

This is only a superficial account of the highly essential relevance of the complementarity of the $G_{\mu\nu} - F_{\mu\nu}$ coordinate systems. More development is given in Chap. 6. Suffice it to simplistically summarize here that the dynamics of the continuous-state SUSY symmetry breaking are key to the ontological properties of this putative model of bulk QC. The G-EM coordinates couple and uncouple fixing one and then the other in a dual seesaw-leapfrogging effect which is like a form of topological wave-particle duality. It is the utilization of this structural-phenomenology as a covariant resonant hierarchy that allows the ontological violation of the Copenhagen regime uncertainty principle. See Chap. 9 for a more complete description of the empirical protocol and Chap. 12 for application to shield technology. The triune geometry of Fig. 11.12a represents the point 0 in 11.12b shown as an Ising lattice array in Fig. 11.11c. This is similar to the vertex in string theory (Fig. 11.11b) able to topologically undergo spin flips of the Riemann sphere from zero to infinity (Fig. 11.20b). In these continuous-state points the Ising vertices as governed by the super quantum potential (unified field) as described by the noetic field equation [26]. There is a foreground and background duality as illustrated in Fig. 11.10 where the *EM* and metrics continuously "leapfrog" in the spacetime backcloth. These factors are imposed on spacetime geometry by the symmetry conditions of noetic cosmology. Traditionally parallel transport of a vector or spinor around a closed path P,Q,R (Fig. 11.13a) or P,Q,R,S (Fig. 11.13b) generally results in a deficit angle, a mass deficit that signifies the amount of curvature at that vertex when the Riemann tensor is $\neq 0$ [30,58,59].

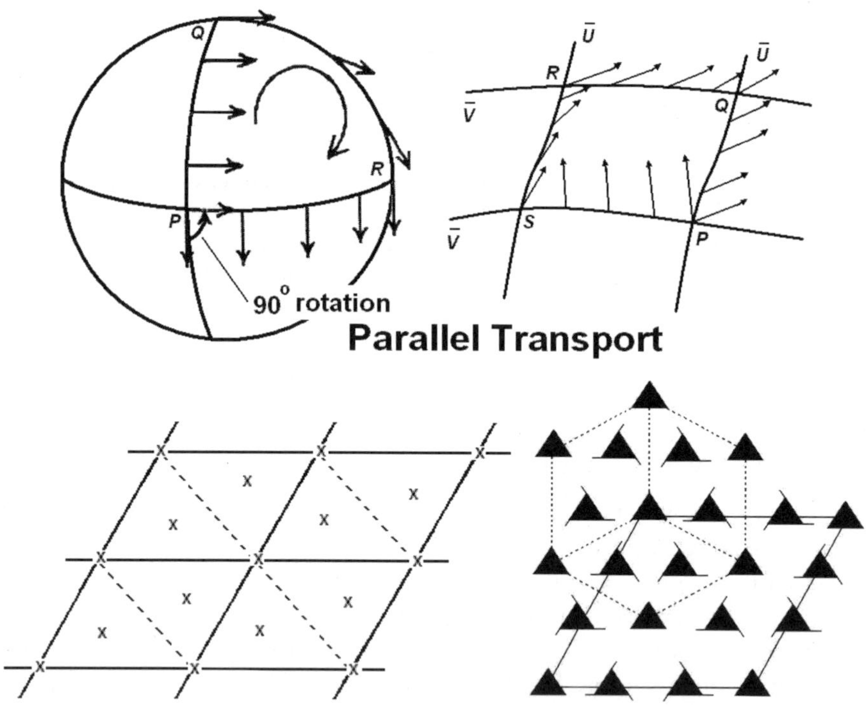

Figure 11.13 (a) 3D and (b) 2D modeling of parallel transport of a vector or spinor around closed paths which by the Regge calculus generally results in a deficit angle which is a gravitational mass defect where the vector does not return to the original position *P*. (c) Tiling of the spacetime backcloth and (d) projective geometry inherent in the Ising model string/brane spacetime backcloth giving rise to the emergence of higher dimensionality.

Tiny loops approximated by a parallelogram of two tangent vectors $\vec{\mu}$ and $\vec{\nu}$ close (no deficit) if $\left[\vec{\mu},\vec{\nu}\right]=0$; then the curvature operator is the commutator of covariant derivatives along $\vec{\mu}$ and $\vec{\nu}$, $R(\vec{\mu},\vec{\nu})=\left[\nabla_{\vec{\mu}},\nabla_{\vec{\nu}}\right]$ [30]. If $\left[\vec{\mu},\vec{\nu}\right]\neq 0$, $\left[\nabla_{\vec{\mu}},\nabla_{\vec{\nu}}\right]$ is subtracted from the commutator, the parallelogram doesn't close and the Riemann tensor is $\neq 0$.

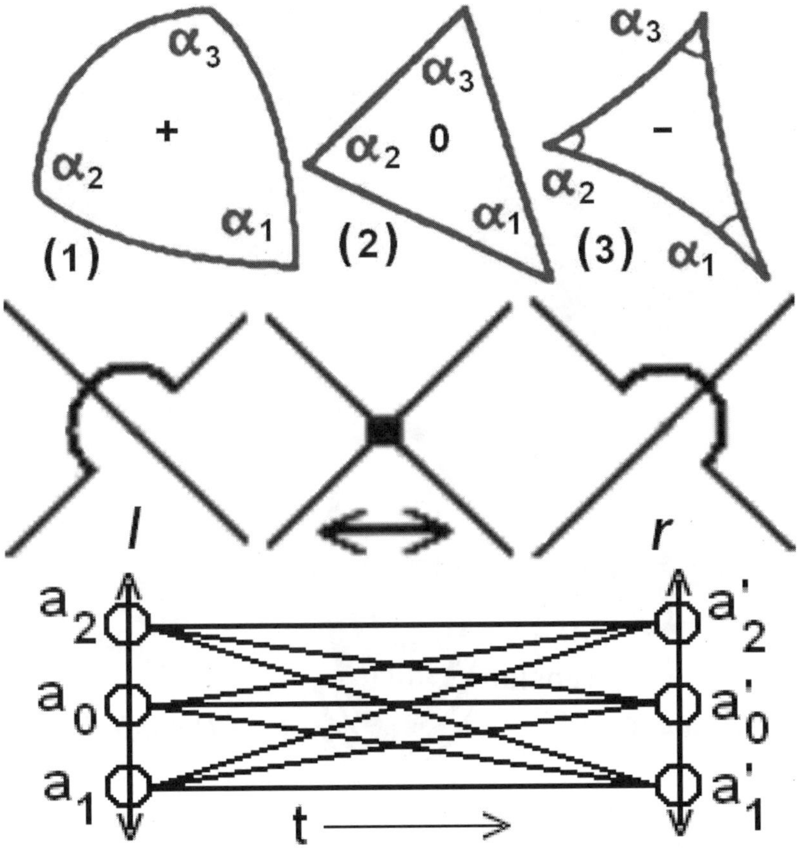

Figure 11.14 (a) Three types of geodesic triangles with Gaussian curvature. 1) Circumsphere with positive curvature, sum of internal angles $>\pi$. 2) Mesosphere, E^3 with zero curvature. 3) Insphere, internal angle sum $<\pi$ so curvature is negative. (b) Chiral properties of a vertex where the coordinate basis topologically switches from fixed to l or r open. (c) Triune elements of an HD transaction in noetic terms where the elements of a least-unit are tertiary.

In Fig. 11.14a the sum of the three internal angles minus π is the Gaussian curvature integral $(\alpha_1 + \alpha_2 + \alpha_3) - \pi \int K dA$ with K the Gaussian curvature. Taking Fig. 11.14 triangle (a) for example on a sphere of radius r with $\alpha_1 = \alpha_2 = \alpha_3 = \pi/2$ the area of the triangle is $(4\pi r^2)/8$ and the Gaussian curvature would be $K = 1/r^2$ which is positive [30].

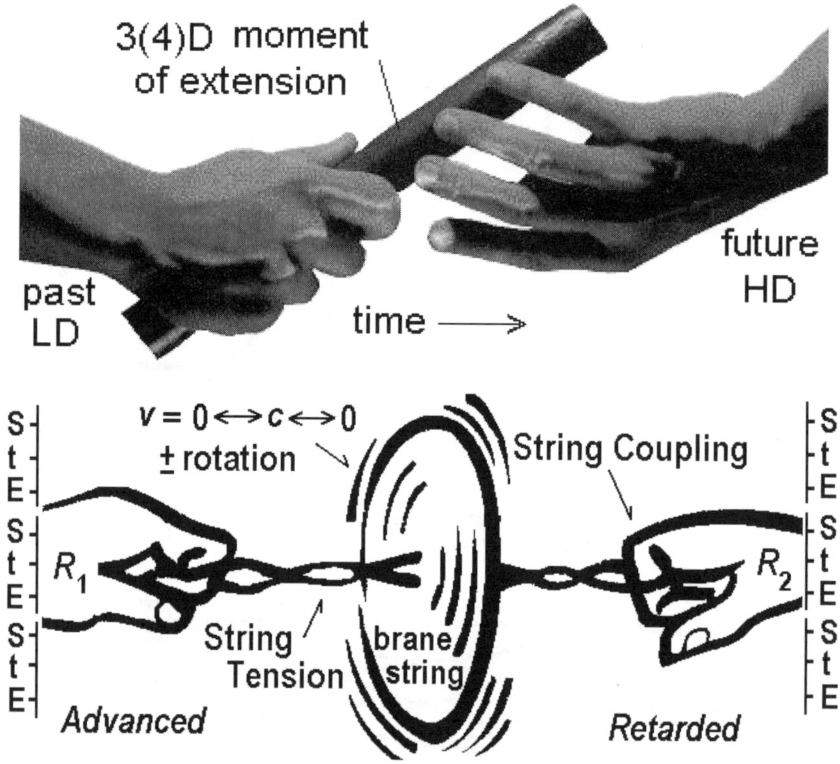

Figure 11.15 Spin Exchange (a) The spin exchange mechanism requires a coupling-decoupling moment between the $c \to q \to u$ components of the spacetime least-units like the passing of a baton in a relay race. (b) The spinning disk toy further illustrates elements of the continuous-state. Imagine an array of disks as in Fig. 11.12b.

The fundamental continuous-state spin-exchange hierarchy process has many components; more are shown in Fig. 11.16.

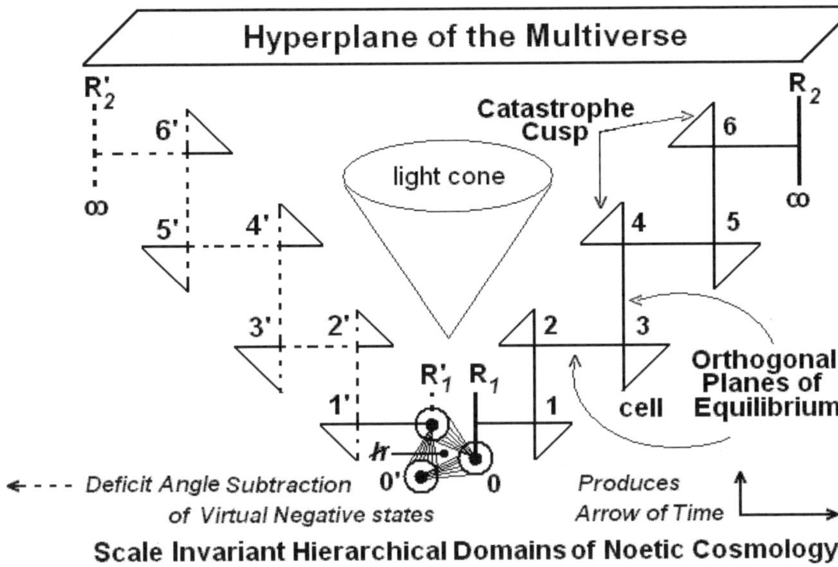

Figure 11.16 Covariant scale invariant hyperplane domains in the hierarchy of noetic superspace.

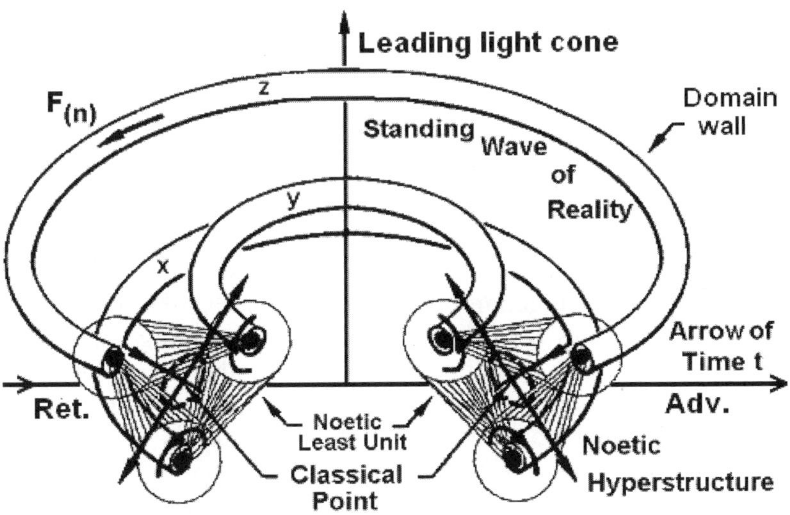

Figure 11.17 Illustration of a single future-past, retarded-advanced domain where the properties illustrated in Figs. 11.10, 11.11, 11.12 and 11.16 interplay to produce the observed macroscopic arrow of time. See Chap. 5.

The dominant view among cosmologists regarding extra dimensions is that if they exist they must be microscopic because they are not observed in the same manner that the 10^{-33} cm Planck scale is not observed. Even though M-Theory is cast in HD, perhaps a majority of physicists are still opposed to dimensionality > 4D because as yet there is no empirical evidence (see Chap. 9 for possibility) In noetic cosmology extra dimensions are macroscopic and take part in the creation and recreation of spacetime, the arrow of time and observed macroscopic reality (Fig. 11.13). This scenario arises during the inherent 'continuous-state spin-exchange dimensional reduction compactification process' by parallel transport (Fig. 11.10) within the additional context of a dual Dirac spherical rotation of the least-unit topology (Fig. 11.8) of subspace elements producing deficit angles during decoupling-coupling allowing relativistic subtraction of supralocal-nonlocal domain components (Figs. 11.10, 11.11, 11.12 and 11.13) producing the arrow of time (Chap. 5). The scaling process begins in the microscopic backcloth without a physical arrow of time and ramps up the helicoid hierarchy (Fig. 11.5) to the virtual standing-wave macroscopic present. Because of its relativistic nature the 'baton' passing (coupling-decoupling) between domains appears smooth to the observer. Figure 11.13 is meant to be synonymous with the lightcone rings of Fig. 11.7 where the leapfrogging domain frequency provides the context for assigning coupling parameters required for utilizing the synchronization backbone for the UQC.

11.8 Geometric Introduction to the Noetic QC Ontology

What are the topological conditions required to achieve a commutative ontology for UQC? Newton's 2nd Law of motion says position and velocity completely determine an observable's 'state', (p, q) at an instant in time. Quantum mechanically an observable has a probabilistic distribution of values (P, Q), with quantization making correspondence between the two [60]; conditions that delineate the uncertainty principle and provide no framework for a pragmatic absolute ontology. In HAM cosmology neither spacetime nor stochasticity is considered fundamental. This is not a different basis than the concept of Heisenberg's potential; so what is required is a new process. Spacetime is a continuous harmonic state comprised of the Amoroso-Vigier dual $EM \rightleftarrows \mu\nu$ metric [53,54] comprised of conventional transverse 'EM' elements described by Maxwell's traditional equations plus longitudinal $\mu\nu$ elements with additional degrees of freedom derived from the Einstein-de Broglie relation $E = h\nu = mc^2$ (with $m = m_0(1 - v^2 / c^2)^{-1/2}$) such that Maxwell's equations do not 'cut off' at the vacuum. Evidence for such a metric is implied by the Casimir, Zeeman & Aharonov-Bohm effects [61]. These two sets of coordinates EM + $\mu\nu$ would generally be exclusive and independent. The aim here is to reveal a framework for their 'continuous-state' integration, not in 4D as previously done [53,54] but in 12D where integration is completed to unitarity.

The close-packed least unit hypersphere tiling of this noetic superspace is a complex self-organized scale invariant anticipatory system. While beyond the scope of this paper, operational interplay of the parameters of the fundamental least unit is discretized macroscopically into perceived reality. Normally local application of an observational RF pulse in the z direction discretizes the uncertainty relations of microscopic quantum states for particles. To avoid production of these uncertainties inherent in the quantum principle, a new set of Noetic transformations beyond the Galilean-Lorentz-Poincaré must be implemented by a cumulative interaction methodology to allow a 'coherent control' transformation of the phenomenology of discretization into an ontological superposition of the information.

To illustrate we apply general mechanical principles for 'pure rolling contact' [62] to the transmission of angular momentum translating through the topology of this HD spin tower (Figs. 11.7 and 11.18), the relative motion of consecutive elements propagate successively in proper order with the elements of parallel axes in the corresponding topological surface. These motions may be ± coupled combinations relative to the center of mass and components of angular momentum that are singular (degenerate), linear, circular, cylindrical, spherical and hyperspherical. This reveals the richness of the cosmological least unit as it undergoes continuous-state spin-exchange (rolling contact) compactification (past orientation) and Ising dimensional flip (future orientation) in quantifiable stages of dimensional jumps from 12D to 0D by superluminal Lorentz boosts [2,63] in cyclic progression $S \rightarrow t \rightarrow E$ and $C \rightarrow Q \rightarrow U$ (space to time to energy; classical to quantum to unitary).

12D, the minimum to describe eternity or escape from the temporal bounds of uncertainty is a result of the dimensional tower (Figs. 11.7, 11.8, and 11.9) where time and E_3 / M_4 is a standing wave subspace of eternity (Figs. 11.3, 11.4). This structure whether Calabi-Yau, dodecahedral, or some M-Theory, F-Theory combination entails a reciprocal spiral topology. In this context we utilize logarithmic, helicoid or cyclotron resonance spirals (Figs. 11.7, 11.8, 11.18) to illustrate new angular momentum commutators. The future/past asymmetry has a Doppler relationship (Figs. 11.16 and 11.17) (relative only to the perception of the 4D observer). *E* & *E'*, (Fig. 11.8d) therefore represent equal Wheeler-Feynman future/past symmetries. The Doppler effect arises because of inherent *E-E'* boosts and compactifications. In this picture the Wheeler-Feynman-Cramer elements [9,10] may be understood conceptually by pairs of logarithmic spirals (Fig. 11.18a) of equal obliquity rolling on a common tangent, *ed* where each coupled point signifies a present spacetime moment; the locus of which (*ed*) is the arrow of time. A radiant of the spiral, *r* is

$$r = ae^{b\theta} \tag{11.23}$$

with a the value of r if $\theta = 0$, e base of the Naperian logarithms and $b = 1/\tan\phi$, with ϕ the constant angle between the tangent to the curve and radiant to the point of tangency [64]. If the value of θ takes uniform increase (quantized values) the radiants, r will be ct = 0,1,2,3…,*n* in geometric progression relative to the hierarchal topology of the space (Fig. 11.18b).

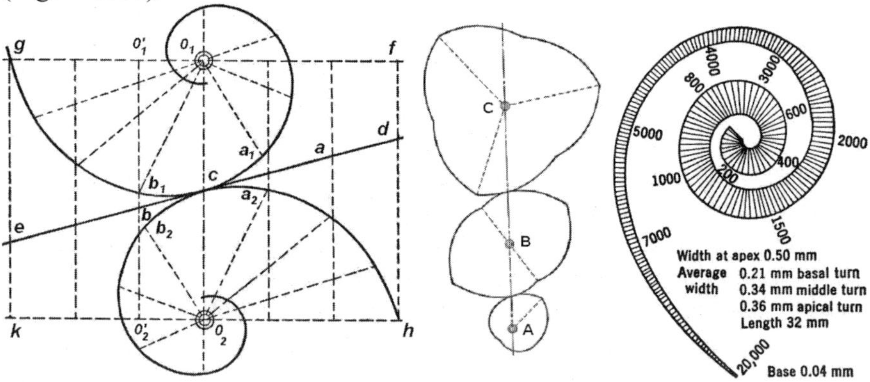

Figure 11.18 (a) Two logarithmic spirals illustrating perfect rolling contact (no slip) that cannot be continuous because of boundary limits. (b) Continuous rolling contact by 1, 2 & 3 lobed spheres (segments of (a)) illustrating how HD hierarchy nodes may be formed. (c) The basilar membrane of the ear is tapered (like radiants) roughly like a logarithmic spiral beginning at a Planck point and widening to the Larmour radius of an atom. Each width is frequency dependent causing sound input of specific frequencies to vibrate more in the location where the radius has the same characteristic resonant frequency.

These log curves are not closed; to adapt to continuous motion, pairs must be utilized. Joining corresponding sections of the spiral form symmetrical unilobed wheels. While sectors needn't be equal or symmetrical, the 'wheels' must be paired with sectors of equal obliquity in contact for pure rolling motion to occur. Wheels may also be bilobed or trilobed etc. up to ND to illustrate the Superspace. A tier of three symmetrical wheels is illustrated in Fig. 11.18b.

The mechanical concept of rolling contact is used to geometrically illustrate the ontological framework for the new noetic commutation rules of angular momentum. A logarithmic spiral coupled to another of the same obliquity undergoes perfect rolling motion (no slippage and constant touching) as long as arcs of the same obliquity coincide. This system of spirals reaches a limit that could be said to be points of Ising flip; but the rotation is not continuous. To make the rolling continuous one must take 2 sections of the logarithmic spiral (Fig. 11.18a) and join them into a spheroid. Then continuous motion may occur. As in Fig. 11.18b this single lobed gear may be made bilobed or trilobed, again for continuous or perfect rolling motion proper obliquity must be maintained.

So here as in the ear metaphor the points of contact correspond to frequencies. If the point of contact corresponds to the *z* axis we have moments of commutation of angular momentum. Leaving one gear set (the spin tower of frequencies) we have a system of close packed spheres of least cosmological units undergoing the noetic mantra (spin-exchange, dimensional reduction, compactification) which means that there are HD moments of commutation in the 12D structure.

Since angular momentum is the resultant of the atomic magnetic moment and (center of mass) harmonic frequencies (as in the cyclotron frequencies of synchrotron radiation) should make these other (x and y) components of angular momentum accessible, In any given discretized (composite) E_3 frame only the *z* axis will commute as per standard quantum theory; but in the complex HD space the E_3 non-commutative parameters commute periodically on rotation through mirror tangent nodes of proper obliquity in the continuous state topology; i.e. in considering all HD hyperplanes, there are periodic simultaneous moments where nodes of commutation may be accessible by synchronizing RF pulses of the proper harmonic cyclotron frequency.

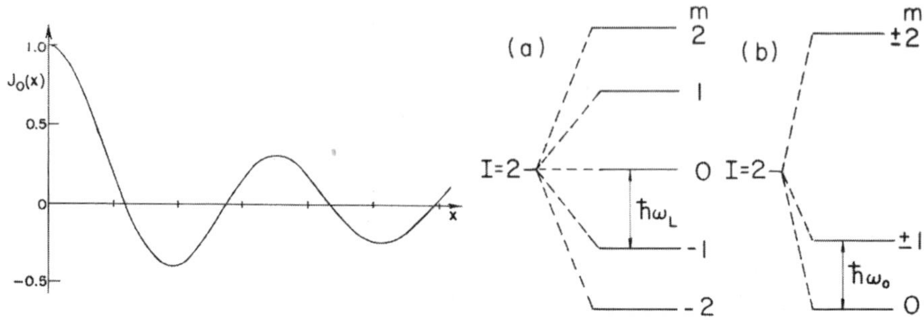

Figure 11.19 (a) Graph of Bessel function $J_0(x)$ showing how J_0 oscillates as x increases. (b) Nuclear energy splitting of substates for total spin 2 in a magnetic field and (c) axially symmetric electric field with quadratic energy splitting, ω is the spin precession frequency.

Bessel functions could be used to manipulate the complex cavity resonance modes. For example in a generalized Cramer event cavity (between future-past topological boundaries) the magnitude of a uniform applied electric field with E_0 constant can be taken as $E = E_0 e^{i\omega t}$. If the frequency increases the electric field flux through any loop Γ_1 produces an oscillating magnetic field $B = i\omega r / 2c^2 \cdot E_0 e^{i\omega t}$ proportional to r, the radius of the cavity. This varying magnetic field, proportional to the rate of change of E and thus ω, effects the electric field so it can no longer be uniform by Faraday's Law and also changes with r [65]. This requires corrections to our original uniform field E_1 such that the corrected field must now be $E = E_1 + E_2 + E_3 ... E_n$ which is best described by the Bessel function J_0 with $x = \omega r / c$

$$J_0(x) = 1 - \frac{1}{(1!)^2}\left(\frac{x}{2}\right)^2 + \frac{1}{(2!)^2}\left(\frac{x}{2}\right)^4 + \frac{1}{(3!)^2}\left(\frac{x}{2}\right)^6 + \ldots \quad (11.24)$$

such that E is now

$$E = E_0 e^{i\omega t} J_0\left(\frac{\omega r}{c}\right). \quad (11.25)$$

11.9 Microphysical Computation Limits: Case of Relativistic Qubits

In the conventional consideration of quantum computing a quantum bit or qubit is any two-state quantum system defined as a superposition of two logical states of a usual bit with complex coefficients that can be mapped to the Riemann sphere by stereographic projection (Fig. 11.20). Formally a qubit is represented as: $\Psi = \xi|0\rangle + \eta|1\rangle$ with each ray $\xi, \eta \in C$ in complex Hilbert space and $\|\Psi\|^2 = \xi\bar{\xi} + \eta\bar{\eta} = 1$, where $|0\rangle$ corresponds to the south or 0 pole of the Riemann sphere and $|1\rangle$ corresponds to the opposite or north or ∞ pole of the Riemann complex sphere. The conventional qubit maps to the complex plane of the Riemann sphere as:

$$\xi\bar{\eta} + \eta\bar{\xi} \to X, \quad \xi\bar{\eta} - \eta\bar{\xi} \to iY, \quad \xi\bar{\xi} - \eta\bar{\eta} \to Z. \qquad (11.26)$$

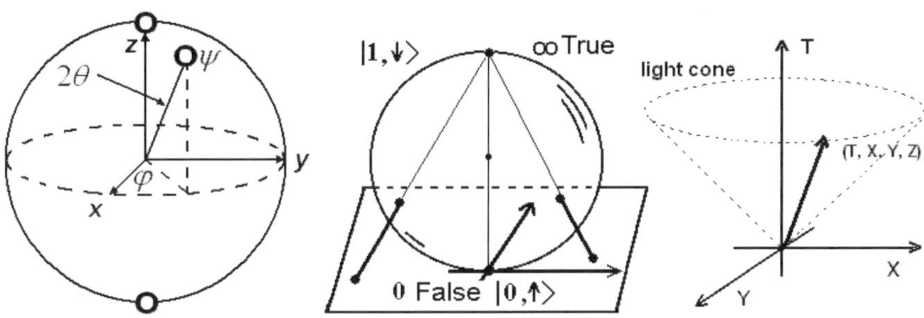

Figure 11.20 The qubit. (a) Block Sphere representation of a qubit, a geometrical representation of the pure state space of a two-level quantum mechanical system. Alternately, it is the pure state space of a 1 qubit quantum register. (b) Stereographic projection model of a qubit on a complex Riemann sphere. (c) Relativistic model of a qubit with interacting quantum fields.

Unitary transformations of a qubit correspond to 3D rotations of the Riemann sphere. Following Vlasov [66] for relativistic consideration of a qubit (r-qubit) an additional 4D parameter is added to equation (11.24):

$$\begin{aligned} &\xi\bar{\eta} + \eta\bar{\xi} \to X, \quad \xi\bar{\eta} - \eta\bar{\xi} \to iY, \\ &\xi\bar{\xi} - \eta\bar{\eta} \to Z, \quad \xi\bar{\xi} + \eta\bar{\eta} \to T \end{aligned} \qquad (11.27)$$

In cartography and geometry, the stereographic projection is a mapping that projects each point on a sphere onto a tangent plane along a straight line from the antipode of the point of tangency (with one exception: the center of projection, antipodal to the point of tangency, is not projected to any point in the Euclidean plane; it is thought of as corresponding to a "point at infinity"). One approaches that point at infinity by continuing in any direction at all; in that respect this situation is unlike the real projective plane, which has many points at infinity.

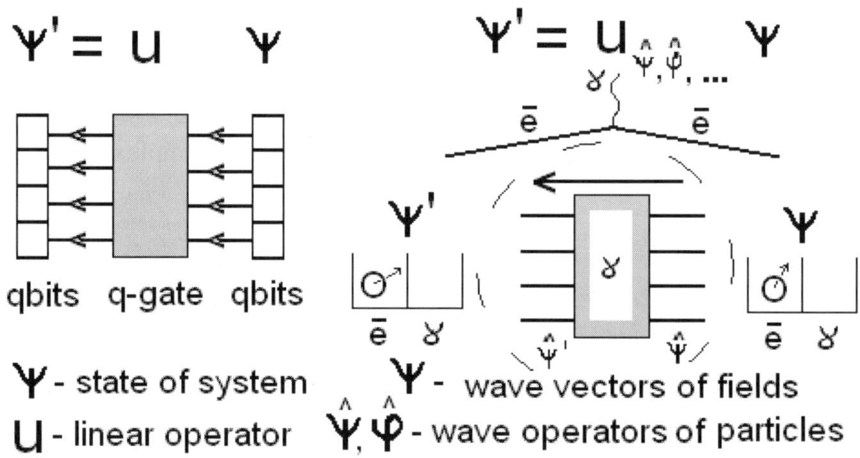

Figure 11.21 (a) Usual q-gate with constant number of states and particles. (b) Relativistic quantum bit (r-qubit) with constant particles but variable or infinite states.

11.10 Essential Parameters of the Incursive Oscillator

The evolution of physical theory from Classical to Quantum changed the fundamental understanding of a point or point particle from continuous – represented in 3D Euclidean space, to discrete fuzzy units with wave-particle duality–represented in 3(4)D Minkowski/Riemann spacetime. As physical cosmology has evolved towards M-Theory it is now realized that neither of these contexts is sufficient or complete. In Multiverse cosmology the nature of a vertex or point changes into a continuous-state 12D superspace. This means there are three regimes existing simultaneously/individually: Classical, Quantum and Unity depending on mode of observation.

Motion of a one-dimensional classical harmonic oscillator is given by $q = A\sin(\omega t + \varphi)$ and $p = m\omega A\cos(\omega t + \varphi)$ where A is the amplitude and φ is the phase constant for fixed energy $E = m\omega^2 A^2 / 2$. For state $|n\rangle$, with $n = 0,1,2,...,\infty$ and with Hamiltonian $E_n = (n+1/2)\hbar\omega$ the quantum harmonic oscillator becomes

$$\langle n|q^2|n\rangle = \hbar / 2m\omega \langle n|(a^\dagger a + aa^\dagger)|n\rangle = E_n / m\omega^2$$

and

$$\langle n|p^2|n\rangle = 1/2(m\hbar\omega)\langle n|a^\dagger a + aa^\dagger = mE_n$$

where the terms a and $a^\dagger$ are the annihilation and creation operators, $q = \sqrt{\hbar/2m\omega}(a^\dagger + a)$ and $p = i\sqrt{m\hbar\omega/2}(a^\dagger a)$. For the 3D harmonic oscillator each equation is the same with energies $E_x = (n_x + 1/2)\hbar\omega_x$, $E_y = (n_y + 1/2)\hbar\omega_y$ and $E_z = (n_z + 1/2)\hbar\omega_z$ [18,21].

In Dubois' notation the classical 1D harmonic oscillator for Newton's second law in coordinates *t* and *x(t)* for a mass *m* in a potential $U(x) = 1/2(kx^2)$ takes the differential form

$$\frac{d^2x}{dt^2} + \omega^2 x = 0 \quad \text{where} \quad \omega = \sqrt{k/m} \tag{11.28}$$

which can be separated into the coupled equations (11.29)

$$\frac{dx(t)}{dt} - v(t) = 0 \quad \text{and} \quad \frac{dv(t)}{dt} + \omega^2 x = 0. \tag{11.29}$$

From incursive discretization, Dubois creates two solutions $x(t+\Delta t)$ $v(t+\Delta t)$ providing a structural bifurcation of the system which together produce Hyperincursion. The effect of increasing the time interval discretizes the trajectory as in Fig. 11.22 [6,67-69]. This represents a background independent discretization of spacetime.

Each mode of a quantum harmonic oscillator is associated with cavity-QED dynamics, hexagon lattices (Fig. 11.22c) of spacetime topology undergoing continuous transitions. *E* is the state of energy for *n* photons. For $n = 0$ the oscillator is in the ground state, but a finite energy

$1/2\hbar\omega$ of the ground state, called the zero-point energy, is still present in the region of the cavity. According to Eq. (11.30), the quantum harmonic oscillator field energy of the photons undergo periodic annihilation and recreation in the periodic spacetime [70].

$$E_n = (n + \frac{1}{2})\hbar\omega . \tag{11.30}$$

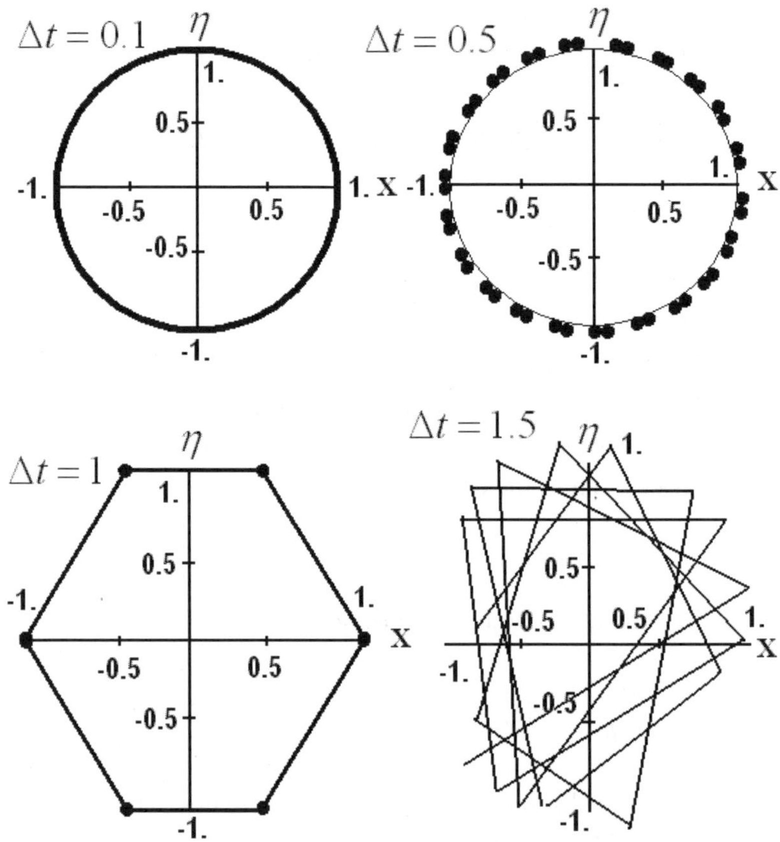

Figure 11.22 Numerical simulation of phase space trajectory for Dubois' *superposed incursive oscillator* based on coordinates and velocities $x_n = 1/2[x_n(1) + x_n(2)]$ $v_n = 1/2[v_n(1) + v_n(2)]$ is shown for values of $\Delta\tau = \omega t$ equal to 0.1, 0.5, 1.0 and 1.5. Initial conditions are $\chi_0 = 1, \eta_0 = 0$ and $\tau_0 = 0$ with total simulation time $\tau = \omega t = 8\pi$. Figure adapted from [6,68].

11.11 Ontological I/O by Superceding Quantum Uncertainty

The critical problem in applying conventional QT to the bulk implementation of QC lies in the accompanying theory of measurement [21]; variables observed change destructively in any interaction between particle and observing apparatus. This phenomenological *force of interaction* is mediated by particle exchange which modifies the Schrödinger equation. In conventional terms 'physical reality is irreducibly quantum' and a qubit resides at a Euclidean, E^3 or Minkowski, M^4 vertex.

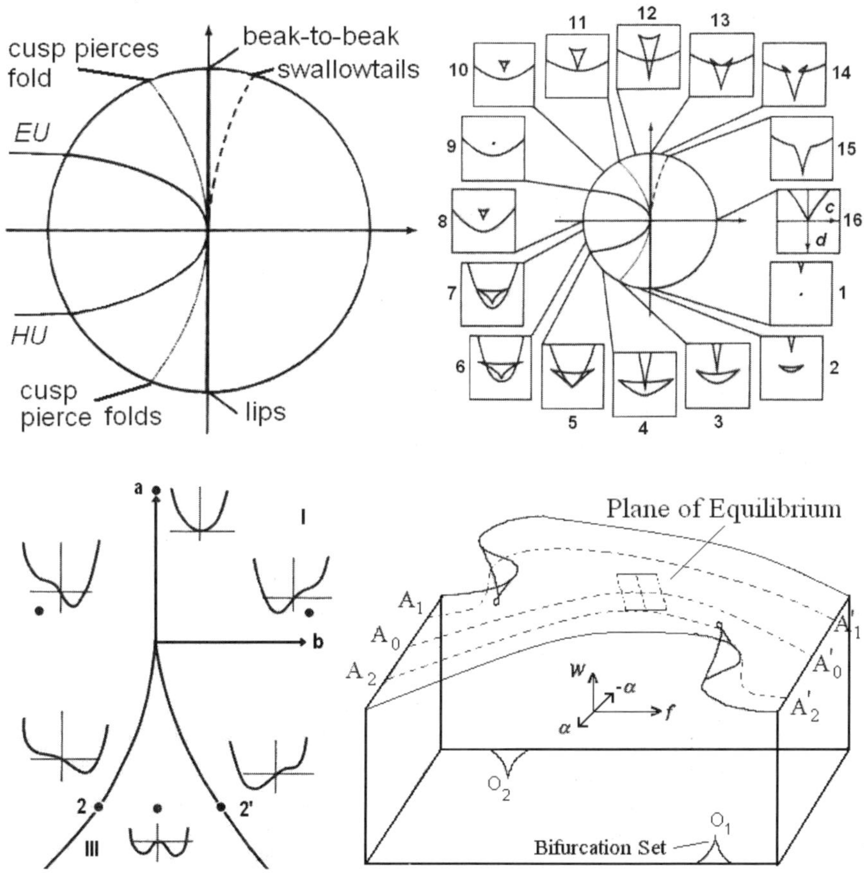

Figure 11.23 A double-cusp catastrophe (DCC) provides a partial geometric and mathematical model of noetic superspace transitions. The inherent Dirac rotation is like a DCC.

All attempts for bulk QC have failed in the Copenhagen regime because measurement destroys the quantum system being measured. To overcome this problem the Dirac equation is hyperdimensionalized utilizing an extension of Cramer's Transactional Model of QT where all off diagonal elements are physically real and conformally invariant. Bulk implementation of UQC requires a new superspace N^{12} without a real vertex where not only is the arbitrarily chosen z-axis of angular momentum accessible; but the x and y components are also real and accessible by a new anticipatory transformation law for ontological evolution utilizing topological switching [15]. This is conceptually elucidated by unfolding a hypercube (Fig. 11.1). Relative to the subspace E_3 the extra square called a satellite is *causally free* of E_3 when carried to 12D unitarity.

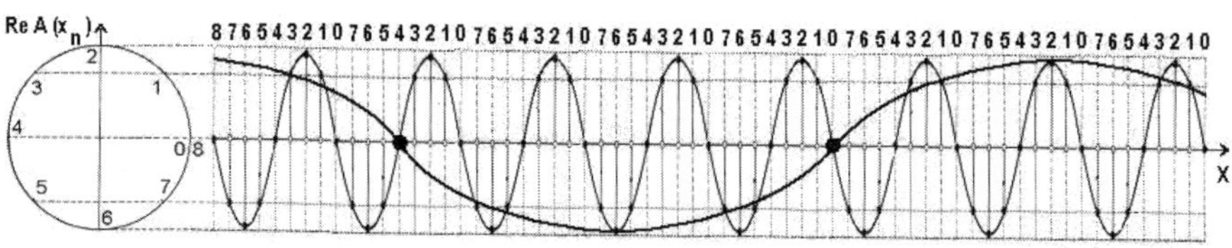

Figure 11.24 Depiction of 2π and 14π harmonic waves coinciding at 2 points on the x-axis corresponding to points 4 and 0,8 on the reference circle. The geometry of a reference circle (2D for simplicity, actually an HD hypersphere) is utilized to set up RF harmonic oscillator π-pulse parameters for phase alignment with the inherent Adv-Ret elements of the spacetime synchronization backbone. The periodicity of the phase points ϕ are aligned to manipulate symmetries of corresponding regimes of commutative and noncommutative modes.

During the HD continuous-state topological transformation of the cosmological form of Dirac spherical rotation, a pinch or twist occurs in the middle of the transform followed by an Ising flip [38] of the close-packed complex Riemann spheres which can be driven by the micromagnetic spintronics [15] of fractional and integer quantum Hall effects because of the highly symmetric topological parameters [2] of driven Micromagnetics [15]. This UQC can be implemented in any sufficient multi-state quantum system, whether solid, liquid, bubble, crystal, dot, network, trap, well, vacuum backcloth, comprised of atoms, molecules, ions, photons, spins, NMR, threads, lines, block walls, domain walls, lattices or arrays able to utilize coherent control of the synchronization backbone [31]. In order to avoid the Copenhagen limitations of collapse and dissipation [71] UQC requires utilization of the hierarchical and recursive properties of complex self-organization inherent in the *whole universe*, not just a portion of its observed parameters. The critical condition is the introduction of a model for evolution of the wave function making correspondence to a new non-collapse (ontological or energyless) version of RQFT.

By a coherent control of Ising spin flips [38] of the noetic spacetime least-units (a topological switching of metrics [15]) domains of discretization ($\Delta x \Delta \rho \cong \hbar$) may be avoided by utilizing periodic nodes in the resonant hierarchy that are commutative because the Riemann curvature tensor equals zero [72]. E_3 is a discretization, a composite of future-past potentials. In HD where the parameters are separated one can manipulate commutative and noncommutative regimes. Another way to illustrate the intended use of coordinated RF sine wave π - pulses (Fig. 11.20) with the geometry of spatial rotations of a pair of common dice to show that some rotations commute, $a \otimes b = b \otimes a$ and others are noncommutative $a \otimes b \neq b \otimes a$.

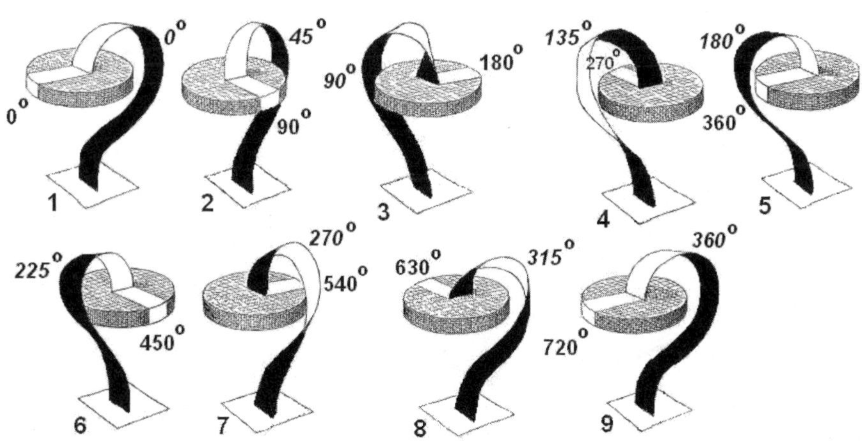

Figure 11.25 Conceptualization of the spinor geometry of Dirac spherical rotation showing the 360° - 720° degree complementarity structure of spin ½ particles without the topological pinch.

11.12 A Twistor Approach to the UQC I/O Ontology

Because of the essential requirement of utilizing an HD form of Dirac spherical rotation to access the inherent synchronization backbone in HAM cosmology; it is suggested that a Penrose twistor approach provides the most efficient methodology for coupling to the resonant hierarchy. We illustrate this only briefly here and leave it to a future paper or other QC researchers to develop more fully.

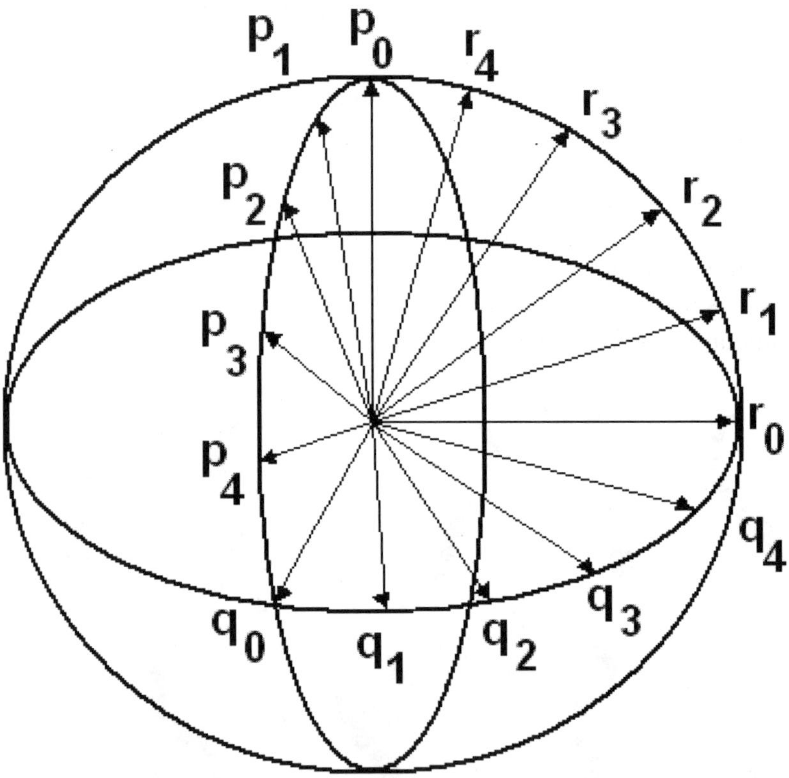

Figure 11.26 Hyperspherical modeling as a visual aid for switching the coordination of phase angles for Dirac spherical rotation.

Given the worldline, $y^a(s)$ and then following Bailey and Penrose [73,74], from the fundamental twistor relation, $Z^\alpha = (\omega^A, \pi_{A'})$ the function $\xi^A(s) = \omega^A - iy^{AA'}(s)\pi_{A'}$ is then defined. Then for the scalar field contours (Fig. 11.23) we define a twistor function $f(Z^\alpha)$ by

$$f(Z^\alpha) = \oint \frac{ds\,\alpha \cdot \beta}{(\alpha \cdot \xi)(\beta \cdot \xi)} \tag{11.31}$$

where α_A and β_A are fixed spinors and $\alpha \cdot \beta = \alpha_A \beta^A$. In this regime the field produced by the unit charge has poles corresponding to advanced and retarded points on the worldline [73,74]. Taking an EM field potential $\Phi AA'(x)$ with left and right handed components given by $\phi A'B' = \nabla^A_{A'\Phi B'A}$ and $\phi AB = \nabla^{A'}_{A\Phi BA'}$ respectively [73,74].

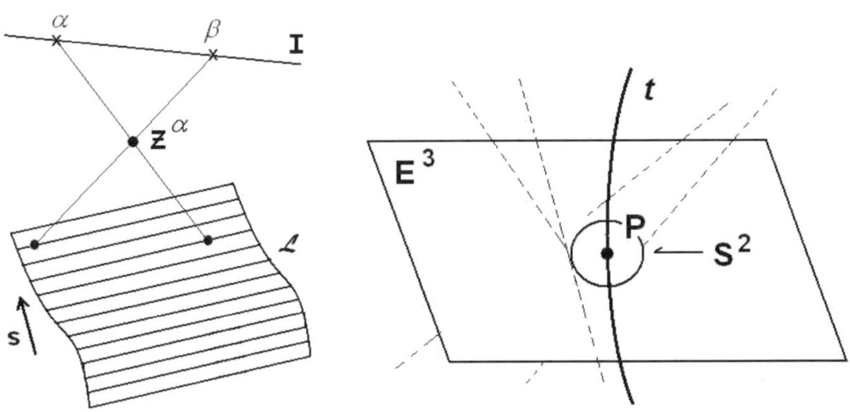

Figure 11.27 (a) A piece of ruled surface L for worldline $y^a(s)$ where each line on the surface represents a point on the complex worldline I. (b) Small sphere, S^2 surrounds E^3 worldline P with null twistors Z^α representing null lines meeting S^2.

Twistor functions describe relative cohomology classes in $\mathbb{PT}$ regions; but the same twistor functions may also be examined geometrically in M^4 [74]. The contour in Eq. (11.31) is a small loop around the $\alpha \cdot \xi = 0$ and $\beta \cdot \xi = 0$ poles (Fig. 11.23). There are two of these, one for advanced and one for retarded solutions. When a singularity is reached (Dirac pinch) one switches from $f(\alpha)$ to $f(\beta)$. In a small neighborhood U of L, U_α, U_β keeps away from the branching singularity of $f(\alpha), f(\beta)$. The process of doing contour integrals gives a well-defined field; choice of contour gives any linear combination of Adv. and Ret. solutions. The α and β spinors represent opposite directions in E^3 but not in the same regions. The contours move continuously from Ret to Adv [73,74].

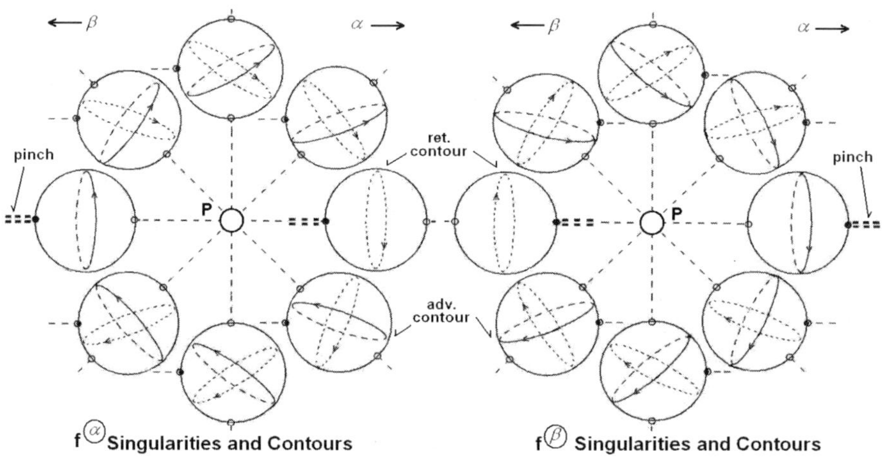

Figure 11.28 Contours and singularities of two linearly independent advanced (solid lines) and retarded (dashed lines) fields that can be computed by contour integration.

Taking the spin structure hierarchy of 1-4 benzosemiquinone (Fig. 11.24) or class II mesoionic xanthines [75] for example and aligning it with the inherent synchronization-backbone of noetic cosmology using the Dirac spherical rotation contour integrals as defined by the Penrose twistor functions in Figs. 11.27 and 11.28 [73,74] as an intermediary we are able to achieve the rolling motion contacts suggested metaphorically in Fig. 11.18. but in the Dirac spherical rotation manner of Fig. 11.25. Why? This is to achieve ontological topological switching with the satellite regime of Fig. 11.3. Noetic theory postulated that this path is only open in the continuous-state leapfrogging of the Vigier-Amoroso coordinates [53,54]. These coordinates fix and unfix; this is a cosmological utility of the Dirac rotation first discovered for the electron.

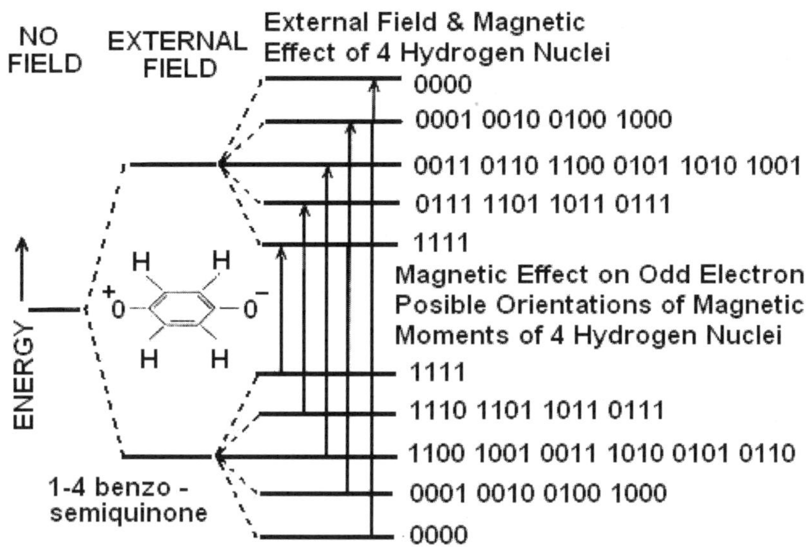

Figure 11.29 The spin states of 1-4 benzosemiquinone, a molecule suitable for testing QC parameters by application of RF fields.

This complexity arises because the Dirac pinch (illustrated in Figs. 11.6c,11.8b and 11.11b) is a fundamental process of reality (not just the electron) since the eternal origin of the unitary field is causally separated from E^3. To comprehend one must hold Fig. 11.6c in ones mind while wrapping it around the context of Fig. 11.21 where the interplay of the three regimes (classical, quantum, unitary) occurs. In order for the I/O pulses to achieve coupling to the proper leapfrogging contours the resonance modes of the RF pulses must align precisely with the inherent beat frequency of the spacetime backcloth, i.e. without the coherent control [76-78] the inherent synchronization backbone provides there can be no cumulative interaction with the Dirac spherical rotation hierarchy and no ontological initialization or processing of the QC registers and the QC remains stuck at the ten qubit limit of the Copenhagen regime.

11.13 Class II Mesoionic Xanthines as Potential 10-Qubit Quantum Computer Substrate Registers

Perhaps better than the 1-4 benzosemiquinone, a molecule are Class II mesoionic xanthines such as *anhydro*-(8-hydroxyalkyl-5-hydroxy-7-oxothia-zolo[3,2-a] pyrimidinium hydroxides) are unique, small atomic weight, stable crystalline organic compounds that can be represented as a combination of ten different resonance structures for each simple xanthine molecule. Each resonance structure contributes a certain percentage to the total resonance of the molecule. This unique resonance represents ten different quantum states of the entire molecule and can thus be exploited as a potential substrate for a ten-qubit register. The number of possible superposition states for such a register in a single molecule is potentially as high as 2^n states or (in this case where $n = 10$) 1,024 complex numbers. In solution the least-unit of this mesoionic crystalline structure is scalable suggesting putative utility for bulk NMR quantum computing. It will be shown that these ten-qubit registers are amenable to standard Deutsch-Jozsa, Shor and Grover algorithms. Additionally, we attempt to formalize I/O techniques for our Class II mesoionic xanthines based on a coherent control RF process of cumulative resonant interaction where by utilizing additional degrees of freedom pertinent to a relativistic basis for the qbit (r-qbit) new HD commutation rules allow decoherence to be ontologically overcome.

Mesoionic purinone analogs, a large and relatively new class of bicyclic heteroaromatic compounds, whose ring systems possess π-electron systems that are isoelectronic with those of the various known purinones, have been synthesized and characterized over the last few decades [79-85]. Class-I mesoionic analogs have been classified and defined as being those that are derived from known five-membered mesoionic ring systems. Class II mesoionic analogs are those that are derived from known six-membered mesoionic ring systems. In 1996, Giandinoto, et al. [86] had synthesized and characterized a number of novel Class II mesoionic xanthine acyclonucleosides as potential anti-neoplastic and antiviral agents. Class-I and Class-II mesoioic purinones have been formulated and examined from a quantum chemical standpoint [87,88]. The generalized structural representation of mesoionic xanthine acyclonucleosides is shown in Fig. 11.30 below.

In particular, the mesoionic xanthine acyclonucleosides where R'= H are especially useful since this moiety is ideal in giving the molecule a handle for attaching it to metallic, organic, polymeric or semiconductor surfaces/substrates such as GaAs, GaN, CdSe/ZnS. The definition of a mesoionic compound is a compound that cannot be adequately represented by any single covalent or single dipolar resonance structure.

These Class II mesoionic xanthines, such as *anhydro*-(8-hydroxyalkyl-5-hydroxy-7-oxothiazolo[3,2a]pyrimidinium hydroxides) cannot be adequately represented by fewer than ten different resonance contributors. Figure 11.31 illustrates these ten resonance forms and all of their possible quantum inter-conversion states. Each resonance structure shown in Figure 11.31 corresponds to an individual quantum state of the total molecule and all ten are required to adequately represent the molecule in its totality of superposed quantum states. In quantum computing, there may be multiple quantum states in superposition. In this particular case where there are ten qubits, the quantum state of superposition would be the following orthonormal basis set $|\psi\rangle = \alpha_i |x_1 x_2 x_3 \ldots x_n\rangle$ for all i=*1-1,024* and for all n=*1-10* where x_n is either 0 or 1.

More succinctly the above may be written: $|\psi\rangle = \sum_{i=1}^{N} \alpha_i |i\rangle$ where $|i\rangle$ is a shorthand notation for an orthonormal basis set of indices $\{i_1, i_2, i_3 \ldots i_j \ldots i_n\}$ where $N = 2^n$.

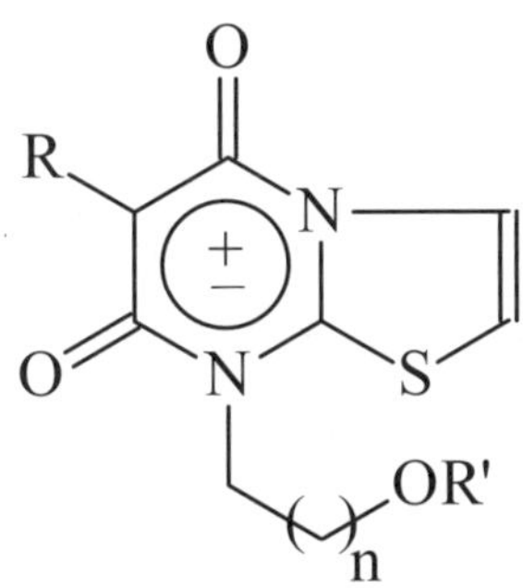

Figure 11.30 Generalized structural representation of Class II mesoionic xanthine acyclonucleosides. n = 1, 2; R = H, CH_3, CH_3CH_2, C_6H_5; R' = H, CH_3.

The Greek letters α_i are referred to as the amplitudes of the register and are complex numbers. In a 10-qubit register, there are therefore 2^{10} or 1,024 complex numbers for the total register. Since the probability ($|\Psi|^2$) of a quantum state or set of quantum superposition entangled states must always be equal to one, the following relationship for the coefficients of the quantum registers must also be true.

$$\sum_{i=1}^{1,024} \alpha_i^2 \equiv |\Psi|^2 = 1. \qquad (11.32)$$

For a X3-qubit register, there exists an 8-dimensional complex vector. For a 10-qubit register there exists a 1,024-dimensional complex vector. In order to initialize this vector space (register) for a quantum computer, an algorithm is necessary. In each step of the algorithm, the vector space is modified by multiplying it with a unitary matrix, which, by definition is a complex matrix.

11.14 Initialization of Mesoionic Xanthine Registers

The mesoionic xanthine molecule, as depicted in Fig. 11.31, represents a molecule that is in a quantum superposition of at least ten distinct and unique quantum states. An efficient scheme for initializing quantum registers with an arbitrary superposed state, without the introduction of additional qubits [89] has been developed by Long & Sun [90]. This scheme begins with the state $|0...0\rangle$ and is then transformed to a general superposed state of the following form: $|\psi\rangle = \sum_{i=0}^{N-1} a_i |i\rangle$. In this particular case, $N = 1,024$ and $|i\rangle$ is the shorthand notation for the basis set $\{i_1, i_2, i_3 ... i_j ... i_n\}$ where $n = \log_2 N$ and where i_j denotes the two possible states (0 or 1) of the jth-qubit. The following diagrams therefore illustrates that $|\psi\rangle$ is a general quantum superposition of N basis states and each basis state is a product state of n qubits. The initialization scheme involves only two types of unitary transformations or gate operations. The first gate operation is a single bit rotation U_θ, $U_\theta \begin{bmatrix} 0 \\ 1 \end{bmatrix} = \begin{bmatrix} \cos\theta & \sin\theta \\ \sin\theta & -\cos\theta \end{bmatrix} \begin{bmatrix} 0 \\ 1 \end{bmatrix}$. This rotation differs from an ordinary rotation because it is an ordinary rotation only for the $|0\rangle$ bit but interjects a minus sign for the $|1\rangle$ bit. The operation thus converts a qubit in the state $|0\rangle$ to a superposition of the two-state $(\cos\theta, \sin\theta)$ and a qubit in the $|1\rangle$ state to the superposition of the two-state $(\sin\theta, -\cos\theta)$. When $\theta = 0$, the state $|0\rangle$ remains unchanged but converts the sign of state $|1\rangle$ (i.e., Pauli-Z gate). When $\theta = \frac{\pi}{4}$, U_θ is simply reduced to the Hadamard-Walsh transformation. When $\theta = \frac{\pi}{2}$ (90° rotation), it acts as the NOT operation (Pauli-X, σ_x) by changing $|0\rangle$ to $|1\rangle$ and $|1\rangle$ to $|0\rangle$.

$$i = \begin{cases} 0 \to \{000...000\} \\ 1 \to \{000...001\} \\ 2 \to \{000...010\} \\ 3 \to \{000...100\} \\ . \\ . \\ . \\ N-1 \to \{111,...111\} \end{cases} \tag{11.33}$$

The second type of gate operation is known as the controlled *k*-operation. This operation is constructed from a string of *k* controlling qubits. The squares represent the controlling qubits $\{i_1, i_2, ...i_k\}$ and the circle is a unitary operation on the target qubit representing an angle of rotation. The uniqueness and power of this operation is that it is a conditional one that is activated only when the controlling qubits hold the respective values indicated in the squares. Controlled *k*-operations may be constructed using $O(k^2)$ standard 1- and 2-bit gate operations [91]. In order to more easily see how these operations are performed we may take a look at the simple example of a two-qubit system.

$$|00\rangle \to \sqrt{|a_{00}|^2 + |a_{01}|^2}\,|00\rangle + \sqrt{|a_{10}|^2 + |a_{11}|^2}\,|10\rangle \qquad \text{Operation}$$

1.) Single bit rotation $\alpha_1, \to |0\rangle\left[a_{00}|0\rangle + a_{01}|1\rangle\right] + |1\rangle\left[a_{10}|0\rangle + a_{11}|1\rangle\right]$ Operation

2.) Two controlled[1]-operations $U_{\alpha 2,i}$ ($i = 0, 1$).

$$= a_{00}|00\rangle + a_{01}|01\rangle + a_{10}|10\rangle + a_{11}|11\rangle$$

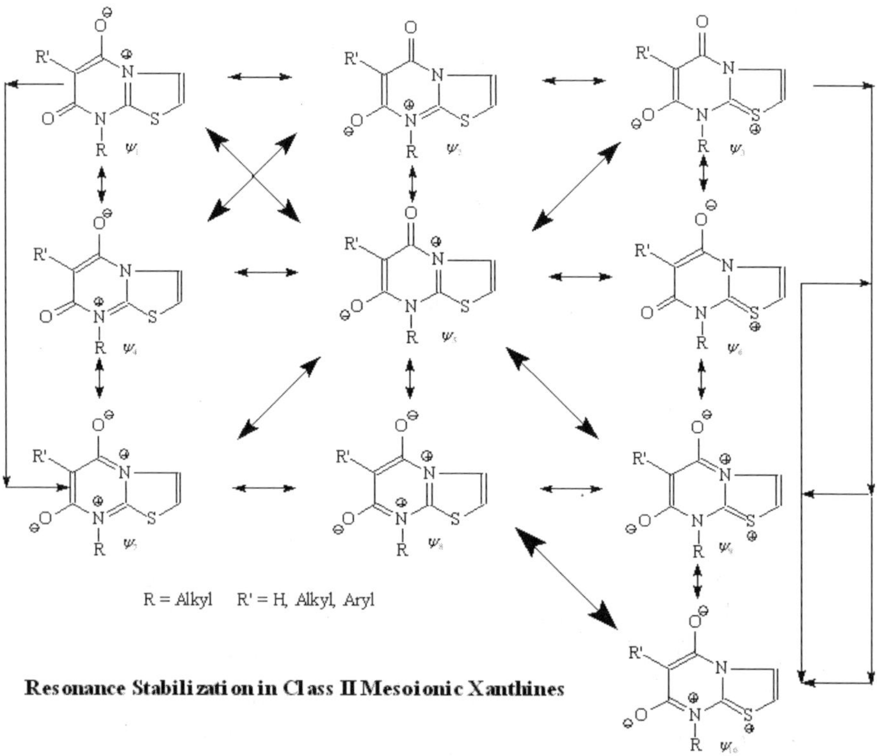

Figure 11.31 Resonance stabilization in Class II Mesoionic Xanthines.

The single bit rotation α_1 is equal to $\tan^{-1}\sqrt{\dfrac{|a_{10}|^2 + |a_{11}|^2}{|a_{00}|^2 + |a_{01}|^2}}$. We may now represent the operations in matrix form as well:

$$U_{\alpha 2,0} = \begin{bmatrix} \dfrac{a_{00}}{\sqrt{|a_{00}|^2 + |a_{01}|^2}} & \dfrac{a_{01}}{\sqrt{|a_{00}|^2 + |a_{01}|^2}} \\ \dfrac{a_{01}^*}{\sqrt{|a_{00}|^2 + |a_{01}|^2}} & \dfrac{a_{00}^*}{\sqrt{|a_{00}|^2 + |a_{01}|^2}} \end{bmatrix} \tag{11.34a}$$

$$U_{\alpha 2,1} = \begin{bmatrix} \dfrac{a_{10}}{\sqrt{|a_{10}|^2 + |a_{11}|^2}} & \dfrac{a_{11}}{\sqrt{|a_{10}|^2 + |a_{11}|^2}} \\ \dfrac{a_{11}^*}{\sqrt{|a_{10}|^2 + |a_{11}|^2}} & \dfrac{a_{10}^*}{\sqrt{|a_{10}|^2 + |a_{11}|^2}} \end{bmatrix} \tag{11.34b}$$

The situation becomes even more interesting when using a larger register such as a 3-qubit register having 8 basis states:

- Starting from the state $|000\rangle$, a single bit rotation is operated on the
- 1st-qubit with the angle $\alpha_1 = \tan^{-1}\sqrt{\dfrac{|a_{100}|^2 + |a_{101}|^2 + |a_{110}|^2 + |a_{111}|^2}{|a_{000}|^2 + |a_{001}|^2 + |a_{010}|^2 + |a_{011}|^2}}$

  transforming the initialized state $|000\rangle$ to the state

$$\sqrt{|a_{000}|^2 + |a_{001}|^2 + |a_{010}|^2 + |a_{011}|^2}\,|000\rangle + \sqrt{|a_{100}|^2 + |a_{101}|^2 + |a_{110}|^2 + |a_{111}|^2}\,|100\rangle .$$

- Then, two controlled-rotations with angles

$$\tan^{-1}\sqrt{\frac{|a_{010}|^2 + |a_{011}|^2}{|a_{000}|^2 + |a_{001}|^2}} \text{ and } \tan^{-1}\sqrt{\frac{|a_{110}|^2 + |a_{111}|^2}{|a_{100}|^2 + |a_{101}|^2}}$$

operate on the 2nd qubit.

The resulting superposed state vector therefore becomes:

$$\sqrt{\left|a_{000}\right|^2+\left|a_{001}\right|^2}\left|000\right\rangle+\sqrt{\left|a_{010}\right|^2+\left|a_{011}\right|^2}\left|010\right\rangle+ \\ \sqrt{\left|a_{100}\right|^2+\left|a_{001}\right|^2}\left|100\right\rangle+\sqrt{\left|a_{110}\right|^2+\left|a_{111}\right|^2}\left|110\right\rangle .$$

3. Finally, 4 controlled-unitary transformations operate on the 3rd-qubit to generate the superposed state:

$$a_{000}\left|000\right\rangle+a_{001}\left|001\right\rangle+a_{010}\left|010\right\rangle+a_{100}\left|100\right\rangle+ \\ a_{011}\left|011\right\rangle+a_{101}\left|101\right\rangle+a_{110}\left|110\right\rangle+a_{111}\left|111\right\rangle .$$

These 4 controlled-unitary transformations are:

$$U_{\alpha 3,00}=\begin{bmatrix} \dfrac{a_{000}}{\sqrt{\left|a_{000}\right|^2+\left|a_{001}\right|^2}} & \dfrac{a_{001}}{\sqrt{\left|a_{000}\right|^2+\left|a_{001}\right|^2}} \\ \dfrac{a_{001}^{*}}{\sqrt{\left|a_{000}\right|^2+\left|a_{001}\right|^2}} & -\dfrac{a_{000}^{*}}{\sqrt{\left|a_{000}\right|^2+\left|a_{001}\right|^2}} \end{bmatrix}$$

$$U_{\alpha 3,01}=\begin{bmatrix} \dfrac{a_{010}}{\sqrt{\left|a_{010}\right|^2+\left|a_{011}\right|^2}} & \dfrac{a_{011}}{\sqrt{\left|a_{010}\right|^2+\left|a_{011}\right|^2}} \\ \dfrac{a_{011}^{*}}{\sqrt{\left|a_{010}\right|^2+\left|a_{011}\right|^2}} & -\dfrac{a_{010}^{*}}{\sqrt{\left|a_{010}\right|^2+\left|a_{011}\right|^2}} \end{bmatrix}$$

$$U_{\alpha 3,10}=\begin{bmatrix} \dfrac{a_{100}}{\sqrt{\left|a_{100}\right|^2+\left|a_{101}\right|^2}} & \dfrac{a_{101}}{\sqrt{\left|a_{100}\right|^2+\left|a_{101}\right|^2}} \\ \dfrac{a_{101}^{*}}{\sqrt{\left|a_{100}\right|^2+\left|a_{101}\right|^2}} & -\dfrac{a_{100}^{*}}{\sqrt{\left|a_{100}\right|^2+\left|a_{101}\right|^2}} \end{bmatrix}$$

$$U_{\alpha 3,11}=\begin{bmatrix} \dfrac{a_{110}}{\sqrt{\left|a_{110}\right|^2+\left|a_{111}\right|^2}} & \dfrac{a_{111}}{\sqrt{\left|a_{110}\right|^2+\left|a_{111}\right|^2}} \\ \dfrac{a_{111}^{*}}{\sqrt{\left|a_{110}\right|^2+\left|a_{111}\right|^2}} & -\dfrac{a_{110}^{*}}{\sqrt{\left|a_{110}\right|^2+\left|a_{111}\right|^2}} \end{bmatrix}$$

For notation purposes we use an "angle" to label a controlled k-operation. If the coefficients are all real, it reduces to an ordinary rotation angle. The notations of angles of the controlled k-rotations, the first subscript designates the target qubit order number and the subscripts following the comma designate the quantum states of the controlling qubits. For example, the 3 in $\alpha_{3,11}$ refers to the target qubit and the subscripts (11 in $\alpha_{3,11}$) refer to the controlling qubits. In the initialization, operations for the first $n-1$ qubits are controlled rotations where each rotation depends only on a single real parameter. The rotation angles take on the following general expressions. In the first qubit there is a 1-qubit rotation. The rotation angle is: $\alpha_1 = \tan^{-1}\sqrt{\frac{\sum_{i_2 i_3 \ldots i_n} \left|a_{1 i_2 i_3 \ldots i_n}\right|^2}{\sum_{i_2 i_3 \ldots i_n} \left|a_{0 i_2 i_3 \ldots i_n}\right|^2}}$. In the 2nd-qubit, there are two controlled-rotations:

$$\alpha_{2,0} = \tan^{-1}\sqrt{\frac{\sum_{i_3 i_4 \ldots i_n} \left|a_{01 i_3 i_4 \ldots i_n}\right|^2}{\sum_{i_3 i_4 \ldots i_n} \left|a_{00 i_3 i_4 \ldots i_n}\right|^2}} \text{ and } \alpha_{2,1} = \tan^{-1}\sqrt{\frac{\sum_{i_3 i_4 \ldots i_n} \left|a_{11 i_3 i_4 \ldots i_n}\right|^2}{\sum_{i_3 i_4 \ldots i_n} \left|a_{10 i_3 i_4 \ldots i_n}\right|^2}}.$$

In general, in the jth-qubit, there are 2^{j-1} controlled-rotations, with each of them having j - 1 controlling qubits labeled as $i_1 i_2 \ldots i_{j-1}$. The rotation angle in the jth-qubit ($j \neq n$) is given by:

$$\alpha_{j, i_1 i_2 \ldots i_{j-1}} = \tan^{-1}\sqrt{\frac{\sum_{i_{j+1} \ldots i_n} \left|a_{i_1 i_2 \ldots i_{j-1} 1 i_{j+1} \ldots i_n}\right|^2}{\sum_{i_{j+1} \ldots i_n} \left|a_{i_1 i_2 \ldots i_{j-1} 0 i_{j+1} \ldots i_n}\right|^2}}. \quad (11.35)$$

For the last qubit, where $j = n$ we have 2^{n-1} controlled unitary transformations where:

$$U_{\alpha_{n,i_1 i_2 \ldots i_{n-1}}} = \begin{bmatrix} \dfrac{A_0}{\sqrt{|A_0|^2+|A_1|^2}} & \dfrac{A_1}{\sqrt{|A_0|^2+|A_1|^2}} \\ \dfrac{A_1^*}{\sqrt{|A_0|^2+|A_1|^2}} & -\dfrac{A_0^*}{\sqrt{|A_0|^2+|A_1|^2}} \end{bmatrix}$$

where $A_0 = a_{i_1 i_2 \ldots i_{n-1} 0}$ and $A_1 = a_{i_1 i_2 \ldots i_{n-1} 1}$.

If A_0 and A_1 are real, the operation is simply a rotation and the angle is given by:

$$\alpha_{n,i_1 i_2 \ldots i_{n-1}} = \tan^{-1}\left(\frac{A_1}{A_0}\right). \qquad (11.36)$$

We are now ready to initialize quantum superposition registers of three different types starting from the state $|0\ldots0\rangle$:

1. The evenly distributed state $|\psi\rangle = \sum_i |i\rangle$ is the most common state in quantum computing. The Hadamard-Walsh gate operation on each qubit generates this form of superposition from the state $|0\ldots0\rangle$. In this particular case, all of the rotation angles are $\pi/4$. In each qubit, the controlling qubits use up all possible combinations and therefore the 2^{j-1} controlled Hadamard-Wash gate operations are reduced to a single Hadamard-Walsh transformation in the jth-qubit.

2. The Greenberger-Horne-Zeilinger or GHZ state is the maximally entangled state with the superposition $\frac{1}{\sqrt{2}}(|0\ldots0\rangle \pm |1\ldots1\rangle)$. Suppose we would like to transform the state $|0000\rangle$ to the state $\frac{1}{\sqrt{2}}(|0000\rangle + |1111\rangle)$. The circuit below shows this diagrammatically:

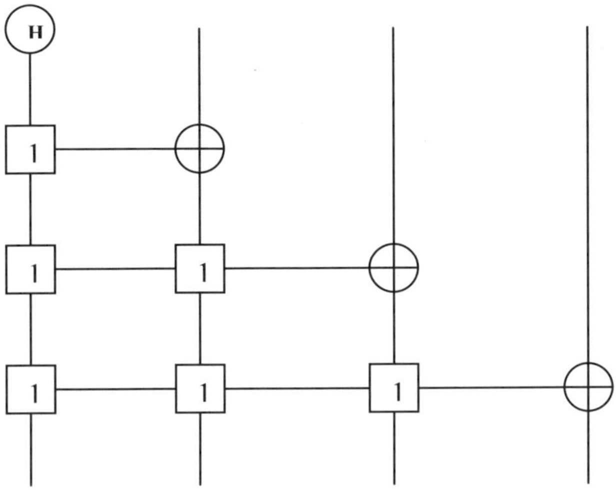

Figure 11.32 Quantum Circuit for the GHZ state.

The rotation in the 1st-qubit is the Hadamard-Walsh transformation. There are two controlled operations $\alpha_{2,0} = 0$ in the 2nd-qubit that are equal to the identity operation and so does nothing to the qubit. However $\alpha_{2,1} = \frac{\pi}{2}$ corresponds to the CNOT operation, so effectively, there is only one controlled-NOT gate in the 2nd-qubit. There are four gate operations in the 3rd-qubit. $\alpha_{3,11} = \frac{\pi}{2}$ is the $|11\rangle$-CNOT gate and $\alpha_{3,00}$ is the identity operation. $\alpha_{3,01}$ and $\alpha_{3,10}$ are undetermined angles equal to $\frac{0}{0}$. Upon closer examination, however, these angles are equal to 0 and are therefore equal to the identity operation. Therefore, the only gate operation in the 3rd-qubit is the $|11\rangle$-CNOT operation. Similarly, there is only the $|111\rangle$-CNOT operation in the 4th-qubit. Should the circuit contain more than four qubits, the same analysis applies until the last qubit. For the last qubit, the rotation is either $\frac{\pi}{2}$ for the state $\frac{1}{\sqrt{2}}\left(|0...0\rangle+|1...1\rangle\right)$ or $-\frac{\pi}{2}$ for the state $\frac{1}{\sqrt{2}}\left(|0...0\rangle-|1...1\rangle\right)$.

3. In the Grover search algorithm [24], the state vector is built up in a two-dimensional space spanned by the so-called "marked" state $|\tau\rangle$ and the "rest" state $|c\rangle=\sum_{i\neq\tau}|i\rangle$. At any step in the search, the state vector has the form $|\psi\rangle=\sin\theta|\tau\rangle+\cos\theta|c\rangle$. In order to initialize such a superposed state, we let $|\tau\rangle=|i_1 i_2 ... i_n\rangle$ be the marked state. We may now construct the state $|\psi\rangle$ from $|0...0\rangle$. The amplitudes a_i of the basis states $|\psi\rangle=\sum_{i=0}^{N-1}a_i|i\rangle$ are $a_\tau=\sin\theta$ and $a_i=\cos\theta/\sqrt{N-1}$ for $i\neq\tau$. According to the following equation,

$$\alpha_1=\tan^{-1}\sqrt{\frac{\sum_{i_2 i_3 ... i_n}\left|a_{1 i_2 i_3 ... i_n}\right|^2}{\sum_{i_2 i_3 ... i_n}\left|a_{0 i_2 i_3 ... i_n}\right|^2}}$$

, the rotation angle in the 1st-qubit is:

$$\alpha_1=\begin{cases}\tan^{-1}\Omega_1, \text{ if } i_1=1\\ \tan^{-1}\frac{1}{\Omega_1}, \text{ if } i_1=0\end{cases}$$

$$\text{where } \Omega_1=\sqrt{\frac{(N-2)\cos^2\theta+2(N-1)\sin^2\theta}{N\cos^2\theta}}.$$

In the kth-qubit, the angle for the $|i_1,i_2...i_{k-1}\rangle$-controlled rotation is therefore:

$$\alpha_{k,i_1,i_2...i_{k-1}}=\begin{cases}\tan^{-1}\Omega_k, \text{ if } i_k=1\\ \tan^{-1}\frac{1}{\Omega_k}, \text{ if } i_k=0\end{cases}$$

$$\text{where } \Omega_1 = \sqrt{\frac{\left(N-2^k\right)\cos^2\theta + 2^k\left(N-1\right)\sin^2\theta}{N\cos^2\theta}} \tag{11.37}$$

A viable organic molecule, a Class II Mesoionic Xanthine, has been introduced as a potential 10-qubit register substrate for scalable quantum computing. We have shown that the ground state of this xanthine molecule exists in a superposition of ten unique wave functions. These unique wave functions can form the basis of 10-qubit registers for quantum computation. Additionally a formalism was devised whereby these registers may be efficiently initialized, subsequently read into and transformed via standard unitary algorithms. We propose that polar solutions of the mesoionic xanthines or small crystalline quantum dots may be suitable for I/O techniques. Furthermore, these solutions or quantum dots may be RF laser pulsed at a certain set of frequencies to produce a cumulative resonant interaction within the xanthines to exploit higher degrees of freedom resulting from new HD commutation rules. Relaxation of the numerous excited states via these HD commutation rules are putatively a vehicle to ontologically overcome the decoherence problem associated with QC applications [92,93]. This ability overcomes the major obstacle for bulk quantum computing.

11.15 Conclusions

The debate over the completeness of quantum theory has raged for nearly one hundred years. There is more to do; but in this volume we believe we have brought it to its endgame. Completing QT to find a method for empirically surmounting the uncertainty principle has been no easy task. We have stated that bulk QC cannot be achieved within the limits of Bigbang cosmology or the bounds described by the Copenhagen regime. Here we have produced a rudimentary path for the completion of QT through a model for the implementation of bulk QC.

We doubt one can understand the ontology without sufficiently comprehending the new cosmology and have perhaps overdone the metaphors in hoping to facilitate this. We can only guess how difficult it will be to build a prototype utilizing our methodology. One could like Edison try 10,000 filaments (multiphase concatenation of resonant hierarchy coupling modes) and expect to achieve success with sufficient effort. Although this is not supposed to be necessary if our protocol is correct. Any style of sufficiently broad quantum system should be able to provide a vehicle for bulk implementation. DiVincenzo [78] has suggested five requirements the physical implementation of quantum computation:

- A physical system with scalable qubits
- Ability to initialize the qubit states
- Long decoherence times, longer than gate operation times
- Universal set of quantum gates
- Qubit measurement capability

We believe we have met these requirements and await the appearance of universal bulk quantum computing.

As a suggestion we have included what we believe to be a viable candidate organic molecule, that of the Class II Mesoionic Xanthine, because it has a potential scalable 10-qubit register substrate. Our general approached is based on a HD form of Dirac Spherical rotation in the context of a completed form of quantum theory able to ontologically surmount uncertainty. A formalism could just as readily be designed around the nomenclature of the spacetime dynamics of M-Theory. Also our method could just as easily be translated into a form of Topological Quantum Field Theory (TQFT), not addressed here but which the brane-world closely resembles. This is also illustrated in the work of L.H. Kauffman, editor-in-charge of this series, in papers such as [94] where he also integrates TQFT with knot theory for quantum computing.

References

[1] Moore, G.E. (1965) Cramming more components onto integrated circuits, Electronics, 38:8.
[2] Amoroso, R.L. & Amoroso, P.J. (2004) The Fundamental Limit and Origin of Complexity in Biological Systems: A New Model for the Origin of Life, in D.M. Dubois (ed.) CP718, Computing Anticipatory Systems: CASYS03-6th International Conference, Liege, Belgium, August 11-16, New York: American Inst. Physics 0-7354-0198-5/04.
[3] Amoroso, R.L. (2002) Developing the cosmology of a continuous state universe, in R.L. Amoroso et al (eds.) Gravitation & Cosmology:

From The Hubble Radius to the Planck Scale, Dordrecht: Kluwer.
[4] Amoroso, R.L. (2006) Paradigm for the cosmology of a continuous state holographic conscious megaverse, in R.L Amoroso, B. Lehnert & J-P Vigier (eds.), Extending the Standard Model: The Search for Unity in Physics, Orinda: The Noetic Press.
[5] Amoroso, R.L. (2008) Defining a context for the cosmology of awareness, in R.L. Amoroso & K.H. Pribram (eds.) The Complementarity of Mind and Body: Realizing the Dream of Descartes, Einstein and Eccles, Cambridge: MIT Univ. Press.
[6] Antippa, A.F. & Dubois, D.M. (2004) Anticipation, orbital stability and energy conservation in discrete harmonic oscillators, in D.M. Dubois (ed.) Computing Anticipatory Systems, AIP Conf. Proceedings Vol. 718, pp.3-44, Melville: American Inst. of Physics.
[7] Randall, L. (2005) Warped Passages, Unraveling the Mysteries of the Universe's Hidden Dimensions, New York: Harper-Collins.
[8] Amoroso R.L. (2003) Awareness: physical cosmology of the fundamental least unit, Noetic Journal 4:1, 1-15.
[9] Cramer, J.G. (1986) The Transactional Interpretation of Quantum Mechanics, Rev. Mod. Phys 58, 647-687.
[10] Wheeler, J. & Feynman, R. (1945) Interaction with the Absorber as the Mechanism of Radiation, Rev. Mod. Phys. 17, 1578.
[11] Holland, P.R. (2000) The Quantum Theory of Motion: An Account of the de Broglie- Bohm Causal Interpretation of Quantum Mechanics, Cambridge: Cambridge Univ. Press.
[12] Vigier, J-P and Bohm, D., (1955) Model of the causal interpretation of quantum theory in terms of a fluid with irregular fluctuations, Phys. Rev., 96:1; 208-17.
[13] Kaku, M. (1998) Introduction to Superstrings & M-Theory, 2nd ed., New York: Springer.
[14] Amoroso, R.L., How to Build a Conscious Quantum Computer: The Immanent Era of Conscious Technologies, Submitted to MIT Univ.Press.
[15] Kotigua, R.P. & Toffoli, T. (1998) Potential for computing in micromagnetics via topological conservation laws, Physica D, 120:1-2, pp. 139-161.
[16] Penrose, R. (2005) The Road to Reality, New York: Knopf.
[17] Peat, F.D. (1988) Superstrings and the Search for the Theory of Everything, Chicago: Contemporary Books.
[18] Messiah, A. (1999) Quantum Mechanics, Mineola: Dover.

[19] Atkins, P.W. (1994) Molecular Quantum Mechanics, Oxford: Oxford Univ. Press.
[20] Brink, D.M. & Satchler, G.R. (1968) Angular Momentum, Oxford: Oxford Univ. Press.
[21] Bohm, D. (1989) Quantum Theory, New York: Dover.
[22] Pauli, W. (1927) *Zeitschrift Physik*. 43: 601.
[23] Schwinger, J. (1992) Casimir energy for dielectrics, Proc. Nat. Acad. Sci. USA 89, 4091-3.
[24] Schwinger, J. (1993) Casimir light: The source, Proc. Nat. Acad. Sci USA 90, 2105-6.
[25] Schwinger, J. (1994) Casimir energy for dielectrics: spherical geometry, Proc. Nat. Acad. Math. Psych. 41:64-67, San Francisco: W.H. Freeman.
[26] Amoroso, R.L. (2002) The Physical Basis of Consciousness: A Fundamental Formalism, Part 1 Noesis, XXVI, Romanian Acad.
[27] Jibu, M, & Yasue, K. (2005) Quantum Brain Dynamics and Consciousness, Amsterdam: John Benjamins.
[28] Schiff, L.I. (1955) Quantum Mechanics, New York: McGraw-Hill.
[29] Peebles, P.J.E. (1992) Quantum Mechanics, Princeton: Princeton Univ. Press.
[30] Hatfield, B. (1992) Quantum Field Theory of Point Particles and Strings, Reading: Addison Wesley.
[31] Feynman, R P. (1948) Spacetime approach to non-relativistic quantum mechanics, Rev. Mod. Phys. 20:367-385.
[32] Accardi, L. (1988) Foundations of quantum mechanics: A quantum probabilistic approach, in G. Tarozzi & A. van der Merwe (eds.) The Nature of Quantum Paradoxes, Dordrecht: Kluwer Academic.
[33] Green, M.B., Schwarz, J.H. & Witten, E. (1987) Superstring Theory V-1, Cambridge: Cambridge Univ. Press.
[34] Amoroso, R.L., Giandinoto, S. & Rauscher, E.A. (2008) Derivation of the string tension formalism, and super quantum potential as inherent parameters of a holographic conscious multiverse cosmology (HAM), in Cs. Varga, I. Dienes & R.L. Amoroso (eds.) Unified Theories, Oakland: The Noetic Press.
[35] Rauscher, E.A. and Amoroso, R.L. (2008) The Schrödinger equation in complex Minkowski space, nonlocality and anticipatory systems, in D. Dubois (ed.) Proceedings of CASYS07, IJCAS Vo. 22. Liege, Belgium, August 2007.

[36] Borne, T., Lochak, G. & Stumpf, H. (2001) Nonperturbative Quantum Field Theory and the Structure of Matter, Dordrecht: Kluwer Academic.
[37] Chu, M-Y,J. & Amoroso, R.L. (2008) Empirical Mediation of the Primary Mechanism Initiating Protein Conformation in Prion Propagation, in D. Dubois (ed.) Proceedings of CASYS07, August, 2007, Vol. 22, Liege, Belgium.
[38] Witten, E. (1996) Reflections on the fate of spacetime, Physics Today (April), pp. 24-30.
[39] Wheeler, J.A. (1955) Geons, Phys. Rev., 97:2, 511-536.
[40] Amoroso, R.L., Burns, J. & Rauscher, E.A. A relativistic approach to the arrow of time utilizing large size extra dimensions, in preparation.
[41] Battey-Pratt, E.P. & Racey, T.J. (1980) Geometric model for fundamental particles, Int. J. Theor. Phys., 19:6; 437-475.
[42] Wolff, M. (1990) Exploring the Physics of the Unknown Universe, Manhattan Beach: Technotran Press.
[43] Wolff, M. (2002) Cosmology, the quantum universe and electron spin, in R.L. Amoroso, G. Hunter, M. Kafatos & J-P Vigier (eds.) Gravitation and Cosmology: From Hubble radius to the Planck scale, Proceedings of a symposium in Honour of the 80th Birthday of Jean-Pierre Vigier, Dordrecht: Kluwer Academic Publishers.
[44] Franck, G. (2000) Time and presence, in R.L. Amoroso, R. Antunes, C. Coehlo, M. Farias, A. Lieto & P. Soares (eds.) Science and the Primacy of Consciousness: Intimation of a 21st Century Revolution, pp. 187-189, Orinda: The Noetic Press.
[45] Evans, M.W. & Vigier, J-P (1994) The Enigmatic Photon; The Field $B^{(3)}$, Dordrecht: Kluwer Academic.
[46] Stevens, H.H. (1989) Size of a least-unit, in M. Kafatos (ed.) Bell's Theorem, Quantum Theory and Conceptions of the Universe, Dordrecht: Kluwer Academic.
[47] Meusnier, J. B. (1776) Mémoire sur la courbure des surfaces, *Mém. des savans étrangers* 10, 477-510, 1785.
[48] Grotz, K. & Klapdor, H.V. (1990) The Weak Interaction in Nuclear, Particle and Astrophysics, New York: Adam Hilger.
[49] Coles, P. & Lucchin, F. (2002) Cosmology, The Origin and Evolution of Cosmic Structure, Chichester: J. Wiley & Sons.
[50] Narlikar, J.V. (2002) An Introduction To Cosmology (3rd Edition) Cambridge: Cambridge University Press.

[51] Peebles, P.J.E. (1993) Principles of Cosmology, Princeton: Princeton Univ. Press.
[52] Puthoff, H.E. (2002) Polarizable vacuum approach to General Relativity, in R.L. Amoroso, G. Hunter, M. Kafatos & J-P Vigier (eds.), Gravitation & Cosmology: From the Hubble Radius to the Planck Scale, pp. 431-446, Dordrecht: Kluwer Academic.
[53] Vigier, J.-P., and Amoroso, R.L. (2002) Can one unify gravity and electromagnetic fields?, in R.L. Amoroso, G. Hunter, M. Kafatos & J-P Vigier (eds.) Gravitation and Cosmology: From the Hubble Radius to the Planck Scale, Dordrecht: Kluwer Academic.
[54] Amoroso, R.L. and Vigier, J.-P. (2004) Toward the unification of gravity and electromagnetism, in V.V. Dvoeglazov & A.A. Espinoza Garrido (eds.) Relativity, Gravitation, Cosmology, New York: Nova Science.
[55] Dirac, P.A.M. (1973) New ideas of space and time, *Naturwissenschaften* 32:6; 529-531.
[56] Jackson, J.D. (1975) Classical Electrodynamics, 1st edition, New York: Wiley & Sons.
[57] Baum, C.E. & Kriticos, H.N. (eds.) (1995) Electromagnetic Symmetry, Washington: Taylor & Francis.
[58] Misner, C.W., Thorne, K.S., & Wheeler, J.A. (1972) Gravitation, San Francisco: Freeman & Company.
[59] Schutz, B. (1980) Geometrical Methods of Mathematical Physics, Cambridge: Cambridge University Press.
[60] Accardi, L., (1988) Foundations of quantum mechanics: A quantum probabilistic approach, in G. Tarozzi & A. van der Merwe (eds.) The Nature of Quantum Paradoxes, Dordrecht: Kluwer Academic.
[61] Aharonov, Y. & Bohm, D. (1959) Phys. Rev. 115, 485.
[62] Shigley, J.E., and Uicker, J.J. Jr. (1980) Theory of Machines and Mechanisms, New York: McGraw-Hill.
[63] Rauscher, E. A. (1983) Electromagnetic Phenomena in Complex Geometries and Nonlinear Phenomena, Non–Hertzian Waves and Magnetic Monopoles, Millbrae: Tesla Books; (2007) 2nd edition, Oakland: The Noetic Press.
[64] Schwamb, P., Merrill, A.L. & James, W.H. (1930) Elements of Mechanism, New York: J. Wiley & Sons.
[65] Feynman, R.P. (1966) The Feynman Lectures on Physics, Reading: Addison-Wesley.
[66] Vlasov, A.Y. (1996) Quantum theory of computation and relativistic

physics, in T. Toffoli, M. Biafore & J. Leao (eds.) Physcomp96, Cambridge: New England Complex Systems Institute; and additional material from private communication.
[67] Dubois, D.M. (2001) Theory of incursive synchronization and application to the anticipation of delayed linear and nonlinear systems, in D.M. Dubois (ed.), Computing Anticipatory Systems: CASYS 2001-Fifth Intl Conference, Am Inst of Physics: AIP Conf. Proceedings 627, pp. 182-195.
[68] Antippa, A.F. & Dubois, D.M. (2008) The synchronous hyperincursive discrete harmonic oscillator, proceedings of CASYS07, preprint.
[69] Dubois, D.M. (2008) The quantum potential and pulsating wave packet in the harmonic oscillator, proceedings of CASYS07, preprint.
[70] Loudon, R. (1994) The Quantum Theory of Light, Oxford: Clarendon Press.
[71] Kak, S. (1999) The initialization problem in quantum computing, Found, Phys., 29:2, 267-279.
[72] Carruthers, P. & Nieto, M.M (1968) Phase and angle variables in quantum mechanics, Rev. Mod. Phys., 40:2; 411-440.
[73] Bailey, T.N. (1982) Relative cohomology and sources on a line I & II, Twistor Newsletter, No. 14, pp. 9-15, 19-21.
[74] Penrose, R, (1982) Real-space topology for advanced/retarded twistor functions, Twistor Newsletter, No. 14, pp. 16-18.
[75] Giandinoto, S. & Amoroso, R.L. (2008) Class II Mesoionic Xanthines as Potential Ten-Qubit Quantum Computer Substrate Registers, in D. Dubois (ed.) Proceedings of CASYS07, IJCAS Vol. 22, Liege, Belgium.
[76] Garcia-Ripoll, J.J., Zoller, P., & Cirac, J.I. (2005) Coherent control of trapped ions using off-resonant lasers, Phys Rev A 71, 062309.
[77] Nielsen, M.A. & and. Chuang, I.L. (2000) Quantum Computation and Quantum Information, Cambridge: Cambridge University Press.
[78] DiVincenzo, D.P. (2008) The Physical Implementation of Quantum Computation, arXiv:quant-ph/0002077v3.
[79] Mbagwu, G. O., Bass, R. G. and Glennon, R. A., J. Heterocyclic Chem., 22, 465 (1985).
[80] Mbagwu, G. O., Bass, R. G. & Glennon, R. A. (1983) Org. Magn. Reson., 21, 527.
[81] Rogers, M. E., Glennon, R. A., Bass, R. G., Mbagwu, G. O., Smith, J. D., Boots, M. R., Nanavati, N., Maconaughey, J. E., Aub, D. &

Thomas, (1981) S., J. Med. Chem., 24, 1284.
[82] Mbagwu, G. O. (1982) Diss. Abs. Int., B42 4428.
[83] Mbagwu, G. O., Bass, R. G. & Glennon, R. A., Virginia (1980) J. Sci. 31, 113.
[84] Mbagwu, G. O., Bullock, B., & H. O., Greer, (1988) J. Heterocyclic Chem., 22, 475.
[85] Mbagwu, G.O. & Garland, R. (1988) J. Heterocyclic Chem., 22, 571.
[86] Giandinoto, S., Mbagwu, G. O., Robinson, T. A., Ferguson, C., Nunez, J., (1986) J. Heterocyclic Chem., 33, 1839.
[87] Coburn, R. A., (1971) J. Heterocyclic Chem., 8, 881.
[88] Coburn, R.A., Carapellotti, R.A., & Glennon, R.A. (1973) J. Heterocyclic Chem., 10, 479.
[89] Ventura, D. & Martinez, T. (1999) Found. Phys. Lett.,12, 547.
[90] Gui-Lu Long & Yang Sun, (2007) Efficient scheme for initializing a quantum register with an arbitrary superposed state, arXiv:quant-ph/0104030v1.
[91] Barenco, A., et. al., (1995) Phys. Rev. A52, 3457; and Proc. R. Soc. London, A449, 679.
[92] Grover, L. K. (1997) Phys. Rev. Lett., 79, 325.
[93] Amoroso, R.L. (2007) Universal quantum computing: anticipatory parameters predicting bulk implementation, in D. Dubois (ed.) Proceedings of CASYS07, IJCAS Vol. 22, Liege, Belgium.
[94] Kauffman, L.H. & Lomonaco, S.J. (2007) q-Deformed spin networks, knot polynomials and anyonic topological quantum computation, arXiv:quant-ph/0606114v3.

Chapter 12

Practical Matter-Wave Antiballistic Defense Shield Technologies

Utilizing a new concept of a static (albeit relativistic) de Broglie matter-wave resonance hierarchy to coherently putatively control a highly symmetric F-Theory model of SUSY structural-phenomenology of the 'total' regime of spacetime, practical matter-wave antiballistic defense shield technologies appear feasible in the near term. The model although obvious to us is not based on 'politically correct' theory and has therefore been missed by the scientific community's rigid adherence to myopic views of quantum theory and cosmology. In simple terms the vacuum (Dirac type), not just the superficial surface equated with the zero-point field and quantum stochasticity, but the complete HD structural regime of spacetime itself. This requires a completed form of quantum theory able to manipulate causality (surmount uncertainty through an ontological form of scale invariant conformal invariance) and utilize unitary field parameters (topological) to control constructive interference of the matter-wave resonant hierarchy. Another key element is to discard the belief that nucleons have been created at a primordial Big Bang era. Operation of the shield technologies relies on the central premise of de Broglie-Bohm modeling of quantum theory (albeit radically extended in several ways) that matter is continuously created, annihilated and recreated as physically real stationary waves imbedded in the local fabric of spacetime. We anticipate matter-wave antiballistic defense shield technologies to appear in three stages or generations: 1) Simple coherently controlled constructive interference that might strengthen aluminum to the density of depleted uranium. 2) Full incorporation of the HD SUSY properties of spacetime. 3) Mature manipulation of spacetime, nanoscale programmable matter, probable antimatter phase configurations, energy efficiency and incursive nonlinear control of the nonlocal 'coherence-length'. From our vantage point we envision no physical reason why the 3rd generation device cannot withstand nuclear ordinance as soon as the principles outlined here are incorporated in the design of shield technologies. In this respect this chapter is a review of the various principles required for implementing matter-wave defense shield technologies rather than a complete engineering manual.

12.1 Introduction – Current Status of Shield Technology

Until this writing the concept of 'shields' only existed as a construct of science fiction media dating from about 1920. Such shields usually take the form of a force field designed to protect against a variety of weapons by deflecting or absorbing their impact. The field is projected along the surface or directed into the space around a spaceship, planet, moon, space station or building. Some are small enough to shield a soldier in combat, or from radiation or biological contaminants. These shields are often invisible or appear as translucent surfaces that glow when struck.

A variety of shield technologies are already under development by organizations such as NASA's Institute for Advanced Concepts which is currently exploring several types of active electrostatic defensive energy force-field radiation shield technologies [1-3]. These efforts are for the design of deflector shields for protecting spacecraft traveling beyond the safety of the earth's magnetic field from high energy charged particles like solar protons and electrons and galactic center cosmic ray particles from striking or penetrating a ship [4,5]. These 'active shields' deflect charged particles through the Lorentz force on a point charge

$$F_L = qE + qv \times B \tag{12.1}$$

where q is the charge of a single radiation particle, v the particle velocity, E is the electric field of the shield, and B the magnetic field component of the shield. The type of shield is determined by the field type:

- Electrostatic shield – only the electric field (time-independent)
- Magnetic shield – only the magnetic field (time-independent)
- Plasma shield – both electric and magnetic fields.

One popular design being considered by NASA utilizes an array inflatable spheres about five meters in diameter placed atop forty meter poles. For repelling protons smaller negatively charged spheres are placed around the periphery with somewhat larger positively charged spheres are placed in the center. The spheres are charged with about 100 megavolts. A mesh underneath the array acts as a ground plane. Other designs consider magnetic fields and plasma arrays [1-3]. See Fig. 12.1. These are not the type of defense shield to be considered here although the antiballistic shields we have in mind would readily repel radiation.

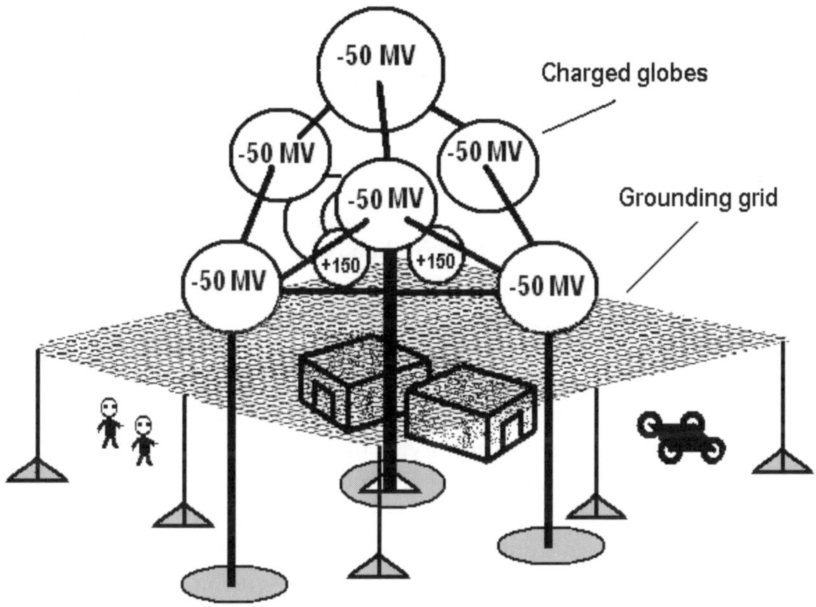

Figure 12.1 Schematic of a popular NASA radiation shield design. Figure redrawn from [1].

12.2 Overview of New Theoretical and Physical Requirements

Trying to address the vacuum has been like confronting an irresistible force and immoveable object. The vacuum is thought to contain infinite energy but none of it is considered accessible beyond virtual quanta for a duration of the Planck time. It seems if this were not so the fabric of the cosmos and our perception of reality would unravel; so that we may not really 'dip a ladle in' and draw any soup out. How then may we utilize or engineer it for various purposes? The 'vacuum' is everything and nothing, everywhere and nowhere; it is infinite potentia. In the midst of this our virtual holographic regime of spacetime. Relative to us as observers the surface topology of spacetime is a stochastic quantum foam of virtual quanta called the zero-point field believed to be governed by the Copenhagen uncertainty principle. To get at the vacuum we must reach beyond or through this surface barrier. We cannot 'tug' on it as conceptualized in the Chinese finger puzzle in Fig. 12.2 below.

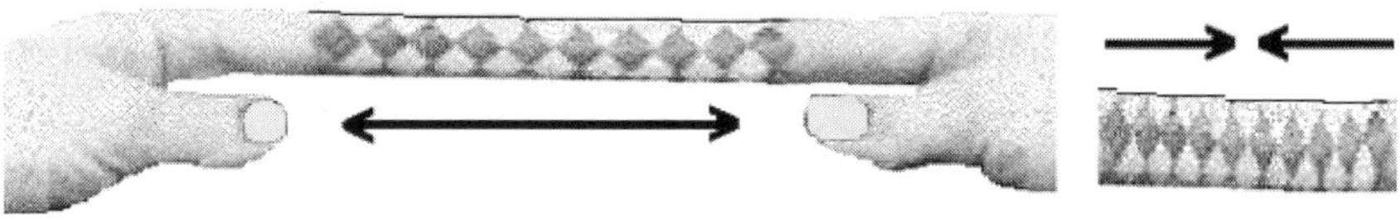

Figure 12.2 Chinese finger cuff – a) The harder one pulls the tighter one gets stuck. b) When kept relaxed the fingers slide out easily. Perhaps this is like the operation of string tension and coupling.

This loosening and tightening illustrated by the Chinese finger cuff in Fig. 12.2 is reminiscent of the 'leapfrog' cycle of coordinate fixing and unfixing (Chap. 6) inherent in the continuous-state standing-wave that is programmable in HAM cosmology. This is the putative element that makes or breaks the whole shield technology gambit. Our approach is opposite to the high energy bombardment currently used in pair production [6,7]; it is a 'gentle' approach combining the phenomenology of the standard model of quantum theory with a new ontological energyless approach that surmounts the uncertainty principle (see Chaps. 9 and 11) by application of a coherently controlled covariant resonance hierarchy. For this we need a special definition of the vacuum, which for the most part exists, but has been ignored in the literature as unpopular because it introduces photon mass, m_γ which is erroneously assumed to violate Gauge principles. We have addressed this vacuum to varying degrees in Chaps. 7 and 9. The features of the covariant Dirac vacuum of interest here is that it is polarizable because for extended EM theory (Proca equation) Maxwell's equations do not cut off at the vacuum but are continuous into or through it. This is key for resonant manipulation; not the usual 4D context of the required Causal Stochastic Interpretation of quantum theory, but its extended-completed HD version as inherent in HAM cosmology. Once we are conceptually over that critical hurdle (difficult for those adhering to the status quo) it is a straightforward matter to utilize the protocols of Chaps. 9 and 11 to ballistically program the vacuum in terms of the additional features introduced and extended here for amorphous Ising model lattice-gas cellular automata following the seminal work of Toffoli and from other parameters found in the literature such as non-Newtonian fluid mechanics and various nanotechnology techniques [8-14].

We see the main inertia to developing defense shield technologies is adamant adherence to the Copenhagen interpretation of quantum theory and associated limits inherent in the parameters of a Big Bang cosmology. Some GUT theories postulate proton decay with a half-life of 10^{36} years [15,16] according to $p \to e^{+} + \pi^{0}; \quad \pi^{0} \to 2\gamma$. The conundrum is that this lends some support to the Copenhagen/Big Bang scenario of nucleosynthesis occurring near the time of the original singularity. The lifetime of a proton is not the concern; it is its true quantum nature in terms of relativistic quantum field theory. However what we need to do is extend the de Broglie-Bohm point of view that suggests matter is a form of HD complex standing-wave that is continuously annihilated and recreated with the quantum wave function piloted by a unitary anthropic action principle tantamount to a 'super-quantum potential'.

Einstein postulated that Planck's quantization rule applied to an atom oscillating about its equilibrium point in a solid [17]. For shield technology we must extend this principle to spherical 'standing de Broglie matter-waves' in spacetime the internal motion of which obeys the Lorentz transformation. Simultaneous points produced by the wave are wavelet centers according to Huygens' principle [18] that reinforce their common envelope (of the main wave) because if the waves are parallel they summate. Another important consideration in this regard that all of the energy of the particle is focused at one point for all observers [19]. This is aligned with M-theory where all elementary particles are not discrete points but comprised of fundamental resonance modes of vibrating strings of a fixed tension [20].

If we remind ourselves of the $360^{\circ} - 720^{\circ}$ Dirac spherical rotation of the electron spinor suggesting HD topological components for the fundamental structure of matter the idea is not as difficult to explore. The intuitive reader will realize this additional topology is almost a proof, as shown by the representation of physical quantum states in abstract mathematical spaces like Hilbert space, phase space or configuration space, of the incompleteness of quantum theory which when complete will entail a physically real space (Chaps. 9 and 11) for the manipulation of observable properties of operators. The probability amplitude is not considered real, but in Cramer's transactional interpretation [21] an event is considered to be a physically real standing wave with all off diagonal elements also real so that an extended or completed form of quantum theory would be manifest in a physically real conformal space. The great import of this situation is that the phase elements of quantum mechanics are physically real which has far reaching consequences and is required for our development of shield technology.

In Chaps. 9 and 11 we have confined our discussion for the most part to atomic or molecular quantum systems. Here it is essential to program spacetime itself [22], so as Copenhagen has ignored the physicality of this HD topology; we must in a sense take the perspective of an HD observer and ignore the 'particle-in-a-box' and deal with the HD topological structural-phenomenology, which I suppose is a round about way of saying we must move well beyond non-relativistic quantum mechanics to an extended or dualistic form of relativistic quantum field theory. By dualistic we mean taking our fundamental basis not from the resultant particulate matter as a stage for sequencing the locus of the evolution of quantum states for present events but from the mirror symmetric standing-wave elements from which the resultant particle arises which can be thought of as two Calabi-Yau 3-forms (Fig. 12.3). This is an HD extension of Cramer's transactional interpretation [21], a standing-wave model requiring a pair of Dirac equations, one for R1 and one for R2 – the future-past advanced-retarded components of the virtual discretized present instant forming our perceived virtual reality.

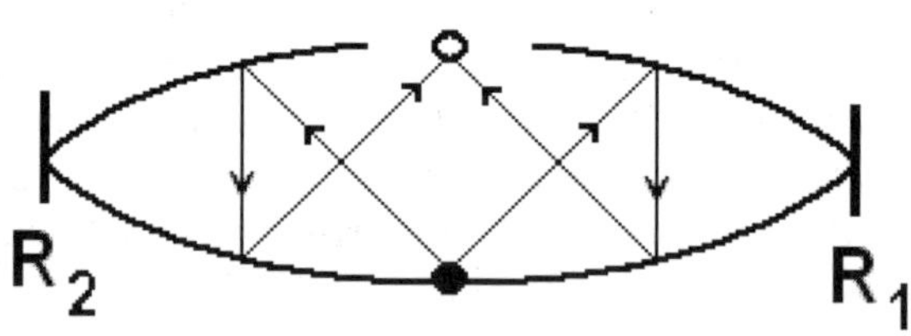

Figure 12.3 2D rendition of an HD holographic process. An object (black circle) placed inside two parabolic mirrors produces a virtual image (white circle). It is suggested that our virtual holographic reality is produced in a similar fashion by Cramer future-past standing-wave parameters of spacetime.

Until now physicists have generally accepted the existence of de Broglie matter waves, $\lambda = h / \rho$ for accelerated particles; but with little utility other than to demonstrate their existence through a variety of diffraction experiments applied to elementary particles, atoms and molecules in order to confirm the wave-particle duality of all matter [23-31]. De Broglie waves are physically real matter-waves associated with any moving particle traveling in a surrounding material medium [32,33].

$$h\nu_0 = m_0 c^2 . \qquad (12.2)$$

This limited view of matter-waves has little bearing on the development of matter-wave defense shields. The development of matter-wave shield technologies depends on a new view of cosmology, quantum theory and its associated new view of the nature of matter and a reconsideration of the nature of relativistic de Broglie wave mechanics.

Initially we were very timid in considering that a design for defense shields could be presented that could withstand atomic bombs. But with continued contemplation and research especially in reading De Broglie's original paper where he mentions that "...the quantum principle suggests associating this internal energy with a simple periodic phenomenon of frequency, ν_0 such that, $h\nu_0 = m_0 c^2$" [30]. This is the same amount of energy that occurs in an atomic bomb, so ideally once the model is sufficiently mature there is more than adequate energy hanging around to offset an incident amount tantamount to a nuclear explosion when non-linear conditions are setup properly. This however would only be the stage two shield technology indicative in the cry of Star Trek's Capt. Kirk "More power to the shields Scotty" which only utilizes intermediate vacuum engineering. But it wasn't so much the reminder of the inherent energy, it was the manner we were struck by de Broglie's statements about the periodicity of the internal energy, a notion left out of most textbooks and is a key element of the Vigier causal interpretation of quantum theory [34,35] that we felt could be applied to cellular automata. Toffoli gives the metaphor of a band leader conducting the various instrument players in a band; this is the usual Turing machine programming. Then he relates that the programming of cellular automata is internal for each unit allowing nonlinear ballistic computation [8,9].

Simplistically if one tears out a tiny segment of a holographic film the whole image may still be reproduced only much less bright and with much less resolution. Continuing the metaphor, then if one considers the tiny piece the usual energy associated with a coordinate point in the vacuum zero-point field, one must now utilize or set up a back-reaction potential that asymptotically approaches infinity as the mean-free-path ballistic coherence length increases [36,37]. In reverse (the zero-point becomes the infinity point), by a coherent control cumulative interaction process, the wave structure of matter may undergo a power-factored constructive interference when set up in a Mirror symmetry-T-duality spacetime resonance hierarchy. The tricky part at the limit of our current comprehension is a manipulation of the arrow of time essential to creating a focused array.

To achieve this result a new basis for the standard model of particle theory is required, one cast in an F-Theory Holographic Anthropic Multiverse (HAM) with a commensurate form of Relativistic Quantum Field Theory utilizing Dirac spherical rotation parameters and de Broglie matter-waves. The effectiveness of a shield is directly proportional to the amount of vacuum energy available to it. This vacuum energy is not utilized directly. The vacuum efficiently conserves itself at an apparent level of give and take one would expect if an immovable object met an irresistible force. The substance of the matter is that one does not wish to present the coherence length for the conservation parameters of a virtual photon if one is confronted with the explosive capacity of a 20 megaton nuclear device. It would not be practical to try to develop a shield technology where one would apply the amount of energy tantamount to what would be required for a shield even if a zepto (10^{-21}) or yocto (10^{-24}) second switch could turn on the power at the instant of impact. What is needed instead is to move from the current Copenhagen phenomenological viewpoint to the extended (completed) de Broglie-Bohm 'ontological' point of view as the basis for programming the vacuum. What we have failed to realize from the Copenhagen view is that we are the vacuum and to operate our devices from the perspective that it also is the vacuum rather than from the local perspective where so-called collapse of the wave-function and the uncertainty principle rule the day. What we are trying to delineate is that nothing new needs to be created or built; it already exists. We need merely to uncouple the perspective of operation and recouple it to the required resonant phase modality in the continuous-state hierarchy. See Chaps. 9 and 11.

As stated we are required to utilize a concept of static de Broglie matter-waves rather than the customary point of view of an associated wave with projectiles. The associated material medium for our 'static' de Broglie matter-waves [38] is the vacuum of spacetime itself which we will look at as a programmable tessellation of Ising model lattice gas cellular automata arising from the continuous-state parameters of the close-packed cosmological least-units inherent to HAM cosmology which tile the spacetime backcloth. Conformally correlated with this vacuum regime we couple resonantly and program a Nanoscale kinematic matrix substrate of amorphous programmable matter to facilitate an ontological phase cascade. The basis for developing a de Broglie ballistic defense shield arises from the Einstein-de Broglie stability conditions [30].

$$\begin{gathered}\int_0^{T_r}\left(p_x dx + p_y dy + p_z\right) = \\ \int_o^{T_r} \frac{m_0}{\sqrt{1-\beta^2}}\left(p_x dx + p_y dy + p_z\right) dt = \\ \frac{m_0 \beta^2 c^2}{\sqrt{1-\beta^2}} T_r = nh .\end{gathered} \tag{12.3}$$

12.3 Critical Philosophical Considerations on the Limits of Potentia

The sub-quantum domain has been called a stochastic foam, a regime within which time asymmetry is considered more fundamental than quantum theory; and that time emerges from a more fundamental unitary domain organizing the structure of and guiding the evolution of events in local reality [39,40]. We consider this a regime of infinite potentia the utility of which is essential to the defense shield technology. This usage is beyond the usual meaning applied to Heisenberg potentia because it only refers to the body of probabilistic states of the wave function before a measurement is taken. We do wish to align with those who claim nothing exists before a measurement, but to an even greater degree in that reality itself does not exist either other than for the basis of the observer. This is a multilevel process; first the boundary conditions forming the foundation of reality are created, then the quantum stochasticity of matter as its upper bound. Observed reality evanesces from a central hysteresis loop of this action. We don't think anyone has suggested this before, that reality is like an intermediate continuous-state collapse, a 'Dirac twist' collapse as a stage for all the rest of what is considered the microscopic evolution of the quantum wave function to rest on. We see this pretty much as if the film in an analog movie projector is a 2D or 3D hologram strip and the bulb in the projector an anthropic laser producing the perceived 3D images on the screen perceived by the observer seated in the theatre (Figs. 5.2 and 5.3). This is not a popular view because it represents a dualist-interactionist model of awareness [41] and gives an inherent importance to the nature and role of the observer. We believe this is correct and have presented empirical models to support it [42,43]; the protocols delineated in Chaps. 9 and 11 are related. Could all this mumbo-jumbo be skipped for the purposes here? Would the reader be satisfied if we merely postulated a deeper Dirac sea rather than the usual thinner Planck surface of the stochasticity of the zero-point field? We look at the zero-point field as the 'fog over the ocean'; whereas we require the utility of the full depth of the ocean. We wish to stick with something that suggests a domain that is truly like a hologram in an HD sense because it seems the most efficient manner to operate an anthropic multiverse.

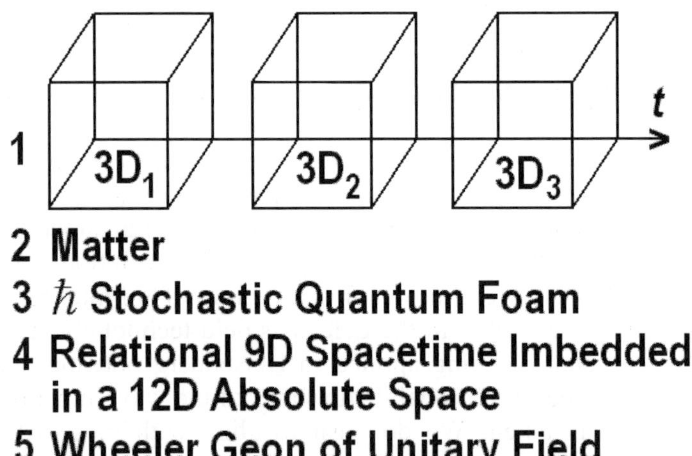

Figure 12.4 Reductionist hierarchic levels of HAM reality from the local standing-wave future-past eternal present to the atemporal geon of unitarity.

We insist up front that this shield technology is impossible to any degree of power without Gödelizing outside the limiting domain of Copenhagen quantum theory into this 5th regime of Fig. 12.4 beyond spacetime to a degree where even the de Broglie-Bohm version is also unsatisfactory and needs further extension to the point of full ontological completion. One must get 'under' or 'beyond' spacetime in order to engineer or program the required full Ising lattice rotations that are able to utilize the 'infinite' power inherent in the vacuum by 'ontologically becoming the vacuum'. The other reason this Gödelization [44] is so important is the requirement not just to summate the phase of stationary de Broglie matter-waves (they are only level 2-3 on Fig. 12.4), but to also coherently control the phases of the topological hierarchy so the mean-free-path will ballistically compute [45,46] in a sufficiently HD regime. The full Gödelization process controls the symmetry of the arrow of time. The ontological foray into level 5 achieved by programming the geometric information of spacetime is before time at the level of the unitary field. This is key to controlling the mean-free-path because it is this manipulation that allows the complete control of the Ising model hypersphere spin flips in a manner able to 'reflect the infinity' of the vacuum and be able to withstand nuclear ordinance.

12.4 The Shield Vacuum

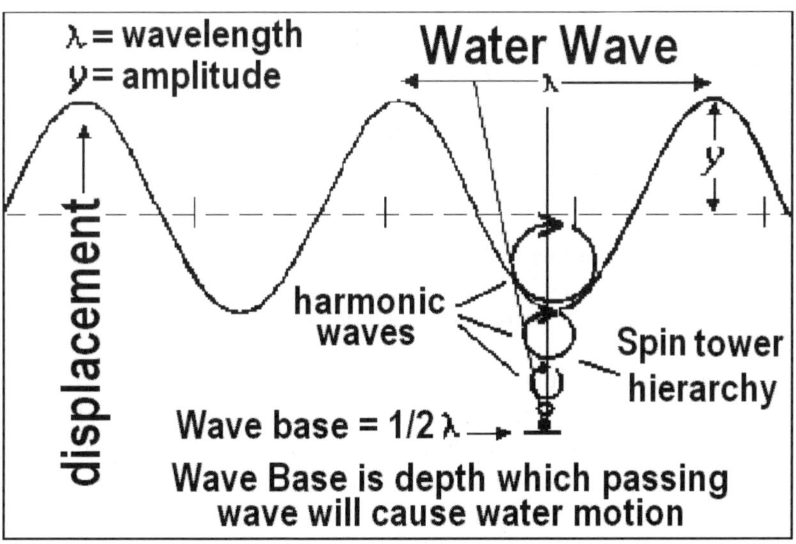

Figure 12.5 Dynamics of water-waves. Water remains stationery as waves pass through. Ocean surface waves are a combination of transverse and longitudinal waves; thus surface points follow orbital paths as in the HAM spacetime model.

The essential shield vacuum is considered to be a form of the well-known covariant polarized Dirac vacuum extended to include a deep structure of close-packed HD least cosmological units introduced by the HAM cosmological paradigm in this volume. This regime of deep structure includes an infinite domain of programmable holographic potential. For purposes of delineating our shield technology, we model the least-unit topology of the 'Dirac sea' with properties characteristic of water waves. When the 'ocean depth' becomes shallow near the shore the waves summate. We use this effect later to model standing-wave boundary conditions of the resonant summation hierarchy in conjunction with conditions mechanically called 'perfect rolling motion' (Fig. 12.6).

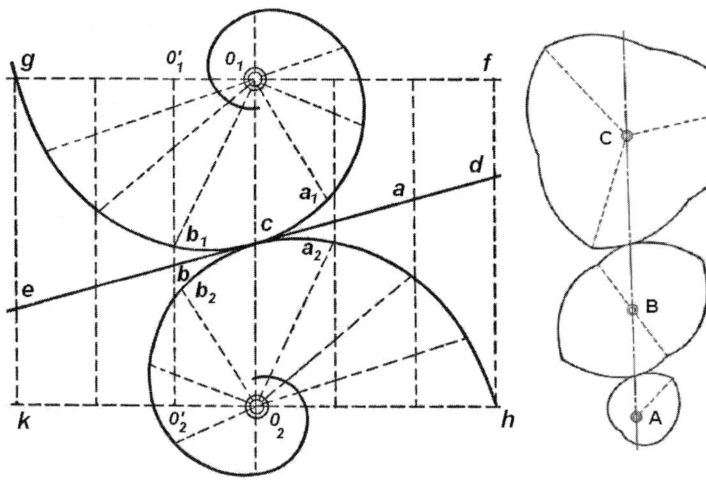

Figure 12.6 Perfect rolling motion allows a resonance hierarchy to be set up at the points of contact that are in phase. Here logarithmic spirals are used to conceptually illustrate the hierarchical coupling concept.

Another way for illustrating resonant hierarchy properties more akin to spacetime topology is the genus-1 helicoid parking-garage structure in Fig. 12.7 which also symbolizes the Calabi-Yau duality/mirror symmetry; or the Kaluza-Klein spin tower which model the Cramer-like standing-wave structure of virtual 3(4)D reality. Maybe we overdo these horrendous explanatory concatenations in our wish to convey how to align and couple spin-spin modes coherently in the HD hierarchy. The approach is to consider the vacuum as a cellular automata topology and program it with coherent control methods.

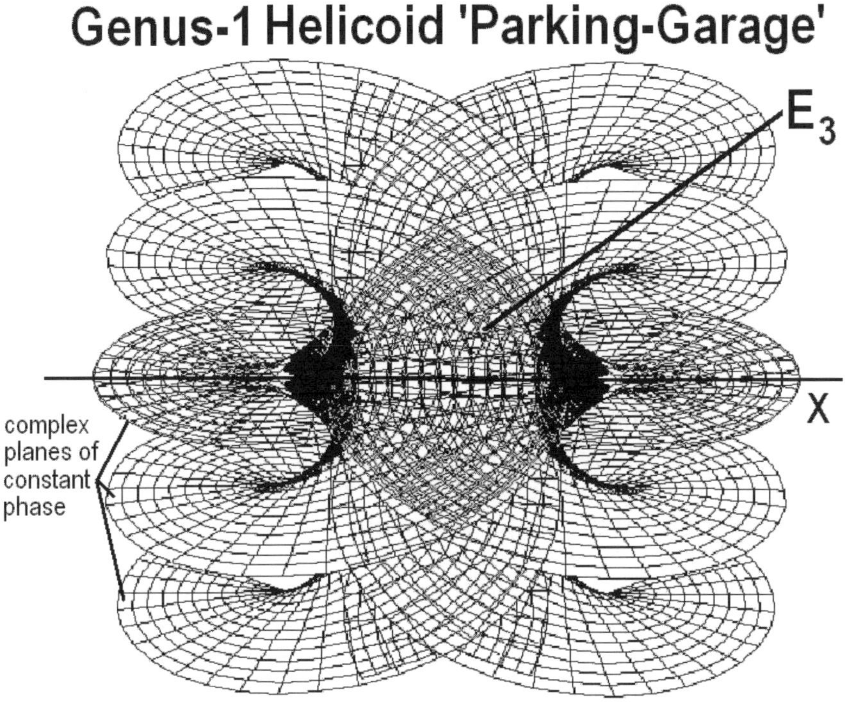

Figure 12.7 Genus-1 helicoid 'parking garage' hierarchy representing the advanced-retarded future-past symmetry of a mirror symmetric Calabi-Yau dual 3-form K-K spin tower inherent in the continuous-state fabric of spacetime.

12.5 What are the Required Vacuum Parameters?

Surprisingly, not just the dynamics of the zero-point field, such as the Van der Waal forces, the Casimir effect, *Zitterbewegung,* Zeeman-Stark effects and such, are open to manipulation by the application of EM fields for example, but the whole structure of the fabric of spacetime itself should be considered amenable to vacuum engineering [22,23,49-51]. It is fully utilizing this degree of accessibility that is required to develop practical matter-wave antiballistic defense shield technologies. This is not evident from within the domain described by the Copenhagen interpretation of quantum theory, or a Big Bang oriented string theory. The plausibility arises only under the auspices of a radical new cosmological perspective which we here call HAM cosmology. The current view of the de Broglie-Bohm [35] and Cramer interpretations [21] go halfway; and string theory (M-theory) is perhaps 80% there in available parameters but heretofore not organized in a helpful manner. De Broglie-Bohm-Vigier and Cramer need extension to the HD regime of a 12D F-Theory [20]; and string theory needs to step away from the Big Bang's limiting insight into the symmetry conditions to align SUSY symmetry breaking with the unique vacuum afforded by the anthropic unitary principles driving the hierarchical structure of cosmology as a complex self-organized system.

The details of these radical new symmetry conditions are somewhat daunting at first bite especially since they are 'not politically correct'; and we hardly claim to muster a complete understanding at this writing ourselves leaving little gap to be filled by the inspiration of those that follow. Before presenting the substantive details of shield technology, it is of passing interest to note for example:

- A .357 Magnum Handgun firing a 150 gram slug at 400 meters per second would have an impact of ~500 Joules.
- A chunk of space debris in low Earth orbit (LEO) with a velocity of ~ 16 km/s would be a projectile with an impact of ~ 130 Mega Joules per kilogram.
- A mature matter-wave antiballistic defense shield technology able to deter, for example, a 20 kiloton nuclear explosion would require the ability to repel an energy of ~80 Tera Joules.

A primitive prototype 'test of concept' shield could be constructed by using just a focused constructive interference of de Broglie matter-waves which might be like increasing the strength of an aluminum sheet to the tensile strength of depleted uranium. This initial 'foray' would arise from a more primitive or superficial utilization of the extended form of quantum theory. Engineers could conceivably get stuck at this level even with a full blown completion of quantum theory if an insufficient understanding of the vacuum structural-phenomenology remained.

Recent work called 'sparking the vacuum' at SLAC has with high energy photon beams produced e^-, e^+ pairs [47,48]. This 'head on' approach requires a lot of energy because it attacks the vacuum surface head on which remains sealed. Although in reverse, this is like the Chinese finger puzzle (Fig. 12.2) - the harder one pulls the more stuck the finger becomes, if relaxed the fingers slip out easily. Our process is like that, it manipulates the periodic coupling moments of string tension leaving the vacuum lattice open. If the usual Copenhagen collapse model is like the view of an orchard/vineyard where from some periodic positions one sees into infinity and from others the trees or particulate positions block the view. The HD model is like the view from a helicopter where the whole programmable array is open to view.

If the close-packed cosmological least-units tiling the spacetime backcloth are considered to have properties like an Ising model lattice-gas Bloch sphere cellular automata array then space becomes programmable as has been suggested [22,49]. A Bloch sphere is a form of Riemann sphere here purported to fill the spacetime raster as Calabi-Yau dual 3-forms where relativistic, r-qubits become physically real rotatable transformable Ising lattice-gas Riemann spheres to which when perfect rolling motion resonance techniques are applied cascade transformations for ballistic computing can be set up.

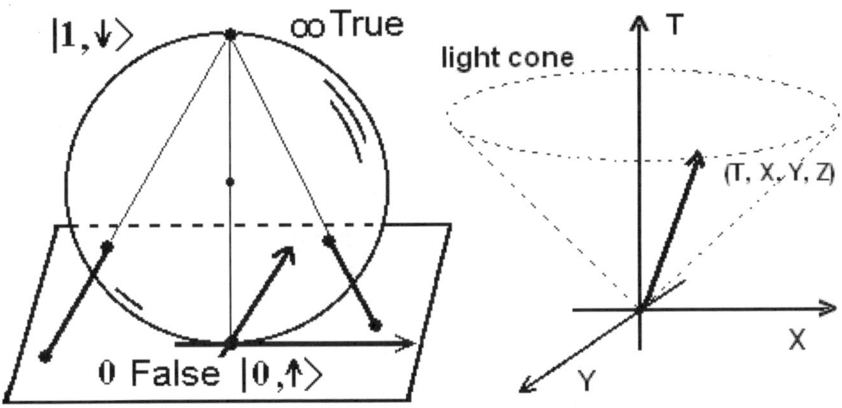

Figure 12.8 a) Block sphere rendition of a qubit. b) Relativistic qubit or r-qubit with more degrees of freedom.

Construction materials of the bunker, vehicle or personnel shield must have a special layer specifically for shield material or be completely constructed out of shield materials that contain alloys with amorphous nanoscale programmable matter [8,9,14,51] as the site where the cellular automata Ising model lattice gas programming occurs. Following Smolin [52,53] for ideas he developed for loop quantum gravity further consider these programmable Bloch spheres as a raster of complex spacetime s[in networks.

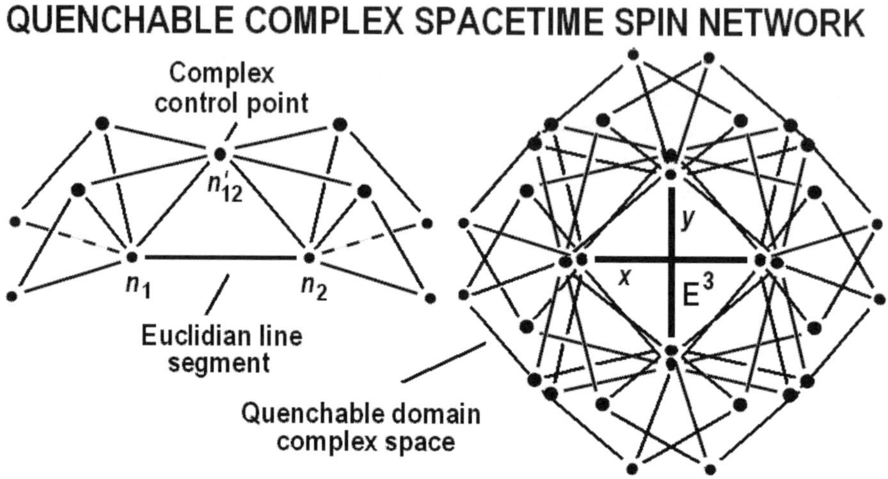

Figure 12.9 Spin networks of the programmable matter array with periodic control points for setting up the nanotech programmable matter substructure able to implement energy cascades in conjunction with ballistic programming.

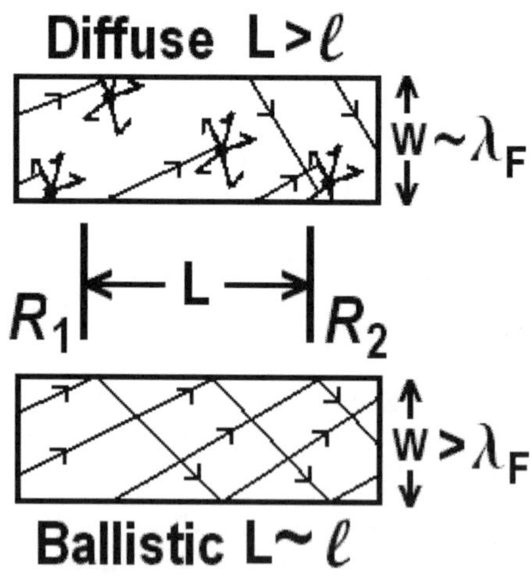

Figure 12.10 By resonant phase coherence the basis for a ballistic transport avalanche may be programmed into the spacetime topology using amorphous nanotech materials that simulate or map to the structural-phenomenology of spacetime. L is the coherence length of the resonator, and l the mean free path.

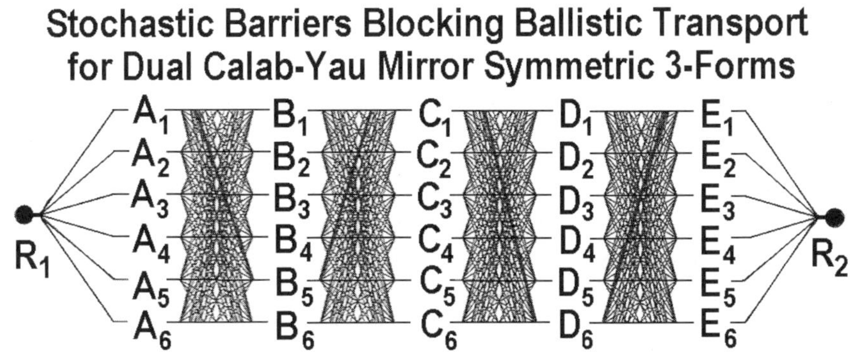

Figure 12.11 A Quantum Calabi-Yau spacetime brane array with 6^5 possible locked paths acting as a barrier to ballistic transport by stochastically disrupting the mean-free-path. The array must be programmed as a harmonic oscillator resonance hierarchy to order the topology coherently.

In Fig. 12.12 application of the proper resonant field at nodes R1-R2 will prepare the mean-free-path for a ballistic transport 'avalanche' for quenchable shield parameters in the 1st order and unquenchable or infinite recursion in the asymptotic stepwise infinite limit as in Fig. 12.9.

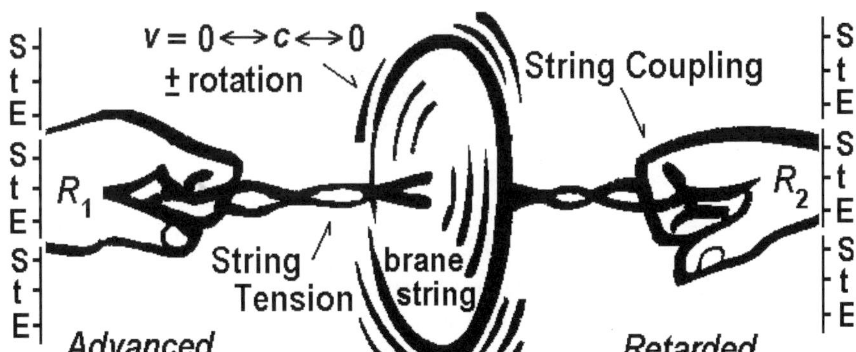

Figure 12.12 Conceptualization of the spin tension-coupling dynamics of a 2-brane spacetime element representing one of the 6^5 paths of Fig. 12.11.

In the HAM cosmology as stated the spacetime raster is self-organized and thus has all the properties of complex self-organized systems such as incursion and evolution controlled by an external action principle [41,54]. With a strike on a grid, 2D *x,y* for simplicity, the points parameters are updated. If the force is below threshold, i.e. quenchable, the neighboring elements of the array help to maintain equilibrium; but for a $z(x,y) \to z(x,y)+1$ threshold an avalanche occurs [55-57]. This has an associated asymptotic power law that fractally (Figure 12.13) propagates with a domino effect of varied stepwise levels and thresholds of quenchable and infinite ballistic transport parameters mediated by the ability of the algorithm to program the amorphous nanoscale material for coherently control of the spacetime hierarchy.

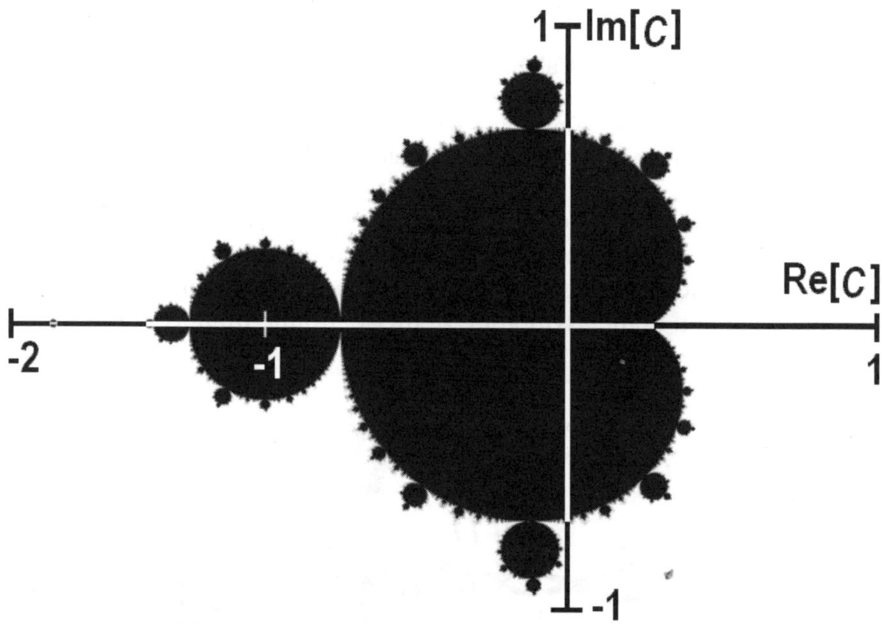

Figure 12.13 Map of the Mandelbrot fractal set from Eq. (12.4). We use it to illustrate the continuous fractal-like incursion through the HD hierarchy that occurs when ballistic computing of the spacetime topology is achieved.

The Mandelbrot fractal set can be mathematically produced by the Feigenbaum fractal generator, $F_C(x) = x^2 + C$ which produces an iteration fulcrum with a period-doubling bifurcation cascade by repeated iteration of, F_C which is a family of complex polynomials from the critical point, x_0

$$
\begin{aligned}
x_0 &= x \\
x_1 &= F_C(x) = x^2 + C \\
x_2 &= F_C\left(F_C(x)\right) = \left(x^2 + C\right)^2 + C \\
x_3 &= F_C^3(x) \\
&\vdots \\
x_n F_C n(x)
\end{aligned}
\tag{12.4}
$$

where C is a complex number and for $C = i$ the sequence is 0,i,(-1+i),-i,(-1+i),-1... The map may escape to infinity or stay within the Mandelbrot set of a disk with infinite radii [58]. In contrast the Mandelbrot fractal set generator, $F_C(Z) = Z^2 + C$ where $Z = x - iy$ and $c = c_1 + ic_2$ maps a subset of the complex plane (Fig. 12.13) for values of c whose orbits don't escape to infinity by

$$
Z^2 + C = \left(x + iy\right)^2 + \left(c_1 + ic_2\right) = x^2 + y^2 + c_1 + \left(2xy + c_2\right) \tag{12.5}
$$

12.6 Domain Wall Boundary Conditions and Emission Absorption Loci for Advanced-Retarded Waves

We shall consider a static thick domain wall constructed by a scalar field with self-interaction in the Schwarzshild black hole spacetime [59,60].

$$
g = -\left(1 - \frac{2M}{R}\right) dt^2 + \left(1 - \frac{2M}{R}\right)^{-1} dR^2 + R^2\left(d\vartheta^2 + \sin^2\vartheta d\varphi^2\right) \tag{12.6}
$$

The metric of the background Schwarzschild black hole is written in terms of the isotropic coordinates, t, r, ϑ, φ, where the new radial coordinate, r is defined by

$$
R = r\left(1 + \frac{M}{2r}\right)^2. \tag{12.7}
$$

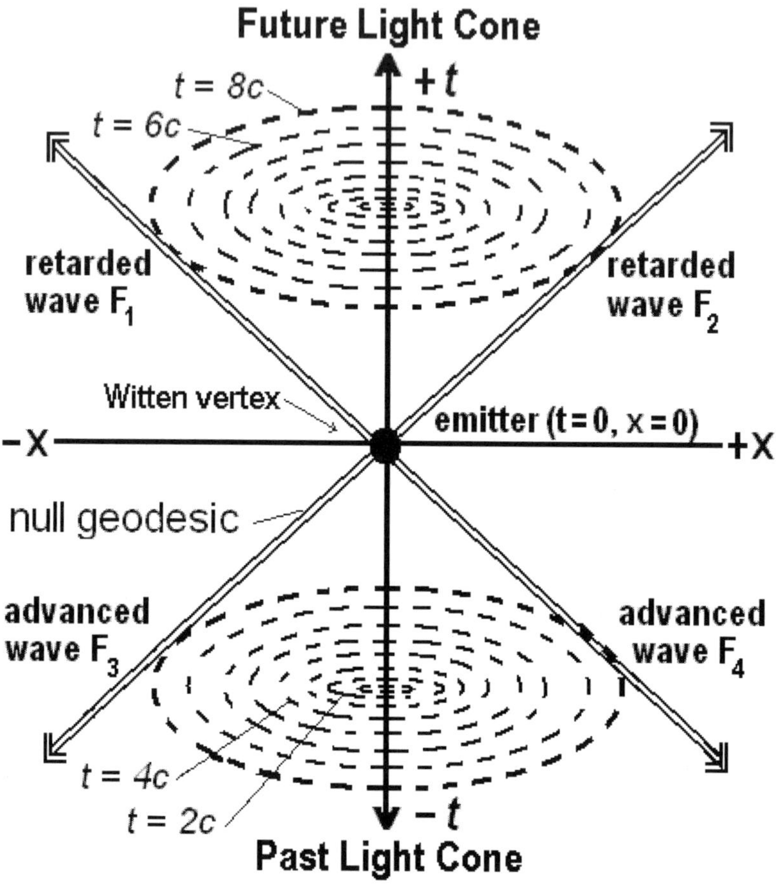

Figure 12.14 Cramer transaction emission locus at x,t = 0,0. We are concerned with the boundary conditions in the region outside the event horizon, where r ≥ M/2 which are of interest even though here applied to a black hole because it might reflect scale invariant principles.

The scalar equation in spherical coordinates of wave motion in spacetime which has spherical symmetry [61,62]

$$\nabla^2\Phi - \frac{1}{c^2}\partial^2\frac{\Phi}{\partial t^2} = 0 \tag{12.8}$$

where Φ is the wave amplitude. The equation has two solutions

$$\Phi_{out} = \frac{1}{r}\Phi_{\max}\exp\left(i\omega t - ikr\right)$$
$$\Phi_{in} = \frac{1}{r}\Phi_{\max}\exp\left(i\omega t + ikr\right) \quad (12.9)$$

which for the programming of spacetime can be applied to the propagation of Cramer's advanced retarded waves from an emission locus at $x,t = 0,0$ by Eqs. (12.9) and (12.10) and Fig. 12.14.

$$F_{1-Ret} = F_0 e^{-ikx} e^{-2\pi ift}, \quad F_{2-Ret} = F_0 e^{ikx} e^{-2\pi ift}$$
$$F_{3-Adv} = F_0 e^{-ikx} e^{2\pi ift}, \quad F_{4-Adv} = F_0 e^{ikx} e^{2\pi ift} \quad (12.10)$$

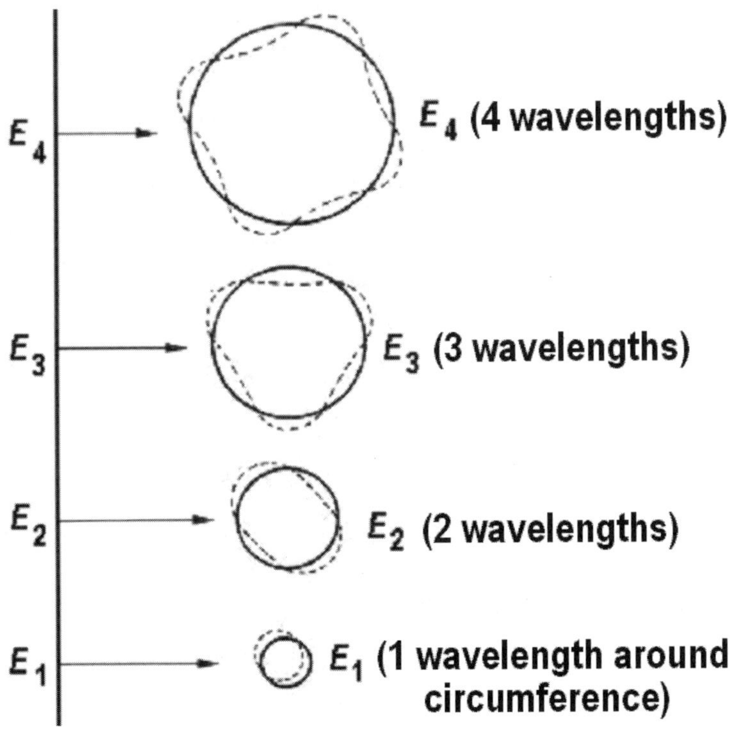

Figure 12.15. A Ring may vibrate with n standing wavelengths depending on the relationship of the circumference to the multiple number of whole wavelengths. Simplified here, it is suggested that the topology of spacetime and matter vibrate on and as hyperspherical surfaces.

Traditionally electron standing-waves oscillate about the atomic nucleus. Here we attempt to expand the wave nature of matter itself as static waves centered on the locus of least spacetime units as it is annihilated and recreated in the arrow of time relative to the observer. This requires a conversion of the de Broglie wave equation, $mvr = n(h/2\pi)$ to a static form amenable to the parameters of continuous-state cosmology [19,38]. For Hyperspherical Representation the magnitudes of the radial coordinates of a two-state wave function, $\psi(\vec{r}_1, \vec{r}_2)$ in hyperspherical representation are replaced by the hyperspherical radius, R and the hyperspherical angle, α such that

$$R \equiv \left(r_1^2 + r_2^2\right)^{1/2} \quad \text{and} \quad \alpha \equiv \arctan\frac{r_2}{r_1} \tag{12.11}$$

in order that the symmetries may be more clearly shown. The hyperspherical radius, R represents the size of the two-state system and the hyperspherical angle, α is a measure of the radial correlation of the two-state system [63]. It is critical to note that when $\alpha = \pi/4$, $r_1 = r_2$; and when $\alpha = 0$ or $\pi/2$ one of the states is at a greater distance from the least-unit vertex than the other.

12.7 Energy Increase from Ising Model Lattice-Gas Properties

In terms of the SUSY spacetime lattice represented by close-packed least units functioning as a Riemann 3-sphere Ising model spin lattice, where total energy, $E_T\{s_i\}$ is a function of the spin hysteresis loop

$$E_T\{s_i\} = \sum_i e_i(s_i) = E_0 - \sum_i h_i s_i \tag{12.12}$$

where $e_i(s_i)$ is the energy of an isolated individual least unit, E_0 the ground state and h_i the energy from spin orientation from the external field that allows coherent control of the Ising spin lattice [64]. The external field is the unitary action driving the evolution of the spacetime lattice structure as a putative self-organized complex system.

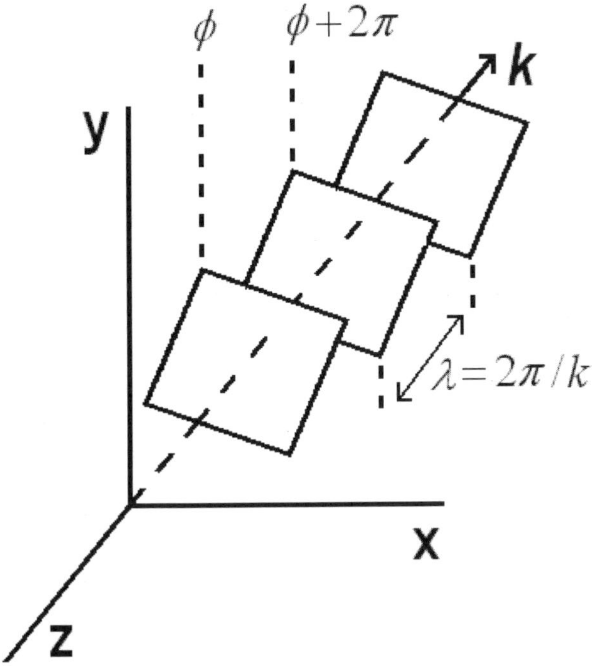

Figure 12.16 Surface of constant phase, in this case to represent orthogonal standing reality waves.

A surface of constant phase, $k \cdot r - \omega t = k_x x + k_y y + k_z z - \omega t =$ *constant* is a wavefront [17]. For a surface of constant phase if any wave equation has a time harmonic (sinusoidinal) solution of the form $Ae^{i\phi}$ where A is the amplitude and the phase, ϕ a function of position with (x,y,z) constant and phase difference 2π separated by wavelength, $\lambda = 2\pi / k$. The direction cosines of the planes of constant phase are proportional to k and move in the direction of k equal to the phase velocity where

$$\mu = \frac{\omega}{k} = \frac{\omega}{\sqrt{k_x^2 + k_y^2 + k_z^2}} . \tag{12.13}$$

Where $\lambda = 2\pi / k = 2\pi\hbar / p = h / p$ is equivalent to the de Broglie matter wave relations, $E = \hbar\omega$, $\mathbf{p} = \hbar k$ [65].

12.8 Programming Matter Through Cellular Automata

Programmable matter is defined as a material that locally adjusts its response to external inputs through programmed control. Amorphous Ising model lattice-gas cellular automata can be used for programming spacetime if designed to mirror the spacetime structure utilized. Each independent computational element in the amorphous or stochastic (accepting all) medium is identically programmed on a topological surface which in this case conforms to the least-unit tori of spacetime. There are too many units to program individually so programming is achieved by neighbor connectedness. Toffoli formed a metaphor to describe this neighbor model [8,9]. Usually a marching band has a leader, this will not work for cellular automata where local self assembly is internalized for each individual unit which acts as it own agent. This is a fundamental requirement for a massive ballistic response. The nanostructure of the defense shield materials must contain a computing substrate that is composed of fine-grained computing nodes distributed throughout space which communicate using only this nearest neighbor type of interactions [8,9,12,13,66]. According to Drexler [66] the closely packed computational units may be constructed to simulate a fractal system that for us would mean has the required incursive properties.

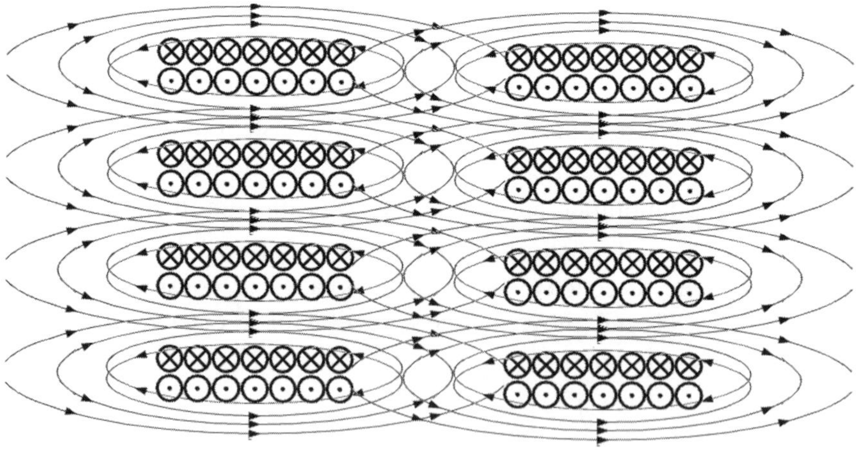

Figure 12.17 Nanoscale programmable matter substrate acting as receptors of modulated cascades to be built into the construction materials to act as a transducer of static de Broglie matter-waves resonating from the cellular automata into the hierarchical structure of spacetime.

12.9 Introduction to de Broglie Matter-Waves

De Broglie by considering a material moving object of restmass, m_0 for a stationery observer suggested that a phase wave, or 'pilot' wave, accompanies a particle because the principle of inertia said it should possess an internal energy equal to m_0c^2 [30]. This phase wave arises as an inevitable consequence of de Broglie's assumption of the internal periodic phenomenon of the particle and the Lorentz transformation laws of the special theory of relativity

$$h\nu_0 - m_0c^2, \tag{12.14}$$

with $\nu - \beta c,\ (\beta < 1)$ for total energy $\nu - m_0c^2 / h\sqrt{1-\beta^2}$. De Broglie's result arose from a combination of the principle of Einstein's special relativity and the quantum relationship for the observer which he initially applied to a photon of nonzero restmass, $m_\gamma\ (< 10^{-50} g)$ which because of its associated internal motion he associated with a piloting phase wave of frequency, ν at each point in space.

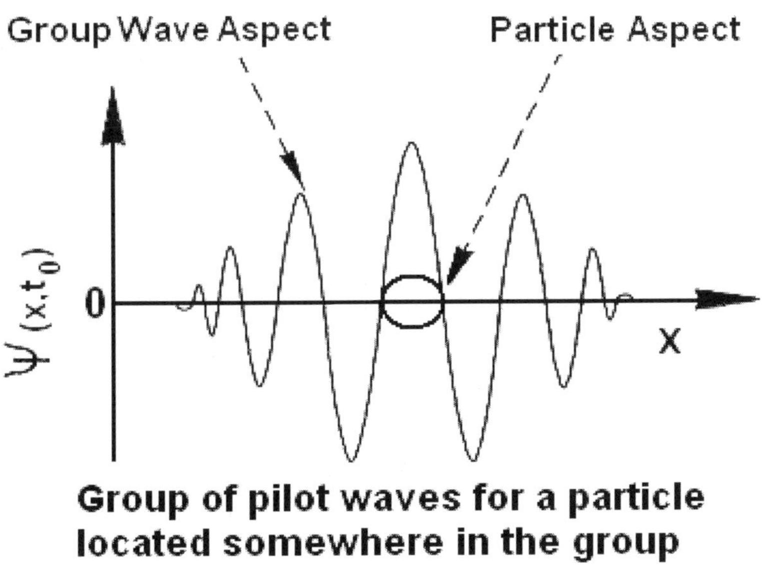

Figure 12.18 The group velocity of de Broglie waves is associated with the velocity of a particle.

MacKinnon [19,38,67] described the de Broglie wave packet for stationery states and nondispersive wave packets of a free particle. He states that the nondispersive wave packet, ψ is a solution of

$$\Box\psi = 0 \tag{12.15}$$

where

$$\Box = \nabla^2 - \frac{1}{c^2}\frac{\partial^2}{\partial t^2}. \tag{12.16}$$

From this MacKinnon shows that the nondispersive wave packet for a particle relative to the observer has the form

$$\psi = \sin(kr/kr)\exp\left[i(\omega t - k_0 x)\right] \tag{12.17}$$

where

$$k = m_0 c/\hbar, \quad r = \left\{\frac{\left(x - vt^2\right)}{1-\left(v^2/c^2\right)} + y^2 + z^2\right\}^{1/2}, \tag{12.18}$$

$$\omega = mc^2/\hbar \quad k_0 = mv/\hbar.$$

Equation (12.17) is a spherically symmetric solution to Eq. (12.15) after being subjected to the Lorentz transform as initially obtained by de Broglie.

Of critical interest to us is MacKinnon's work to set up a de Broglie wave packet for a stationery state. Although we are interested in relativistic waves, it is not the de Broglie waves for the usual particles in coordinate motion, but for the de Broglie waves for stationery matter with internal 'continuous-state' relativistic effects.

Consider two identical particles moving in opposite directions relative to an observer at x^* and t^*

$$\psi_1^* = A\cos\left(\omega t^* - kx^*\right), \quad \psi_2^* = A\cos\left(\omega t^* + kx^*\right) \tag{12.19}$$

which represent standing waves when solved by the Schrödinger equation for a particle in a box and cannot depend on the reference frame

[38]. MacKinnon concludes that these stationery states are static and for which Bohm postulated a quantum potential to account for it. MacKinnon carries this point further [19] to suggest that:

> The motion of a particle in spacetime does not depend on the motion relative to it of any observer or any frame of reference [and] if the particle has an internal vibration of the type hypothesized by de Broglie, the phase of that vibration at any point in spacetime must appear to be the same for all observers...Each observer or reference frame will have its own de Broglie wave for the particle. The phase of the particle's vibration must, by definition, be the same as that for all possible de Broglie waves at the point where the particle is. By superimposing all these possible de Broglie waves, a [nondispersive] wave packet is formed centered in space on the particle.

In his original work de Broglie was not able to properly form a wave packet that could localize the particle; MacKinnon was able to construct a wave packet from de Broglie's original wave phenomena that is also nondispersive [19].

12.10 Coherent Control of Standing Matter-Waves

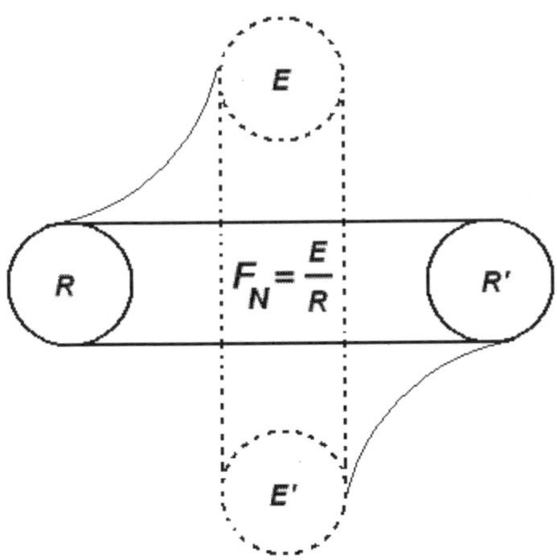

Figure 12.19 Ultimately the control mechanism for controlling standing de Broglie waves depends on applying the noetic field equation to the other programming parameters for the ballistic programming of cellular automata.

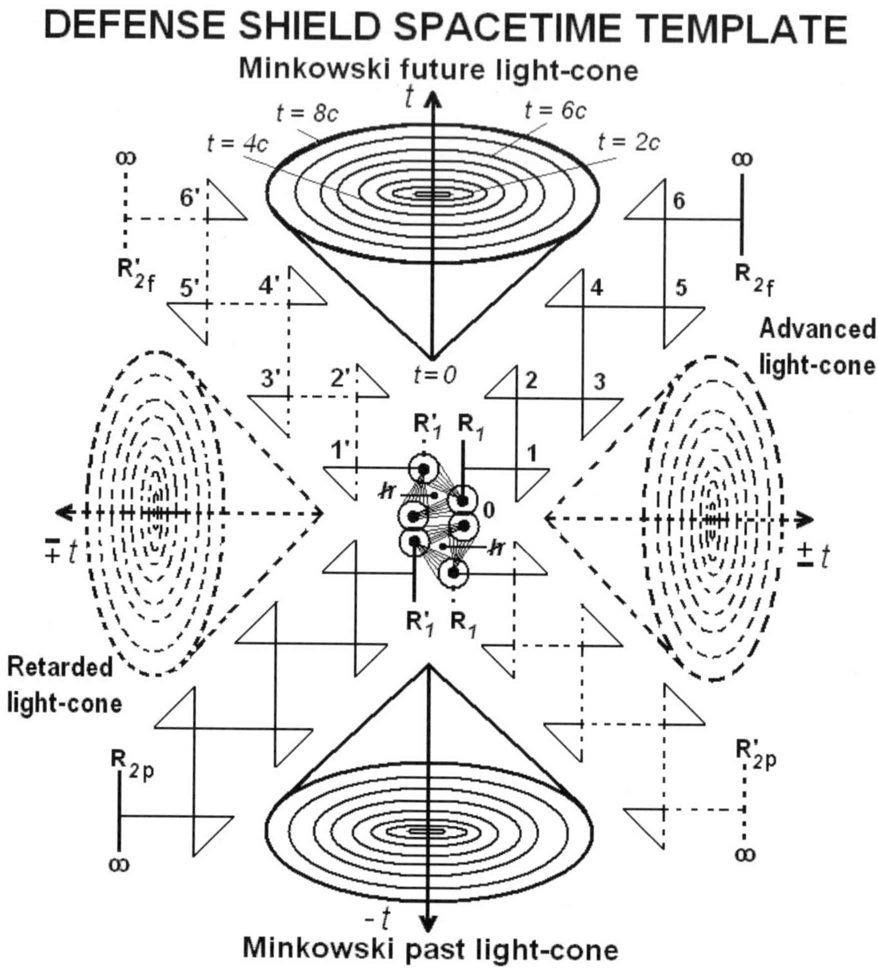

Figure 12.20 Conceptualized schema of the underlying spacetime structure utilized as a template for modulating the matter-wave resonance hierarchy mimicked in the programmable matter of the shield construction materials.

If it were possible to conceptually summarize everything required to develop a de Broglie matter-wave antiballistic defense shield it is illustrated in Fig. 12.20 above which is an exploded conformal scale-invariant view of the continuous-state wave-particle seesaw leapfrog dynamics inherent in the topology of spacetime shown as a template within a brane topology hierarchy amenable to application of resonance.

12.11 Afterward

When gazing out the window of one's mind it is hard to imagine that tangible objects like moons, mountains or cannon balls from a certain perspective can be as gossamer as the essence of love. The solid surfaces we walk on are made of relativistic holographic oscillations of tiny electron waves that quantum mechanically are everywhere and nowhere at the same time. If the nucleus of an atom were the size of the Earth the electron orbitals would be further away than the moon. That's a lot of empty space. We are made out of and imbedded in these materials and unaware of just how virtual reality is. It's the planes of constant quantum phase that make it so for us. A very complex self-organized dynamic holographic image process just for the observers benefit. By $E = mc^2$ there is a lot of energy in a pinhead; a baseball size clump of plutonium can level a city. Remembering a cartoon seen some decades ago depicting a couple of astronauts just as they were rounding the far side of the moon and the surprised looks they had when they saw that the moon was just a painted billboard; it's quite a challenge to accept a reality of that form. But this is the form that the principles here are based on.

We have provided a preliminary introduction for constructing matter-wave antiballistic defense shield technologies. The 1st prototype or test of the concept may be no more than constructive interference of stationary de Broglie waves that could in practice give aluminum the strength of depleted uranium for example. This would have some immediate utility in various applications. However what we predict is that a true shield, perhaps the 3rd generation technology, would utilize all of the deep structure of spacetime and be able to withstand a nuclear blast. Internal power consumption is required to operate the programmable matter substrate, in line with the mundane Star Trek cry by Captain Kirk: 'more power to the shields Scotty!'. The mature antiballistic defense shield technology would have these same subelements, i.e. coherently controlled constructive interference of matter-waves, nanoscale programmable matter substrate, but also the leading edge or wave envelope would not merely be a Huygens wave front but be programmed with an antimatter spin structure which would asymptotically increase the shields effectiveness. Also with sufficiently versatile programming this 'surface' would not create a percussive back-reaction but annihilate or damp the phases of the incoming matter and shock waves to attenuation by destructive rather than constructive interference techniques.

12.12 Summary of the Defense Shield Design Parameters

We have given a model for antiballistic matter-wave defense shield technology. It is not an engineering blueprint; the first prototype will require a little more effort. We take a moment to summarize the salient features and requirements:

- Observed reality is like a virtual HD standing-wave of future-past advanced-retarded parameters.
- All current thinking confined to the limits of any/all standard models, i.e. particle physics, cosmology or quantum theory is insufficient and we therefore can safely emphasize that a de Broglie matter-wave defense shield cannot be built from within these confines.
- Most particularly matter in the extended de Broglie-Bohm-Vigier causal stochastic interpretation of quantum theory, the wave function is physically real as are both 'wave and particle' which may exist simultaneously. The properties of the Dirac equation is extended from the original concept of matter to include both spacetime and domain walls of the reality of the observer, all of which are created-annihilated and recreated in a covariant continuous-state scale-invariant process.
- The Dirac Polarized vacuum is a programmable 'ocean' of potentia, part of which we treat as a backcloth of 'close-packed' least cosmological units with Ising model properties like a cellular automata.
- In this general context the key to a de Broglie matter-wave antiballistic defense shield is simply to ballistically program the mean-free-path of this HD spacetime array, not in the usually considered linear path but for all coordinates simultaneously in a minimum of 6 spatial dimensions. Six-D may not turn out to be adequate; the three temporal and three unitary (for quantum potential or piloting) may also need to be addressed. We 'guess' three may drop out and just a 9D matrix will be required.
- We see three generations of shield technology:
    1. Limited HD programming – constructive interference of matter-waves giving aluminum the strength of depleted uranium.
    2. Full shield that could withstand nuclear ordinance but would require energy input for operation ala Star Trek 'more power to the shields, Scotty'.

3. Mature 3rd generation shield technology with all refinements. Antimatter topological configurations of cellular automata programming. Utilizes the infinite energy of the vacuum or even energy taken from the projectile with no energy input required.

- Some form of Noetic Transformation (Chap. 5) is probably required in the programming. Possibly the unique identifier term for a person to receive a transcendent insight can be omitted; but perhaps with a many-body addition to handle the 'ubiquity factor' whereas the individual the requirement is more like a singularity which wouldn't require ballistic transport conditions.

FORMS OF THE NOETIC TRANSFORMATION

1. Observer receives transcendent information form HD, ⇓
2. Subject S_1 and S_2 have open channel ⇔ between them.
3. Ballistic spacetime programming for HD, ⇕↖↘↙↗⇔
4. Combination of 2. & 3. plus imbedding information
    a. Structural-phenomenological for spacetime info.
    b. Data content – like imbedding actual qualia in quantum computer music

Acknowledgement

In the spirit of the recent tercentenary of one of our heroes, early American printer, journalist, publisher, author, philanthropist, abolitionist, public servant, scientist, librarian, diplomat, statesman and inventor Benjamin Franklin, who put all of his myriad patents, such as the Franklin stove, bifocals, the medical catheter, lightning rod, swim fins, and the odometer in the public domain; in this same spirit we would like to offer our insights into defense shield technology as a gift toward world peace. Franklin believed:

> As we benefit from the inventions of others, we should be glad to share our own...freely and gladly.

We are Americans who consider ourselves highly patriotic, but also good world citizens. Many Americans are ashamed at how poorly our country has been run recently. If we wish to continue to 'police the planet' and present diplomatic, scientific and democratic leadership; we need to do a lot better. Case in point regarding the content of this chapter; the US Department of Defense (DOD) created DARPA, the Defense Advanced Research Projects Agency in 1958 by order of then US President Dwight D. Eisenhower in response to the surprise Russian launch of Sputnik. Eisenhower's guidance was clear: 'find and quickly develop advanced technology for the Armed Forces so the United States would never again suffer a technological surprise by another nation'.

In direct contact with DARPA management, we were told DARPA was not interested in our shield technology proposals, 'that they knew of no experiment....'; we said we knew this, we wanted to present the experiment, if they wouldn't fund that at least let us write a 'white paper' describing it. We were politely told to 'come back in twenty years'. We also tried the DARPA BAA research programs; but our institute wasn't considered large enough to pass the type of 'Dunn & Bradstreet' screening the BAA system required. Finally we attempted to get a NATO advanced projects grant but could not find a willing NATO Mediterranean or Eastern European partner which was part of the application requirements. So there you have it...Of course we commit no treason here as theory of any kind apparently is not considered a threat to national security. But just in case we gave no blueprint.

What are the remaining enlightened person's priorities – God, country, world, family, self? American prophet Brigham Young said, 'all scientific discovery comes as revelation from God'. So we give this technology back to God! Let the arms race, no, let the peace race begin...

References

[1] Levy, R.H. & Fren, F.W. (1967) The plasma radiation shield: concept, and applications to space vehicles, Prepared under NASA Contract No. NAS 8-20310 NASA CR-61176.
[2] Buhler, C.R. & Wichmann, L. (2005) Analysis of a Lunar Base Electrostatic Radiation Shield Concept, Phase I Final Report, Kennedy Space Center, Florida: ASRC Aerospace Corporation.
[3] US Army Corps of Engineers (1946) Fundamentals of Protective Design.

[4] Davisson, C.J. & Germer, L.H. (1928) Reflection of electrons by a crystal of nickel, PNAS, v 14, pp 317-323.
[5] Thomson, G.P. (1927) The diffraction of cathode rays by thin films of platinum, Nature 120, p. 802.
[6] Hubbell, J.H. (2006) Electron positron pair production by photons: A historical overview, Radiation Physics and Chemistry 75:6; 614–623.
[7] Agger, C.K. (1996) The Theory of Photon-Impact Bound-Free Pair Production and Applications to Relativistic Heavy-Ion Collisions, www.modspil.dk/agger/speciale.pdf.
[8] Toffoli, T. Programmable Matter: An introduction, Cambridge: MIT Univ. Press (in preparation).
[9] Toffoli, T. & Margolus, N. (1987) Cellular Automata Machines; A New Environment for Modeling, Cambridge: MIT Univ. Press; Russian translation (1991) *Mashiny Kletochnykh Avtomatov, Izdatelstvo 'Mir'.*
[10] Ionicioiu, R., Amaratunga, G. & Udrea, F. (2000) Quantum computation with ballistic electrons, arXiv:quant-ph/0011051v1.
[11] Viswanathan, G.M. & Viswanathan, T.M. (2008) Spontaneous symmetry breaking and finite time singularities in D-dimensional incompressible flow with fractional dissipation, arXiv:0807.1563v1 [physics.flu-dyn].
[12] Kodama, T. & Koide, T. (2008) Memory effects and transport coefficients for non-Newtonian fluids, arXiv:0812.4138v1 [hep-ph].
[13] Abelson, H. et al. (2000) Amorphous computing, Communications of the ACM, 43:74-82.
[14] Wolfram, S. (1994) Cellular Automata and Complexity, Reading: Addison-Wesley.
[15] Sreekantan, B.V. (1984) Searches for proton decay and superheavy magnetic monopoles, J Astrophysics & Astronomy, vol. 5, p. 251–271.
[16] Hagiwara, K. et al. (2002) Particle Data Group current best estimates of proton lifetime, Phys. Rev. D 66, 010001.
[17] Peebles, P.J.E. (1992) Quantum Mechanics, Princeton: Princeton Univ. Press.
[18] Huygens, C. (1690) *Traite de la Lumiere*, (Treatise on Light).
[19] McKinnon, L. (1978) A nondispersive de Broglie wave packet, Found. Phys., 8:3-4; 157-176.
[20] Kaku, M. (1998) Introduction to Superstrings & M-Theory, 2nd ed., New York: Springer.

[21] Cramer, J.G. (1986) The Transactional Interpretation of Quantum Mechanics, Rev. Mod. Phys. 58, 647-687.
[22] Zizzi, P. (2005) Spacetime at the Planck scale: The quantum computer view, in R.L Amoroso, B. Lehnert & J-P Vigier (eds.), Beyond the Standard Model: Searching for Unity in Physics, Oakland: Noetic Press.
[23] Kulhánek, J. (1965) Propagation of de Broglie waves in space-time, Il Nuovo Cimento, 38:3; 1955-1965.
[24] Croca, J.R., Garuccio, A., Lepore, V.L. & Moreira, R.N. (1990) Quantum-optical predictions for an experiment on the de Broglie waves detection, Found. Phys. Lett. 3:6; 557-564.
[25] Feoli, A. & Scarpetta, G. (1998) De Broglie Matter Waves from the Linearized Einstein Field Equations, Found. Phys. Lett. 11:4; 395-403.
[26] Lepore, V.L. (1992) Homodyne detection of de Broglie waves, Found. of Physics Letters, 5:5; 469-478
[27] Croca, J.R.., Ferrero, M., Garuccio, A. & Lepore, V.L. (1997) An experiment to test the reality of de Broglie waves, Found. Phys. Lett. 10:5; 441-447.
[28] Barut, A.O. (1990) Quantum theory of single events: Localized de Broglie wavelets, Schrödinger waves, and classical trajectories, Found. Phys. 20: 10.
[29] Rauch, H. (2008) Test of quantum mechanics by neutron interferometry, Euro. Phys. J. - Special Topics, V.159, No.1.
[30] de Broglie, L. (1923) Radiation, waves and quanta, *Comptes Rendus*, Vol. 177, pp. 507-510.
[31] Davisson, C.J. & Germer, L.H. (1929) A test for polarization of electron waves by reflection, Phys. Rev. 33: 5; 760-772.
[32] Dewdney, C., Kypeianidis, A., Vigier, J.-P., Garuccio, A. &. Gueret, Ph. (1984) Time-dependent neutron interferometry: Evidence in favour of de Broglie waves, *Lett. Al Nuovo Cimento*, 40:16; 1971-1985.
[33] Croca, R. (1987) Neutron interferometry can prove (or refute) the existence of de Broglie's waves, Found. Phys. Vol. 17, No. 10.
[34] Rybakov, Y.P. (1974) On the causal interpretation of quantum mechanics, Found. Phys.Vol. 4, No. 2.
[35] Vigier, J.-P. (2000) selected papers, in S. Jeffers, B. Lehnert, N. Abramson, & L. Chebotarev (eds.) Jean-Pierre Vigier and the Stochastic Interpretation of Quantum Mechanics, Montreal: Aperion.
[36] Yang, S., Hammack, A.T., Fogler, M.M. & Butov, L.V. (2006) Coherence length of cold exciton gases in coupled quantum wells,

arXiv:cond-mat/0606683v1.
[37] Kleinert, H. (2005) Vortex origin of tricritical point in Ginzburg-Landau theory, arXiv:cond-mat/0509430v1.
[38] McKinnon, L. (1979) The de Broglie wave packet for a simple stationery state, Found. Phys., 9:9-10; 787-791.
[39] Barbour, J. (1999) *Then end of time - The next revolution in physics*, Oxford Press, Oxford, U.K.
[40] Leibniz, G.W. (1768) *Opera Omnia.* 6 volumes, Louis Dutens, ed. Geneva
[41] Amoroso, R.L. & Pribram, K.H. (eds.) (2010) The Complementarity of Mind and Body: Realizing the Dream of Descartes, Einstein and Eccles, Cambridge: MIT Univ. Press.
[42]] Chu, M-Y.J. & Amoroso, R.L. (2008) Empirical mediation of the primary mechanism initiating protein conformation in prion propagation, in D. Dubois (ed.) Partial Proceedings of CASYS07, IJCAS, Vol. 22, Univ. Liege Belgium.
[43] Amoroso, R.L. (1996) The production of Fröhlich and Bose-Einstein coherent states in in vitro paracrystaline oligomers using phase control laser interferometry, Bioelectrochemistry & Bioenergetics, 41:1, pp.39-42.
[44] Smullyan, R.M. (1992) Gödel's Incompleteness Theorems, Oxford: Oxford University Press.
[45] Heiblum, M., Nathan, M.I., Thomas, D.C. & Knoedler, C.M. (1985) Observation of Ballistic Transport in GaAs, Phys. Rev. Lett. 55:20; 2200-03.
[46] Javey, A., Guo, J., Paulsson, M., Wang, Q., Mann, D., Lundstrom, M. & Dai, H. (2003) High-field, quasi-ballistic transport in short carbon nanotubes, arXiv: 0309/0309242.
[47] Hubbell, J. H. (2006) Electron positron pair production by photons: A historical overview, Radiation Phys.& Chem. 75:6; 614-623.
[48] Schwinger, J. (1950) Physical consequences of vacuum polarization, Phys. Rev. 78:135; 105.
[49] Lloyd, S. (2006) Programming the Universe, New York: Knoph.
[50] Marshall, T. (2002) Engineering the vacuum, in R.L. Amoroso, G. Hunter, M. Kafatos & J-P Vigier (eds.) Gravitation and Cosmology: From the Hubble Radius to the Planck Scale, Dordrecht: Kluwer.
[51] McCarthy, W. (2003) Hacking Matter, New York: Basic Books.
[52] Smolin, L.(1997) The future of spin networks, arXiv:gr-qc/9702030v1.

[53] Smolin, L. (2002) Quantum gravity with a positive cosmological constant, arXiv:hep-th/0209079v1.
[54] Amoroso, R.L & Amoroso, P.J. (2004) The fundamental limit and origin of complexity in biological systems: A new model for the origin of life, in D. Dubois, (ed.) Computing Anticipatory Systems, Proceedings of CASYS04, 6th Int, Conf., Liege, Belgium, 11-16 Aug, 2003, AIP Proceedings Vol. 718, pp. 144-159.
[55] Jensen, H.J. (1998) Self-Organized Criticality - Emergent Complex Behavior in Physical and Biological Systems, Cambridge Univ. Press.
[56] Krink, T. & Tomsen, R. (2001) Self-organized criticality and mass extinction in evolutionary algorithms, Proc. IEEE int. Conf, on Evolutionary Computing, pp. 1155-1161.
[57] Bak, P., Tang, C. & Wiesenfeld, T.(1987) Self-organized criticality: an explanation of 1/f noise, Phys. Rev. Lett. Vol. 59, No. 4.
[58] Feigenbaum, M.J. (1980) Universal behavior in nonlinear systems, Los Alamos Science v1, p4-27.
[59] Morisawa, Y., Ida, D., Ishibashi, A.& Ken-ichi Nakao, K-I (2002) Thick domain walls around a black hole, arXiv:gr-qc/0209070v2.
[60] Morisawa, Y., Yamazaki, R., Ida, D., Ishibashi, A. & Nakao, K-I (2000) Thick domain walls intersecting a black hole, arXiv:gr-qc/0005022v1.
[61] Wolff, M. (2002) Cosmology, the quantum universe and electron spin, in R.L. Amoroso, G. Hunter, M. Kafatos & J-P Vigier (eds.) Gravitation and Cosmology: From the Hubble Radius to the Planck Scale, pp. 517-524, Dordrecht: Kluwer.
[62] Wolff, M. (2008) Schrödinger's Universe and the Origin of the Natural Laws, Parker: Outskirts Press.
[63] Starace, A.F. (1988) Hyperspherical description of two-electron systems, in Briggs, J.S., Kleinpoppen, H. & Lutz, H.O. (eds.) Fundamental Processes of Atomic Dynamics, pp. 235-258, NY: Plenum.
[64] Harding, S.L., Miller, J.F. & Rietman, E.A. (2006) Evolution in materio: Exploiting the physics of materials for computation, arXiv:cond-mat/0611462v1.
[65] L. De Broglie, *Théorie générale des Particules à Spin (Méthode de Fusion)*, p. 93-116. Paris, Gauthier-Villars, 1943.
[66] Drexler, K.E. (1992) Nanosystems : Molecular Machinery, Manufacturing and Computation, New York : Wiley & Sons.
[67] MacKinnon, L. (1981) A fundamental equation in quantum mechanics? Lett. Al Nuovo Cimento, 32:10; 311-316.

Chapter 13

Is a Different Search Protocol Required for Success in SETI Research?

In an Anthropic Multiverse intelligent civilizations are purported to be the rule rather than the statistical exception, providing circumstantial evidence that the current SETI search protocol to find civilizations similar to our own is inadequate since the SETI program is over fifty years old. From the current cosmological perspective researchers are content to perform straightforward searches for narrow-band electromagnetic transmissions from technologically based extra-solar civilizations. Here it is suggested that the scenario in an Anthropic Multiverse requires a radical new observational approach.

13.1 On the Horns of a Dilemma

If the physicality of the universe is not a Big Bang cosmology following **a** Friedmann-Robertson-Walker (FRW) metrical solution to Einstein's gravitational field equations as the current vogue suggests; it could have a radically different basis with an anthropic principle guiding its temporal evolution as we've done our best to delineate in this volume. In this context we explore the possibility that the Search for Extraterrestrial Intelligence (SETI) which formally began in 1959 (originally called the Order of the Dolphin) will require a radically different search protocol in order to have any hope of achieving success.

The highly ordered symmetry conditions of anthropic multiverse cosmology suggest that conformal scale-invariance is an inherent property. This could mean that "other worlds" exhibit the property of the 1st person 3rd person barrier. This conjecture suggests that the current SETI strategy will fail because current search protocols are based on Big Bang naturalism in the context of the Copenhagen interpretation of quantum theory which is governed by the uncertainty principle (The observer can only view one complement of an observation at a time because of the Pauli exclusion principle). This exclusive property according to the principle of invariance would apply to both other minds and other worlds. The current cosmological perspective does not include any putatively required addressable anthropic parameters of 'consciousness' relative to the physics of an observer. To repeat in order to ameliorate any surprise: What is suggested is that other worlds have properties similar to 'other minds' relative to an Earth observer which keeps conscious content blocked or hidden from detection by the limit of 4D Minkowski space based instrumentation by the quantum uncertainty principle. Why anthropic properties are required is explored. An outline for implementation of an alternative strategy based on new cosmological principles inherent in unitary field parameters as represented in a Holographic Anthropic Multiverse (HAM) is delineated. Design for an inter-dimensional radio Q-telescope utilizing entanglement with an HD 'ontological wave function of the universe' that HAM cosmology might require is presented in a preliminary manner.

Sometimes scientific enquiry follows lines of inductive reasoning just to explore the spurious conclusions. This chapter is a logical extension of anthropic multiverse parameters in penultimate form; if its SETI speculation is wrong it remains an interesting diversion to follow the curious thread of logical deductions. If it turns out correct then we are prophets dreaming of the day we can watch *Bronson Beta* TV or listen to *Ethnarch Gamma* radio! We cross the line of usual conservatism in order to leave no stone unturned, especially as SETI researchers have stated that under the current methodology it will probably take several generations to achieve success. With the pace of improving technology this would only take about two generations to check every star in the Milky Way Galaxy.

> The first episode of *I Love Lucy* was broadcast on October 15, 1951. About 0.0002 seconds later, the signal glided over the rooftops of the farthest city suburbs, and headed into space. It's still going. Every day, that first installment passes through an additional 4 thousand trillion trillion trillion cubic kilometers of the cosmos. Given that stars in our galactic neighborhood are separated by about 4 light-years, it's easy to figure that roughly 10 thousand star systems have been exposed to "I Love Lucy" in the past five decades [1].

When first researching this chapter the ideas seemed outrageous even to us; we suspect this may initially be the case for you also dear reader. The endeavor was initially conducted more as a recreational exercise to see what tinkering provided rather than as a serious avenue of scientific inquiry. We still remain a bit dumbfounded at the final result but have learned to embrace it in view of the broad explanatory power of other aspects of the new paradigm. Readers will find this by far the most unusual chapter in the volume. We do not wish to belittle it by our sarcasm or by passing it off as errant twaddle; but we want to make it clear we realize the challenges it presents within the currently popular Big Bang mindset and wish to state this obvious fact. Finally this chapter is a radical blend of theoretical anthropic cosmology, philosophy of science and epistemological theology. We hope you will enjoy at least the exercise of trying to keep an open mind as we have done; which is after all supposed to be the guiding light of scientific inquiry!

13.2 SETI Epistemology from the Anthropic Perspective

Currently two standard techniques are utilized in the search for extra-solar planets. So far most planets have been discovered by observing tiny wobbles in a stars radial velocity due to the gravitational tug of an orbiting planet. This method favors the detection of very massive Jupiter class planets. The other technique, called the *transit method*, observes the slight dimming in a stars light when a planet passes over its disk. Technological improvements in the sensitivity of digital cameras and spectrometers continue to enhance both techniques significantly.

It seems reasonable to assume that if extraterrestrial intelligence exists at a technological level sufficient to broadcast radio signals into space, an Earth based search could be conducted by searching the sky with radio telescopes of sufficient sensitivity. This appears inherently obvious from the perspective of a naturalistic Big Bang cosmology. But by 2050 the SETI program will be considered a failure if negative results remain; proponents have stated this search protocol would then have to be rethought.

Is the situation different in an anthropic cosmology? If it is, thinking metaphorically as queried above, aiming a radio antenna at the sky may be no different than aiming a microscope or telescope at someone's head in order to read the content of their mind. This is another epistemological conundrum. Cognitive psychologists who currently consider the brain tantamount to mind contend that forthcoming improvements in f-MRI brain scans will be able to resolve consciousness. Therefore an f-MRI enhanced radio telescope should be sufficient to resolve a signal. We have addressed this issue in detail elsewhere [2-6]. The suggestion explored here is that from the anthropic multiverse perspective SETI protocols must consider an additional parameter for resolving the signal because 'other world' will be caught under the 'other minds' 1st person 3rd person barrier that entails causal and therefore information separation. This is a stretch even for us to accept; future empirical tests will resolve the issue. Quantum theory discovered the nonlocal Einstein, Podolsky, Rosen (EPR) correlation. A completed form of quantum theory with an HD SUSY form of EPR entanglement that can be used to design a radical new type of inter-dimensional Q-telescope for manipulating and accessing causally separated anthropic domains. This relates to the nature of the observer, a challenge considered distasteful and therefore ignored so far in physical theory and epistemology in general [2].

Before delving into this thesis we wish to discuss some additional epistemological considerations. Theoretical physicist P. Rowlands states in volume 41 of this series,

> Physics appears to be the only source of fundamental knowledge about the natural world. No other system of thought has shown any of its systematic explanatory or predictive power [7].

The tools of physical science have been the pragmatic use of logic, reductionism and empiricism. But this can now be considered a myopic view because currently physics is devoid of underlying principles of 'consciousness' associated with not just the role, but also the fundamental nature of the observer. We have ignored the fact that the observer is imbedded in and made out of the same materials as the instrumentation. We have postulated that perceived reality itself is an intermediate collapsed or limited state. The time has come to intricately inspect the bias any inherent limitations the virtual nature of the observers reality intrudes on the pragmatic foundations of empiricism. This means to complete epistemology the tools associated with transcendence must be included. Recall that thousands of years ago Greek philosopher Plato stated that 'no matter how great ones intelligence, or how great ones wisdom, noetic insight (transcendent) from the cosmos is still greater' [8]. We are suggesting that in order to complete the tools of epistemology, transcendence or at least the role of anthropic properties should be utilized as a tool in guiding scientific theory formation [2].

In contrast to the methods of scientific inquiry, theology also attempts to describe the nature of reality utilizing faith and revelation or transcendence. Scientific inquiry by definition should not discount one any more quickly than the other. Myopic scientific interpretation has historically made as many horrendous mistakes as rigid bias in theology. We suggest that Theology and Science are not mutually exclusive as they became around the time of Galileo because of both the failure of sound logic as in the case of gravity for different weight objects and the narrow mindedness of the contemporary ruling theocracy, but opposite ends of a long continuum of human epistemology. We are not suggesting theology should replace science or that the empirical stance should be weakened in any way, only that theology and transcendence can be used as a viable tool to temper scientific theory formation and interpretation in some form of 'empirical metaphysics' [2,9]. Case in point, the principles introduced here could have been utilized in the year 2,000 rather than after the proposed 2,050 change in the SETI protocol, for a possible savings of many thousands of man hours and many millions of dollars. Not to mention that some future generation instead of our generation gets to enjoy the wonders of extrasolar intelligence.

SETI researchers have suggested three possible reasons for the failure:

1) Technical intelligence is very rare,
2) Technical societies are short lived,
3) Failure occurred because the correct search strategy is not utilized. (The one we address here)

We use this chapter as an axiomatic test case for future hindsight into the viability of an 'empirical metaphysics [2,9] in evaluating the utility of the putative integration of science and theology, but not to return to the bias shown in the following quotation:

> "For the first time since the Dark Ages", physicists Paul Ginsparg and Sheldon L. Glashow wrote 12 years ago, "we can see how our noble search may end, with faith replacing science once again" [10].

13.3 The Drake Equation

The Drake Equation [11-13], first formulated by radio astronomer Frank Drake in 1961 purports to estimate the number of communicative civilizations in the Milky Way Galaxy. Drake's initial solution to the Drake Equation estimates 10,000 communicative civilizations in our galaxy of about 250 to 400 billion stars. But Drake's equation is based on various assumptions. If different values to the assumptions are utilized, intelligence can just as readily be suggested to be the general rule not the statistical exception. At the time of writing about 400 extra-solar planets have been discovered; most of which are Jupiter size found by the minute stellar wobble their gravitational mass creates on the star [14]. But on 7 March 2009 NASA launched the Kepler extrasolar planet finder which will search the Milkyway for Earth-size planets utilizing the transit method. Kepler's photomultiplier sensitivity allows searches for Earth-size planets around stars up to 3,000 light years distance.

The Drake Equation:

$$N = R^* f_p n_e f_l f_i f_c L, \tag{13.1}$$

where

- **N** = The number of communicative civilizations
- **R*** = The rate of formation of suitable stars (such as our Sun)
- $\mathbf{f_p}$ = Fraction of those stars with planets. (Current evidence indicates planetary systems may be common for stars like the Sun.)
- $\mathbf{n_e}$ = The number of Earth-like worlds per planetary system
- $\mathbf{f_l}$ = The fraction of those Earth-like planets where life actually develops
- $\mathbf{f_i}$ = The fraction of life sites where intelligence develops
- $\mathbf{f_c}$ = The fraction of communicative planets (those on which electromagnetic communications technology develops)
- **L** = The "lifetime" of communicating civilizations

Using one of the interactive sites in [12,13] one may experiment with the parameters of the Drake equation to explore the range of possibilities.

Following the principle that life is the rule in an anthropic cosmology we propose corollary §13.1

§13.1 All Stars Have An Anthropic Zone.
Every star has an anthropic zone associated with it. Since 'gravity is caused by the movement of spirit' [15] or the unitary field if you prefer; the anthropic action principle optimizes the formation of planets compatible with life in this region; but of course this doesn't necessarily mean a star has planets in this zone.

13.4 Brief Review of Anthropic Multiverse Parameters

As gleaned from the introductory chapters in Part-1, HAM cosmology is not a naturalistic Big Bang cosmology but a form of continuous-state eternal/atemporal timeless multiverse derived by extending Einstein's original static universe model [16] to include HD SUSY parameters. HAM cosmology has properties reminiscent of Kant's antinomy of spacetime which he proposed as a solution to the argument between Newton and Leibniz as to whether the universe was open or closed [17]. A fundamental HAM premise is that the observed Hubble radius, H_R the Einstein 3-sphere of our perceptual reality is closed and finite temporally; and open and infinite atemporally. Thus the topology of HAM cosmology is like a continuously transforming HD hyper-Klein bottle with the possibility for an infinite number of Hubble type spheres 'holographically' nested within it, each of which might have a fine-tuned variance of the laws of physics [18]. The paradigm was developed by extending the Wheeler-Feynman-Cramer radiation/transactional interpretation models [19,20] and the de Broglie-Bohm ontological models to an HD regime commensurate with our version of SUSY-M-Theory parameters [2-6], but not interpreting Everett's many worlds condition as a duplicate parallelism but as additional and unique in their own right.

For Kant there are four antinomies connected with

1. the limitation of the universe in respect of space and time,
2. the theory that the whole consists of indivisible atoms (whereas, in fact, none such exist),
3. the problem of freedom in relation to universal causality
4. the existence of a necessary being

all of which by pure reason contradict the empirical, as thesis and antithesis. This was part of Kant's critical program of determining limits to science and philosophical inquiry [17].

The inherent purpose of an Anthropic Multiverse is the evolution of life and consciousness, therefore intelligent life is the rule not the accidental exception as often considered by Big Bang cosmologists. The difference in evolution for an Anthropic Multiverse is that evolution is not random but guided by a teleological action principle. A major premise of HAM cosmology in this respect is that cosmology is a self-organized complex system. Such systems are driven by an external action principle which in this case is the anthropic teleology. This action of course applies to each nested Hubble sphere. Thus the horizon of knowledge moves outside the temporal bounds of H_R where theory is still silent about other aspects of the Multiverse; but since the model is empirically testable we hope to soon begin to unravel new mysteries.

The difference in the two scenarios demands radically different search protocols for detecting extra-solar intelligence.

13.5 Does SETI Require a Different Strategy for Success?

Even if an EPR[1] correlation were somehow employed; under current methodologies EPR superposition is unable to teleport detailed information from one domain to the other. The EPR condition only arises under the simultaneous emission of photons (photon pairs) which as a result are correlated in time. It is suggested that neither people nor planets naturally maintain such a correlation and the current interpretation of quantum theory is not sufficient to develop such a superposition from causally separated entities by any known method of parametric conversion. The EPR paradox therefore illustrates both why a Big Bang SETI strategy will fail and as will be illustrated below leads to the explanation of why a HAM SETI strategy holds more promise.

This anthropic condition for SETI is an issue of both philosophy of science and theology. For example although cognitive brain scientists remain hopeful, the myriad forms of brain scan technology cannot achieve the analogous result of resolving consciousness. Even utilizing

---

[1] EPR – From a thought experiment devised by Einstein, Podolsky and Rosen which later in relation to experimental tests proved the existence of 'quantum nonlocality', a condition of long range entanglement or superposition.

the highly anticipated new MRI or fMRI advances will fail because 'mental information' is not contained in the energetic atomic or molecular states of brain neural biochemistry. Consciousness is not only a brain state, but a teleological cosmological condition [2-6]. For the purposes here the proper criteria for a successful SETI program must be aligned with a Cartesian dualistic model of mind-body interactionism [2].

This is the historical debate over the sufficiency of *Biological Mechanism*[2] versus the need for an additional life principle [2,9]. At the moment this issue remains highly controversial and many would insist that the 'telescope/microscope' metaphor applies neither to the cosmological issues pertaining to SETI research nor especially not to information content in the brain/mind. The critique revolves around the philosophical issue of Biological Mechanism. To insist at the present level of scientific progress that the mechanistic model of mind or the Big Bang cosmology is incorrect is highly unpopular. Unfortunately at this moment nearly every scientist or SETI researcher would therefore summarily label the premises given here as nonsense, believing that solar systems or planets are nothing like brains; and to even consider such a claim would be absurd. No current standard model of science (of which there are up to a dozen depending on how one counts them) can predict this conundrum because they remain naturalistic and the problems addressed here therefore outside this scope of influence.

In HAM cosmology the Hubble radius is an observational limit based on a 'tired light' redshift [21-31] rather than a Doppler redshift. Hubble discovered redshift not expansion of the universe. With the alternative interpretation the Universe is not limited merely to the ~15 billion light year radius Hubble sphere of the Big Bang model that we observe. The Multiverse has the potential for an infinite number of holographically nested Hubble-type spheres in causal separation each with their own laws of physics [18].

Here is where the conceptual problem arises for SETI as it is currently employed. In a Conscious or anthropic universe it appears that each intelligent system (by this we mean planetary system with intelligent life) has its own natural laws that are out of phase with the cosmological conditions of other intelligent systems (The same might be said for people) making oscillations of the background matter, not the mental content (like the quantum fluctuations of brain chemistry explored by

[2] Biological Mechanism – The philosophical position that the principles of Chemistry and Physics are sufficient to describe all life; no additional life principle is required.

fMRI) the only material viewed. So the discussion here is centered on the problem of how to get at the 'Qualia' or endogenous 'conscious light' of intelligence rather than just the thermodynamic energetics of the atomic structure of the rocks and gasses there.

13.6 Theological Arguments – Adam given his Reckoning versus the Cosmological Age of the Earth

The putative age of the Earth based on scientific radiometric dating has an upper limit of about 4.567 billion years based on rock in the earths crust which is also compared to moon rock and Martian meteors. The best-known absolute dating technique is carbon-14 dating, which archaeologists prefer to use. However, the half-life of carbon-14 is only 5730 years, so the method cannot be used for materials older than about 70,000 years. Radiometric dating involves the use of isotope series, such as rubidium/strontium, thorium/lead, potassium/argon, argon/argon, or uranium/lead, all of which have very long half-lives, ranging from 0.7 to 48.6 billion years. Subtle differences in the relative proportions of the two isotopes can give good dates for rocks of any age. Radiometric dating is based on solidified rock. Should this be the valid origin of dating? It is purported that the Martian core took about .5 billion years to solidify. The Earth's core has not solidified because of its larger mass. Thus extrapolating the age of the Earth as a molten ball before the crust solidified could also add a billion years to the age of the Earth, or do we not call the molten ball before crust solidification Earth? I make this analysis to contrast the scientific age of the Earth with the opposition to the 'young earth' theological doctrine with a different type of theological calculation making correspondence to the scientific model. Although not a rock hard theory it at least gives suggestive support to the Anthropic cosmology put forward here.

Theologically, after the creation delineated in the Book of Genesis of the Judeo-Christian Bible/Torah, Adam (the 1st man) was 'given his reckoning' in the Garden of Eden. This is interpreted to mean that instead of existing in God's timeless eternal frame; Adam was given a different Earthly temporal clock or reckoning where "A day with the Lord is a 1000 years with man" [32] and a lifespan of "four score years". HAM cosmology suggests that Adam's 'clock' is relative only to Earth's intelligence and therefore the fine tuning relative to Earthly laws of physics, so that both God and other cosmic civilizations are out of phase (and therefore invisible and causally separated) as it were with both the clock and geometric structure of our reality for the Earth observer.

It is possible to make a simple 'age of the Earth' calculation to illustrate this in contrast to the myopic view of the young-earth creationists who claim the Earth is only 7,000 years old by utilizing the Judeo-Christian premises:

- A day with the Lord is a 1000 years with man
- God created the world in seven days
- The creation occurred first spiritually and second temporally.

Therefore there is a modicum of credibility of doing a simple calculation using 'Gods time'. Therefore 7 days x 1,000 man years per day = 7,000 years which is the usual point of view taken for the period of man's existence. But Adam was not given his reckoning until after he was placed on the Earth so instead if we consider the creation was 1,000 God years, multiplying 365.4 Man days per year x 7000 God days x 1,000 = 2,557,800,000 billion Man years x 2 because of the spiritual and temporal creations = 5,115,600,000 we come up with a proper order of magnitude for the scientific age of the Earth. Oh you noticed the number appears to be a half billion years off. You forgot that the scientific reckoning date begins from the radio-dating time when rock solidified. It seems logical that it took half a billion years of cooling to solidify. We agree the calculation is somewhat silly and sloppy, but is it just a coincidence that it arrives at the correct answer of ~ 4.6 billion years?

Now lets fudge in a different way. If we keep the spiritual creation as the 7,000 God years and reduce the temporal to 6,000 we get an age of the Earth of 4,750,200,000 billion years well within acceptable error for the putative radiometric dating method. But how can we justify this concatenation? Easy. We go back now to make correspondence to the 'young Earthers' in terms of Adam given his reckoning and interpret the omitted 1,000 years as the age of man for the fulfillment of a society of Gods people. Adam to Abraham to Moses to Christ to present age being the 6,000 years with the 7th to come as the Millennium. No offense if the reader wishes to call this 'hogwash'. It has as much valid logic as any other consideration; and its purpose is only to be thought provoking and foster debate on the utility of some form of the anthropic principle in cosmology.

This has more pertinent meaning in terms of the first line of Genesis: "In the beginning God created the Heaven and the Earth" – Genesis 1:1.

This is taken to mean God created this earth and its heaven. (A pocket universe in the Multiverse?) We realize that moot theological dogma is not considered scientific; but it can be used as a philosophical basis for developing a science of cosmology. Thus in a Holographic Multiverse this implies room for an infinite number of nested Hubble spheres each with their own fine-tuned laws of physics [18].

Four Theological Arguments can be used to support failure of the current SETI search protocol.

1. Eternal HAM cosmology versus Temporal Big Bang Cosmology. In an anthropic universe a teleological action principle guides evolution. The argument is that even quantum theory is insufficient for success. A model utilizing the noetic unified field is required.

2. Genesis 1:1 states that God created[3] both 'the Earth and its Heaven' suggesting the laws of physics may be different for each civilization because the nested Hubble spheres are in causal separation requiring a different method for SETI.

3. Scripture also states that 'Adam is given his reckoning' preaching that Man's time is not God's time suggesting his time rate is different than that of other Hubble spheres (In a holographic anthropic multiverse a Hubble sphere represents the observational limit of a particular civilization).

4. Anthropic Principle (AP) arguments exist against Steady State cosmology. They suggest that ETI should have already filled a timeless universe and currently be in our solar system [33]. The argument is logical for a naturalistic Big Bang cosmology; but not for HAM cosmology. Earthlings are barely able to coexist with each other. By putative teleological considerations the organizer of intelligence may have a planets millennium (judgment day) begin before a civilization is able to use 'warp drive' space travel to visit other solar systems and interfere with their societies natural evolution. This argument may seem unacceptable to some but one is forced to admit that it is as logically valid as AP arguments against Steady State cosmology.

---

[3] Not *ex nihilo* (out of nothing) creation, but 'organized' from existing raw materials.

There is one scripture supporting this argument albeit from obscure Mormon theology:

> Is not the reckoning of God's time, angel's time, prophet's time, and man's time, according to the planet on which they reside? ... Yes. But there are no angels who minister to this earth but those who do belong or have belonged to it [34].

This suggests that God will not allow external interference in the evolution of other planetary societies and could be a reason other stars are so many light years distant from ours.

13.7 Seemingly Far-Out Absurd Pseudo-Scientific Arguments

Now some very radical theoretical inference is introduced. The renowned physicist J.A. Wheeler proposed what he called the geon concept, a ball of photonic light of sufficient size to gravitationally self cohere into a sphere (not a star with matter, but a ball of pure light). The propagation of photons in the group would be in circular rays such that there would be no loss [35]. According to the special theory of relativity [36], in terms of the laws of conservation of energy and momentum, even though photons are considered to have zero rest mass; they have an equivalent *moving-mass*, $m = E / c^2$ that is able to facilitate the cohesion of the Wheeler geon [35].

French physicist Louis de Broglie proposed that the same wave properties exhibited by electromagnetic phenomena applied also to all material particles [37-39]. Recall first that in dualistic models of consciousness the mind is like the light (rather than the usual electricity) running an optical computer. If the de Broglie wave of the Earth is considered to be like a Wheeler geon representing the light of all of Earth's intelligence over all time, then the mass of the Earth according to Einstein's formula for special relativity equating matter and energy $E = mc^2$, when converted to energy could represent the geon energy mass for all life on Earth over the approximately 5 billion years of its existence.

Our moment to moment view of reality is in time. This is like the centroid of a wave or internal spinning mass-particle moment of a propagating wave according to the standard concept of wave-particle duality. The envelope (wave) is timeless and the internal motion (particle) is temporal. Therefore the total mass-energy of the wave can be used to theoretically represent the total mass-energy of planetary intelligence, which can be considered reminiscent of Jung's Collective Unconscious [40] which most transpersonal psychologists consider to be physically real.

The 14 billion light-year Hubble radius, H_R is the limit of observation. In Big Bang Cosmology this radius is said to be the current limit of expansion of the universe in terms of a Doppler redshift. In HAM cosmology the interpretation is different and based on a so-called 'tired light' mechanism. During the centroid moment of photon propagation the internal motion causes an inertial gravitational coupling to the spacetime vacuum. This coupling is responsible for the observed cosmological Redshift. This means that at the Hubble radius all visible light has attenuated to zero energy signifying the limit of observation. This 'mass of intelligence' would correlate with a classical 3(4)D gravitational force causing a perceptual collapse of the 11(12)D eternal space of HAM cosmology reducing it to the more limited temporal reality we observe as the external world. What I'm trying to say is that the photonic mass of the Earth's geon of intelligence provides a force collapsing observational space to the Hubble radius for the temporal based Earth observer which is a limited virtual reality.

Let me try to explain this more clearly. When an EPR experiment is performed, simultaneously produced photon pairs are required to provide the correlated system. These photons are propagating through space at the speed of light. Imagine instead a stationery system with an eternal present [41] as might be viewed by a HD observer from the point of view of Einstein's pre-relativistic thought of 'what it would be like to ride on a photon' – Time would stand still, i.e. one could circumnavigate the whole universe without the passage of time. In HAM cosmology both the multiverse and human beings contain an inherent duality of temporality and eternity. In this duality the invisible core is a 12D node of the unitary field from which our 3(4)D view continuously flows as a standing wave subspace.

The Pauli exclusion principle is illustrated in the uncertainty principle by the fact that components of the quantum wave function do not commute in the lesser Copenhagen regime. Correlated photon pairs are produced by what is called parametric down conversion [42]; although theoretically described science currently has no method for parametric up conversion [43] to correlate unpaired photons. EPR superposition is the most primitive form of these correlations. The point I am trying to make is that a higher level unitary correlation is required to obtain 'conscious' information like through telepath or 'revelation from God'. This information is scale invariant so that what I am trying to describe at the photonic and human level is also postulated to occur at the cosmological scale. All the information is contained in the hologram but by the 1st person 3rd person barrier macroscopic uncorrelated systems are not able to share internal information. This is a form of collapse of the wave function at the level of reality itself which our awareness is naturally coupled to.

This is a converse of the observational limit in physical cosmology where the ~14 billion light year radius of the Hubble sphere H_R is caused by light **redshifted to complete attenuation**. This relies on an alternative interpretation of Hubble's 1940 discovery. Big Bangers interpret his discovery of the redshift as an indicator of expansion of the universe, when actually all he discovered is a redshift distance effect that can just as reasonably be interpreted to be non Doppler as Doppler. In summary, the gravitational mass of Earthly intelligence over all time represents the force that gives the Hubble sphere its observed radius, i.e. in a conscious universe U_R is infinite, but the innate force of the observer creates in the continuous creation of his reality, the observational limit which for Earthlings is ~14 billion light years. This means that within another HAM Hubble type nested sphere elsewhere in the multiverse that mass of intelligence would be different, the laws of physics different by fine tuning and therefore the Hubble radius different for the denizens there! This is the crux of our premise for the failure of the current SETI protocol. These other intelligent domains would be 'out of phase' with our reality and therefore not observable with a standard telescope or microscope. This is considered a standard feature of wave mechanics. This is similar to making a quantum measurement; only a portion of the information is observable and only ontological versions propose methods for obtaining complete information beyond uncertainty. (See Chap. 9.) This is part of an ongoing debate as to whether the Schrödinger wave function provides a complete description of reality.

Therefore in this model the Barnard's star geon of intelligence would have a radically different gravitational mass (statistically highly improbable that they would be the same; representing the highly improbable only case where the current SETI protocol could work[4]. It still would not work because this would be like an EPR experiment that cannot be used to teleport information) creating the Hubble type sphere created by the denizens living there. This would relate to the 'reckoning' given to the 'Adam' there, which must be taken into account for our reception of Alpha Centauri technological broadcasts according to the theoretical model presented here. There are several kinds of coordinate transformation, the Galilean transformation for small velocities, the Lorentz-Poincaré transformations for relativistic velocities and a newly proposed noetic transformation for ontological considerations.

13.8 The Anthropic Principle (AP)

The development of HAM cosmology aligns with arguments for the Anthropic Principle which states that the observed universe is designed to accommodate intelligent beings [44-46]. Fine tuning also seems to play an important part. For example:

- A star must remain on the main sequence for about 10 billion years in order for there to be sufficient time for planet formation and 'guided' evolution of intelligent life; which for Earth as an average case took about 4.5 billion years.
- If the Earth spins much faster the atmosphere will fly off.
- If the Earth spins much slower opposite sides will alternately burn and freeze.
- If the mass of the Earth is very large gravitational forces will be too strong to support life as we know it. The linear increase in body size and mass requires a quadratic increase in bone strength.
- If the Earth were too small (like Mars for example) the molten core would solidify, the dynamo would stop rotating attenuating the geomagnetic field causing the atmosphere to fly off into space since there would be no charge to trap the gasses.

It's interesting to speculate on possible environments of extrasolar planets. Living conditions would depend on the distance of a planet from

---

[4] Because simultaneously created entangled EPR photon pairs would be required – this is why the Q-telescope postulated here is required to produce such a correlation.

a star and the stars spectral qualities. For a binary star system which is very common, if a dim massive red giant had a smaller bright star orbiting it, the red giant might be far enough away from a planet orbiting the smaller star to have little effect on it. If one assumed that the second star had a significant effect on the planet, it would probably result in the seasons being based on the proximity of the secondary star. Summer would occur while the planet was between the two stars, and daylight would be continuous. Winter would occur when the planet was on the far side of the primary star, and there would be the usual darkness at night. The habitable world of such a system would have a highly elliptical orbit that was much further away from the primary star during the summer because of the gravity of the secondary star. The elliptical orbit would regulate the annual temperature: the heat of two stars would be tolerable because the planet stays farther away from one of the heat sources [47].

In order for life to be the rule rather than the exception, these fine-tuned details unique to each star system would be far easier to meet with an anthropic action principle guiding planet and star formation. As seen in Chap. 10 soon enough we will have sufficient data for Titius-Bode profiles for the planets of numerous star systems.

The Weak Anthropic Principle (WAP) [33] asks where and when are good conditions for intelligent life realized. The Strong Anthropic Principle (SAP) extends the AP to explain the actual nature of physical constants. Some suggest that the SAP leads to teleology or belief in a 'Divine Purpose'. The SAP can be extended to apply to an ensemble of Hubble spheres in a multiverse with varied laws of physics for each Hubble-like domain. In this respect the SAP puts constraints on the physical laws of the universe.

Darwinian style evolutionary biologists argue that man is probably unique in the universe because of the number of improbable evolutionary steps required to produce a Homo Sapiens. There is also the space travel argument against ETI in our galaxy. If ETI existed in our galaxy, by the Principle of Mediocrity (Earth evolution being typical) [33], ETI would have evolved sufficiently to be in our solar system because within 300 million years they would have explored the whole galaxy. AP arguments against Steady-State cosmology, where time is meaningless, state that ETI would also fill the whole universe [33].

However another plausible theological argument follows from anthropic teleology. After thousands of years of the evolution of human consciousness history remains continuously littered with the destruction of man getting along with man. One may make three reasonable assumptions by anthropic design.

1) Other stars are far away. Proxima Centauri, the closest known star to Earth is 4.2 light years away, requiring some form of advanced Star Trek style warp drive technology which for us is still science fiction.
2) In Judeo-Christian theology one might assume the prophesized millennium begins before a civilization develops a warp drive capacity taking human intelligence off planet. (Based on non-intervention and scripture.)
3) Fully developed intelligence has no need for physical travel. A form of inter-dimensional superposition would be tantamount to providing an ontological presence invisible to us but not to them.

13.9 Calculations for a Holographic Anthropic Multiverse

We assume axiomatically from earlier work [18] that the universe is scale invariant from microcosm to macrocosm (i.e. no change in physical laws). But like most regimes there is a limit; because of fine-tuning parameters in the Multiverse, such a limit also applies to 'our' Hubble sphere. Also from prior work a formula for the noetic force of consciousness, F_N (an anthropic potential) [48] was derived as $F_N = E / R$ in general unexpanded form, where E equals the energy of a conscious noetic field in Einstein's (the physical unit defined as a mole of photons) and R represents a complex or rotational radius derived from momentum[5]. The formalism becomes complex when expanded and applied to a variety of SUSY domains in its range of application; for example E/R becomes alternating dynamic and static Casimir boundary conditions [49-51] in spacetime brane domains of continuous-state symmetry breaking. This noetic formula is utilized to perform a proportionality order of magnitude calculation to see if the mass of the universe, M_U calculated from the philosophy of standard physics can be considered commensurate to the alternative derivation of mass from the

---

[5] Note that this noetic formalism takes the precise form as the fundamental equation for string tension ($T = E/L$) in M-theory and could be said to represent an alternative manner for its derivation. See Chap. 4.

HAM model. The purpose of the calculation is to demonstrate the possibility that the collapse of the universe from the 11(12) dimensional eternal realm to the subspace of our observed temporal reality of 3(4) dimensions occurs by or is a property of the action of consciousness.

We make the axiomatic assumptions that:

1. All spirit is matter; and that all matter under proper conditions may evolve to become spirit. (The latter being the task of the Earth.)
2. That the mass of the Earth, M_E is equivalent to the mass of intelligence for all life living during all time (~ 5 billion years) of the Earth's existence.

The mass of the universe, M_U is not nearly as precise as M_E and is derived from the Eddington and Dirac 'large number hypotheses' [18,52] for the number of nucleons $N_p \sim 10^{80}$ in the universe. Since the mass of an average nucleon is $1.6\times10^{-24}\,g$ multiplication tell us $M_U \approx 10^{57}\,g$. This will be utilized as the standard M_U for our H_R to be compared below with the mass derived from a conscious universe, M_{UC}.

The first part of the alternative calculation is to derive the energy of Earth's intelligence I_E from its mass m_E by using Einstein's well known mass energy relation formula $E = mc^2$. With simple multiplication in *cgs* units using M_E as $6\times10^{27}\,g$ and the speed of light *c* as $3\times10^{10}\,cm/\sec$ we arrive at a magnitude of 5.4×10^{128} for I_E.

Next we find the radius of the Earth E_R from its circumference $C = 2\pi R$. If the circumference of the earth is taken to be ~25,000 miles, converting to cgs we find $E_R = 1.6\times10^9\,cm$.

Finally from the Hubble radius, H_R of the universe ~14 billion light years; we convert to *cm* to again be in proper *cgs* form for our purposes. Since the speed of light, *c* is $3\times10^{10}\,cm/\sec$, and 60s x 60m x 24h x 365.4d seconds are in a year, we find H_R to be $9.14\times10^{17}\,cm$.

Using the noetic universe proportionality formula:

$$\frac{I_U}{H_R} \propto \frac{I_E}{E_R}, \quad I_U = \frac{H_R \times I_E}{E_R} \tag{13.2}$$

we find for the intelligence of the universe

$$I_U = \frac{9.14\times10^{17}\,cm \times 5.4\times10^{128}\,cm}{1.6\times10^{9}\,cm} = 3.1\times10^{137}. \tag{13.3}$$

Using Einstein's mass-energy equivalency formula $E = mc^2$ again we convert to the complex mass

$$m = \frac{E}{c^2} = \frac{3.1\times10^{137}}{3\times10^{10}\cdot 3\times10^{10}} = 3.33\times10^{116} \tag{13.4}$$

Finally we assume this is a complex mass related to the wave function of the universe from the Wheeler-DeWitt equation (13.5) [53-56].

If we interpret result (13.4) in terms of Cramer's transactional model of quantum theory [20] where all off diagonal advanced and retarded components of the wave function are considered to be physically real standing wave future-past elements; we may use the standard quantum equation $\Psi = \psi\psi^*$ and take the square root of the complex mass, M_U of the $M_U^{'}$. This gives the result for real Euclidean/Minkowski space the mass $M_U \sim 10^{58}\,g$ which, within acceptable limits of error for cosmological numbers, is the same mass of the universe calculated by conventional cosmological means!

We use this phantasmagoric albeit simple calculation to suggest that for an anthropic multiverse a quantum condition of our virtual reality acts as a force of collapse of the wave function (reality is like a continuous measurement) such that H_R is determined by the mass of intelligence, I_U or I_E. This is the basis for fine-tuning in nested holographic anthropic pockets where each pocket has its own laws of physics based on this anthropic principle.

13.10 Wave Function of the Universe (WFU)

A number of authors have postulated a wave function of the universe satisfying the Wheeler-DeWitt equation with the general unexpanded form

$$H\Psi = 0 \tag{13.5}$$

which is a gravitational form of the Schrödinger equation suggesting the universe is described by a quantum wave function rather than a classical spacetime [53-57]. This equation has become the basis for theories of quantum gravity. Cosmologically this wave function, Ψ depends on the action of gravity and matter on a 3-torus for a closed universe. There are three main proposals – The Hartle-Hawking wave function [53,57], the Linde wave function [58] and the tunneling wave function [59,60]. While this avenue has spurred years of creative thinking, it remains insufficient to solve the problem. Many scientists currently believe we live in a quantum universe. This is not true; we do not live in a quantum universe any more than we live in a Newtonian universe. We live in a multiverse that is a continuous anthropic state (like a complex HD standing wave) of complementary classical, quantum and unitary parameters [19,20] requiring an ontological interpretation of the WFU [61]. Feynman has said that their might not be a quantum gravity [62] and HAM cosmology also suggests that this is true. The quantum regime has a limit in the same way the Newtonian regime does. The quantum world and gravity do not integrate with each other, but with unitarity. The purpose of the quantum regime is to separate the reduced or limited Newtonian AS from the new HD absolute space. This separation allows our temporal reality to 'surf' as it were on the face of eternity; and is why for Earthly observers reality appears 3D.

13.11 Subtractive Interferometry

The technology exists today to implement subtractive interferometry; but it will probably wait some twenty-five years before NASA-ESA will budget such a program. For SETI research a wide baseline interferometer is set up with telescopes on opposite sides of the Sun. Such a telescope array is able to subtract out stellar coronas which normally block 'M-class' planets from view which orbit too close to the star to be observed by conventional telescopes.

Subtractive interferometry is required for two reasons. M-class planets are much too small to cause a stellar wobble which is the method used to discover the roughly three hundred and fifty Jupiter sized extra-solar planets. When the stellar corona is subtracted out by interferometry spectroscopy may then be used to detect water and chlorophyll. It is these planets that are of paramount interest and where SETI searches should be concentrated for highest efficiency.

The technology to construct a pair of precision telescopes able to remain in formation while maintaining the required stability for observations is an engineering challenge, but researchers at NASA and the European Space Agency (ESA) have programs to meet the challenge. The ESA is developing a formation-flying interferometer, called Darwin, planned for a joint NASA-ESA mission before 2020.

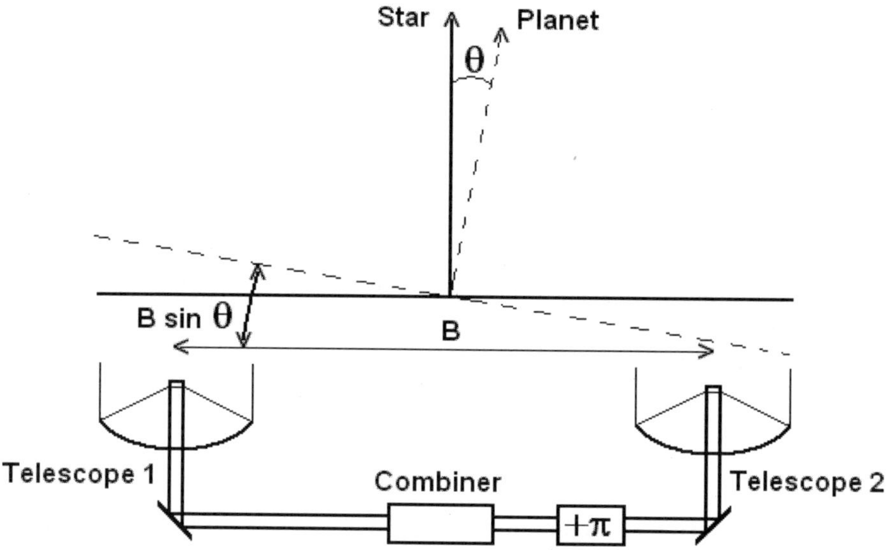

Figure 13.1. Two telescope space-based nulling interferometer, simplest configuration for producing a dark fringe in the line of sight when properly phased. This destructive interference enables planet detection by subtracting stellar coronas when focusing the null region on the stars disk. Redrawn from [63].

13.12 New SETI Technology - The Interdimensional Q-Telescope

HAM cosmology proposes that a new type of interdimensional Q-Telescope is required for an anthropic multiverse in order for SETI to be successful. The Q-Telescope requires the completed 12 dimensional version of Quantum Theory that allows utilization of the Noetic unified field to obtain the information of other planetary civilizations. This information is lost to temporal 4D reality because it is only a subspace of the higher reality and some information is lost. Analogously this is the same as solving the problem of the 1st person 3rd person barrier in philosophy of mind and is also similar to the EPR paradox.

The new technology proposed is named a Q-Telescope (Q to represent both the mental Qualia and the Q charge of radio) to be designed to take these new considerations into account and perform a new type of advanced EPR measurements on Proxima Centauri. Currently EPR correlated photons are only created or entangled by simultaneous emission of photon pairs from excited atoms such as mercury. We currently do not have the ability to entangle uncorrelated photons in an interferometer for example by any known method of parametric upconversion [42,43] to obtain entanglement in the nonlocal eternal present with no temporal separation. What is the relation between the two types of photons?

> **§13.2** In an anthropic multiverse because of inherent conformal invariance, the 1st person – 3rd person barrier said to exist between other minds is postulated to exist between extra-solar societies and our own because of the phase variation of fine-tuning making this information inaccessible in the usual 4D reality of standard measurement protocols.

How can this be logical since we can observe these planets? In the same manner we can observe other people but not observe their minds through telescope or microscope, only the matter they are made of.

It is now technologically feasible (although NASA says it will be at least 30 years before budgeting) to deploy a dual satellite system to utilize what is called ‘subtractive interferometry’. This entails putting a satellite on each side of the sun. Currently astronomers have discovered over 100 extra-solar Jupiter class planets by observing the slight wobble their orbits cause in the stars they circle. Earth type planets, the ones deemed most likely to produce life as we know it, are too small to create stellar wobbles and close enough to their stars to be unobservable because they are hidden behind the stellar coronas. Subtractive Interferometry has the ability to cancel out these stellar coronas so that powerful telescopes can search for Earth class planets. Once Earth type planets are observed; and it is speculated that they are common, spectroscopic observations can look for water and bioluminescence.

Earth class planets with water are currently anticipated to be the most likely venues to produce technological societies.

So rather than broadband radio frequency searches across the whole sky as currently being performed by SETI programs; only the Earth class planets discovered by subtractive Interferometry would be investigated. These planets would be searched with the radically different approach using the new Q-Telescope technology. The Q-Telescope would be designed to utilize

1. a temporal adjustment accounting for the 'Adam's reckoning' and
2. phase resonance protocols to couple to the 'Intelligent geon' state of the system.

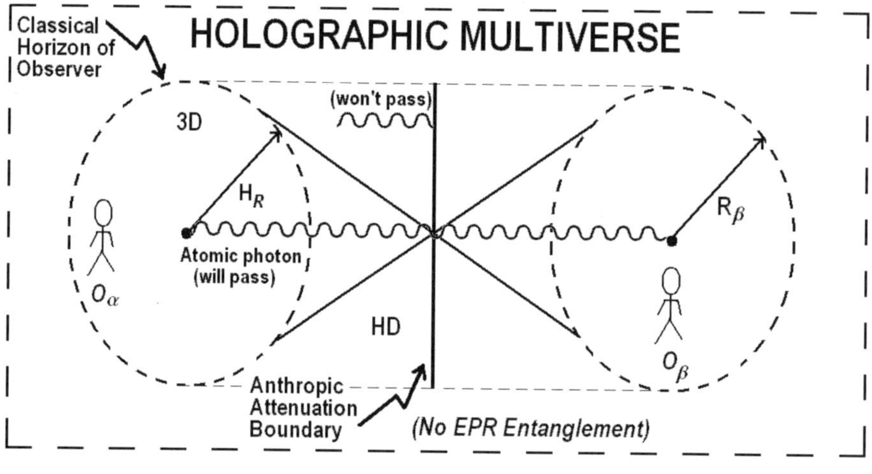

Figure 13.2 In an Anthropic Multiverse causally separated nested Hubble spheres may have no EPR correlation because of the different basis for the temporal 'reckoning' of the denizens so that atomic photons will pass but anthropic photons will not.

According to Fig. 13.2, while 'material' or atomic light in M_4 passes between the two causally separated observers; because each observer's awareness has a different 'reckoning' [32] consciousness related phenomena like TV or radio program waves will be out of phase and require an HD parametric upconversion EPR process to synchronize the reckoning times as in the entangled state of simultaneously emitted photons in parametric down conversion by Hg atoms [42,43]. We admit this seems ridiculously strange to current thinking but we get halfway there by the clock paradox of special relativity where one twin space traveling relativistically away from Earth may age only 5 years while his Earthbound twin could age hundreds of years [64].

> Understanding holography in space–times other than Anti de Sitter space, e.g. flat space and cosmological space–times. This throws up new conceptual challenges. For example, the boundary of Anti de Sitter space is time-like, but the boundary of cosmological space–times is often space-like and for flat-space it is null. Extending holography to these cases, will surely be worthwhile [65].

Because of the manner in which the skeptical scientific mind works, I suppose SETI-I is doomed to take some decades to fail before many will pay serious attention to the views presented here [66]. But instead of beginning to design a Q-Telescope 30 years from now when NASA and SETI programs might be ready to listen; perhaps if scientists start thinking about it now Q-Telescope designs could be ready at the same time. Because I for one don't want to have to wait for Proxima Centauri radio or Barnard's star TV any longer than necessary!!

13.13 Conclusions Contusions and Cowardice

On a lighter note, even to the authors (guilt by association) the concepts and postulates introduced here seemed utterly preposterous at first. It took many months to get used to them. Curiously though while in the process of habituating to the 'strangeness', we noticed a relationship that solved a medical research problem we were working on which gave us a new and deeper appreciation of the conformal scale invariant properties of this curious cosmological concatenation. This engendered the initial decision to outline the SETI framework and follow it to its logical contusions for the purpose of an entertaining and momentary escape from the drudgery of engaging in more complex conservative problems like developing a protocol for surmounting the uncertainty principle. We laughed at times thinking no one would ever believe such a concatenation of horrendous gewgaw. But now, years later we realize how wonderfully this marvelous heresy fits together in ethereal bliss. The test of time will tell if it's ephemeral fantasy or not...

Still if the SETI-I protocol fails as anticipated, perhaps because of the explanatory power of other aspects of the theory someone may get around to empirical tests. Then we will see what merit remains for the 'Earth being the center of the universe'... The Lord certainly does seem to work in mysterious ways because if we had not delved into this whimsical work (speaking now in terms of conformal scale invariance for certain physical principles) we would not have realized until much much later that these strange cosmological principles also apply directly to the microscopic scale in terms of the new noetic class medical paradigm we have been developing...

References

[1] Shostak, S. (2006) Listening for ET's Television, www.space.com/searchforlife/seti_shostak_tv_061109.html.
[2] Amoroso, R.L. & Pribram, K.H. (2009) The Complementarity of Mind and Body: Realizing the Dream of Descartes, Einstein and Eccles, Cambridge: MIT Press.
[3]Amoroso, R.L. (2006) The Holographic Conscious Multiverse, in R.L. Amoroso, B. Lehnert & J-P Vigier (eds.) Beyond the Standard Model: The Search for Unity in Physics, Orinda: Noetic Press.
[4] Amoroso, R.L. (2003) The Fundamental Limit and Origin of Biological Systems, Noetic Journal 4:1; 24-32.
[5] Amoroso, R.L. & Amoroso, P.J. (2004) The Fundamental Limit and Origin of Complexity in Biological Systems: A New Model for the Origin of Life, in D.M. Dubois (ed.) Computing Anticipatory Systems, AIP Conf. Proceedings V. 718, pp. 144-159, Melville: Am Inst. Phys.
[6] Amoroso, R.L. (2002) Developing the cosmology of a continuous state universe, in R.L. Amoroso, G. Hunter, M. Kafatos & J-P Vigier (eds.), Gravitation & Cosmology: From the Hubble Radius to the Planck Scale, Dordrecht: Kluwer Academic.
[7] Rowlands, P. (2007) Zero to Infinity: The Foundations of Physics, Singapore: World Scientific.
[8] Plato (1989) The Collected Dialogues of Plato: Including the Letters, E. Hamilton & H. Cairns (eds.) L. Cooper (trans.) Princeton: Bollingen Series LXXI.
[9] Meehl, P.E. (1966) The compleat autocerebroscopist: A thought experiment on professor Feigel's mind-body identity thesis, in P.K.

Feyerabend, & G. Maxwell, Mind, Matter and Method; Essays in Philosophy and Science in Honor of Herbert Feigl, pp. 103-180.
[10] Musser, G. (1998) String Instruments, Sci. Am., V.10, pp. 17-19.
[11] Drake, F. & Sobel, D. (1992) Is Anyone Out There? New York: Delacorte Press.
[12] Beyond the Drake Equation (does calculations) www.station1.net/DouglasJones/drake.htm#dolphin
[13] www.as.utexas.edu/astronomy/education/drake/drake.html; or www.classbrain.com/artmovies/publish/article_50.shtml; or www.pbs.org/wgbh/nova/origins/drake.html; http://frombob.to/drake.html.
[14] http://exoplanet.eu
[15] Smith, J. (1974) The Doctrine and Covenants, D&C 88: 11-13, Salt Lake City: The Church of Jesus Christ of Latter-Day Saints.
[16] Einstein, A.(1917) *Kosmologische Betrachtungen zur allgemeinen Relativitätstheorie, Sitzungsberichte der Preussischen Akademie der Wissenschaften*,142–152; Also Cosmological considerations on the general relativity theory, pp. 177–188; (1923) in The Principle of Relativity, a collection of original memoirs on the special and general theory of relativity, H.A. Lorentz, A. Einstein, H. Minkowski, & H. Weyl, with notes by A. Sommerfeld; W. Perrett & G. B. Jeffery (trans.) London: Methuen & Co.; reprinted (1990); also (1952) New York: Dover.
[17] Kant, E. (1787) *Kritik der reinen Vernunft*, Critique of Pure Reason, (1996) W. Pluhar (trans.), Indianapolis: Hackett.
[18] Kafatos, M, Roy, S & Amoroso, R. (2000) Scaling in cosmology and the arrow of time, in R. Buccheri, V. di Gesu & M. Saniga (eds) Studies on the Structure oF Time, Dordrecht: Kluwer Academic.
[19] Wheeler, J.A. & Feynman, R. (1945) Rev. Mod. Phys. 17, 157.
[20] Cramer, J. (1985) The Transactional interpretation of Quantum Mechanics, Rev. Mod. Phys 58, 647-687.
[21] Bass, L. & Schrödinger, E. (1955) Proc. R. Soc. Lon A 232,1-6.
[22] Einstein, A. (1917) Ann. Phys., vol. 18, p. 121.
[23] Zwicky, F. (1929) Proc. Nat. Ac. Sc., Washington, 15, 773.
[24] Finlay-Freundlich, E. (1953) *Göttinger Nachrichten*, 7, 95-102.
[25] Finlay-Freundlich, E. (1954) Philosophical Magazine, 45, 303-319.
[26] Finlay-Freundlich, E. (1954) Proceedings of the Physical Society A 67, 192-193.
[27] Born, M. (1954) *Nachr. Ak. Wiss. Göttingen*, 7, 102.

[28] Pecker, J-C., Roberts, A.P., & Vigier, J-P (1972) Non-velocity redshifts and photon-photon interactions, Nature, v. 237, pp. 227-229.
[29] Pecker, J-C & Vigier, J.-P. (1988) A possible tired-light mechanism, Apeiron, No. 2, pp. 19-23.
[30] Vigier, J-P (1990) Evidence for nonzero mass photons associated with a vacuum-induced dissipative redshift mechanism, IEEE Trans Plasma Sci, 18:1;64-72.
[31] Amoroso, R.L., Kafatos, M. & Ecimovic, P. (1998) The origin of cosmological redshift in spin exchange vacuum compactification and nonzero rest mass photon anisotropy, in G. Hunter, S. Jeffers & J-P Vigier (eds.) Causality & Locality in Mod. Physics, Dordrecht: Kluwer.
[32] Bible, New Testament, King James Version, 2 Peter 3:8.
[33] Barrow, J.D. & Tipler F.J. (1996) The Anthropic Cosmological Principle, Oxford: Oxford Univ. Press.
[34] Smith, J. (1974) The Doctrine and Covenants, D&C 130: 4-5, Salt Lake City: The Church of Jesus Christ of Latter-Day Saints.
[35] Wheeler, J.A. (1955) Geons, Phys. Rev., 97:2, 511-536.
[36] Shadowitz, A. 1968, Special Relativity, New York: Dover.
[37] De Broglie, L. (1927) *J. Physique, 6e serie* 8, 225.
[38] Kulhánek, J. (1965) Propagation of de Broglie waves in space-time, Il Nuovo Cimento, Volume 38:3; 1955-1965.
[39] Feoli, A. & Scarpetta, G. (1998) De Broglie Matter Waves from the Linearized Einstein Field Equations, Found. Phys. Lett. 11:4; 395-403.
[40] Jung, C. (1970) Collected Works of C. G. Jung, Princeton: Princeton Univ. Press..
[41] Franck, G. (2000) Time and presence, in Science and The Primacy of Consciousness, R.L. Amoroso et al, (eds.) Orinda: Noetic Press.
[42] Marshall, T.W. (2002) Engineering the vacuum, pp. 459-468, in R.L. Amoroso, G. Hunter, M. Kafatos & J-P Vigier (eds.), Gravitation & Cosmology: From the Hubble Radius to the Planck Scale, Dordrecht: Kluwer Academic
[43] Marshall, T.W. (1998) Parametric up conversion of the vacuum; http://arXiv.org/PS_cache/quant-ph/pdf/9803/9803054.pdf.
[44] Sagan, K & Drake, F. (1975) The search for extraterrestrial intelligence, Scientific American, May
[45] Sagan, K. (1994) Search for extraterrestrial life, Sci Am, October.
[46] Tarter, J.C. & Chyba, C.F. (1999) Is there life elsewhere in the universe?, Scientific American, 281:6, pp. 118-123.
[47] www.answerbag.com/q_view/506418.

[48] Amoroso, R.L (2000) Derivation of the fundamental equation of consciousness, Part I, Boundary conditions, Noetic J 3:1, pp. 91-99.
[49] Schwinger, J. (1992) Casimir energy for dielectrics, Proc. Nat. Acad. Sci. USA 89, 4091-3.
[50] Schwinger, J. (1993) Casimir light: The source, Proc. Nat. Acad. Sci. USA 90, 2105-6.
[51] Schwinger, J. (1994) Casimir energy for dielectrics: spherical geometry, Proc. Nat. Acad. Math. Psych. 41:64-67, SF: W.H. Freeman.
[52] Carter, B., 1974, Large number coincidences and the Anthropic Principle, in M.S. Longair (ed.), Confrontation of Cosmological Theory with Astronomical Data, Dordrecht: Reidel.
[53] Hartle, J.B. & Hawking, S.W. (1983) Phys. Rev. D28, 2960
[54] Halliwell, J.J., 2000, Trajectories for the wave function of the universe from a simple detector model; arXiv:gr-qc/0008046.
[55] Vilenkin, A. (1998) The quantum cosmology debate, arXiv:gr-qc/9812027.
[56] Smith, Q. (1997) The Ontological Interpretation of the Wave Function of the Universe, The Monist, 80:1,160-185.
[57] Penrose, R. 2005, The Road to Reality, New York: Knopf.
[58] Linde, A.D. (1984) Lett. Nuovo Cimento, 39, 401
[59] Vilenkin, A. (1984) Phys. Rev. D30, 509.
[60] Vilenkin, A. (1986) Phys. Rev. D33, 3560.
[61] Callender, C. & Weingard, R. (1994) The Bohmian Model of Quantum Cosmology, PSA, vol. 1, D. Hull et al. (eds.) pp. 218-27.
[62] Feynman, R.P. (1971) Lectures on Gravitation, Pasadena: California Inst of Technology.
[63] Velusamy, T. et al (2001) Single and double bracewell nulling interferometer in space, Proc. Toward Other Earths: DARWIN/TPF; and the Search for Extrsolar Terrestrial Planets, Heidelberg, Germany. 22-25 April 2003 (ESA SP-539, October 2003).
[64] Bohm, D. (1965) The Special Theory of Relativity, New York, W.A. Benjamin.
[65] Trivedi, S.P. (2001) Holography, black holes and string theory, Current science, 81:12; 1582 -1590.
[66] Basalla, G. (2006) Civilized Life in the Universe: Scientists on Intelligent Extraterrestrials, Oxford: Oxford Univ. Press.
[67] Amoroso, R.L. (2010) A seminal introduction to immanent parameters of noetic medicine, in W. Carvalho & Pelizoli, M. (eds.) Proc. 1st Symp.for 'Cure and Consciousness' Nov. 27-29, Recife, Brazil.

Index

SERIES ON KNOTS AND EVERYTHING

Editor-in-charge: Louis H. Kauffman *(Univ. of Illinois, Chicago)*

The Series on Knots and Everything: is a book series polarized around the theory of knots. Volume 1 in the series is Louis H Kauffman's Knots and Physics.

One purpose of this series is to continue the exploration of many of the themes indicated in Volume 1. These themes reach out beyond knot theory into physics, mathematics, logic, linguistics, philosophy, biology and practical experience. All of these outreaches have relations with knot theory when knot theory is regarded as a pivot or meeting place for apparently separate ideas. Knots act as such a pivotal place. We do not fully understand why this is so. The series represents stages in the exploration of this nexus.

Details of the titles in this series to date give a picture of the enterprise.

Published:*

Vol. 1: Knots and Physics (3rd Edition)
by L. H. Kauffman

Vol. 2: How Surfaces Intersect in Space — An Introduction to Topology (2nd Edition)
by J. S. Carter

Vol. 3: Quantum Topology
edited by L. H. Kauffman & R. A. Baadhio

Vol. 4: Gauge Fields, Knots and Gravity
by J. Baez & J. P. Muniain

Vol. 5: Gems, Computers and Attractors for 3-Manifolds
by S. Lins

Vol. 6: Knots and Applications
edited by L. H. Kauffman

Vol. 7: Random Knotting and Linking
edited by K. C. Millett & D. W. Sumners

Vol. 8: Symmetric Bends: How to Join Two Lengths of Cord
by R. E. Miles

Vol. 9: Combinatorial Physics
by T. Bastin & C. W. Kilmister

Vol. 10: Nonstandard Logics and Nonstandard Metrics in Physics
by W. M. Honig

Vol. 11: History and Science of Knots
edited by J. C. Turner & P. van de Griend

*The complete list of the published volumes in the series, can also be found at
http://www.worldscibooks.com/series/skae_series.shtml

Vol. 12: Relativistic Reality: A Modern View
edited by J. D. Edmonds, Jr.

Vol. 13: Entropic Spacetime Theory
by J. Armel

Vol. 14: Diamond — A Paradox Logic
by N. S. Hellerstein

Vol. 15: Lectures at KNOTS '96
by S. Suzuki

Vol. 16: Delta — A Paradox Logic
by N. S. Hellerstein

Vol. 17: Hypercomplex Iterations — Distance Estimation and Higher Dimensional Fractals
by Y. Dang, L. H. Kauffman & D. Sandin

Vol. 18: The Self-Evolving Cosmos: A Phenomenological Approach to Nature's Unity-in-Diversity
by S. M. Rosen

Vol. 19: Ideal Knots
by A. Stasiak, V. Katritch & L. H. Kauffman

Vol. 20: The Mystery of Knots — Computer Programming for Knot Tabulation
by C. N. Aneziris

Vol. 21: LINKNOT: Knot Theory by Computer
by S. Jablan & R. Sazdanovic

Vol. 24: Knots in HELLAS '98 — Proceedings of the International Conference on Knot Theory and Its Ramifications
edited by C. McA Gordon, V. F. R. Jones, L. Kauffman, S. Lambropoulou & J. H. Przytycki

Vol. 25: Connections — The Geometric Bridge between Art and Science (2nd Edition)
by J. Kappraff

Vol. 26: Functorial Knot Theory — Categories of Tangles, Coherence, Categorical Deformations, and Topological Invariants
by David N. Yetter

Vol. 27: Bit-String Physics: A Finite and Discrete Approach to Natural Philosophy
by H. Pierre Noyes; edited by J. C. van den Berg

Vol. 28: Beyond Measure: A Guided Tour Through Nature, Myth, and Number
by J. Kappraff

Vol. 29: Quantum Invariants — A Study of Knots, 3-Manifolds, and Their Sets
by T. Ohtsuki

Vol. 30: Symmetry, Ornament and Modularity
by S. V. Jablan

Vol. 31: Mindsteps to the Cosmos
by G. S. Hawkins

Vol. 32: Algebraic Invariants of Links
by J. A. Hillman

Vol. 33: Energy of Knots and Conformal Geometry
by J. O'Hara

Vol. 34: Woods Hole Mathematics — Perspectives in Mathematics and Physics
edited by N. Tongring & R. C. Penner

Vol. 35: BIOS — A Study of Creation
by H. Sabelli

Vol. 36: Physical and Numerical Models in Knot Theory
edited by J. A. Calvo et al.

Vol. 37: Geometry, Language, and Strategy
by G. H. Thomas

Vol. 38: Current Developments in Mathematical Biology
edited by K. Mahdavi, R. Culshaw & J. Boucher

Vol. 39: Topological Library
Part 1: Cobordisms and Their Applications
edited by S. P. Novikov and I. A. Taimanov

Vol. 40: Intelligence of Low Dimensional Topology 2006
edited by J. Scott Carter et al.

Vol. 41: Zero to Infinity: The Fountations of Physics
by P. Rowlands

Vol. 42: The Origin of Discrete Particles
by T. Bastin and C. Kilmister

Vol. 43: The Holographic Anthropic Multiverse
by R. L. Amoroso and E. A. Ranscher